Applying AutoCAD®

A Step-by-Step Approach

for AutoCAD Release 14

by
Terry T. Wohlers
Wohlers Associates, Inc.

*This work-text was created using AutoCAD
Release 14 and was fully tested with
Windows 95 and Windows NT 4.0.*

New York, New York   Columbus, Ohio   Woodland Hills, California   Peoria, Illinois

Except where otherwise credited, all CAD drawings within this book were developed with AutoCAD. Cover design was provided courtesy of Pudik Graphics, Inc; images on the front cover were provided courtesy of Gary J. Hordemann, Gonzaga University.

Applying AutoCAD is a work-text for those who wish to learn how to use the AutoCAD software. AutoCAD is a computer-aided drafting and design package produced by Autodesk, Inc. For information on how to obtain the AutoCAD software, contact Autodesk at (800) 964-6432.

Applying AutoCAD is not an Autodesk product and is not warranted by Autodesk. Autodesk, AutoCAD, AutoLISP, AutoCAD Development System, 3D Studio, the Autodesk, logo and AutoShade are registered trademarks of Autodesk, Inc. ACAD, AutoCAD SQL Extension, Autodesk Animator Clips, Autodesk Device Interface, DXF, Kinetix, ObjectARX, and *WHIP!* are trademarks of Autodesk, Inc.

Glencoe/McGraw-Hill

A Division of The McGraw·Hill Companies

Send all inquiries to:
Glencoe/McGraw-Hill
3008 W. Willow Knolls Drive
Peoria, IL 61614-1083

Eighth Edition
ISBN 0-02-667636-2 (Work-text)
ISBN 0-02-667637-0 (Instructor's Guide)
ISBN 0-02-667638-9 (Diskette)

Printed in the United States of America

3 4 5 6 7 8 9 10   009   02 01 00 99 98

Dedication

The author dedicates this edition of *Applying AutoCAD* to Gary Hordemann of Gonzaga University and Robert Pruse of Fort Wayne Community Schools for their many contributions.

Acknowledgments

The author wishes to thank the editorial, marketing, and production staff at Glencoe/McGraw-Hill. A special thank-you goes to Trudy Muller and Debie Baxter, both of Glencoe/McGraw-Hill, for their help and guidance.

A very special thank-you to Jody James of James Editorial Consulting for her contribution to this book. Her speed, organization, and attention to detail helped to keep the book on schedule. Thanks also to Chad Wohlers for copying the revised pages and preparing them for editing.

Thanks to Don Sanborn of Unique Solutions for lending his software development expertise. Thanks to Jim Quanci of Autodesk for answering countless questions and providing critical input. Finally, thanks to Autodesk's technical support staff for promptly fielding technical questions.

Advisory Board

The author and publisher gratefully acknowledge the contributions of the Advisory Board members, who reviewed *Applying AutoCAD* and provided valuable input for its revision.

Table of Contents

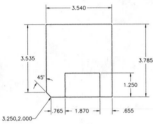

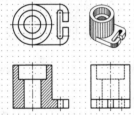

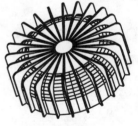

Unit 16: AutoCAD's Magnifying Glass **170**
Objective: To practice the ZOOM, REGEN, and VIEWRES commands

Unit 17: Getting from Here to There **183**
Objective: To apply the PAN and VIEW commands, scrollbar panning, dynamic zooms and pans, and Aerial View

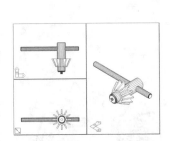

Unit 18: Viewports in Model Space **195**
Objective: To learn how to use viewports in model space

Unit 19: AutoCAD File Maintenance **203**
Objective: To practice using the Select File dialog box and the AUDIT and RECOVER commands

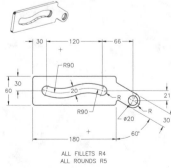

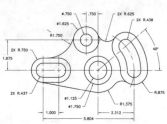

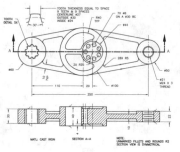

x

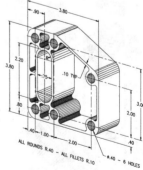

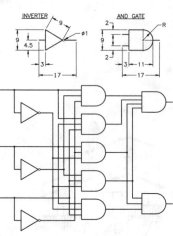

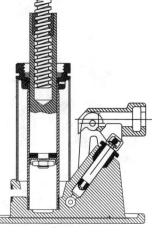

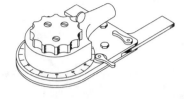

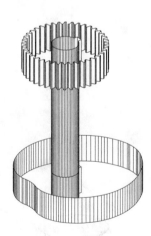

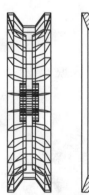

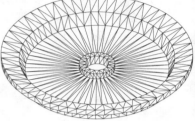

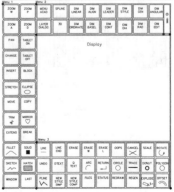

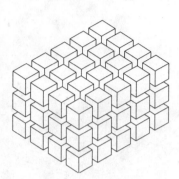

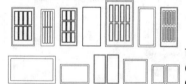

AutoCAD at Work Stories

Illustrations in the Table of Contents provided courtesy of the following:

Autodesk, Inc., p. xiii
CAD Northwest, p. xiii
Dan Cowell, Red Bud High School, p. xx
Alan Fitzell, Central Peel Secondary School, p. x
Gary J. Hordemann, Gonzaga University, pp. vi, vii, viii, ix, xi, xii, xv, xvi
Steve Huycke, Lake Michigan College, p. xvii
Lendert Kersten, University of Nebraska—Lincoln, p. xii
Robert Pruse, Fort Wayne Community Schools, p. x
Lansing Pugh, Architect, p. viii
Tim Smith, Hyland Design, p. xi
Joseph K. Yabu, Ph.D., San Jose State University, pp. v, vi

Introduction

Applying AutoCAD: A Step-By-Step Approach is a work-text based on the AutoCAD Release 14 computer-aided drafting and design software. It is designed primarily for new users of AutoCAD, though experienced users will also find it helpful for reference and review. Through step-by-step instruction, the book takes students from the beginning to the advanced level. Along the way, they are encouraged to experiment, create, and learn firsthand the power, depth, and breadth of AutoCAD.

Applying AutoCAD is not restricted to one discipline. Rather, it serves all areas that require methods of drafting, design, and engineering. These areas include architecture, civil engineering, mapping, landscaping, electricity/electronics, mechanical, product design, tooling design, structural engineering, facilities planning, and interior design. Less common but potentially productive areas include theater set and lighting design, museum display design, and even archaeology.

Operating System

Applying AutoCAD: A Step-By-Step Approach is fully compatible with personal computers running Microsoft® Windows 95 or Windows NT 4.0.

The AutoCAD software is available from Autodesk, Inc. Contact Autodesk at 1-800-964-6432 for the name of a local AutoCAD dealer. Autodesk can also provide you with a list of the minimum hardware requirements for AutoCAD.

Format

The contents of this book are formatted in a straightforward, simple-to-use manner so that instructors, regardless of their background, can easily adapt the book to their existing courses. The book's structure in fact lends itself to picking and choosing units and problems as instructors see fit. Therefore, instructors should by no means feel obligated to use the entire work-text to accomplish their course objectives.

For educators beginning a new course using AutoCAD, the book provides a foundation for developing the course. The units are sequenced in a logical order for learning AutoCAD, so instructors are encouraged to use the book's table of contents as their course outline.

The book contains enough exercises for an entire semester course. Some instructors have used the earlier editions for introductory, intermediate, and advanced courses spread over several semesters. The instructor can adjust the pace and assignments according to the level of the learner group and the number of hours the students receive on the AutoCAD stations.

Features

- Seventy clearly defined units guide students in their progress from basic to advanced levels. Progress is easy to see, and review is simple.

- In addition to the fundamentals of AutoCAD, the book presents topics of special interest, including:

  - symbol libraries
  - attributes and bills of materials
  - 3D modeling using surfaces of revolution, ruled surfaces, tabulated surfaces, Coons surface patches, and basic 3D polygon meshes
  - external references (xrefs)
  - paper space and multiple-view plotting
  - tablet, pull-down, and image tile menu development and customization
  - AutoLISP applications and programming
  - slide shows and slide libraries
  - digitizing and plotting
  - hard disk organization and system management
  - advanced object selection and editing using grips
  - dialog boxes for performing many AutoCAD functions
  - photorealistic rendering, entirely within AutoCAD

  - two-dimensional region modeling, involving the union, subtraction, and intersection Boolean operations
  - "what-you-see-is-what-you-get" (WYSIWYG) plot previews and multiple output device configuration
  - importing and exporting DXF, SAT, STL, DWF, EPS, 3DS, and other file formats
  - importing and editing of raster image files, such as TIF, BMP, JPG, TGA, PCX, and others
  - TrueType fonts, text editing, and spell checking
  - ANSI Y14.5 geometric dimensioning and tolerancing (GD&T)
  - Non-uniform rational B-spline (NURBS) curves and surfaces
  - Boolean solid modeling using the ACIS modeling engine
  - Object linking and embedding (OLE)

- Hint sections throughout the units help students effectively tap the full power of AutoCAD.

- Questions and problems at the end of each unit ensure mastery of AutoCAD. Challenge Your Thinking questions encourage students to use higher level thinking to expand their knowledge.

- Optional Problems section challenges and motivates advanced students.

- Appendices on topics such as AutoCAD toolbars help students expand their knowledge of AutoCAD when they do not have access to an AutoCAD system.

- "AutoCAD at Work" stories help students understand how AutoCAD is used in business and industry.

What's New in Release 14

This edition of *Applying AutoCAD* includes new and enhanced commands and features of AutoCAD Release 14. All units have been updated, and new ones have been added to cover new commands and features. The following is a sampling of what's new:

- Start-up dialog box, including Windows standard "wizards" to streamline the drawing setup process

- Drawing templates that replace the prototype drawings used in previous releases

- AutoSnap, a new object snap feature that enables you to preview and confirm snap point candidates visually

- Improved layer and linetype management

- Improved object property controls

- Right-button shortcut menus

- Command-line editing

- Real-time zoom and pan

- Improved plot previewing

- Dramatically improved text editor

- Photorealistic rendering

- Semi-automatic production of orthographic views, sections, and auxiliary views from solid models

- External reference (xref) enhancements

- Hybrid raster/vector drawings and raster image support

Instructor's Guide

An instructor's guide is available as a companion to the *Applying AutoCAD* work-text. It provides instructors with a range of helpful teaching supplements, such as a sample course syllabus, instructional tips and advice, optional group activities, and transparency masters. For purchasing information, phone 1-800-334-7344.

Applying AutoCAD Diskette

An optional companion diskette is available for use with *Applying AutoCAD*. The diskette contains many useful files. For example, AutoLISP routines (including a special parametric program) and menu files presented in this book are contained on the diskette, saving you the time and effort of accurately entering them manually.

If you want to experiment with DXF, PostScript (EPS), 3D Studio (3DS), and SAT files created by other systems, several are contained on the diskette so that you can see how they import to AutoCAD. Also available on the diskette are several drawing files for use with many of the exercises presented throughout *Applying AutoCAD*. These files save you drawing preparation time when completing the exercises.

The diskette also contains a sample slide show, as well as a program that enables you to create a bill of materials. For additional information, phone Glencoe/McGraw-Hill at 1-800-334-7344.

To the Student

By following the step-by-step exercises in this book, you will learn to use AutoCAD to create, modify, store, retrieve, and manage AutoCAD drawings and related files. For review and practice, questions and problems have been provided at the end of each unit. There is a section of more challenging problems following Unit 70.

In order to derive the full benefit of this book, you should be aware of the following:

- *Notational conventions.* Computer keyboards differ. In this book, you will find many references to the ENTER key. On your keyboard, this key may be marked RETURN.

  In the step-by-step instructions, user input is in boldface type. For example, the instruction "enter the **LINE** command" means that you should pick the Line icon or type the LINE command at the keyboard and press the ENTER key. Command names are usually shown in uppercase letters, but you can type them in either upper- or lowercase letters.

  On the computer screen, AutoCAD default values are often displayed within angle brackets like this: < >. You can select the default value by simply pressing the ENTER key or the space bar.

- *Icon reference.* On many pages of this book, you will see icons in the right margin of the page. These icons can help you find and use the many icons AutoCAD provides for performing drawing and editing tasks. The name of the toolbar in which the icon can be found appears above the icon in the margin.

- *Acad.dwt template file.* As you work with AutoCAD, you will learn about the AutoCAD defaults and how these modes and settings are stored in AutoCAD's default template file named acad.dwt.

- *End-of-unit questions.* The questions at the end of each unit are intended to help you review the material in the unit. If you enjoy more challenging questions or want to expand your knowledge, try the "Challenge Your Thinking" questions. In order to answer many of these questions, you may need to work on the computer or refer to the manuals or on-line references that accompany the AutoCAD software.

About the Author

Since 1977, Terry Wohlers has focused his education, research, and practice on design and manufacturing. In 1983, he taught one of the nation's first university credit courses on AutoCAD. Since then, Terry has published more than 200 books, articles, reports, and technical papers on engineering and manufacturing automation. He holds several advisory board positions with publishers and conference organizers in the United States, Europe, and Asia.

Over the past decade, Terry has been given the opportunity to share his views with thousands of engineers and managers in many North American, European, and Asian nations and has been a keynote speaker at major industry events. His award-winning presentation in 1986 in London, England, accurately predicted the explosive growth of desktop-based systems for computer-aided design and drafting.

In 1992, Terry led a group of 14 individuals from industry and academia to form the first association dedicated to rapid prototyping. In 1993, the association joined the Society of Manufacturing Engineers (SME) to become the Rapid Prototyping Association (RPA) of SME. Today, RPA/SME is the world's largest individual member association on rapid prototyping. Terry is also a founder of the CAD Society, a non-profit CAD membership organization.

Prior to leaving his teaching and research position at Colorado State University in 1986, Terry formed Wohlers Associates, Inc., a firm that works closely with organizations to streamline product development using modern approaches to design, modeling, prototyping, and tooling. The company has counseled many organizations on CAD/CAM, rapid prototyping, and reverse engineering developments and trends. Terry earned a master's degree (1982) and a bachelor's degree (1980) in industrial sciences from Colorado State University and the University of Nebraska, respectively.

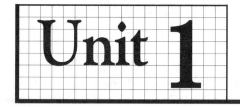

Unit 1   Straight to AutoCAD

■ OBJECTIVE:

To understand the purpose of the components found in the AutoCAD graphics screen and to learn the purpose of the NEW, OPEN, SAVE, SAVEAS, QSAVE, and QUIT commands

The AutoCAD graphics screen is where AutoCAD users spend most of their time. From the graphics screen, you can open, change, view, and plot drawings. It is therefore necessary for you to understand the purpose of each component found in the AutoCAD graphics screen.

Taking a Look Around

1. Start AutoCAD, and you will find yourself in the AutoCAD graphics screen, as shown below.

HINT:

From Microsoft® Windows®, start AutoCAD by double-clicking the AutoCAD R14 icon. Double-clicking means to position the pointer on the item and then press the left button on the pointing device twice very quickly. If the AutoCAD R14 icon is not present, pick the Windows Start button, move the pointing device to Programs, move it to AutoCAD R14, and pick the AutoCAD R14 option.

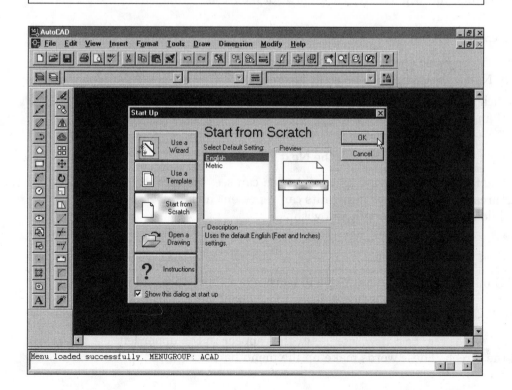

The exact configuration of the AutoCAD screen elements may vary depending on how they were left by the last person to use the AutoCAD software. Also, the display hardware may cause the appearance of the screen elements to vary. For example, at lower resolutions, the icons at the top of the screen may extend across the entire screen. At higher resolutions, they may extend only about halfway across the screen.

2 Pick the **Use a Wizard** button on the **Start Up** dialog box.

HINT: Position the pointer over the Use a Wizard button and press the left button on the pointing device. This is called the *pick button*.

A wizard steps you through a sequence; in this case, a setup sequence.

3 In the middle of the dialog box, pick **Quick Setup**. (It may be picked already.)

Read Wizard Description. Note that many of the settings are based on the acad.dwt template file. If you were instead to pick the Start from Scratch button, all of the settings would come from this template file. In many of the future units, you will use the Start from Scratch button.

4 Pick the **OK** button.

Step 1: Units focuses on the unit of measurement. Decimal is the default selection. Notice the sample in the right area of the box.

5 Pick the other units of measurement and watch how the sample changes.

6 Pick **Decimal** and then pick the **Next** button.

Step 2: Area focuses on the drawing area. As you can see, the default drawing area is 12×9 units. These units can represent inches, feet, meters, kilometers, or any length you wish.

7 Pick the **Done** button.

AutoCAD displays a grid of dots. You will learn more about this grid in Unit 9. Also, in future units, you will learn more about setup wizards and the overall setup process.

Notice the words File, Edit, View, Insert, etc., that appear in the upper left area of the screen. These words make up the menu bar.

⑧ Select **File**, causing a pull-down menu to appear.

The File pull-down menu contains items that enable you to open, save, export, and print AutoCAD drawings.

⑨ With the pointing device, pick either the **Save** or **Save As...** menu item.

The Save Drawing As dialog box appears as shown in the following illustration. If you are a user of other Windows applications, the dialog box should look familiar to you.

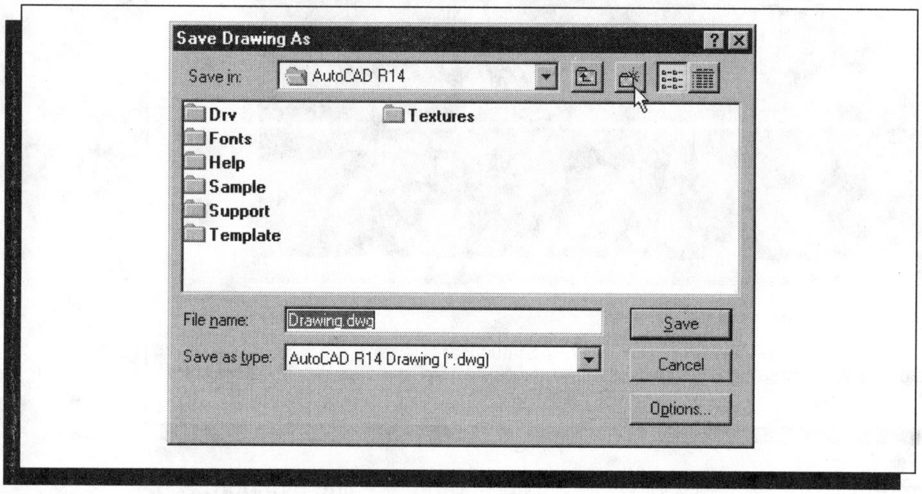

⑩ Pick the **Create New Folder** icon (as shown), type your first name, and press **ENTER**.

This creates a new folder.

⑪ Double-click this new folder to make it current.

⑫ In the File name text box, highlight **Drawing.dwg** by double-clicking it, and enter **stuff**.

⑬ Pick the **Save** button to create a file named stuff.dwg.

Notice that the drawing file name appears at the top of the screen.

NOTE:

You can also enter the SAVE, QSAVE, and SAVEAS commands at the Command prompt (located near the bottom of the screen) to save your work. See "Maneuvering AutoCAD Files" later in this unit for more information about these commands.

The following illustration describes the parts that make up the graphics screen.

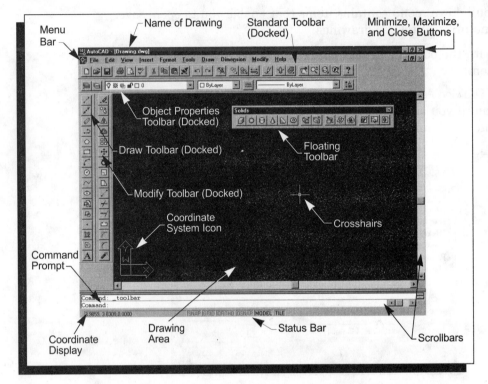

Under the menu bar (File, Edit, View, . . .) you should see the Standard toolbar docked horizontally across the top of the screen. Below it, you should see the Object Properties toolbar, also docked horizontally. A toolbar is said to be *docked* when it appears to be a part of the top, bottom, left, or right border of the drawing area. When a toolbar is docked, the name of the toolbar does not appear on the screen.

NOTE:

If the Standard and Object Properties toolbars are not present, that's okay. You will learn how to display and position them later in this unit, in the section titled "Docked Toolbars."

Notice the Command prompt near the bottom of the screen. Keep your eye on this area because this is where you receive important messages from AutoCAD. Also, you can enter AutoCAD commands at the Command prompt.

At the bottom of the screen is the status bar. The status bar tells you the coordinates of the screen crosshairs and the status of various AutoCAD modes.

The rest of the screen is the drawing area.

Floating Toolbars

Toolbars include icons that can save you time when working with AutoCAD.

Let's focus on displaying and positioning toolbars. Toolbars that are not attached are called *floating* toolbars. The Solids toolbar in the previous illustration is an example.

1 If any other floating toolbars are present, close them also.

NOTE:

If you are using Windows 95, you may need to pick the × in the upper right corner to close the toolbar. If you are running Windows NT 4.0, you may need to pick the dash in the upper left corner to close the toolbar.

2 Select the **View** pull-down menu.

Notice that several menu items contain a small arrow pointing to the right. This means that when you point to them, they will display another menu called a *cascading menu.*

3 Move the pointing device to any one of the menu items containing a small arrow and allow the pointer to rest for a moment.

A cascading menu appears, displaying another set of menu options.

Menu items that contain ellipsis points (three dots or periods), as in the Toolbars... item, display a dialog box when picked.

4 Pick the **Toolbars...** item.

This displays a list of toolbars that AutoCAD makes available to you, as shown in the illustration on the next page.

5

5 Scroll down the list until Solids appears and click the check box.

The Solids toolbar appears.

6 Pick the **Close** button on the dialog box.

7 Move the Solids toolbar by clicking on the bar located at the top of the toolbar and dragging it. (This bar contains the word Solids.)

As you can see, it's easy to move floating toolbars to a new location. As you work with AutoCAD, you should move the toolbars to a location that is convenient, yet does not interfere with the drawing.

Changing a Toolbar's Shape

AutoCAD makes it easy to change the shape of toolbars.

1 Move the pointer to the bottom edge of the Solids toolbar, slowly and carefully positioning it until a double arrow appears.

2 When the double arrow is present, click and drag downward until the toolbar changes to a vertical shape and then release the pick button.

3 Move the pointer to the right edge of the Solids toolbar, positioning it until a double arrow appears.

4 When the double arrow is present, click and drag to the right until the toolbar changes back to a horizontal shape.

5 Close the toolbar. (See the note on page 5 if you need help.)

Docked Toolbars _____

A docked toolbar is locked into position at the edge of the graphics screen.

1 If the Standard and Object Properties toolbars are *not* present on the screen, skip to Step 5.

2 Move the pointer to the Object Properties toolbar and click and drag the outer edge (border) of the toolbar, moving the toolbar down into the drawing area.

3 Repeat Step 2 with the Standard toolbar.

4 Close both of these toolbars.

Now let's dock the Standard and Object Properties toolbars into their standard positions at the top of the drawing area.

5 Pick the **View** pull-down menu, the **Toolbars...** item, and check the **Object Properties** and **Standard** toolbars.

6 Pick the **Close** button.

7 Click and drag the Standard toolbar and carefully dock it under the menu bar (File, Edit, View, . . .).

The toolbar locks into place.

8 Click and drag the Object Properties toolbar and carefully dock it under the Standard toolbar.

9 If the Draw and Modify toolbars are not present and locked into place, position them now as shown on page 1.

You should now see four docked toolbars on the screen.

_____ NOTE: _____

In future units, *Applying AutoCAD* assumes that these four toolbars are present on the screen in a configuration similar to the one on page 1. This book has been written to work with AutoCAD straight from the installation. It assumes that the screen color, crosshairs, and other elements of AutoCAD have not been customized.

10 If you did not complete Steps 2 through 4, do so now. Also, be sure to complete Steps 5 through 9.

Picking Icons _____

It can be faster to pick an icon than to enter a command at the keyboard or from a pull-down menu.

_____ NOTE: _____

> The *AutoCAD User's Guide* sometimes refers to icons as *tool icons*.

1 Position the pointer on the Line icon located in the upper left corner of the Draw toolbar, but *do not* pick it.

The word Line appears after about one second, as shown below. This word assists you in understanding what function the icon performs.

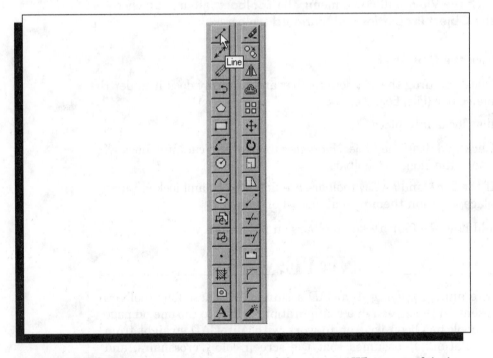

Also, notice the status bar at the bottom of the screen. When a tooltip is displayed, the status bar changes to provide more detailed information about the icon.

2 Slowly position the pointer on top of other icons and read the information that displays.

Draw

③ Pick the **Line** icon.

HINT: Notice the icon at the right of Step 3. This is provided as a visual aid to help you more easily locate and select the icon. The name of the toolbar in which the icon is located appears above the icon for reference. If you need more help, refer to Appendix H, "Toolbars."

The icons appear in the margin throughout the unit in which they are introduced and in the subsequent three or four units. They also appear occasionally in later units when the icon has not been used recently, to refresh your memory.

The following text displays on the Command line.

Command: _line From point:

You have just entered the LINE command. AutoCAD is asking where you want the line to start.

④ Pick a point anywhere on the screen using the pick (left) button on the pointing device; then pick a second point to form a line.

⑤ Press the **ENTER** key to terminate the LINE command.

⑥ Pick another icon (any one of them) from the Draw toolbar.

⑦ Press the **ESC** key to cancel the last entry.

Standard

⑧ From the **Standard** toolbar, pick the **Save** icon.

This is equivalent to picking Save from the File pull-down menu.

⑨ After you review the information in the following two sections, exit AutoCAD by picking **Exit** from the **File** pull-down menu or by picking the × button located in the upper right corner of the AutoCAD window.

NOTE:

It is very important that you exit AutoCAD at the end of each unit. This resets the icons in the toolbars to their default settings.

Maneuvering AutoCAD Files

The commands described below are basic ones that you will use in almost every unit of this book. They include commands for opening a drawing file, storing a drawing to disk, and exiting AutoCAD. It is therefore important for you to understand what each command does.

NEW	presents the Create New Drawing dialog box, from which you can set up a new drawing
OPEN	opens an existing drawing; automatically converts older AutoCAD drawings to Release 14
SAVE	requests a file name and saves the drawing
SAVEAS	identical to SAVE, but also renames the current drawing
QSAVE	saves a named drawing without requesting a file name
QUIT	prompts you to save the current drawing and exits AutoCAD

You will have many opportunities to practice these commands throughout this book.

What If I Enter the Wrong Command?

As you work with AutoCAD, you will be entering commands by picking icons, selecting them from menus, or typing them at the keyboard. Occasionally you might accidentally select the wrong command or make a typing error. It's easy to correct such mistakes.

If you catch a typing error *before* you press ENTER . . .	use the backspace key to delete the incorrect character(s). Then continue typing.
If you select the wrong icon or pull-down menu item . . .	press the ESC key to clear the Command line.
If you type the wrong command . . .	press the ESC key to clear the Command line.
If you accidentally pick a point or object on the screen . . .	press the ESC key twice.

Questions

1. Describe one of two ways of starting AutoCAD.

   _____

   _____

2. Explain the overall purpose of the AutoCAD graphics screen.

   _____

   _____

3. How do you display a toolbar that is not on the screen?

   _____

   _____

4. How do you move a floating toolbar?

   _____

   _____

5. How do you make a floating toolbar disappear?

   _____

   _____

6. Briefly explain the basic function of the AutoCAD Command prompt line.

   _____

   _____

7. Explain the purpose of the following AutoCAD commands:

   OPEN _____

   SAVE _____

   SAVEAS _____

   QSAVE _____

   QUIT _____

8. On the lines that follow this illustration, write the names of items A
through O.

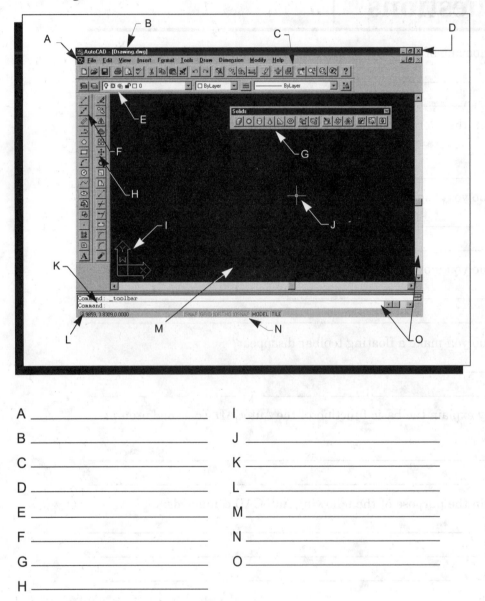

A _____   I _____

B _____   J _____

C _____   K _____

D _____   L _____

E _____   M _____

F _____   N _____

G _____   O _____

H _____

Challenge Your Thinking: Questions for Discussion

1. AutoCAD provides ways to change the appearance and colors of the
graphics screen. For example, if you prefer to work on a gray surface,
you can change the drawing area from black to gray. Find out how
to customize AutoCAD's appearance and colors. Then write a short
paragraph explaining how to customize the graphics screen and why
it might sometimes be necessary to do so.

Problems

1-3. Using the features presented in this unit, rearrange the toolbars so that they are similar to the ones in the following illustrations.

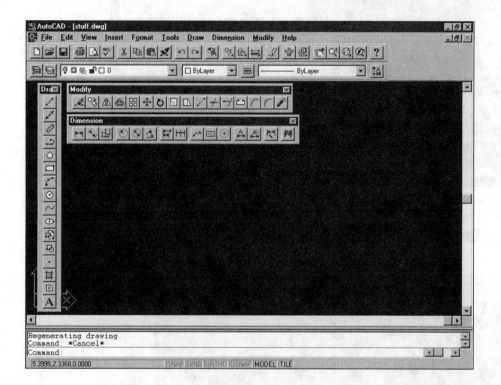

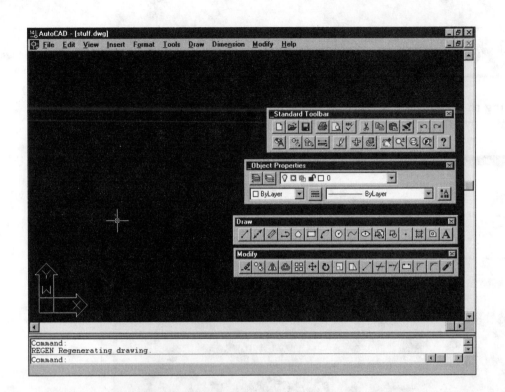

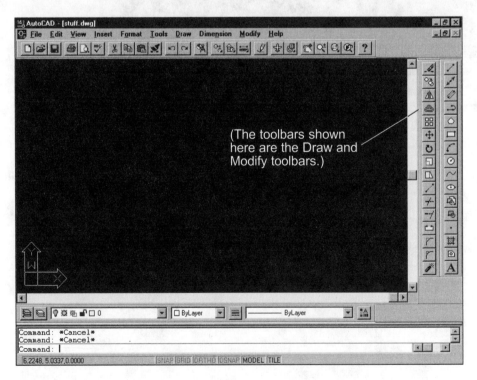

(The toolbars shown here are the Draw and Modify toolbars.)

4. Change the positions of the toolbars back to their standard appearance as shown on page 1.

AUTOCAD® AT WORK

You're Expecting How Many?

Have you ever planned a party or meeting and needed to figure out how many chairs or tables were required? Did you have to decide whether there would be room for all the people you wanted to invite? Imagine what would happen if that party included thousands of people, and you discovered at the last minute that they would not fit into the space.

That was the problem frequently faced by the Pine Bluff Convention Center in Pine Bluff, Arkansas, before the introduction of AutoCAD. The staff would be setting up the tables and chairs and then learn, to their horror, that the room set aside for the event would not hold even half the people who would be arriving in a few hours. It was a planner's nightmare.

Then the convention center created a computer graphics department for the purpose of drawing convention setups to scale and calculating the correct amount of space and equipment needed. Using AutoCAD, the department currently turns out about 25 drawings per week, the same number that formerly took about three months to complete. Most important, the seating and floor space estimates are accurate. The staff can draw setups within setups and keep closer control over equipment inventories.

Says the center's technical director, Steve Barnett, "The application of AutoCAD has given us a greater ability to show creativeness in our setups, has reduced part-time labor costs due to incorrect hand drawings, and has provided us with a powerful advertising tool for securing clients."

The center hosts 1,500 events each year and accommodates about

350,000 visitors. It is the most fully computerized convention center in the world. Not only are floor plans done by computer, but heating and cooling, accounting, payroll, and scheduling and booking of events are also handled by the center's mainframe. Computerization has proved so successful that the center has begun offering its services to other convention centers across the country.

Unit 2

The Line Forms Here

■ OBJECTIVE:

To apply the LINE, MULTIPLE, POLYGON, and RECTANG commands, and to use the Select File dialog box and command aliases

The LINE command is one of the most often used AutoCAD commands. There are a number of ways to produce lines. Some ways are simple; others are more involved. The following exercise uses the simplest approach to producing lines.

Opening Files

1 Start AutoCAD.

It is very important that you start (or restart) AutoCAD at the beginning of each unit. This ensures that the icons in the toolbars are set to their default settings.

2 Pick **Start from Scratch**, **English**, and **OK** in the Start Up dialog box.

NOTE:

Unless otherwise noted, all units in this book assume that the AutoCAD graphics screen is configured to look like the illustration on page 1 in this book. If AutoCAD currently looks different, reconfigure the toolbars before proceeding to Step 3. Unit 1 explains how to display, position, and close toolbars.

3 Pick the **Open** icon from the Standard toolbar or select **Open...** from the **File** pull-down menu.

Standard

The following dialog box appears.

![Select File dialog box showing Look in: AutoCAD R14, with folders Drv, Fonts, Help, Sample, Support, Template, Textures, YourName; File name field; Files of type: Drawing (*.dwg); Open, Cancel, Find File, Locate buttons; Open as read-only and Select Initial View checkboxes; Preview pane.]

16

The same dialog box appears when you enter OPEN at the keyboard.

4 Double-click the folder named after yourself that you created in Unit 1.

5 Select **stuff.dwg** and pick the **Open** button.

HINT:

You can also double-click stuff.dwg to select the drawing. This is the same as clicking on it once and then picking Open or pressing ENTER, but faster.

You have just opened the drawing file you created in Unit 1. It contains one line.

Abbreviated Command Entry _____

AutoCAD permits you to issue commands by entering the first one or two characters of the command. Command abbreviations such as these are called *command aliases*.

1 Type **L** in upper or lower case and press **ENTER**.

This enters the LINE command. You can also enter other frequently used commands, such as those listed here.

Command	Entry	Command	Entry
ARC	A	LAYER	LA
CIRCLE	C	MOVE	M
COPY	CO	PLINE	PL
DVIEW	DV	REDRAW	R
ERASE	E	ZOOM	Z
LINE	L		

2 Press the **ESC** key to cancel the LINE command.

3 Enter the commands listed above using the alias method. Press **ESC** to cancel each command.

Command aliases are defined in the AutoCAD file acad.pgp, contained in the \Programs Files\AutoCAD R14\support directory. You can create additional command aliases by modifying this file using a text editor.

 With the pointing device, draw two of the polygons shown below. (See the following hint.)

HINT: After you've completed one polygon, press ENTER or the space bar to terminate the LINE command. To construct the next polygon, issue the LINE command again by pressing ENTER or the space bar. This enters the most recently used command—a real shortcut and timesaver. (On most systems, the right button on the pointing device can also be used for this purpose.)

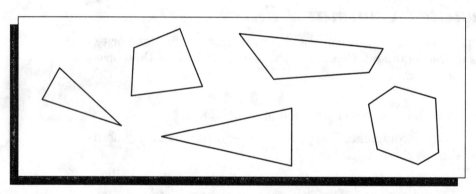

5 Create more polygons using the **LINE** command. (See the following hint.)

HINT: To automatically and precisely close a polygon and terminate the LINE command, type the letter C and press ENTER prior to constructing the last line segment of the polygon. Be sure to press ENTER or the space bar to reenter the LINE command.

MULTIPLE Command _____

The MULTIPLE command is used to reenter other commands multiple times. For example, entering MULTIPLE and then LINE causes AutoCAD to repeat the LINE command automatically until you cancel (ESC).

 Type **MULTIPLE** and press **ENTER.**

Notice that the Command prompt changes to Multiple command.

 Type **L**, press **ENTER**, and create several polygons.

Notice that the LINE command remains entered.

3 Press **ESC** to cancel the LINE command.

4 Practice drawing additional polygons using the automatic close feature as well as the other shortcuts just described.

 ## *LINE Undo Option* _____

From time to time, it's necessary to back up or undo one or more line segments. Suppose you have drawn three lines in creating a polygon, and you're about to enter the fourth line when you realize the third line you drew is incorrect. The easiest and fastest way of correcting this is to pick the Undo icon or type U (short for Undo) and press ENTER.

1 Enter the **LINE** command. ("Enter" means to type the command or command alias and press **ENTER**, or pick the equivalent icon.)

2 Create three connecting line segments of any length and orientation.

Draw

3 Pick the **Undo** icon from the Standard toolbar or type **U** and press **ENTER**.

Standard

The last line segment is replaced by a "rubberband" segment attached to the crosshairs. You can now redefine the line segment by picking a new endpoint.

 Continue to pick the **Undo** icon or enter **U** until all the line segments are gone. Then draw new line segments.

POLYGON Command _____

The POLYGON command enables you to create regular polygons with 3 to 1024 sides. A regular polygon is one with sides of equal length. Let's try a pentagon.

Draw

1 From the Draw toolbar, pick the **Polygon** icon.

Notice the <4> at the end of the Command line. This is the default value, meaning that if you were to press ENTER, AutoCAD would enter 4 for Number of sides.

2 Type **5** in reply to Number of sides and press **ENTER**.

AutoCAD now needs to know if you want to define an edge of the polygon or select a center point. Let's use the default, Center of polygon.

3 With the pointing device, pick a point in an open area on the screen in response to the <Center of polygon> default. This will be the center of the polygon.

4 Press **ENTER** to select the I (Inscribed) default value.

AutoCAD now wants to know the radius of the circle within which the polygon will appear.

5 With the pointing device, move the screen crosshairs from the center of the polygon and notice that a pentagon begins to form.

6 Pick a point or enter a numeric value, such as **.5**, at the keyboard. (Entering a 0 before the decimal point is optional.)

7 Draw several more regular polygons using the remaining options provided by the POLYGON command. Press the space bar to reenter the POLYGON command.

RECTANG Command _____

You can quickly create basic rectangles using the RECTANG command.

Draw

1 From the Draw toolbar, pick the **Rectangle** icon.

As you can see at the Command prompt, AutoCAD is asking for the first corner of the rectangle.

2 In reply to First corner, pick a point at any location on the screen.

3 Move the pointing device in any direction and notice that a rectangle begins to form.

4 Pick a second point at any location to create the rectangle.

5 Create a second rectangle. Since the RECTANG command was just entered, reenter it by pressing the space bar or **ENTER**.

6 Create a third rectangle.

As you experiment with the different methods of entering commands (*i.e.,* icons, the keyboard, pull-down menus, and eventually tablet menus), you may prefer one method over another. For example, picking the Line icon may be faster than entering L at the keyboard. Bear in mind, however, that there is no right or wrong method of entering commands. Experienced users of AutoCAD use a combination of methods.

Selecting Objects by Mistake

Occasionally, you might inadvertently pick an object on the screen.

1 Without a command entered on the Command line, pick any polygon on the screen.

The line segments that make up the polygon appear broken. Also, small blue boxes appear at each endpoint of each line segment. In Unit 15, you will learn why they appear.

2 Press **ESC** twice to return the polygon to its original status and make the blue boxes disappear.

In the future, if you select an object by mistake when a command is not entered, press **ESC** twice.

3 Pick the **Save** icon from the Standard toolbar, or select the **File** pull-down menu and pick **Save** to save your work.

Standard

Notice that either action enters the QSAVE command at the Command prompt. You could have also entered QSAVE at the keyboard.

4 Select **Exit** from the **File** menu to exit AutoCAD.

Always be sure to save your work before selecting Exit.

Questions

1. Describe the purpose of the Open icon and the Open... item contained in the File pull-down menu.

   _____

   _____

2. Explain the relationship between the pointing device and the screen crosshairs.

   _____

   _____

3. What is the fastest and simplest method of reentering the previously entered command?

   _____

4. What is the fastest method of closing a polygon when using the LINE command?

   _____

5. Describe the purpose of the MULTIPLE command.

   _____

   _____

6. Explain the use of the LINE Undo option.

   _____

   _____

7. How can you enter commands such as LINE, ERASE, and ZOOM quickly at the keyboard?

   _____

   _____

8. Name two ways of entering the RECTANG command.

   _____

■ *Challenge Your Thinking: Questions for Discussion*

1. Experiment with the @ option of the LINE command. What is the purpose of this option?

2. Experiment with and explain each of the following POLYGON command options:
   Edge
   Inscribed in circle
   Circumscribed about circle

3. AutoCAD provides two ways to create lines and other objects in a drawing file. Discuss the advantages and disadvantages of having more than one way to issue a drawing command.

How to Save Your Problems

Most units of this work-text conclude with some problems for you to complete. You'll probably want to save your problems, so start a new file for each one. Code the file by unit and problem number. For example, for the problems in this unit:

Standard

1. Begin a new drawing by picking the **New** icon or entering the **NEW** command.

2. For the first problem, save the file with the name **prb2-1**. (AutoCAD adds the DWG extension automatically. The full name of the file is now prb2-1.dwg.)

3. When you are finished with that problem, save your work.

When you are ready to start prb2-2.dwg, repeat Steps 1 through 3. When you finish the last problem, save it and exit AutoCAD.

NOTE:

When drawings are small and simple, you may want to include more than one drawing in a file. This will save time and disk space, and you will end up with fewer files to manage.

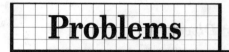

Problems

Create one drawing file for each numbered problem.

1. Using the LINE, POLYGON, and RECTANG commands, draw the
   following objects. Don't worry about exact sizes, but do try to
   make them look as much like the ones below as possible. Practice
   the shortcuts and various options covered in this unit.

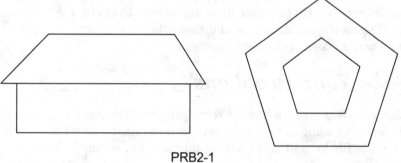

PRB2-1

2. Use the LINE, MULTIPLE, POLYGON, and RECTANG commands to create
   this sawhorse.

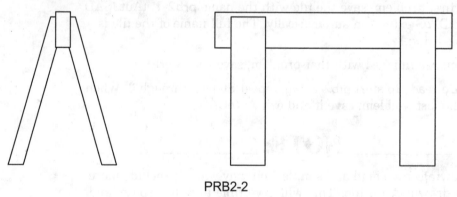

PRB2-2

3. Use the appropriate AutoCAD command to create the following
   objects.

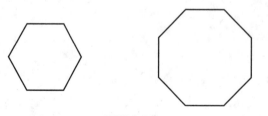

PRB2-3

4. Create the object shown on the right by first drawing a six-sided polygon with the POLYGON command. Then draw the six-pointed star using the LINE command.

PRB2-4

5. Create the same object again, but draw the star using two three-sided polygons.

6. How would you draw a five-pointed star? Draw it.

7. Draw a block with a rectangular cavity using the RECTANG and LINE commands.

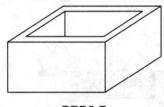

PRB2-7

8. The two objects shown below are composed entirely of equal-sided and equal-sized polygons with common edges surrounding a central polygon. Can this be done with equal-sided and equal-sized polygons of any number of sides? Answer this question by trying to draw such objects using the POLYGON command with polygons of five, six, seven, and eight sides.

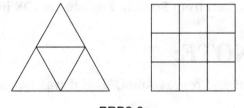

PRB2-8

Problem 2 courtesy of Joseph K. Yabu, Ph.D., San Jose State University
Problems 4 through 8 courtesy of Gary J. Hordemann, Gonzaga University

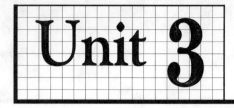

Unit 3 And Around We Go

■ **OBJECTIVE:**

To apply the CIRCLE, DRAGMODE, ARC, ELLIPSE, and DONUT commands

The purpose of this unit is to experiment with the AutoCAD commands that allow you to produce arcs and circular objects.

The following race car is typical of the extent to which drawings contain round and curved lines. Later, you may wish to create a simplified version of this car.

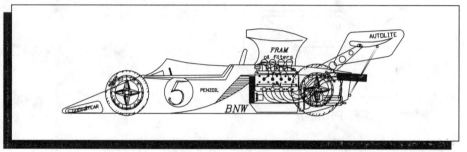

AutoCAD drawing courtesy of BNW, Inc.

CIRCLE Command _____

AutoCAD makes it easy to draw circles and curved lines. For example, to draw wheels, follow these steps.

1 Start AutoCAD and pick **Start from Scratch**, **English**, and **OK** in the Start Up dialog box.

_____ NOTE: _____

As a reminder, be sure to exit and restart AutoCAD if it was loaded prior to Step 1.

Standard

2 Pick the **Save** icon from the Standard toolbar, or select **Save As...** from the **File** pull-down menu.

This causes the Save Drawing As dialog box to appear.

3 Double-click the folder with your first name and double-click **Drawing.dwg**.

4 Type **car** in upper- or lowercase letters and pick **Save** or press **ENTER**.

This creates and stores a new drawing file named **car** in your named directory.

Draw

5 Pick the **Circle** icon from the Draw toolbar, or enter **C** (the command alias for CIRCLE) at the keyboard.

26

6 Draw the larger (outer) circle of the wheel first. Use your pointing device to pick the center point and then the radius, as shown in the following illustration. AutoCAD will complete the circle. Don't worry about exact location or size.

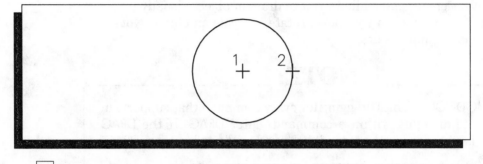

7 Next, draw the smaller circle.

HINT: Press the space bar to reenter CIRCLE. Pick the approximate center point of the circle. Notice that Radius is the default value, so pick a point for the radius of the small circle.

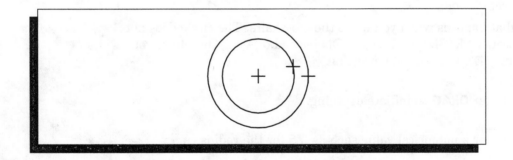

DRAGMODE Command

DRAGMODE enables and disables dynamic dragging of objects.

1 Enter **DRAGMODE** at the keyboard.

Note the prompt line that appears on the screen.

2 If the DRAGMODE default value is <On> or <Auto>, enter **OFF** to turn it off. If DRAGMODE is <Off>, enter either **ON** or **Auto**.

3 Draw the second wheel, again using the **CIRCLE** command.

What is the difference between drawing circles with DRAGMODE ON (or Auto) and drawing them with DRAGMODE OFF?

As you have seen, dragging allows you to drag an object visually into place. It also applies when you move, rotate, or scale an object. Not every command supports dragging.

NOTE:

DRAGMODE ON enables dragging for every command that supports it. To initiate dragging within a command, enter DRAG. If the DRAG request is embedded in the menu item, you do not need to enter DRAG.

DRAGMODE Auto enables dragging for every command that supports it, without having to enter DRAG.

When DRAGMODE is OFF, no dragging of objects can occur, even when DRAG is embedded in menu items.

To learn more about how dragging works, try the following.

4 With DRAGMODE **ON**, enter the **CIRCLE** command and pick a center point.

Note what happens when you move the crosshairs. The circle does not drag because DRAGMODE is set to ON instead of Auto. Therefore, you must enter DRAG to initiate dragging.

5 Enter **DRAG** to initiate dragging.

6 Enter a numerical value such as **.75** for the radius.

7 Experiment with other ways of drawing circles, such as 3P (3 Points), 2P (2 Points), and TTR (Tan, Tan, Radius).

HINT:

After issuing the CIRCLE command, enter 3P, 2P, or TTR.

ARC Command

Now let's focus on the ARC command.

1 Enter the **DRAGMODE** command and the **Auto** option.

Draw

2 Pick the **Arc** icon from the Draw toolbar, or enter **A** at the keyboard.

3 Pick three consecutive points.

4 Reenter the **ARC** command and produce several additional arcs.

ARC Options

Let's experiment with other methods of creating arcs.

Draw

1 Enter the **ARC** command.

2 Pick a start point and then enter **C** for Center.

3 Pick a center point.

As you move the crosshairs, notice what occurs. An arc forms in the counterclockwise direction.

4 Pick an endpoint.

5 Reenter the **ARC** command and specify start and center points.

6 Enter **A** for Angle and enter a number (positive or negative) up to 360. (The number specifies the angle in degrees.)

7 Reenter the **ARC** command and pick a start point.

8 Enter **E** for End and pick an endpoint.

9 Enter **R** for Radius and enter a numerical value for the radius.

HINT: If you see *Invalid* at the Command line, try Steps 7 through 9 again, entering a larger number for the radius.

ARC Continue _____

Next, let's produce the following curved line, which is really a series of arcs.

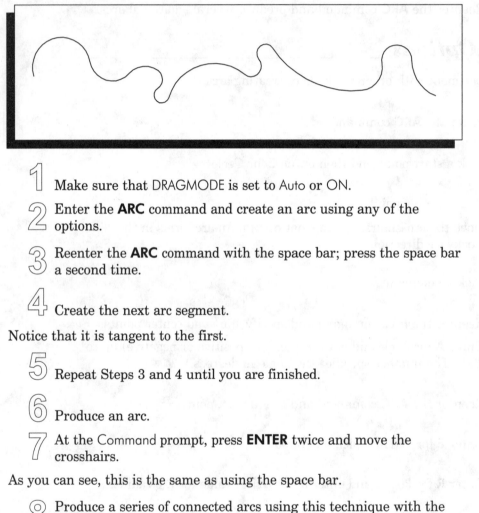

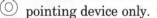 Make sure that DRAGMODE is set to Auto or ON.

2 Enter the **ARC** command and create an arc using any of the options.

3 Reenter the **ARC** command with the space bar; press the space bar a second time.

4 Create the next arc segment.

Notice that it is tangent to the first.

5 Repeat Steps 3 and 4 until you are finished.

6 Produce an arc.

7 At the Command prompt, press **ENTER** twice and move the crosshairs.

As you can see, this is the same as using the space bar.

8 Produce a series of connected arcs using this technique with the pointing device only.

ELLIPSE Command _____

The ELLIPSE command enables you to create mathematically correct ellipses.

1 Make sure DRAGMODE is set to Auto.

Draw

Draw

2 From the Draw toolbar, pick the **Ellipse** icon.

AutoCAD enters the ELLIPSE command.

3 Pick a point for Axis endpoint 1 as shown in the following illustration.

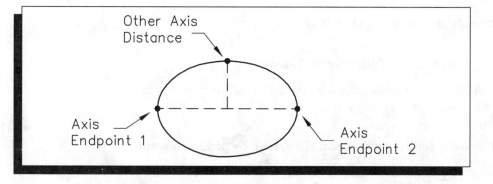

4 Pick a second point for Axis endpoint 2 directly to the right of the first point.

5 Move the crosshairs and watch the ellipse develop. Pick a point, or enter a numeric value, and the ellipse will appear.

6 Experiment with the Center option on your own.

Elliptical Arcs

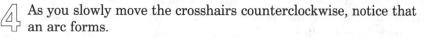

The ELLIPSE Arc option permits you to produce elliptical arcs.

Draw

1 Pick the **Ellipse** icon and enter **A** for Arc.

2 Pick three points similar to those shown in the previous illustration.

A temporary ellipse forms.

3 In reply to <start angle>, pick a point anywhere on the ellipse.

4 As you slowly move the crosshairs counterclockwise, notice that an arc forms.

5 Pick a point to form an arc.

DONUT Command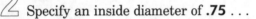

With the DONUT command, it's possible to create solid-filled donuts.

1 From the **Draw** pull-down menu, pick **Donut**, or enter **DONUT** at the Command prompt.

2 Specify an inside diameter of **.75** . . .

3 . . . and an outside diameter of **1.3**.

The outline of a donut locks onto the crosshairs and is ready to be dragged and positioned by its center.

4 Place the donut anywhere on the screen by picking a point.

5 Move the crosshairs away from the new solid-filled donut and notice the prompt line at the bottom of the screen.

6 Place several donuts on the screen.

7 Press **ENTER** to terminate the command.

That's all there is to the DONUT command.

8 Save your work and exit AutoCAD.

Standard

Time Out for a Discussion on Storing Your Drawings

In future units, you will not be instructed to store the files you create in any particular directory. This leaves it up to you to decide the best location for them. Once you've decided, be sure to specify the appropriate directory when you create and store new files. If you don't, the files will be stored in the default directory, a practice that is not recommended.

From this point forward, the AutoCAD directory is assumed to be AutoCAD R14. If the directory on the machine you're using is different, substitute the appropriate directory name for that shown in this book. Also note that, throughout this book, the terms "directory" and "folder" are used interchangeably.

HINT:
Back up your files regularly. The few seconds required to make a backup copy may save you hours of work. Experienced users back up faithfully because they have experienced the consequences of not doing it.

Questions

1. When drawing circles, arcs, and ellipses, what happens when the DRAGMODE command is set to Auto?

   _____

   _____

2. What happens when the DRAGMODE command is OFF?

   _____

   _____

3. Briefly describe the following methods of producing circles.

   2 Points _____

   3 Points _____

   TTR _____

4. In what AutoCAD toolbar are the ARC and CIRCLE commands found?

   _____

5. What function does the ARC Continue feature serve?

   _____

   _____

6. Explain the purpose of the DONUT command.

   _____

   _____

7. Describe the recommended procedure of storing AutoCAD files.

   _____

   _____

   _____

■ *Challenge Your Thinking: Questions for Discussion*

1. Review the information in this unit about specifying an angle in degrees. How might you be able to create an arc in a clockwise direction? Try your method to see if it works, and then write a paragraph describing the method you used.

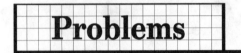

Problems

1-5. Using the commands you've just learned, complete the following drawings. Don't worry about text matter or exact shapes, sizes, or locations, but do try to make your drawings look similar to the ones below. You do not need to enter the text for "Lake AutoCAD."

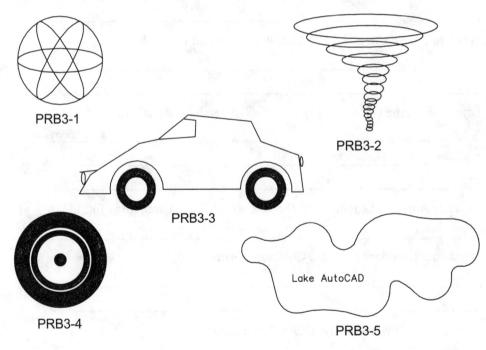

PRB3-1

PRB3-2

PRB3-3

PRB3-4

Lake AutoCAD

PRB3-5

6. Use the LINE, POLYGON, CIRCLE, and ARC commands to create this hex bolt.

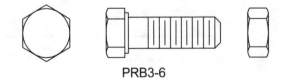

PRB3-6

7. Draw the eyebolt below using all of the commands you've learned.

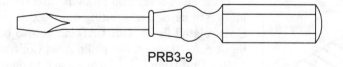

PRB3-7

8. Using the LINE, ARC, and CIRCLE commands, draw the following two views of a hammer head.

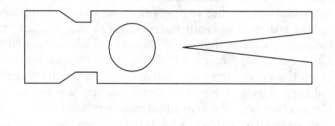

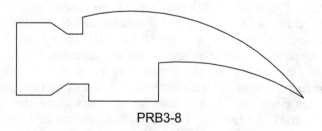

PRB3-8

9. Use the ARC and LINE commands to create this screwdriver.

PRB3-9

Problems 6 through 9 courtesy of Joseph K. Yabu, Ph.D, San Jose State University

AutoCAD® at Work

CAD Lab Becomes Red Bud V.I.P.

Red Bud High School's drafting program was in need of a revision. We were offering strictly board drafting, and we used no computers. Although industry had been supportive of our program in the past, we found that we were placing fewer students because they lacked entry-level computer skills.

We sent several proposals to the board for purchasing computers, but costs were prohibitive. As we researched our options, we discovered that working hand in hand with the private sector and promoting economic links with other educational and governmental agencies are two very important approaches to promoting our educational program.

Our goal was to begin instructing drafting students using the latest technology so that they could obtain good jobs in related fields. Our first big break came when V.I.P. (Vocational Instructional Practicum) provided the opportunity for Ron Weseloh, a faculty member, to work in industry during a summer period. As he worked at Snyder General Corporation and Red Bud Industries, a relationship formed. These companies agreed to match our funds to purchase the needed CAD equipment. VICA (Vocational Industrial Clubs of America) raised funds for the project.

After the first station was purchased and the first year's training took place, the computer, along with the CAD student who had trained on it, was placed in industry for summer employment. The first placement was very successful, and the results that followed were quite positive. Local industry responded with financial support as well as equipment and software donations. It has been seven years since the first system was obtained. We now have ten complete stations and three plotters. Over the years, students have competed in various CAD contests, including the VICA Skill Olympics in which they have won more than $32,000 worth of CAD software for the school.

Because of increased student interest in the program, more students are seeking post-secondary training in CAD-related fields. Our CAD lab is also being used for a night class through Belleville Area College.

Red Bud High School's approach to obtaining financial support from industry and other outside agencies is easily adoptable by other schools. The first requirement is a strong commitment by the school and its staff to the goal of developing a full CAD program. The other major requirements are flexibility, perseverance, and the desire to equip drafting students with highly marketable job skills.

Feature story courtesy of Ron Weseloh, Red Bud High School

Drawing by Dan Cowell,
Red Bud High School

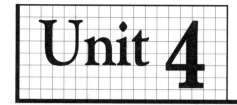

Unit 4   Now You See It . . .

■ **OBJECTIVE:**

To apply the ERASE and OOPS commands and object selection options

This unit shows you how to delete and restore all or part of a drawing. Also, the unit introduces you to AutoCAD entities, called *objects,* and the different ways you can select them.

First, a Word about Objects _____

An object, also called an *entity,* is an individual predefined element in AutoCAD. The smallest element that you can add to or erase from a drawing is an object.

The following list shows examples of object types in AutoCAD.

3D face	leader	shape
3D solid	line	solid
arc	mtext	spline
attribute	point	text
block	polyline	tolerance
circle	polymesh	trace
ellipse	ray	viewport
group	region	xline
hatch		

You'll learn more about these objects and how to use them as you complete the exercises in this book.

Erasing and Restoring Objects _____

 Start AutoCAD and start a new drawing from scratch.

_____ NOTE: _____

In this work-text, all new drawings use English units unless otherwise specified.

2 Using the **LINE** and **CIRCLE** commands, draw the following triangle and circle at any convenient size and location.

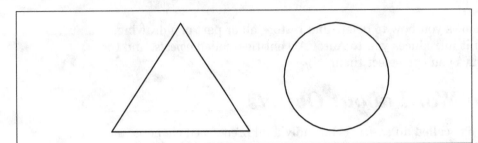

Modify

3 Enter the ERASE command by picking the **Erase** icon from the Modify toolbar or by entering **E** at the keyboard.

The Command prompt changes to Select objects. Notice that the crosshairs have changed to a small box. This is used to pick objects.

4 Type **W** for Window and press **ENTER**.

The crosshairs return.

5 Place a window around the triangle.

HINT: To do this, imagine a box or rectangle surrounding the triangle. Move the crosshairs to any corner of the imaginary box and pick a point. Then move the crosshairs to the opposite corner and pick a point. The triangle must lie entirely within the window.

Notice what happens. The object to be erased is highlighted with broken lines as shown in the following illustration.

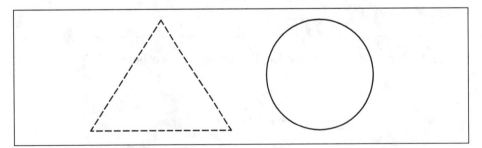

6 Press **ENTER** to make it disappear. (This also terminates the ERASE command.)

What if you erased an object by mistake and you want to restore it?

7 Enter the **OOPS** command at the keyboard.

The triangle reappears.

What if you want to select single objects, such as individual line segments?

Modify

1 Pick the **Erase** icon or enter **E** at the keyboard.

The prompt changes to Select objects.

2 Use the pointing device to pick two of the line segments in the triangle.

3 Press **ENTER** to make the highlighted objects disappear.

HINT:

As you may recall, if you are using a pointing device that has more than one button, pressing one of them, usually the second or third, is the same as pressing the ENTER key. This is usually faster than pressing ENTER at the keyboard.

Now let's try the Last option.

Modify

1 Issue **OOPS** to restore the line segments.

2 Enter the **ERASE** command.

3 Type **L** (for Last) and press **ENTER**.

The last object you drew should highlight.

4 Press **ENTER**.

If you continue to enter ERASE and Last, you will erase objects in the reverse order from which you created them.

Now let's try a new object selection procedure.

Modify

1 Enter **OOPS** to restore the objects or, if necessary, recreate the object(s).

2 Enter the **ERASE** command.

3 Without first entering W, place a window around the triangle. (See the following hint.)

HINT: You can select an object using a window without entering W (for Window). Produce a window by picking two corner points.

4 Type **R** (for Remove) and press **ENTER**.

Notice that the prompt line changes to Remove objects. You can now remove one or more lines from the selection set. The line(s) you remove will *not* be erased.

5 Remove one of the lines by picking one of them.

Note that the line you picked is no longer highlighted with broken lines.

6 Type **A** for Add and press **ENTER** to restore the Select objects prompt.

Notice that the prompt line changes back to Select objects.

7 Select the circle.

8 Press **ENTER** to make the objects shown in broken lines disappear.

So you see, you can select and remove objects as you wish until you are ready to perform the operation. The objects selected are indicated by broken lines. These selection procedures work not only with the ERASE command but also with other commands that require object selection, such as MOVE, COPY, MIRROR, ARRAY, and many others.

9 Enter **OOPS** to restore the objects to the screen.

Let's try another way of backing up or undoing objects you have selected.

1 Enter the **ERASE** command.

Modify

2 Pick two lines from the triangle, but do not press ENTER.

Two lines should now be highlighted.

3 Type **U** (for undo) and press **ENTER**.

Notice what happened to the last line you selected.

4 Type the **U** option again and press **ENTER**.

So you see, you can back up one step at a time with the Undo option.

5 Press **ENTER** to terminate the ERASE command.

Crossing Option

The Crossing option is similar to the Window option, but it selects all objects that cross the window boundary as well as those that lie within it.

Modify

1 Enter the **ERASE** command.

2 Type **C**, press **ENTER**, and pick a point in the center of the screen.

3 Move the crosshairs to form a box and notice that it is made up of broken lines.

4 Form the box so that it crosses over at least one object on the screen and press the pick button in reply to Other corner.

Notice that AutoCAD selected those objects which cross over the box.

5 Press **ENTER** to erase the object(s).

Box Option

The Box option enables you to apply the Window and Crossing options without the need to enter them.

Modify

1 Enter **OOPS** to restore the objects you erased in the previous section.

2 Enter the **ERASE** command.

3 Pick a point in the center of the screen in reply to Select objects.

4 Move the pointing device to the right, and then up or down to form a box, but do not pick a point.

If you were to pick a second point, you would erase all objects that lie entirely within this box. This is equivalent to the Window option.

5 Move the pointing device to the left of the first point, and then up or down to form another box.

Notice that the box is made up of broken lines. If you were to pick a second point, you would erase all objects that lie within or cross the box. This is equivalent to the Crossing option.

6 Pick a point and press **ENTER**.

7 Experiment further with this object selection option.

WPolygon, CPolygon, and Fence Options

The WPolygon, CPolygon, and Fence options are similar to the Window and Crossing options. However, they offer more flexibility.

1 Fill the screen with several objects, such as lines, circles, arcs, rectangles, polygons, ellipses, and donuts.

2 Enter the **ERASE** command.

Modify

3 In reply to Select objects, enter **WP** for WPolygon. (W is for Window.)

4 Pick a series of points that form a polygon of any shape around one or more objects and press **ENTER**.

NOTE:

When forming the polygon, AutoCAD automatically connects the last point with the first point.

5 Press **ENTER** to erase the objects.

The CPolygon option is similar to WPolygon.

Modify

1 Enter the **ERASE** command or press the space bar.

2 In reply to Select objects, enter **CP** for CPolygon. (C is for Crossing.)

3 Pick a series of points to form a polygon. Make part of the polygon cross over at least one object and press **ENTER**. (As you create the polygon, notice that it is made up of broken lines, indicating that the Crossing option is in effect.)

As you can see, CPolygon is similar to the Crossing option, whereas WPolygon is similar to the Window option.

4 Press **ENTER** to complete the erasure.

The Fence option is similar to the CPolygon option except that you do not close a fence as you do a polygon. When you select objects using the Fence option, AutoCAD looks for objects that touch the fence.

Modify

1 Enter **ERASE**.

2 Enter **F** for Fence.

3 Draw a line that crosses over one or more objects and press **ENTER**.

4 Press **ENTER** to complete the erasure.

 All Option _____

The All object selection option permits you to select all objects on the screen quickly.

1 Create several objects.

Modify

2 Enter the **ERASE** command.

3 In reply to Select objects, enter **All**.

AutoCAD selects all objects on the screen.

4 If you want to erase them all, press **ENTER**.

5 Exit AutoCAD. Do not save your work.

In the upcoming units, you will have the opportunity to apply these object selection techniques to all commands that require you to select objects. In Unit 15, you will experiment with advanced object selection and editing techniques. That unit describes the useful grips feature and the noun/verb selection method.

Questions

1. After you enter the ERASE command, what does AutoCAD ask you to do?

   _____

2. Experiment with and describe each of these object selection options.

   Multiple = _____

   Last   = _____

   Previous = _____

   Window   = _____

   Crossing = _____

   Box   = _____

   AUto   = _____

   SIngle   = _____

   Add   = _____

   Remove   = _____

   Undo   = _____

   WPolygon = _____

   CPolygon = _____

   Fence   = _____

   All   = _____

   Group   = _____

3. How do you place a window around a figure during object selection?

   _____

4. If you erase an object by mistake, how can you restore it?

   _____

   Will this method work if you draw something else after erasing the object?

   _____

5. How can you retain part of what has been selected for erasure while remaining in the ERASE command?

_____

6. What is the fastest way of erasing the last object you drew?

_____

 ### *Challenge Your Thinking: Questions for Discussion*

1. AutoCAD offers many object types. Using AutoCAD's Help, find a complete listing of them, along with brief definitions. Hint: Use the Command Reference or Index feature.

Problems

To gain skill in using the object selection options and the ERASE command, try the following exercise.

1. Create a new drawing named select.dwg. Draw polygons *a* through *e* as shown in the following illustration.

2. Enter the ERASE command and use the various object selection options to accomplish the following *without* pressing ENTER.

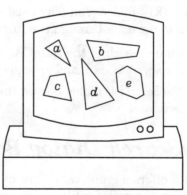

- Place a window around polygon *a*.
- Pick two of polygon *b*'s lines for erasure.
- Use the Crossing option to select polygons *c* and *d*.
- Remove one line selection from polygon *c* and one from polygon *d*.
- Pick two lines from polygon *e* for erasure but then remove one of the lines so it won't be erased.

3. Now press ENTER.

You should have nine objects (line segments) left on the screen.

4. Enter OOPS to restore the drawing to its original form.

5. Save your work and exit AutoCAD.

Unit 5 Help

■ **OBJECTIVE:**

To obtain help when using AutoCAD commands

Working with AutoCAD is easy as long as everything goes smoothly, but sometimes you can get stuck on how to use a command or feature. Help is available, and this unit shows you how to get it.

Drawing Preview

When you open a drawing, AutoCAD provides visual help, making the selection process as fast as possible.

1 Start AutoCAD and start a new drawing from scratch.

Standard

2 Pick the **Open** icon or select **Open...** from the **File** pull-down menu.

3 Locate and highlight the **car.dwg** file. (Car.dwg is the file from Unit 3.)

Notice that the drawing appears in a preview box in the right area of the dialog box. This allows you to find drawings more easily that you or others have created.

4 Find and double-click the **Sample** folder, which is located in the AutoCAD R14 folder. You may need to use the **Up One Level** button to navigate through the folders.

5 Click once on each of the AutoCAD drawing files, including campus.dwg, pipes.dwg, and tower.dwg.

6 Keep the Select File dialog box on the screen and proceed to the next section.

Browse/Search Dialog Box

This dialog box offers a graphical means of browsing a group of files. It also permits you to search for files using specific search criteria.

1 Pick the **Find File...** button.

The Browse/Search dialog box displays the graphical contents of all drawing files in the specified directory, making it easier to find a particular file.

Near the top, you should see two tabs labeled Browse and Search. AutoCAD displays either the browse or search information, depending on how it was left by the last user.

2 If the browse information is not present, pick the **Browse** tab.

3 In the Directories list box, double-click **AutoCAD R14**.

4 In the same list box, double-click **support**.

This displays the contents of the support directory.

5 In the lower right area of the dialog box, change Size to **Medium** using the down arrow. (It may already be set to Medium.)

This displays medium-size images of the drawings.

6 Use the scrollbar to browse the files.

Double-clicking any of these images will open the file. Do not open a file at this time.

7 Change Size to **Small**.

Performing Searches

1 Pick the **Search** tab, causing the search information to appear over the top of the browse information.

This part of the dialog box allows you to search for files according to their type and date of creation.

2 Change the text in the Search Pattern edit box to ***.dwg**, unless it is already set to this.

3 In the Date edit box, change the date to **1-1-97**.

4 In the Search Location area, pick the **Path** radio button and enter **\Program Files\AutoCAD R14** in the Path text box, assuming that this is the path to your AutoCAD directory.

5 Pick the **Search** button.

AutoCAD searches for DWG files in the AutoCAD R14 directory, as well as in all subdirectories in AutoCAD R14.

6 Use the scrollbar to browse the files.

7 Find the file named **WATCH.DWG** and double-click it.

This opens and displays the watch.dwg file.

HELP Command _____

Standard

1 Find and open the **car.dwg** file.

This automatically closes the watch.dwg file.

Several objects should now be on the screen.

Standard

2 Pick the **Help** icon from the Standard toolbar, or enter **HELP** or **?** at the keyboard.

HINT: You can also obtain help by pressing the F1 function key.

The AutoCAD Help dialog box appears as shown in the following illustration.

At this point, you can obtain instructions on how to use a particular command, or you can obtain an alphabetical list of AutoCAD topics.

③ Pick the **Contents** tab unless it is already in the foreground.

④ Double-click **Command Reference**.

⑤ Under Command Reference, double-click **Menus**.

This displays a new set of options.

⑥ Pick **File Menu** and then pick the green text **SAVEAS**.

As you can see, this provides information on the SAVEAS command, which is issued when you select the Save As... menu item.

Perhaps you can see a pattern developing. It's possible to obtain information on any menu item. This technique is commonly referred to as *context-sensitive help using hypertext links*.

⑦ Pick the **Back** button.

AutoCAD backs up to the previous help screen.

⑧ Pick the **Back** button again.

⑨ Close the AutoCAD Help dialog box by clicking the upper right corner (the ✕) or by selecting **Exit** from the File pull-down menu.

Searching _____

AutoCAD's help facility enables you to search for help on specific topics and commands.

① Reopen the **AutoCAD Help** dialog box.

② Pick the **Index** tab.

Standard

The following dialog box appears.

③ Use the scrollbar to scroll down through the list of topics.

Notice a blinking cursor inside the empty entry box.

④ Type **M** in upper or lower case, but *do not* press ENTER.

AutoCAD very quickly finds the first topic in the list box that begins with the letter M.

⑤ After the M, type **OV**, but *do not* press ENTER.

AutoCAD finds the first topic that begins with MOV. In this case, it finds Move (Modify menu).

⑥ Double-click the **MOVE** command.

Selections related to the MOVE command appear.

⑦ Double-click **MOVE command (ACR)**.

This produces information about the MOVE command. ACR stands for *AutoCAD Command Reference*.

⑧ Read the help information. Then pick **See Also** and read this information.

⑨ Pick a point anywhere on the screen.

10 Close the AutoCAD Help dialog box when you're finished.

Obtaining Help After Entering a Command

There is another very useful way of obtaining help.

1 Enter the **PLINE** command at the keyboard.

2 In reply to From point, pick the **Help** icon.

Standard

As you can see, AutoCAD allows you to obtain help in the middle of a command, when you're most likely to need it. When you close the AutoCAD Help dialog box, the command resumes.

3 After closing the AutoCAD Help dialog box, press the **ESC** key to cancel the PLINE command.

4 Open the AutoCAD Help dialog box, pick the **Find** tab, and experiment with it on your own.

5 Exit AutoCAD without saving.

Questions

1. After you enter the HELP command, what does AutoCAD display?

2. How do you obtain a listing of all AutoCAD topics and commands?

3. Suppose you have entered the MIRROR command. At this point, what is the fastest way of obtaining help on the MIRROR command?

4. How do you obtain help on the menu items contained in AutoCAD's pull-down menus?

■ *Challenge Your Thinking: Questions for Discussion*

1. If you forgot how to use AutoCAD's on-screen help, how could you find out how to use it?

2. For some commands, the help information may be long and complicated. Discuss ways to make a permanent hard copy of commonly used help information to keep for reference.

3. Explain the benefit of using the Find capability in AutoCAD Help dialog box. When would you use it instead of the Index?

Problems

1. Obtain AutoCAD help on the following commands and state the basic purpose of each.

   EXPORT _____

   GRID _____

   TIME _____

   SAVETIME _____

   ORTHO _____

2. Locate each of the above commands in the *AutoCAD User's Guide* and read what it says about each of them.

AUTOCAD® AT WORK

AutoCAD, the Ultimate Furniture Mover

As might be expected. AutoCAD makes room design easier. Moving a chair or even a wall here or there and then printing the different results helps designers plan the most efficient use of space. However, what about really *big* areas? What about a facility comprising 1.6 million square feet under one roof?

Well, for big areas, AutoCAD is almost indispensable. The Indiana Convention Center & RCA Dome in Indianapolis makes available to its clients one million square feet of exhibit space, banquet halls, ballrooms, and meeting rooms. The center uses AutoCAD to make the floor plans for upcoming events. The plans show seating, stage and riser locations, rows of booths, even positions of indoor tracks and basketball courts.

Every meeting room, convention hall, ballroom, and even the 60,000-seat RCA Dome is stored in the computer as a block and appears on a customized screen menu. Another customized menu lists the different sizes of stages, risers, podiums, chairs, and tables used. Many of the blocks are drawn as generic rectangles so that a block representing a stage, for instance, might also be used to represent a dance floor. More specialized items, such as pianos, are included in the tablet menu also.

Working from client specifications, the center's event coordinators start with a room from a screen menu and insert major features needed, such as a dance floor or basketball courts. Then, the coordinator adds any standard seating arrangements, such as bleacher sections, that are of a uniform size. If rows of tables or seating are required, the coordinator inserts one item of that type from the existing block, then uses the ARRAY command to generate the required number of units. Afterwards the arrays can be moved, in whole or in part, allowing for staggered table arrangements or aisles between seating sections. An array can be copied or mirrored to create an angled section, to avoid doing the same array again and again. Notes, arrows, and other details, such as the distance between sections of seats, are then added. The floor plan is then ready for the plotter.

One of the benefits of using AutoCAD to plan events is that many events are rescheduled year after year. Only minor changes may need to be made to the previous year's diagram, saving hours of planning time.

Some of the events that have been planned using AutoCAD include the 1987 Pan American Games, Indiana Black Expo, Indianapolis Colts football, the Rolling Stones Voodoo Lounge concert tour, Farm Aid IV, NCAA Indoor Track and Field Championships, and the Jefferson Jackson Banquet, an event President Clinton attended in May 1994.

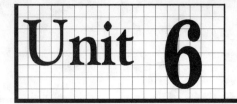
Unit 6

Becoming a Keyboard Artist

■ **OBJECTIVE:**

To enter coordinates using the absolute, relative, polar, and direct distance methods

So far, you have used only the pointing device to enter coordinates. Another method is to use the keyboard to specify coordinates. This method allows you to specify points and draw lines of any specific length and angle. This method also applies to creating arcs, circles, and other object types.

AutoCAD uses a Cartesian coordinate system. (See the illustration below.) The *origin* is the point where the values of x and y are both zero. On the computer screen, the origin is usually located at the lower left corner. AutoCAD calls this system the *world coordinate system* (WCS). Temporary coordinate systems whose origins are specified by the user are also available in AutoCAD. They are called *user coordinate systems* (UCSs) and are explained fully in Unit 42.

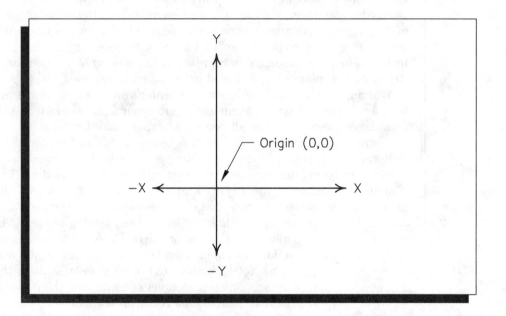

A point is expressed as an *(x,y)* coordinate pair. For two-dimensional (2D) drafting, all work can be done using *(x,y)* coordinates. If your work involves three-dimensional (3D) modeling, the Z axis must be added to locate 3D points using an *(x,y,z)* coordinate triple. The Z axis is perpendicular to the plane defined by the X and Y axes. Refer to Units 41-49 and 51-57 for details on how to apply AutoCAD's 3D capabilities.

Methods of Entering Points _____

Consider the following three ways to specify coordinates when using the LINE command.

Absolute Method

Example:

> LINE From point: 2,3
> To point: 5,8

This begins the line at absolute point 2,3 and ends it at 5,8.

Relative Method

Example:

> LINE From point: 2,3
> To point: @2,0

This draws a line 2 units in the positive X direction and 0 units in the Y direction from point 2,3. In other words, the distances 2,0 are relative to the location of the first point.

Polar Method

Example:

> LINE From point: 2,3
> To point: @4<60

This produces a line segment 4 units long at a 60-degree angle. The line begins at point 2,3.

The polar method is useful for producing lines at a precise angle. Note the following illustration. If you specify an angle of 90 degrees, the line extends upward vertically from the last point.

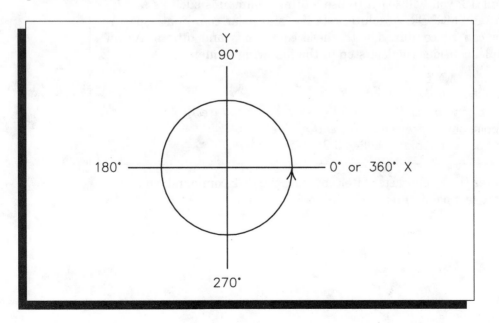

1 Start AutoCAD and start a new drawing from scratch.

2 Enter the **LINE** command and input the following sequence to produce a drawing. Don't forget to press **ENTER** after each entry.

LINE From point: **4,3**
To point: **@3,0**
To point: **@2.5<90**
To point: **C** *(and press **ENTER**)*

These entries produce a triangle, as shown in the following illustration.

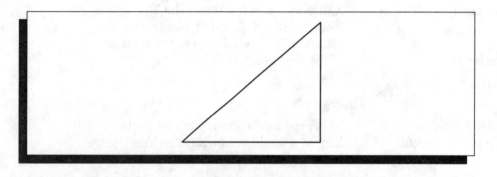

NOTE:

In Unit 12, you will learn to use editing commands such as MOVE, COPY, and MIRROR. Absolute, relative, and polar coordinate specification can be combined with these commands and others. As an example, consider the last step in the following sequence.

Command: MOVE
Select objects: *(select a line)*
Select objects: *(press ENTER to complete the selection set)*
Base point or displacement: *(pick end of line)*
Second point of displacement: @3.25<45

If you enter the @ character after entering the LINE command, you will specify the last point entered. Try it.

Direct Distance Entry

AutoCAD enables you to enter precise line lengths quickly using the pointing device to indicate the direction. This is faster than entering the direction at the keyboard.

Draw

1 Enter the **LINE** command and pick a point anywhere on the screen.

2 Move the crosshairs up and to the right from the first point, but do not pick a second point.

3 At the To point prompt, enter **2**. (Be sure to press **ENTER**.)

AutoCAD creates a line segment 2 units long in the direction you specified with the pointing device.

4 Produce a series of line segments using this method.

Creating a Gasket

1 Erase the objects you just created so that the screen is blank.

Draw

2 Enter the **LINE** command again.

3 Using the keyboard, create the following drawing of a gasket. Don't worry about the exact sizes and locations of the holes, and do not try to place the dimensions on the drawing at this time. However, do make the drawing exactly this size by applying what you've learned thus far in this unit. The top and bottom edges of the gasket are perfectly horizontal.

HINT:
Refer to the previous two pages. Use the U (Undo) option if you need to back up one step but remain in the LINE command.

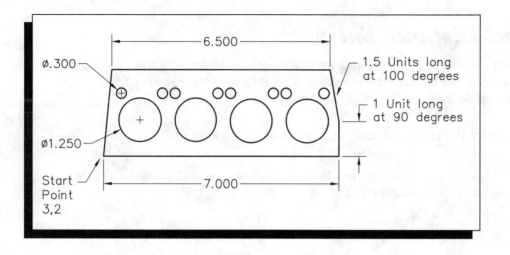

NOTE:

You can enter negative values for line lengths and angles, as shown here.

LINE from point: 5,5
To point: @2<–90

This polar point specification produces a line segment 2 units long downward vertically from absolute point 5,5. This is the same as entering @2<270. Try it and then erase the line.

4 Save your work in a file named **gasket.dwg** and exit AutoCAD.

Standard

Questions

1. Briefly describe the differences between the absolute, relative, and polar methods of point specification.

   _____

   _____

   _____

2. Is there an advantage to specifying endpoints from the keyboard rather than with the pointing device? Explain.

   _____

   _____

3. What is the advantage of specifying endpoints with the pointing device rather than the keyboard?

   _____

   _____

4. How can you back up one step if you make a mistake in specifying line endpoints?

   _____

5. What happens if you enter the @ character at the From point prompt after entering the LINE command?

   _____

6. What is the effect on the direction of the new line segment when you specify a negative number for polar coordinate entry?

   _____

Challenge Your Thinking: Questions for Discussion

1. Why may entering absolute points be impractical much of the time when completing drawings?

2. Consider the following command sequence:

   LINE From point: 4,3
   To point: 5,1

   What polar coordinates could you enter at the To point prompt to achieve exactly the same line?

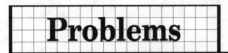

Problems

For drawing problems 1 and 2, list exactly what you would enter when using the LINE command to produce the drawings. Try to incorporate all three methods—absolute, relative, and polar—to enter the points. After completing all of the blanks, step through the sequence in AutoCAD. Note that the horizontal lines in the drawings are perfectly horizontal.

1. Command: LINE From point: _____

   To point: _____

   To point: _____

   To point: _____

   To point: _____

   To point: _____

   To point: _____

PRB6-1

2. Command: LINE From point: _____

   To point: _____

   To point: _____

   To point: _____

   To point: _____

   To point: _____

   To point: _____

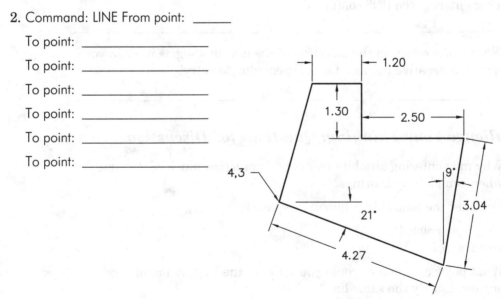

PRB6-2

3. Using @ and <, create the following drawing of a 5¼″ floppy diskette. The slotted hole is optional.

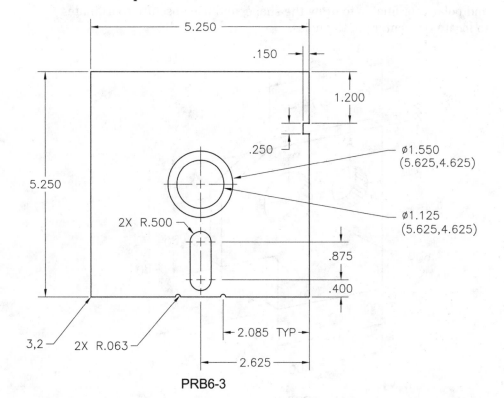

PRB6-3

4. Draw this 3½″ diskette.

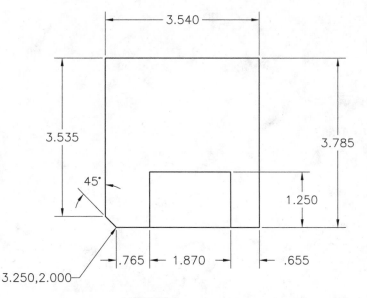

PRB6-4

5. Draw the front view of the spacer plate shown below. Start the drawing
   at the lower left corner using absolute coordinates of 3,3. Use relative
   and polar coordinates to draw the shape, and use absolute coordinates
   to locate the centers of the holes.

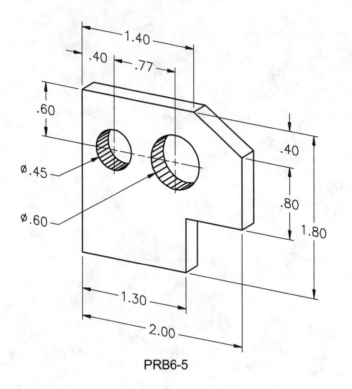

PRB6-5

6. Draw the front view of the locking end cap shown below. Start the drawing at the center of the object using absolute coordinates of 5,5. Use absolute polar coordinates to locate the centers of the four holes, and use absolute coordinates with the RECTANG command to draw the rectangular piece.

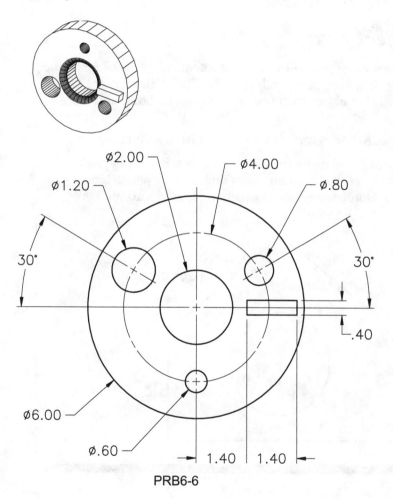

PRB6-6

Problems 3 and 4 courtesy of Joseph K. Yabu, Ph.D., San Jose State University
Problems 5 and 6 courtesy of Gary J. Hordemann, Gonzaga University

Unit 7 Snagging Points

■ **OBJECTIVE:**

To apply the object snap feature

This unit uncovers the powerful object snap capability and shows how to change the settings associated with it.

Using Object Snap

There will be times when you will want to automatically "grab" a specific point in your drawing, such as an endpoint of a line or the center of a circle. With AutoCAD's object snap feature, you can do both, and more.

1 Start AutoCAD and start a new drawing from scratch.

2 In preparation for using the object snap feature, draw the following object. Omit all numbers and, at this point, ignore the words. Don't worry about exact sizes and locations.

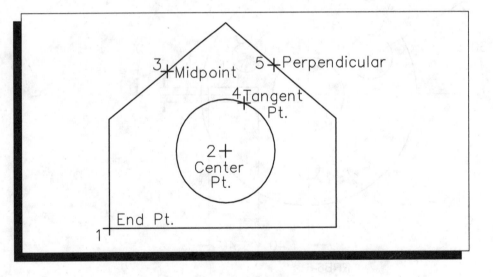

3 Save your work in a file named **objsnp.dwg**.

Let's practice using the object snap feature. Before we begin, we need to review object snap settings, which affect the appearance of the object snaps.

1 Enter the **OSNAP** command.

This displays the Osnap Settings dialog box.

2 Pick the **AutoSnap** tab.

3 If they are not already checked, check **Marker**, **Magnet**, and **SnapTip**. If Display aperture box is checked, uncheck it now, and change Marker color to yellow if it is another color.

You will learn what all of these settings mean later in the unit.

4 Pick the **OK** button to close the dialog box.

5 Enter the **LINE** command.

6 In reply to From point, type **ENDP** to specify the Endpoint object snap mode and press **ENTER**.

The word of appears in the prompt area of the screen.

7 Move the crosshairs so that the the center of it touches the horizontal or vertical line near point 1, and let the crosshairs rest in place.

A small yellow box appears at point 1, along with the word Endpoint. This means that if you were to pick a point at this time, you would snap to the endpoint of the line.

8 Press the pick button to snap to point 1.

Let's experiment with another object snap mode.

9 In response to the To point prompt on the screen, type **CEN** (for Center) and press **ENTER**.

10 Move the crosshairs over the line that makes up the circle. When the small yellow circle appears, press the pick button on the pointing device.

The line snaps to the center of the circle.

11 Press **ENTER** to terminate the LINE command.

Object Snap Toolbar

The Object Snap toolbar can ease the process of entering object snap modes.

1 From the **View** pull-down menu, pick **Toolbars...** and pick the **Object Snap** check box.

The Object Snap toolbar appears, as shown in the following illustration.

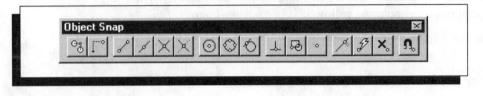

2 Pick the **Close** button in the Toolbars dialog box.

The following is a list of all the object snap modes, their corresponding icons, and their meaning. To enter an object snap mode at the keyboard, type only the capitalized letters.

APParent
intersection — snaps to the apparent intersection of two objects which may not actually intersect in 3D space

CENter — center of arc or circle

ENDPoint — closest endpoint of line or arc

FROm — allows you to use any point on the screen as a reference or base point for coordinate entry

INSertion — insertion point of text, block, or attribute

INTersection — intersection of lines, arcs, or circles

MIDpoint — midpoint of line or arc

NEArest — nearest point on a line, arc, circle, ellipse, polyline, spline, and other object types

NODe — nearest point entity or dimension definition point

OSNAP — presents the OSNAP Settings dialog box, from which you can set or clear object snaps in a single operation

PERpendicular — perpendicular to line, arc, or circle

QUAdrant — quadrant point of arc or circle

QUIck — quickly selects the first snap point found

TANgent — tangent to arc or circle

TRACKing — locates a point using a series of temporary points

NONe — temporarily cancels all running object snaps (for one operation only)

1 Enter the **LINE** command.

2 Pick the **Snap to Center** icon from the Object Snap toolbar.

Object Snap

3 Select the circle.

4 Snap to point 3 using the **Snap to Midpoint** icon. Be sure that a small yellow triangle appears at the midpoint.

Object Snap

Object Snap

5 Snap to point 4 using the **Snap to Tangent** icon.

6 Snap to point 5 using the **Snap to Perpendicular** icon.

Object Snap

7 Press **ENTER** to terminate the LINE command.

Your drawing should now look like the following. If it doesn't, try it again.

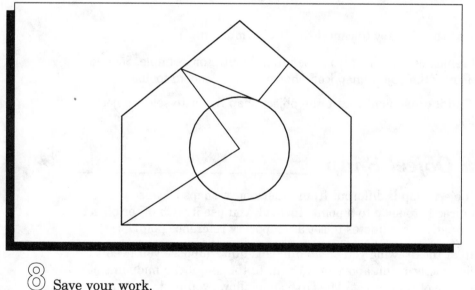

8 Save your work.

9 Experiment with each of the remaining object snap modes.

NOTE:

Even though the Object Snap toolbar may seem convenient, you may choose to type the modes as you need them.

Object Snap Flyout _____

AutoCAD offers a flyout toolbar that provides the full set of object snap mode icons. A flyout appears when you pick an icon that contains a small black triangle in its lower right corner. Each flyout presents icons that are related to the icon that displays it.

1 Enter the **LINE** command.

2 From the Standard toolbar, pick and hold the **Tracking** icon. Hold down the pick button until the flyout appears.

Standard

As you can see, the flyout provides object snap icons that are identical to the ones in the Object Snap toolbar.

Standard

3 Move the pointer down to **Snap to Quadrant** and select it.

4 Move the crosshairs around the circle, causing yellow diamonds to appear at the quadrant points of the circle.

5 Press the **ESC** key to cancel the LINE command.

AutoCAD remembers the last icon used in a flyout. For example, Snap to Quadrant is now the object snap icon shown in the Standard toolbar.

You may find it convenient to use the object snap flyout to select object snap modes.

FROm Object Snap _____

The From object snap is different from other object snaps because you do not use it directly to snap to a point. Instead, you use it with other object snaps and relative coordinates to set a temporary reference point.

Look again at the drawing you created in this unit. Suppose you need to add a circle exactly 1 unit above and 1 unit to the left of the midpoint of the top left line of that object. The From snap allows you to do so quickly and easily.

1 Enter the **CIRCLE** command.

Object Snap

2 Pick the **Snap From** icon in response to the Center Point prompt.

HINT: The Snap From icon is located in the Standard toolbar flyout, as well as in the Object Snap toolbar.

3 At the from Base point prompt, pick the **Snap to Midpoint** icon and pick the top left line of the object.

4 In response to Offset, enter **@–1,1**.

AutoCAD places the center of a circle precisely 1 unit to the left and 1 unit above the midpoint of the line. To finish the circle, you have only to specify the radius or diameter.

5 Specify a radius of **1** unit.

Your drawing should now look like the one below.

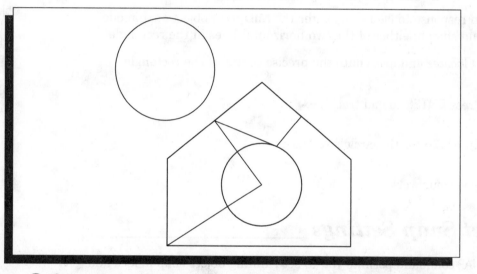

6 On your own, place another circle with its center point exactly **.75** unit below and **1.5** unit to the right of the lower right corner of the object. Specify a radius of **1.25** for the circle.

7 Save your work.

Tracking

You can use tracking to locate points visually relative to other points in the drawing. You can use tracking, for instance, to find the center of a rectangle.

1 Using the **RECTANG** command, create a rectangle.

Suppose you want to create a circle at the rectangle's center point.

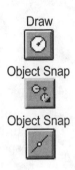

2 Enter the **CIRCLE** command.

3 In reply to <Center point>, enter **tk** at the keyboard or pick the **Tracking** icon from either the flyout or the floating toolbar.

4 In response to First tracking point, enter the **MIDpoint** object snap mode.

5 Snap to either of the two vertical lines in the rectangle.

6 Move the crosshairs to the right and left and notice that AutoCAD forces the line to be perfectly horizontal.

7 In response to Next point, enter the **MIDpoint** object snap mode and snap to either of the two horizontal lines in the rectangle.

AutoCAD locates and locks onto the precise center of the rectangle.

8 Press **ENTER** to end tracking.

9 Enter **.25** for the circle's radius.

10 Save your work.

Object Snap Settings _____

The OSNAP command permits you to set running object snap modes and other settings.

1 Pick the **Object Snap Settings** icon from the toolbar, or enter the **OSNAP** command at the keyboard.

The following dialog box appears.

Osnap Settings

Running Osnap | AutoSnap(TM)

Select settings

☐ ☑ Endpoint   ☐ Insertion
△ ☐ Midpoint   ☐ Perpendicular
○ ☑ Center   ☐ Tangent
⊠ ☐ Node   ☐ Nearest
◇ ☐ Quadrant   ☐ Apparent Int
✕ ☑ Intersection ☐ Quick

Clear all

Aperture size

OK   Cancel   Help

NOTE:

If no running object snaps are currently set, pressing the F3 function key also displays the Running Object Snap dialog box. If running object snaps are selected, the F3 key toggles the selected snaps on and off.

2 If the Running Osnap tab does not appear in the foreground, pick the tab to display the running object snap options.

3 Check the **Endpoint, Midpoint, Center, Quadrant, Intersection,** and **Tangent** check boxes and pick the **OK** button.

4 Enter the **ARC** command and slowly move the crosshairs over the top of the first drawing you created.

Draw

As you can see, the running snap modes permit you to select specific points on the object easily and accurately.

5 With the **ARC** command entered, rest the crosshairs on the circle, and press the **Tab** key several times.

Pressing the Tab key cycles through all of the running snap modes related to the circle. This is useful when an area is densely covered with lines and other objects, making it difficult to snap to a particular point.

NOTE:

You can toggle the running snap modes on and off using the OSNAP button located on the status bar.

6 Snap to any point on the circle and create an arc of any size.

Modify

7 Erase the arc.

Other Osnap Settings

Other settings permit you to customize the object snap feature.

Object Snap

1 Pick the **Object Snap Settings** icon and the **AutoSnap** tab in the dialog box.

Marker, Magnet, and SnapTip should be checked. When Marker is checked, AutoCAD displays geometric shapes at each of the snap points. When Magnet is checked, AutoCAD locks the crosshairs onto the snap point. When SnapTip is checked, AutoCAD displays a tooltip-like flag that describes the name of the snap location.

2 Check **Display aperture box**.

This causes an aperture box to appear at the center of the crosshairs when you snap to an object.

3 Under Marker size, move the slider bar slightly to the right, increasing the size of the markers.

4 Under Marker color, pick the down arrow to display a list of colors and select red.

5 Pick the **Running Osnap** tab.

6 Under Aperture size, slightly increase the size of the aperture using the slider bar.

7 Pick the **OK** button and enter the **LINE** command.

8 Move the crosshairs over the drawing and notice the changes.

NOTE:

Notice that the snap markers do not appear until the aperture box contacts the object. Also, pay attention to the magnet feature.

9 Press the **ESC** key to cancel the LINE command.

The running snap modes are stored with the drawing, not with AutoCAD. The AutoSnap settings, including the display of the aperture box and the color and size of the markers, are stored independently of the drawing files. This means that AutoCAD remembers the changes.

Object Snap

10 Display the **Osnap Settings** dialog box, uncheck **Display aperture box**, reduce the size of the markers, and change the marker color to yellow.

11 Pick the **OK** button.

12 Close the Object Snap toolbar, save your changes, and exit AutoCAD.

Questions

1. Explain the purpose of the object snap modes.

   _____

   _____

   _____

2. In order to snap a line to the center of a circle, what part of the circle must the crosshairs touch?

   _____

3. What is the benefit of using running object snap modes?

   _____

   _____

4. Describe a situation in which you would want to change the aperture box size.

   _____

   _____

   _____

5. Briefly describe the use of each of the following object snap modes.

   Apparent Intersection _____

   _____

   Center _____

   _____

   Endpoint _____

   _____

   Insertion _____

   _____

   Intersection _____

   _____

   Midpoint _____

   _____

   Nearest _____

   _____

   Node _____

   _____

74

⊥ Perpendicular _____

_____

◯ Quadrant _____

_____

⚡ Quick _____

_____

◯ Tangent _____

_____

✕ None _____

_____

6. Name one of two ways to access the Osnap Settings dialog box.

_____

_____

■ *Challenge Your Thinking: Questions for Discussion*

1. Imagine a new object snap mode that does not currently exist in AutoCAD. The new snap mode must provide a useful service. Write a paragraph describing the new object snap. Explain what it does and why it is useful.

2. Explain the difference between the Snap From and Tracking object snap features.

Problems

1. Draw the square on the left. Then use the appropriate object snap modes to make the additions shown on the right.

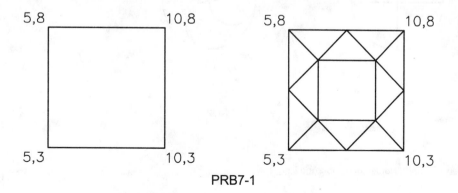

PRB7-1

2. Using object snap, construct the following object. Don't worry about exact sizes and locations.

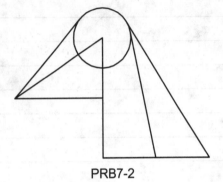

PRB7-2

3. Draw the top view of this object.

PRB7-3

Start by drawing a circle of any size at any location. Then use the Quadrant and Center object snaps to draw the *four* lines as shown in the figure below, left. (You must draw four lines, not two.) Use the ARC command with the Start, Center, End option and Quadrant, Midpoint, and Center snap modes to draw the arcs. Erase the lines.

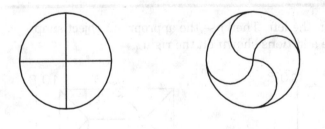

Problem 3 courtesy of Gary J. Hordemann, Gonzaga University

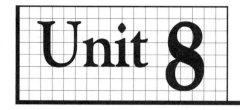

Unit 8 Five Favorable Features

■ OBJECTIVE:

To control certain AutoCAD features, including the coordinate display, ortho, the TIME command, the AutoCAD Text and Command Line windows, and the SAVETIME system variable

This unit describes several special AutoCAD features. You will have an opportunity to practice these features and learn how to apply them in an AutoCAD work environment.

Coordinate Display

The display of coordinate information is a helpful feature. This information, displayed in digital form, is found in the lower left corner of the graphics screen and is part of the status bar. The coordinate display tracks the current position (or coordinate position) of the crosshairs as you move the pointing device. It also gives the length and angle of line segments as you use LINE and other commands.

1 Start AutoCAD and start a new drawing from scratch.

2 Move the crosshairs on the screen by moving the pointing device.

Note how the coordinate display changes with the movement of the crosshairs.

3 Enter the **LINE** command and draw the first line segment of a polygon. Note the coordinate display as you draw the line segment.

———— NOTE: ————

Most personal computer keyboards have function keys called F1, F2, F3, and so on. AutoCAD has assigned many of these function keys to specific functions, such as the coordinate display and the ortho feature.

4 Press the **F6** function key once, draw another line segment, and notice the coordinate display.

It should now be off.

5 Press **F6** again, draw another line segment or two, and watch the coordinate display.

As you can see, this information is similar to specifying line endpoints using the polar method of specifying coordinates.

⑥ While in the LINE command, press the **CTRL** and **D** keys and note the change in the coordinate display as you pick points.

⑦ Create additional line segments, pressing **CTRL D** between each.

As you can see, CTRL D serves the same function as F6. You can also double-click the coordinates in the status bar to toggle the coordinates.

NOTE:

This feature may not work with some display systems.

The Ortho Mode

Now let's focus on an AutoCAD feature called *ortho*. Ortho is a useful feature that allows you to draw horizontal or vertical lines quickly and easily.

① Press the **F8** function key, or double-click the word **ORTHO** in the status bar. (Both actions perform the same function.)

Ortho is on whenever the word ORTHO is present (not grayed out) in the status bar at the bottom of the screen. As you can see, you can toggle ortho on and off by double-clicking the word ORTHO in the status bar.

② Experiment by drawing lines with ortho turned on and then with ortho off. Note the difference. (See the following hint.)

HINT:

Like the coordinate display feature, ortho can be toggled on and off at any time, even while you're in the middle of a command such as LINE.

③ Attempt to draw an angular line with ortho on.

Can it be done?

4 Now, clear the screen if necessary and draw the following plug, first with ortho off and then with ortho on. Don't worry about exact sizes and locations.

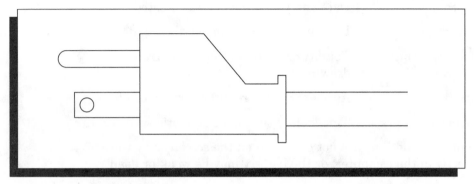

Was it faster with ortho on?

TIME Command

AutoCAD keeps track of time while you work. With the TIME command, you can review this information.

1 Enter **TIME** at the keyboard or select the **Tools** pull-down menu and pick **Inquiry** and **Time**.

Text appears on the screen, providing information similar to the following. Dates and times will of course be different.

```
AutoCAD Text Window                                    _ □ ✕
Edit
To point:
To point:
To point:
To point:
To point:
To point:
To point:
To point:
To point:
To point:
To point:
To point:

Command: time

Current time:              Sunday, June 15, 1997 at 12:22:56:430 PM
Times for this drawing:
  Created:                 Sunday, June 15, 1997 at 12:22:04:300 PM
  Last updated:            Sunday, June 15, 1997 at 12:22:04:300 PM
  Total editing time:      0 days 00:00:52.130
  Elapsed timer (on):      0 days 00:00:52.130
  Next automatic save in:  0 days 01:59:12.050

Display/ON/OFF/Reset:
```

HOURS MINUTES SECONDS MILLISECONDS

Here's what this information means.

Current time:	Current date and time
Created:	Date and time drawing was created
Last updated:	Date and time drawing was last updated
Total editing time:	Total time spent editing the current drawing
Elapsed timer:	A timer you can reset or turn on or off
Next automatic save in:	Time before the next automatic save occurs

If the current date and time are not displayed on the screen, they were not set correctly in the computer, or the battery may be weak or dead.

2 If the date and time are incorrect, minimize AutoCAD (pick the minimize button at the top right corner of the screen), and reset the date and time by picking **Control Panel** and **Date/Time**.

3 If you minimized AutoCAD, maximize it now by clicking the AutoCAD button at the bottom of the screen and enter the **TIME** command.

What information is different? Check the current time, total editing time, and elapsed time.

NOTE:

The times, found after the day, month, and year, are displayed to the nearest millisecond using the 24-hour "military" format. For example, 14:15:00.000 means 2:15 P.M.

4 With AutoCAD's TIME command entered, type **D** for Display, and notice what displays on the screen.

This provides updated time information.

5 Enter the **OFF** option, and display the time information again.

The elapsed timer should now be off.

NOTE:

An example of when you might specify OFF is when you want to leave the computer to take a break. When you return, you turn the timer ON. This keeps an accurate record of the actual time (elapsed time) you spend working on the project.

6 Reset the timer by entering **R**, and display the time information once again.

The elapsed timer should show 0 days 00:00:00.000, with the exception of a second or two that might have passed.

7 Last, turn on the timer and display the time information.

Notice that the elapsed timer has kept track of the time only while the timer was turned on.

Why is all of this time information important? In a work environment, the TIME command can track the amount of time spent on each project or job, making it easier to charge time to clients.

8 Press **ENTER** to terminate the TIME command, but do not close the AutoCAD Text Window.

AutoCAD Text Window

As you are aware, AutoCAD displays the time information in the AutoCAD Text Window. AutoCAD uses this window for various purposes, as you will see in future units.

1 Pick a point in the drawing area but outside the AutoCAD Text Window.

This causes the graphics screen to appear in front of the AutoCAD Text Window, hiding it from view.

2 Press the **F2** function key to make the AutoCAD Text Window come to the front.

3 In the upper right corner of the AutoCAD Text Window, pick the minimize button (the dash).

4 Press **F2** or pick the **AutoCAD Text Window** button at the bottom of the screen to make it full size.

NOTE:

You can press F2 at any time to display the AutoCAD Text Window. F2 serves as a toggle switch, permitting you to toggle the display of the window from the front to the back and vice versa.

5 Using the vertical scrollbar, scroll up the list of text in the AutoCAD Text Window.

Observe that AutoCAD maintains a complete history of your activity. Also, notice that the AutoCAD Text Window offers an Edit pull-down menu.

⑥ Select the **Edit** pull-down menu from the AutoCAD Text Window.

Paste to CmdLine pastes the highlighted text to the Command line. The Copy item permits you to copy a selected portion of the text to the Windows Clipboard. The Clipboard is a memory space in the computer that temporarily stores information (text and graphics). After copying information to the Clipboard, it is easy to paste it into another software application.

Copy History copies the entire contents of the AutoCAD Text Window to the Windows Clipboard. Paste enables you to paste the contents of the Clipboard to the Command line. The Preferences... menu item permits you to display the Preferences dialog box, which allows you to change various settings to meet individual needs. The Preferences dialog box will be covered in a later unit.

⑦ Using the pointer, highlight the most recent time information, as shown in the following illustration.

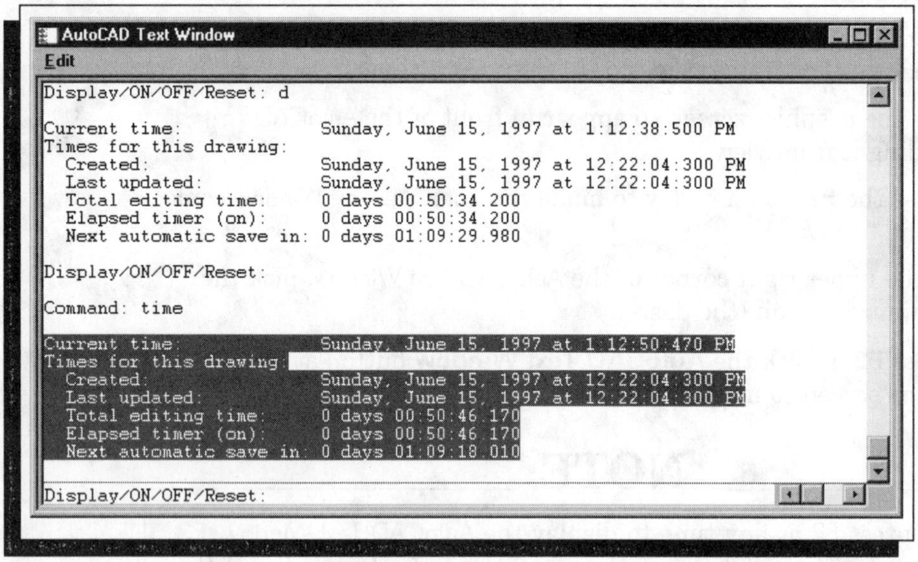

```
▤ AutoCAD Text Window                                    _ □ ✕
E dit
┌────────────────────────────────────────────────────────┬─┐
│ Display/ON/OFF/Reset: d                                 │▲│
│                                                         │ │
│ Current time:            Sunday, June 15, 1997 at 1:12:38:500 PM │ │
│ Times for this drawing:                                 │ │
│   Created:               Sunday, June 15, 1997 at 12:22:04:300 PM │ │
│   Last updated:          Sunday, June 15, 1997 at 12:22:04:300 PM │ │
│   Total editing time:    0 days 00:50:34.200            │ │
│   Elapsed timer (on):    0 days 00:50:34.200            │ │
│   Next automatic save in: 0 days 01:09:29.980           │ │
│                                                         │ │
│ Display/ON/OFF/Reset:                                   │ │
│                                                         │ │
│ Command: time                                           │ │
│                                                         │ │
│ Current time:            Sunday, June 15, 1997 at 1:12:50:470 PM │ │
│ Times for this drawing:                                 │ │
│   Created:               Sunday, June 15, 1997 at 12:22:04:300 PM │ │
│   Last updated:          Sunday, June 15, 1997 at 12:22:04:300 PM │ │
│   Total editing time:    0 days 00:50:46.170            │ │
│   Elapsed timer (on):    0 days 00:50:46.170            │▼│
│   Next automatic save in: 0 days 01:09:18.010           │ │
├────────────────────────────────────────────────────────┴─┤
│ Display/ON/OFF/Reset:                          ◄│ │  │►  │
└──────────────────────────────────────────────────────────┘
```

⑧ Pick the **Copy** item from the **Edit** pull-down menu in the AutoCAD Text Window.

⑨ Close the AutoCAD Text Window by clicking the **Close** button (the ✕) in the upper right corner of the window.

⑩ Minimize AutoCAD by picking the minimize button (the dash) located in the upper right corner.

11 Launch a text editor such as Notepad. (Notepad is a standard Windows application and is a part of the Accessories group of Windows utilities.

12 After launching the text editor, select **Paste** from the **Edit** menu.

This pastes the time information from the Windows Clipboard. At this point, you could name and save this file and/or print this information.

13 Exit the text editor (do not save the file), and click the AutoCAD button at the bottom of the screen to make it full size.

Command Line Window

Now focus your attention on the AutoCAD - Command Line window (the window that contains the Command prompt).

1 Position the pointer so that its tip is touching any part of the AutoCAD - Command Line window's border.

2 Click and drag the window upward into the drawing area until a floating window forms.

You can move the window anywhere on the screen, and you can resize its width and height.

3 Drag the AutoCAD - Command Line window downward and dock it into its original position.

It is possible to copy, paste, and edit text in the AutoCAD - Command Line window.

4 At the right side of the AutoCAD - Command Line window, scroll up using the up arrow.

5 Identify a command (any command) and highlight it by dragging the cursor across it while pressing the pick button on the pointing device.

6 With the pointer inside the AutoCAD - Command Line window, press the right button on the pointing device, causing a pop-up menu to appear.

7 Pick **Paste to CmdLine**.

AutoCAD pastes the highlighted text at the Command line. Later in this book, you may find this feature useful as you work with AutoLISP, AutoCAD's built-in programming language.

8 Press the **ESC** key to cancel.

⑨ Type **TME**, a misspelled version of the TIME command, but do not press ENTER.

⑩ Using the left arrow key, back up and stop between the T and M.

⑪ Type the letter **I**.

NOTE:

The Insert key serves as a toggle to turn on and off the insert mode. If the Insert function is not active, pick the Insert key to turn it on.

⑫ Assuming that TIME is now spelled correctly, press **ENTER**.

⑬ Press **ESC** to cancel.

⑭ On your own, experiment with the following Command line navigation options.

Key	Action
LEFT ARROW	Moves cursor back (to the left)
RIGHT ARROW	Moves cursor forward (to the right)
UP ARROW	Displays the previous line in the command history
DOWN ARROW	Displays the next line in the command history
HOME	Places the cursor at the beginning of the line
END	Places the cursor at the end of the line
INS	Turns on and off the insertion mode
DEL	Deletes the character to the right of the cursor
BACKSPACE	Deletes the character to the left of the cursor
ENTER	Moves the cursor to the end of the line and terminates the command
CTRL+V	Pastes text from the Clipboard

SAVETIME

SAVETIME allows you to preset AutoCAD to save your drawing automatically. This helps to prevent a loss of work due to a power failure, system crash, or some other event that causes your software or hardware to fail. If AutoCAD automatically saves your drawing every 10 minutes, the most you can lose is 10 minutes of work.

1 Enter **SAVETIME** and enter **2** for the new value.

2 Draw an object on the screen.

3 Draw additional objects or sit idle for at least 2 minutes.

4 After 2 minutes have elapsed, add another object or make a change to the existing one. (Watch or listen for hard disk activity.)

AutoCAD displays the message Automatic save to C:\TEMP\auto1.sv$. If you lose your work in the future and need to resort to the auto1.sv$ file, rename it to a drawing file. For example, you could rename it heather.dwg.

5 Enter **SAVETIME** and enter **10**.

6 Exit AutoCAD. Do not save your work.

Questions

1. What key, in conjunction with CTRL, allows you to turn on the coordinate display feature?

_____

2. Of what value is the coordinate display?

_____

_____

3. What's the name of the feature that forces all lines to be drawn only vertically or horizontally?

_____

4. What key, used with CTRL, controls this feature?

_____

5. Of what value is the TIME command?

_____

_____

6. Briefly explain each of the following components of the TIME command.

Current time: _____

Created: _____

Last updated: _____

Total editing time: _____

Elapsed timer: _____

Next automatic save in: _____

7. Which function key toggles the display of the AutoCAD Text Window on and off?

_____

8. Explain how you would copy a few lines from the AutoCAD Text Window to a text editor such as Notepad.

_____

_____

9. Explain the purpose of SAVETIME.

_____

_____

■ *Challenge Your Thinking: Questions for Discussion*

1. Experiment with ortho in combination with various object snap modes using the LINE command. What happens when you try to snap to a point that is not exactly horizontal or vertical to the previous point?

2. Why might you want to copy parts of the AutoCAD Text Window or AutoCAD - Command Line window to the Windows Clipboard?

Problems

Set AutoCAD to save your work automatically every 5 minutes. As you complete the following problems, consult AutoCAD's timer frequently to see when the drawing was last saved and when the next save is scheduled.

1-3. Practice using ortho by drawing the following objects. Create a new drawing to hold them, and use ortho when appropriate. Don't worry about exact sizes. Turn on the coordinate display, and note the display as you construct each of the objects.

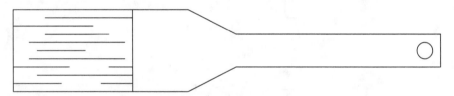

PRB8-1: Paintbrush

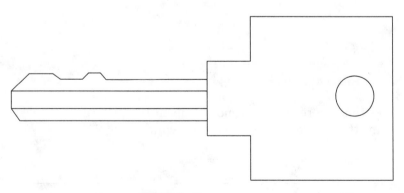

PRB8-2: Key

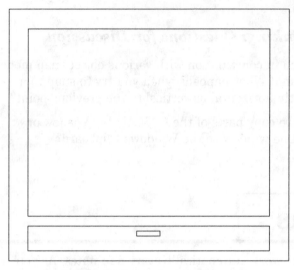

PRB8-3: Computer Monitor

4. Create a new drawing file and draw the following object. Copy everything entered in the AutoCAD - Command Line window to the Clipboard and paste it into a text editor such as Notepad. Then print it. (Read problem 5 before completing this problem.)

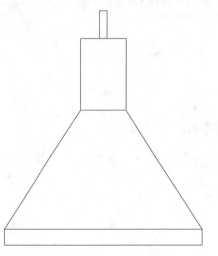

PRB8-4: Lamp Head

5. Use AutoCAD's timer to review the time you spent completing problem 4. Print this information or record it below.

_____   _____

_____   _____

_____   _____

_____   _____

_____   _____

AUTOCAD® AT WORK

LightCAD™ Figures the Foot-Candles

People don't often look up when entering a room. They *always* notice the carpet but seldom give a thought to the ceiling, which is where lighting designers do most of their work. Without lighting, of course, the color of the carpet wouldn't matter.

Lighting design begins with the shape and size of the space to be lit. The hallway of a sleek professional building, for example, would need different lighting than a stairwell in a sports arena. Floor and wall treatments, with different light-reflecting qualities, also factor into the designer's choice of fixture type and location.

Energy consumption is another key consideration. The designer has to answer the question: How much will it cost to keep these lights on? To help designers plan energy-efficient lighting, the Electric Power Research Institute (EPRI) sponsored development of LightCAD, a software package that creates a layer in an AutoCAD drawing to represent lighting.

LightCAD calculates room geometry and reflectance automatically. The designer decides what fixtures to use and where to put them. As the designer specifies the fixtures, LightCAD gives a sidebar readout of how much light (measured in foot-candles) is being supplied in the room and how much power is being consumed.

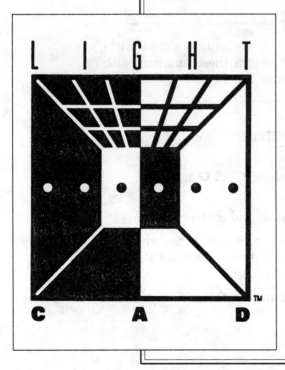

To discover the most energy-efficient plan, the designer tries out different fixtures in LightCAD and compares their illumination levels and costs. Standard fixture information comes from a LightCAD database, but the program can also import files supplied by fixture manufacturers. The ability to use manufacturer-supplied files assures up-to-date information on prices and specifications.

When the job is complete, LightCAD provides a special export file that can be used to show how the lighting plan measures up to the specifications of ASHRAE/IES 90.1, a national lighting standard. According to Tom McDougall of the Weidt Group (Minnetonka, MN), the architecture/energy software organization that produced LightCAD, "this is an advanced lighting layout tool to improve basic commercial lighting."

Ironically, as LightCAD helps lighting designers do their job better so that we can appreciate the color of office carpet at the lowest energy cost, hardly anyone will look up and take notice.

Unit 9 Helpful Drawing Aids

■ OBJECTIVE:

To apply the GRID, SNAP, XLINE, and RAY commands

This unit focuses on four very helpful construction aids. Each of these aids assists the AutoCAD user during the layout and placement of objects.

■ *GRID Command*

The GRID command allows you to set an alignment grid of dots of any desired spacing, making it easier to visualize distances and drawing size. The grid feature is a useful visual aid.

Let's work with this feature.

1 Start AutoCAD and start a new drawing from scratch.

2 Toggle on the grid feature by pressing **CTRL G**.

A grid of dots appears on the screen.

3 Enter the **GRID** command at the keyboard and change the grid spacing to 1.0 unit by entering **1**.

4 Enter the **GRID** command again and change the spacing to **.5** unit.

_____ NOTE: _____

In this book, values less than zero (e.g., .5) are usually shown without a leading zero (e.g., 0.5). AutoCAD accepts these numbers with or without a leading zero, so you may enter the numbers either way.

5 Next, turn the grid off by pressing **CTRL G** . . .

6 . . . and then turn it back on by pressing **CTRL G** again.

7 Press the **F7** function key to toggle the grid off and on.

Notice that the word GRID appears darkened in the status bar when grid is on and grayed out when grid is off. You can also double-click the word GRID in the status bar to toggle the grid on and off.

8 Double-click **GRID** in the status bar to turn the grid off and then back on.

NOTE:

AutoCAD has assigned several additional functions to the function keys. They serve as on/off toggle switches. Press each of them to see what they do.

SNAP Command

The snap feature is similar to the grid feature because it is also a grid, but it is an invisible one. You cannot see the snap feature, but you can see the effects of snap as you move the crosshairs across the screen.

 Press **CTRL B** to turn on the snap feature. The word Snap appears darkened on the status bar when snap is on. If it is not darkened now, press **CTRL B** again.

NOTE:

You can also double-click the word SNAP in the status bar to toggle snap on and off. Or, if you prefer, use the F9 function key.

 Slowly move the pointing device and watch closely the movement of the crosshairs.

The crosshairs jump (snap) from point to point.

3 Enter the **SNAP** command.

4 Enter **.25** to specify the snap resolution.

5 Move the pointing device and note the crosshairs movement.

6 Draw a line.

HINT:

If the grid is not on, turn it on to see better the movement of the crosshairs. Notice the spacing relationship between the snap resolution and the grid. They are independent of one another.

So you see, the snap feature is like a set of invisible magnetic points. The crosshairs jump from point to point as you move the pointing device. This allows you to lay out drawings quickly, yet you have the freedom to toggle snap off at any time. Object snap, on the other hand, permits you to snap to points on objects, such as the midpoint of a line.

There may be times when you want the crosshairs to snap one distance vertically and a different distance horizontally. Here's how to do it:

1 Enter the **SNAP** command.

Snap spacing or ON/OFF/Aspect/Rotate/Style <0.2500>:

2 Enter the Aspect option by typing **A** and pressing **ENTER**.

AutoCAD asks for the horizontal spacing.

3 Enter **.5**.

Next AutoCAD asks for the vertical spacing.

4 For now, enter **1**.

5 Move the crosshairs up and down and back and forth. Note the difference between the amount of vertical and horizontal movement.

Let's experiment with the Rotate option.

1 Enter the **SNAP** command again, and this time enter the **Rotate** option.

2 Leave the base point at 0,0 by giving a null response. (Simply press the space bar or **ENTER**).

3 Enter a rotation angle of **30** (degrees).

The snap, grid, and crosshairs rotate 30 degrees counterclockwise.

4 Draw a small object at this rotation angle near the top of the screen.

5 Return to the original snap rotation of **0**.

Now, let's try drawing another object.

1 Draw the following figure with snap set at **.5** unit both horizontally and vertically. Use the pointing device (not the keyboard) to specify all points.

HINT: Turn on ortho to speed the drawing of the horizontal and vertical lines. Also, use the coordinate display when specifying endpoints.

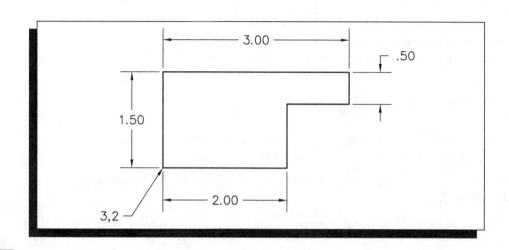

Drawing Aids Dialog Box

The Drawing Aids... selection found in the Tools pull-down menu assists you in reviewing and changing the snap, grid, and other settings. The Drawing Aids dialog box is shown here.

Drawing Aids

Modes	Snap	Grid
☐ Ortho	☐ On	☐ On
☑ Solid Fill	X Spacing `0.5000`	X Spacing `0.5000`
☐ Quick Text	Y Spacing `0.5000`	Y Spacing `0.5000`
☐ Blips	Snap Angle `0`	Isometric Snap/Grid
☑ Highlight	X Base `0.0000`	☐ On
☑ Groups	Y Base `0.0000`	⦿ Left ○ Top ○ Right
☐ Hatch		

OK   Cancel   Help

1 Choose the **Tools** pull-down menu.

2 Select the **Drawing Aids...** item found in this menu.

The Drawing Aids dialog box appears.

NOTE:

You can also make this dialog box appear by entering the DDRMODES command. If you check the Blips check box, construction points called *blips* do not clear automatically from the screen.

3 Examine the items in the dialog box.

As you can see, it permits you to change many different settings by checking boxes and filling in new values. The check boxes serve as on/off toggle switches.

You will learn to use the solid fill, quick text, isometric, and other features in this dialog box in future units.

When the Blips item is checked, temporary construction points, called blips, appear on the screen. They do not appear when the Blips item is not checked.

4 Make several changes to the current settings and pick **OK** to save the changes.

NOTE:

AutoCAD saves these changes with the drawing. Therefore, if you or someone else begins a new drawing, the old settings return.

5 Draw several objects and notice how the new settings have changed the behavior of AutoCAD.

Construction Lines

Construction lines, or "xlines," are lines of infinite length. Their purpose is to help you lay out and create new drawings.

Standard

1 Begin a new drawing.

2 Pick **Start from Scratch**, **English**, and **OK**.

3 From the Draw toolbar, pick the **Construction Line** icon.
This enters the XLINE command.

Draw

4 Pick a point anywhere on the screen.

5 Move the crosshairs and notice what happens.

A line of infinite length passes through the first point and crosshairs.

6 Pick a second point.

A line freezes into place. As you move the crosshairs, notice that a second line forms.

7 Pick another point so that the intersecting lines approximately form a right angle.

8 Press **ENTER** to terminate the XLINE command.

Unlike the snap and grid drawing aids, xlines are AutoCAD objects. You can select them for use with commands, and you can snap to them using the object snap modes.

9 Enter the **LINE** command and use the **Intersection** object snap mode to snap to the intersection of the two lines.

Object Snap

HINT:

Object snap modes are available in the flyout located on the Standard toolbar.

Standard

10 Press **ENTER** to terminate the LINE command.

11 Erase the xlines.

XLINE Options

The XLINE command provides several options for construction lines.

Draw

1 Enter the **XLINE** command and enter **H** for Horizontal.

2 Pick a point anywhere on the screen; then pick a second point and a third.

As you can see, this option produces horizontal construction lines.

3 Press **ENTER** to terminate the XLINE command and press **ENTER** or the space bar to reenter it.

4 Enter **V** for Vertical and pick three consecutive points anywhere on the screen.

This option produces vertical construction lines.

5 Press **ENTER** to terminate the XLINE command and reenter it.

6 Enter **A** for Angle and enter **45**.

7 Place three construction lines.

8 Terminate XLINE.

9 Enter **ERASE** and **All** to clear the screen.

10 Enter **XLINE** and **B** for Bisector.

This option produces a construction line that bisects an angle that you specify.

11 Pick a point for the vertex of the angle.

12 Pick a second point and a third to define the angle.

A construction line freezes at an equal distance between the second and third points.

13 Terminate XLINE, reenter it, and enter **O** for Offset.

14 Enter **1** (for 1 unit), select the construction line, and pick a point on either side of it.

Object Snap

15 Terminate XLINE (press **ESC**), reenter **XLINE**, and enter the **Perpendicular** object snap mode.

16 Pick the last construction line you created and pick another point elsewhere on the screen.

The new construction line is perpendicular to the one you selected.

17 Terminate XLINE and erase all the objects from the screen.

Rays

Rays are similar to construction lines.

1 From the **Draw** pull-down menu, pick **Ray**.

This enters the RAY command.

2 Pick a point in the center of the screen.

3 Move the pointing device in any direction and then pick several points around the first point.

As you can see, these lines extend from a single point into infinity in a radial fashion.

4 Press **ENTER** to terminate the RAY command.

5 Exit AutoCAD, but do not save your work.

Questions

1. What is the purpose of the GRID command? How can the grid be toggled on and off quickly?

_____

_____

_____

2. When would you use the snap feature? When would you toggle off the snap feature?

_____

_____

_____

_____

3. How can you set the snap feature so that the crosshairs move a different distance horizontally than vertically?

_____

_____

4. Explain how to rotate the grid, snap, and crosshairs 45 degrees.

_____

_____

_____

5. How do you make the Drawing Aids dialog box appear?

_____

_____

6. What is the benefit of using the Drawing Aids dialog box?

_____

_____

Challenge Your Thinking: Questions for Discussion

1. Explain the purpose of construction lines and rays. Describe a drawing situation in which you could use construction lines or rays to simplify your work.

Problems

1-3. Draw the objects for problems 1 through 3 using the grid and snap settings provided beside each object.

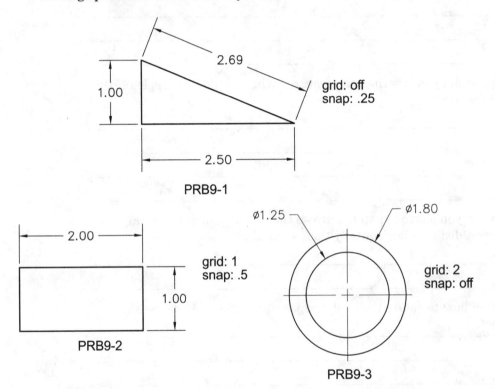

2.69

1.00

grid: off
snap: .25

2.50

PRB9-1

2.00

grid: 1
snap: .5

1.00

PRB9-2

⌀1.25   ⌀1.80

grid: 2
snap: off

PRB9-3

4. Draw three views of this slotted block. Start by drawing the front view using the LINE and CIRCLE commands. Do not worry about drawing the object exactly. Make the task easier by turning ortho on.

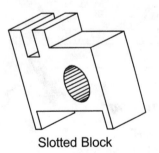

Slotted Block

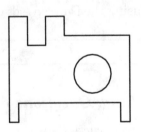

Front View

Use the front view to construct the right-side view. Use the RAY command to extend the edges in the front view into what will become the right-side view. Use the appropriate object snaps to begin the rays at the edges of the front view, as shown below.

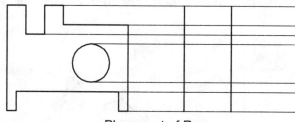

Placement of Rays

To begin drawing the right-side view, add two vertical lines (estimate the space between them), as shown in the previous drawing. Use the construction lines (rays) to draw the visible lines. Draw the hidden lines as a series of very short lines. Later you will learn an easier way to draw dashed lines to represent hidden and center lines.

When you are finished, use the ERASE command with a crossing window to erase all of the construction lines. The drawing should look like the one below.

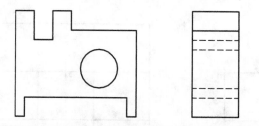

PRB9-4   Finished Front and Right-Side Views

Repeat the process to draw the top view, using the same depth as you used to create the right-side view. The finished drawing should look like the one below. (These three views are known as the orthographic views.)

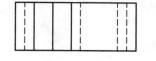

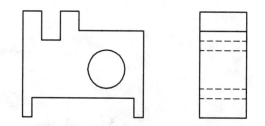

5. Draw the top, front, and right-side views of the block support shown below. Use a grid spacing of .2 and a snap setting of .1. Consider using construction lines or rays to help you position the three views. For now, draw all of the lines as continuous lines, or draw the hidden and center lines using very short lines. You may wish to change the snap spacing to .05.

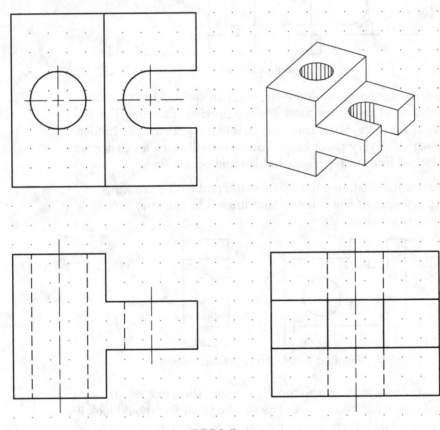

PRB9-5

6. Draw the three orthographic views of the bushing holder below. Use a grid spacing of .2 and a snap setting of .1. Consider using construction lines or rays to help you position the views. The front view is shown as a full section view; *i.e.,* it is shown as if the object has been cut in half. The diagonal lines indicate cut material; you may want to rotate the crosshairs to create these lines. AutoCAD provides a way to insert such hatch patterns easily. You will learn how to do this later.

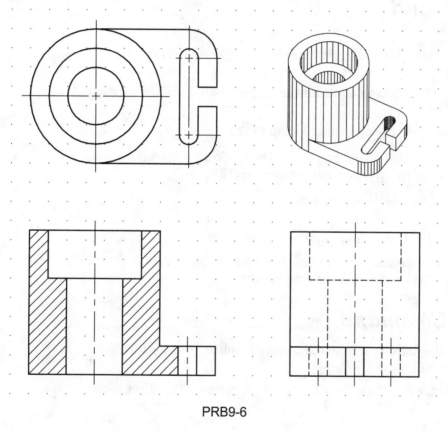

PRB9-6

Problems 4, 5, and 6 courtesy of Gary J. Hordemann, Gonzaga University

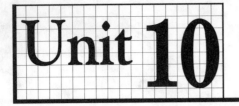 **Unit 10** Undoing What You've Done

■ **OBJECTIVE:**

To apply the U, REDO, and UNDO commands

AutoCAD allows you to back up to any earlier point in an editing session using the commands U, UNDO, and REDO. You can reverse the effect of one Undo if you accidentally go one step too far. Let's work with these commands.

U Command

1 Start AutoCAD and begin a new drawing from scratch.

2 Place a half dozen donuts on the screen at any size and location.

3 Set the grid spacing to **1** and snap to **.25**.

4 Enter the U command by picking the **Undo** icon from the Standard toolbar, or by typing **U** and pressing **ENTER**.

Standard

What happens? AutoCAD undoes your last operation.

5 Enter the **U** command again, and then again.

So you see, the U command backs up one step at a time. With it, you can back up to the beginning of the editing session.

REDO Command

1 Pick the **Redo** icon from the Standard toolbar to enter the REDO command.

Standard

REDO undoes the last Undo. In other words, it reverses the effect of the last Undo.

UNDO Command

The UNDO command is similar to the U command, but UNDO provides several options. Let's try a few.

1 Draw the following objects at any size and location.

2 Enter the **UNDO** command at the keyboard.

The following options appear at the prompt line.

Auto/Control/BEgin/End/Mark/Back/<Number>:

3 Respond to <number> by typing the number **2** and pressing **ENTER**.

By entering 2, you told AutoCAD to back up two steps. UNDO 1 is equivalent to the U command.

Now let's try the Mark and Back options.

Draw

1 Draw a small rectangle at any size and location.

Suppose, for instance, that you want to proceed with drawing and editing, but you would like the option of returning to this point in the session at a later time. This is what you must do.

2 Enter the **UNDO** command.

3 Enter the Mark option by typing **M** and pressing **ENTER**.

AutoCAD has (internally) marked this point in the session.

4 Perform several operations, such as drawing and erasing objects and changing the ortho, snap, and grid settings.

Now suppose you decide to back up to the point where you drew the rectangle in Step 1.

5 Enter the **UNDO** command and then the **Back** option.

You have just practiced two of the most common uses of the UNDO command. Other UNDO options exist, such as BEgin and End. Refer to AutoCAD's on-line help facility or the *AutoCAD User's Guide* for more information on these options.

Control Option

The Undo feature can use a large amount of disk space and can cause a "disk full" situation when only a small amount of disk space is available. You may want to disable the U and UNDO commands partially or entirely by using the UNDO Control option.

Let's experiment with UNDO Control.

1 Enter the **UNDO** command and then the **Control** option.

2 Enter the **One** option.

3 Draw two arcs anywhere on the screen.

4 Enter the **UNDO** command.

UNDO now permits you to undo only your last operation.

5 Press **ENTER**.

With UNDO Control set at One, AutoCAD stores only a small amount of Undo information. This minimizes the risk of a "disk full" situation.

6 Enter the **UNDO Control** option again.

7 This time, enter **None**.

8 Perform a couple of AutoCAD operations.

9 Now enter **UNDO**.

Since UNDO Control is set at None, AutoCAD does not give you the option of undoing. Your only option is to change the setting of UNDO Control.

Standard

10 Press **ESC** and try the **U** command.

The U command is also disabled.

11 Enter **UNDO** and **All**.

Now the AutoCAD Undo feature is once again fully enabled.

12 Enter **UNDO** again and note the complete list of options.

13 Experiment further with AutoCAD's UNDO command.

14 Exit AutoCAD. Do not save your changes.

Questions

1. Explain the differences between the U and UNDO commands.

   _____

   _____

   _____

2. What is the purpose of the REDO command?

   _____

   _____

3. Explain how you would quickly back up or undo your last five operations.

   _____

   _____

4. Explain the use of the UNDO Mark and Back options.

   _____

   _____

   _____

5. Describe the three UNDO Control options.

   All _____

   None _____

   One _____

1. Is the U command enabled or disabled when UNDO Control is set at One? Why?

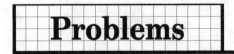

Problems

1. Draw this simple house elevation at any size. Prior to drawing the roof, use UNDO to mark the current location in the drawing. Then draw the roof.

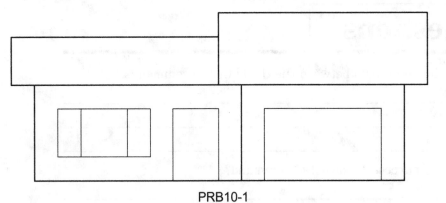

PRB10-1

With UNDO Back, return to the point prior to drawing the roof. Draw the following roof in place of the old roof. Use the U and UNDO commands as necessary as you complete the drawing.

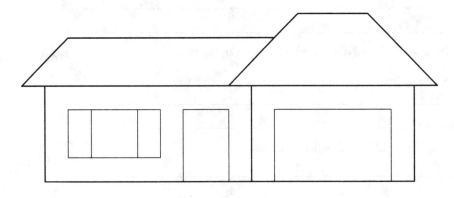

2. Draw the front view of the tube bulkhead shown below. Before drawing the holes, use the UNDO command to mark the current location in the drawing. After completing the bulkhead, use UNDO to replace the two holes with two new ones. Draw both of the new holes with a diameter of 1.00. Change the center-to-center distance from 1.40 to 1.60.

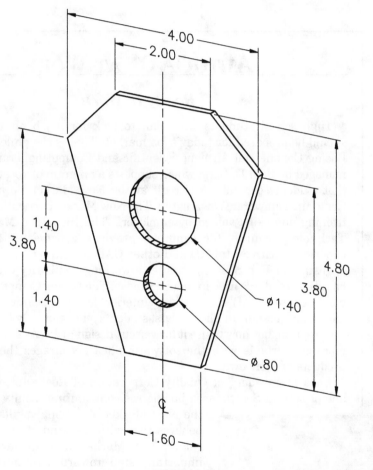

PRB10-2

Problem 2 courtesy of Gary J. Hordemann, Gonzaga University

AUTOCAD® AT WORK

Ready to Work

High school students who plan to look for a job after graduation should be prepared for today's job market. That is the philosophy at the Boeing Commercial Airplane Scientific and Computing Training Group, managed by Paul D. Berg, which supports a program of vocational classes that students can take for credit at the New Market Vocational Skills Center in Tumwater, Washington. The New Market curriculum focuses on training and skills valued by employers. The Industrial Manufacturing Technology course, for example, provides a semester of hands-on experience with AutoCAD and other CAD programs.

Instructor Ron Shea emphasizes a working familiarity with AutoCAD, based on standards used in industry today. The Boeing Company, for example, uses AutoCAD for airplane interior layout, facilities design, and mechanical design. In Shea's class, students learn not only how to operate the program, but how to use it for various design tasks. As students become more comfortable with the software, Shea encourages them to design products of their own.

Shea is president of Quality Corporation of Rochester, Washington, one of several Seattle area businesses contributing faculty members to the New Market Vocational Skills Center. He believes that with the kind of experience they get at New Market, students are taking an important step toward a quality career in drafting, design, engineering, or architecture.

New Market students may or may not be hired by the Boeing Company. Many will go on to work for companies that serve or supply Boeing. By taking part in the program at New Market, and setting real-world standards for the curriculum, the Boeing Company is helping to improve the quality of the region's work force and promoting a brighter future for today's students.

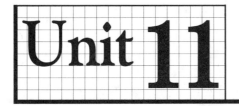 **Unit 11** **Altering and Adding Objects**

■ OBJECTIVE:

To practice the CHAMFER, BREAK, FILLET, OFFSET, and MLINE commands

The CHAMFER and FILLET commands produce chamfered and rounded corners, respectively. The BREAK command removes pieces from entities, while the OFFSET and MLINE commands produce parallel lines.

CHAMFER Command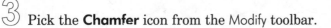

The CHAMFER command enables you to place a chamfer at the corner formed by two lines.

1 Start AutoCAD and open the drawing named **gasket.dwg**.

2 **ZOOM All**. (Enter **Z** and **A** at the keyboard. The ZOOM command is covered fully in a later unit.)

3 Pick the **Chamfer** icon from the Modify toolbar.

Modify

4 Enter **D** for Distance.

5 Specify a chamfer distance of **.25** unit for both the first and second distances.

6 Enter the **CHAMFER** command again (press the space bar or **ENTER**) and place a chamfer at each of the corners of the gasket by picking the two lines that make up each corner.

When you're finished, the drawing should look similar to the one below, with the possible exception of the holes.

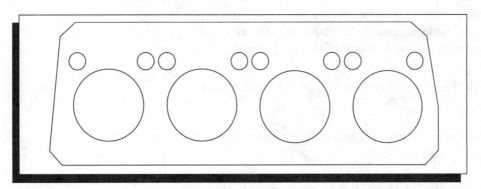

7 Obtain AutoCAD on-line help to learn about the other options offered by the CHAMFER command.

BREAK Command

Now let's remove (break out) sections of the gasket so that it looks like the drawing below.

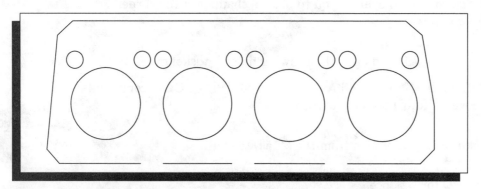

As you know, the bottom edge of the gasket was drawn as a single, continuous line. Therefore, if you were to use the ERASE command, it would erase the entire line since the line is a single object. The BREAK command, however, allows you to "break" certain objects such as lines, arcs, and circles.

Modify

1 From the Modify toolbar, pick the **Break** icon.

This enters the BREAK command.

2 Pick a point on the line where you'd like the break to begin. Since the locations of the above breaks are not dimensioned, approximate the location of the start point.

3 Pick the point where you'd like the break to end.

Did a piece disappear?

Let's break out two more sections of approximately equal size as shown in the above gasket drawing.

4 Repeat Steps 1 through 3 to create each break.

5 Experiment with the other BREAK options on your own.

Draw

⑥ Insert arcs along the broken edge of the gasket, as shown in the illustration on the next page.

Object Snap

HINT: Use the ENDPoint object snap mode to place the arcs accurately.

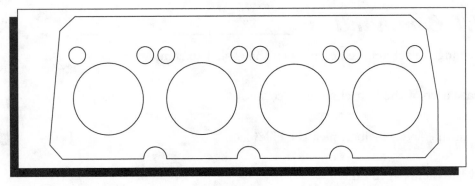

Let's break out a section of one of the holes in the gasket, as shown in the following illustration.

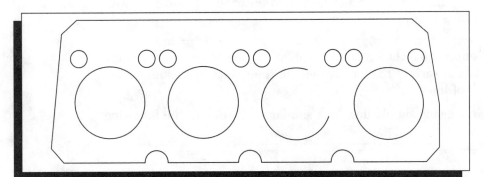

Modify

① Enter the **BREAK** command.

② Pick any point on the circle.

③ Instead of picking the second point, enter **F** for first point.

④ Pick the first (lowest) point on the circle.

⑤ Working counterclockwise, pick the second point.

A piece of the circle disappears.

6 Enter **OOPS**.

The broken piece of the circle does not reappear. Why not? OOPS works only in conjunction with the ERASE command. It does not work with BREAK.

Standard

7 Enter **U** twice.

The broken piece should reappear. If it doesn't, enter U again.

FILLET Command

Now let's change the chamfered corners on the gasket to rounded corners.

1 Erase each of the four chamfered corners.

2 From the Modify toolbar, pick the **Fillet** icon.

Modify

3 Set the radius at **.3** unit.

HINT: Enter R for Radius and then enter .3. Be sure to press ENTER after each entry.

4 Reenter the **FILLET** command (press the space bar) and produce fillets at each of the four corners of the gasket by picking each pair of lines.

The gasket drawing should now look similar to the one in the following illustration.

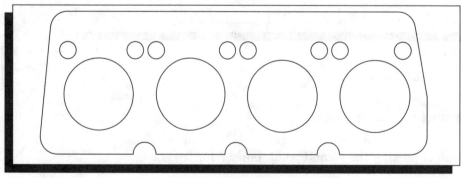

NOTE:

You can also fillet any combination of two lines, arcs, or circles.

5 Save your work.

Let's move away from the gasket and try something new.

1 Above the gasket, draw lines similar to the ones in the illustration below. Omit the numbers.

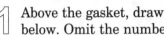

2 Set the fillet radius at **0**.

3 Reenter the **FILLET** command and select lines 1 and 2.

The two lines extend to form a sharp corner.

This technique works with the CHAMFER command, too.

4 Experiment with the remaining FILLET options on your own. If necessary, obtain on-line help.

OFFSET Command _____

The OFFSET command provides a method of offsetting curves, circles, and lines.

1 From the Modify toolbar, pick the **Offset** icon.

This enters the OFFSET command.

2 For the offset distance, enter **.2**.

3 Select one of the circles in the gasket drawing.

4 Pick a point inside the circle in reply to Side to offset?.

Did another circle appear?

5 Select another circle and pick a side to offset.

6 Press **ENTER** to terminate the command.

Let's try the Through option.

1 Using the **LINE** command, draw a triangle of any size.

Modify

2 Pick the **Offset** icon and enter **T** (for Through).

3 Pick any one of the three lines that make up the triangle.

4 Pick a point a short distance from the line and outside the triangle.

The offset line appears. Notice that the line runs through the point you picked.

5 Do the same with the remaining two lines in the triangle so that you have an object similar to the following. Press **ENTER** when you're finished.

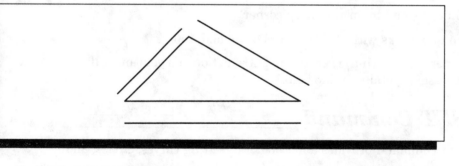

Modify

6 Enter **CHAMFER** and set the first and second chamfer distances at **0**.

7 Enter **CHAMFER** again (press the space bar) and pick two of the new offset lines.

8 Do this again at the remaining two corners to complete the second triangle.

9 Clean up the drawing so that only the gasket remains and save your work.

Multilines

The OFFSET command is useful for adding offset lines to existing lines, arcs, and circles. If you want to produce up to 16 parallel lines, you can do so easily—and simultaneously—with the MLINE command.

Standard

Draw

1 Begin a new drawing from scratch.

2 From the Draw toolbar, pick the **Multiline** icon.

This enters the MLINE command.

3 Create a large triangle by picking three points and entering **C** to close the triangle.

Notice that the corners of the triangle meet perfectly.

The triangle is a single object—an mline object.

4 Erase the object.

Multiline Styles Dialog Box

The Multiline Styles dialog box permits you to change the appearance of multilines and save custom multiline styles. Let's use this dialog box to produce the following drawing.

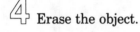

1 Set snap at **.25** unit and turn ortho on.

2 From the **Format** pull-down menu, pick the **Multiline Style...** item.

The Multiline Styles dialog box appears.

```
┌─────────────────────────────────────────┐
│ Multiline Styles                      [X] │
│ ┌─ Multiline Style ──────────────────────┐│
│ │ Current:   [STANDARD              ▼]   ││
│ │ Name:      [STANDARD          ]        ││
│ │ Description: [                    ]     ││
│ │  [ Load... ] [ Save... ] [ Add ] [ Rename ] ││
│ │                                        ││
│ │  ┌──────────────────────────────────┐  ││
│ │  │                                  │  ││
│ │  └──────────────────────────────────┘  ││
│ │                                        ││
│ │       [ Element Properties ... ]       ││
│ │       [ Multiline Properties ... ]     ││
│ │                                        ││
│ │   [ OK ]    [ Cancel ]    [ Help ]     ││
│ └────────────────────────────────────────┘│
└─────────────────────────────────────────┘
```

3 In the box located to the right of Name, double-click the word **STANDARD**.

4 Type **S1**.

5 In the box located to the right of Description, pick a point and type **This style has end caps**.

6 Pick the **Save...** button.

7 In the Save Multiline Style dialog box, pick the **Save** button to store the S1 style in the acad.mln file.

Notice that the current style is still STANDARD.

8 Pick the **Load...** button.

The Load Multiline Styles dialog box appears.

9 Select **S1** and pick **OK**; pick **OK** again.

10 Redisplay the **Multiline Styles** dialog box and pick the **Multiline Properties...** button.

11 Pick the **Start** and **End** check boxes located at the right of Line and pick the **OK** button.

This causes the example multiline in the Multiline Styles dialog box to change. Notice that both ends of the multiline are now "capped" with short connecting lines.

12 Pick the **OK** button.

When you create new multilines, the S1 style will apply.

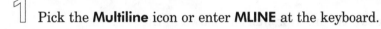

1 Pick the **Multiline** icon or enter **MLINE** at the keyboard.

Draw

2 Enter **S** for Scale and enter **.25**.

This specifies parallel lines that are .25 unit apart.

3 With **MLINE** entered, create the drawing shown in the illustration on page 115, approximating its size.

4 Press **ENTER** to terminate MLINE.

5 Save your work in a file named **multi.dwg**.

Let's add the interior walls shown in the following drawing.

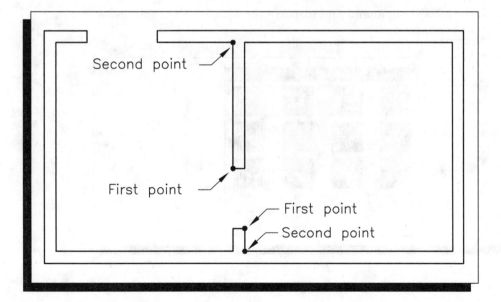

1 Display the **Multiline Styles** dialog box.

2 Create a new multiline style named **S2** with the following description: **This style has a start end cap.** After you save it, be sure to load it.

3 Pick **OK** to exit the Multiline Styles dialog box, if you haven't already.

4 Redisplay the same dialog box and pick the **Multiline Properties...** button.

5 Uncheck the **End** check box located to the right of Line and pick **OK**; pick **OK** again.

6 Using **MLINE**, draw the interior walls as shown in the previous illustration. (Press **ENTER** to complete each wall.)

You will edit the wall intersections in the following section.

7 Save your work.

Draw

Standard

Editing Multilines

Using the Multiline Edit Tools dialog box, it's possible to edit the intersection of multilines.

1 From the **Modify** pull-down menu, pick **Object** and then **Multiline...** from the cascading menu.

This enters the MLEDIT command and displays the Multiline Edit Tools dialog box.

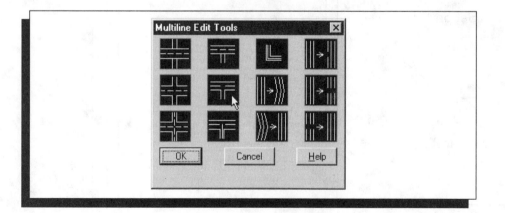

2 Pick the image tile located in the second row and second column, as shown in the illustration on the previous page.

Open Tee appears in the lower left corner.

3 Pick the **OK** button.

Focus your attention on either of the two interior walls.

4 In reply to Select first mline, **pick one of the two interior walls.**

5 In reply to Select second mline, **pick the adjacent exterior wall.**

This causes AutoCAD to break the exterior wall. If it did not break in the correct location, . . .

6 . . . undo and repeat Steps 4 and 5.

7 Edit the second interior wall by repeating Steps 4 through 6.

8 Press **ENTER** to terminate the MLEDIT command.

9 On your own, explore the remaining options found in the Multiline Styles and Multiline Edit Tools dialog boxes.

10 Save your work and exit AutoCAD.

Questions

1. What is the function of the CHAMFER command?

_____

_____

2. How is using the BREAK command different from using the ERASE command?

_____

_____

3. If you want to break a circle or arc, in which direction do you move when specifying points: clockwise or counterclockwise?

_____

4. In what toolbar is the Fillet icon found?

_____

5. How do you set the FILLET radius?

_____

_____

6. Will the FILLET radius change or stay the same after you save your work and exit AutoCAD?

_____

7. What can be accomplished by setting either FILLET or CHAMFER to 0?

_____

8. Explain the purpose of the OFFSET command.

_____

_____

9. How might the MLINE command be useful? Give at least one example.

_____

10. Explain the purpose of the Multiline Styles and Multiline Edit Tools dialog boxes.

_____

_____

■ *Challenge Your Thinking: Questions for Discussion*

1. Explore the Angle option of the CHAMFER command. How is it different from the Distance option? Discuss situations in which each option (Angle and Distance) may be useful.

2. To help potential clients understand floor plans, the architectural firm for which you work draws its floor plans showing the walls as solid gray lines the thickness of the wall. Exterior walls are 6″ thick, and interior walls (those that make up the room divisions) are 5″ thick. Explain how you could achieve these walls using MLINE and the Multiline Styles dialog box.

3. AutoCAD allows you to save the multiline styles and then reload them later. Explore this feature and write a short paragraph explaining how to save and restore multilines.

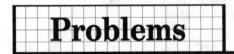

Problems

1. Create the first drawing shown below, left. Don't worry about exact sizes and locations, but do use snap and ortho. Then use FILLET to change it to the second drawing. Set the fillet radius at .2 unit.

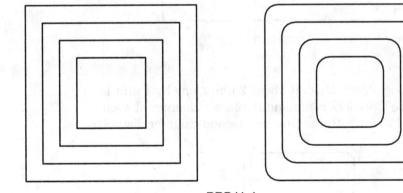

PRB11-1

2. Create the triangle shown below. Then use CHAMFER to change it into a hexagon. Set the chamfer distances at .66 unit.

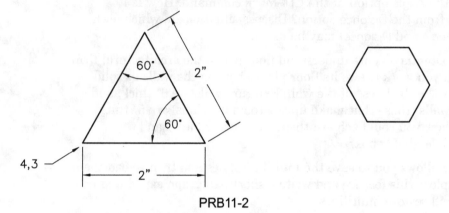

PRB11-2

3. Draw the following object; don't worry about specific sizes. Use the FILLET command to place fillets in the corners as indicated.

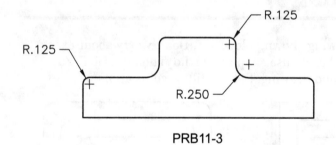

PRB11-3

4. Draw the following object. Make it about 2 units long by 1 unit in diameter. Use the CHAMFER command to place a chamfer at each corner. Specify .125 for both the first and second chamfer distances.

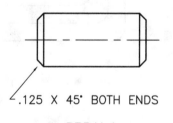

PRB11-4

5. Using the ARC and CHAMFER commands, construct the following object.

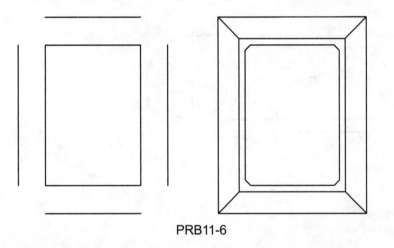

.250

.500

PRB11-5

6. Create the following picture frame drawing. First, create the drawing on the left using the LINE and OFFSET commands. Then modify it to look like the drawing on the right using the CHAMFER command.

PRB11-6

7. Create the following A.C. plug drawing using the MLINE, MLSTYLE, CHAMFER, and FILLET commands. The larger fillet radius is .125, and the smaller fillet has a diameter of .125.

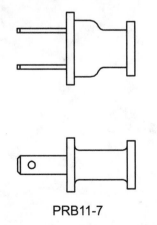

PRB11-7

8. Draw the front view of the rod guide end cap shown below. Use the OFFSET command to create the outer circular shapes and the inner horizontal lines. Use the BREAK and FILLET commands to create the inner shape with four fillets.

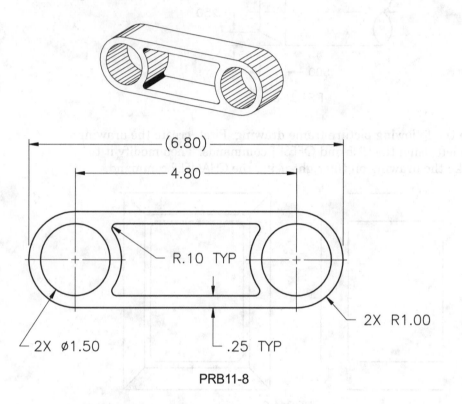

PRB11-8

9. Using AutoCAD's multiline feature, create a basic floor plan of a garage or shed. Be creative.

10. Draw the front view of the ring separator shown below. Start by using the CIRCLE and BREAK commands to create a pair of arcs. Then use OFFSET to create the rest of the arcs.

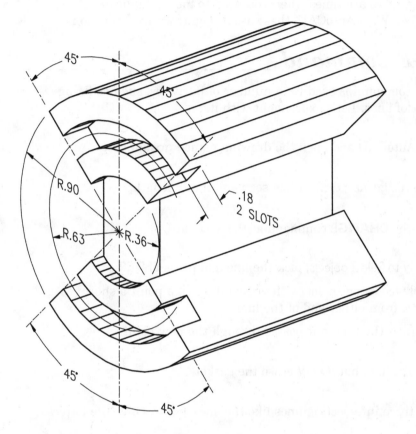

PRB11-10

Problems 6 and 7 courtesy of Joseph K. Yabu, San Jose State University
Problems 8 and 10 courtesy of Gary J. Hordemann, Gonzaga University

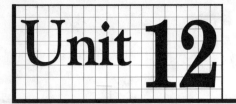

Unit 12 Moving and Duplicating Objects

■ **OBJECTIVE:**

To apply the CHANGE, MOVE, COPY, and MIRROR commands

When drawing, there are times when you need to move, duplicate, or change an object. With AutoCAD, these are straightforward operations.

CHANGE Command

The CHANGE command is used for a number of purposes, such as fixing the placement of lines. Let's experiment with it.

1 Start AutoCAD and open the drawing named **gasket.dwg**.

2 Above the gasket, draw a line segment of any length.

3 Enter the **CHANGE** command at the keyboard.

4 In reply to Select objects, pick the line and press **ENTER**.

5 In reply to Change point (with ortho off), pick a point a short distance from either end of the line.

AutoCAD redraws the line so it passes through the new point.

1 Erase the line, but *do not* erase the gasket.

2 Draw three intersecting lines like the ones below. Omit the letters.

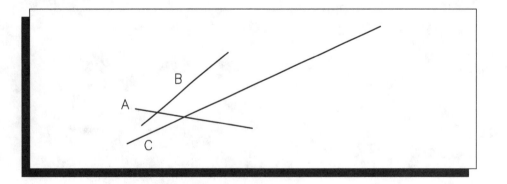

126

Using the CHANGE command, let's fix these lines to form a perfect arrow as shown in the illustration below.

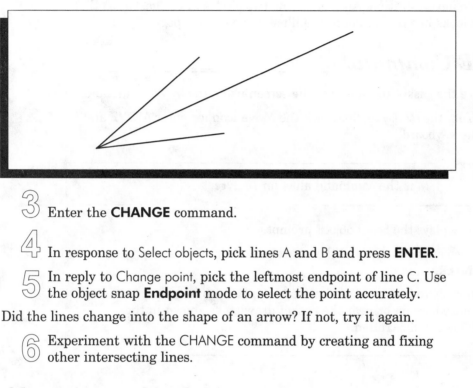

③ Enter the **CHANGE** command.

④ In response to Select objects, pick lines A and B and press **ENTER**.

⑤ In reply to Change point, pick the leftmost endpoint of line C. Use the object snap **Endpoint** mode to select the point accurately.

Did the lines change into the shape of an arrow? If not, try it again.

Object Snap

⑥ Experiment with the CHANGE command by creating and fixing other intersecting lines.

Changing a Circle

You can also use the CHANGE command to change the radius of a circle. Let's do it.

① Draw a small circle of any radius.

② Enter the **CHANGE** command, pick the circle, and press **ENTER**.

③ Pick a point near the circle.

The circle's radius changes through this point.

You can invoke drag by pressing ENTER just prior to specifying a change point. You can then drag the new radius into place. DRAGMODE must be set at Auto. Try it.

We'll explore other ways of using the CHANGE command at a later time.

 For now, erase the arrow and circle so that you'll have room for the next operation. Also, erase the two circles in the gasket so that it looks like the gasket in the illustration on this page.

MOVE Command

Let's move the gasket to the top of the screen using the MOVE command.

 From the Modify toolbar, pick the **Move** icon, or enter **MOVE** at the keyboard.

Modify

HINT:
M is the command alias for MOVE.

AutoCAD displays the Select objects prompt.

Place a window around the entire gasket drawing and press **ENTER**.

In reply to Base point or displacement, place a base point somewhere on or near the gasket drawing as shown in the following illustration.

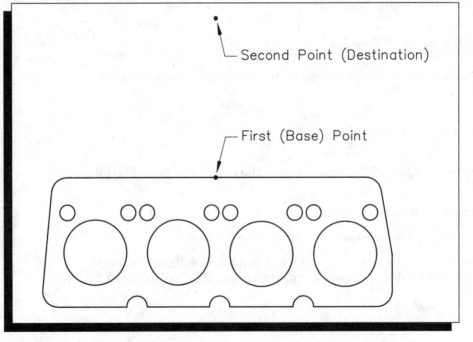

Move the pointing device in the direction of the second point (destination).

5 Pick the second point.

Did the drawing move as illustrated below?

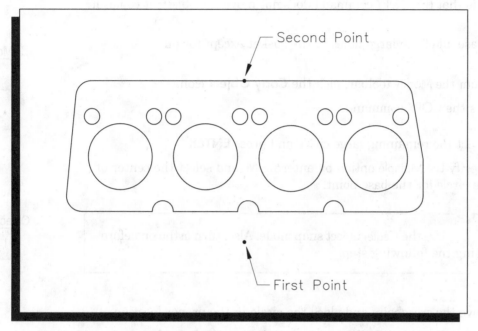

6 Practice using the MOVE command by moving the drawing to the bottom of the screen.

Now let's drag two of the holes in the gasket to a new location.

Modify

1 Enter the **MOVE** command.

2 In response to the Select objects prompt, pick two of the four large holes (circles) and press **ENTER**.

3 Specify the first point (base point). Place the point anywhere on or near either of the two circles.

4 In reply to Second point of displacement, move the pointing device.

5 After you decide on a location away from the gasket, pick that location with the pointing device.

6 Drag the circles back into the gasket as they were before.

COPY Command

The COPY command is almost identical to the MOVE command. The only difference is that the COPY command does not move the object; it copies it.

 1 Erase all of the large holes in the gasket except for one.

2 From the Modify toolbar, pick the **Copy Object** icon.

This enters the COPY command.

Modify

3 Select the remaining large circle and press **ENTER**.

4 Specify the Multiple option by entering **M**, and select the center of the circle for the base point.

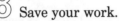
HINT:
Use the Center object snap mode. Also, turn ortho on before completing the following step.

Object Snap

5 Move the crosshairs and place the circle in the proper location.

6 Repeat Step 5 until all four large circles are in place; then press **ENTER**.

7 Practice using the COPY command by erasing and copying the small circles.

8 Save your work.

MIRROR Command

There are times when it is necessary to produce a mirror image of a drawing, detail, or part. A simple copy of the object is not adequate because the object being copied must be reversed, as was done with the butterfly in the following drawing. One side of the butterfly was drawn and then mirrored to produce the other side.

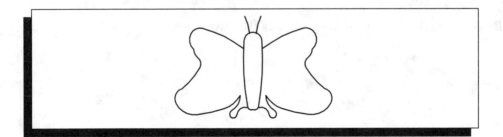

The same is true if the engine head gasket we developed is to be reproduced to represent the opposite side of an eight-cylinder engine.

1 Move the gasket either to the top or bottom of the screen to allow space for another gasket of the same size.

Modify

2 From the Modify toolbar, pick the **Mirror** icon.

This enters the MIRROR command.

3 Select the gasket by placing a window around it, and press **ENTER**.

4 Create a horizontal mirror line near the gasket by selecting two points on a horizontal plane as shown below. Ortho should be on.

5 The prompt line asks if you want to delete old objects. Enter **N** for No, or press **ENTER** since the default is No.

The gasket should have mirrored as shown below.

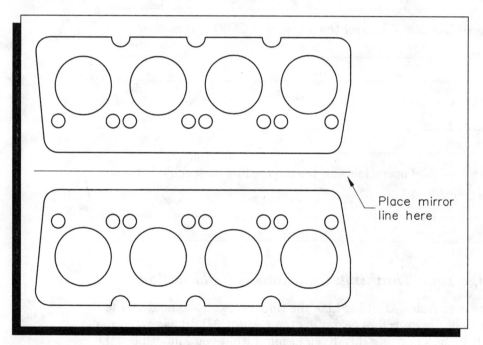

Place mirror line here

You can also create mirrored images around an axis (mirror line) other than horizontal or vertical.

Modify

6 Draw a small triangle and mirror it with an angular (*e.g.*, 45°) mirror line.

7 Erase the triangles, save your work, and exit AutoCAD.

Questions

1. Describe a situation in which CHANGE would be used for changing endpoints of lines.

   _____

   _____

2. In what toolbar is the Move icon located?

   _____

3. Explain how the MOVE command is different from the COPY command.

   _____

   _____

   _____

4. How does DRAGMODE affect the MOVE and COPY commands?

   _____

   _____

5. Describe a situation in which the MIRROR command would be useful.

   _____

   _____

   _____

6. During a MIRROR operation, can the mirror line be specified at any angle? Explain.

   _____

   _____

   _____

■ Challenge Your Thinking: Questions for Discussion

1. Experienced AutoCAD users take the time necessary to analyze the object to be drawn before beginning their AutoCAD drawing. Discuss the advantages of doing this. Keep in mind what you know about the AutoCAD commands presented in this unit.

132

Problems

1-2. In problems 1 and 2, follow each step to create the objects. Use the ARC Continue option at Step 2 of problem 1. Use the MIRROR and COPY commands to complete Steps 3 and 4. Your final work should show all four steps. Use the COPY command to copy your work from one step to the next.

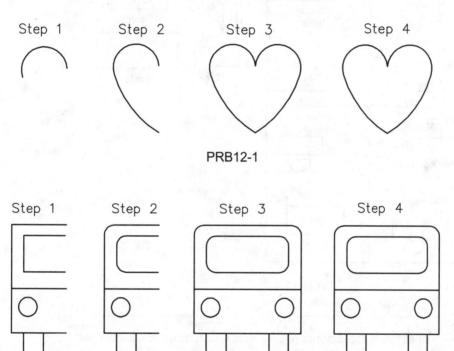

Step 1   Step 2   Step 3   Step 4

PRB12-1

Step 1   Step 2   Step 3   Step 4

PRB12-2

3. Draw the objects and room as shown. Then use the MOVE and COPY commands to move the office furniture into the room. Omit the lettering on the drawings.

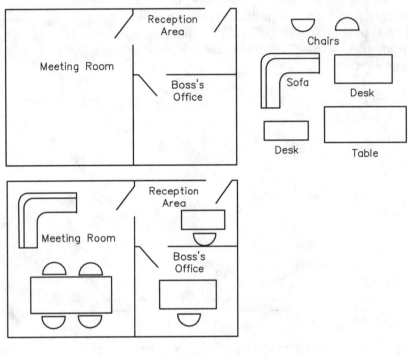

PRB12-3

HINT: In certain situations, MOVE and COPY may work better with the ortho and snap modes off.

4. In the drawing you made for problem 3, use the OFFSET command to produce double-line walls for the office. Set the offset distance at .2 and use other AutoCAD commands, such as CHAMFER and BREAK, to complete and clean up the wall constructions. If necessary, make the drawing larger for easier editing. Save your changes as prb12-4.dwg.

5. Start a new drawing named prb12-5.dwg. Recreate the floor plan you made for problem 3 using AutoCAD's multiline feature. Set the width at .15 unit.

6. Draw the chisel below using the ARC, FILLET, LINE, COPY, and MIRROR commands.

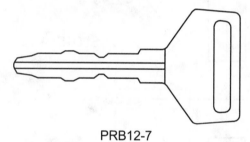

PRB12-6

7. Draw the key below using the ARC, FILLET, LINE, COPY, and MIRROR commands.

PRB12-7

8. Apply the commands introduced in this unit to create the following piping drawing.

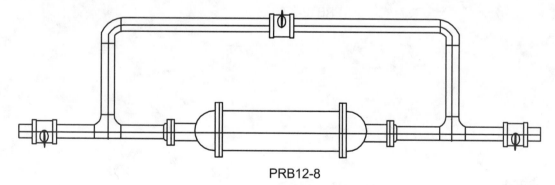

PRB12-8

9. Draw the front view of the bar separator shown below. Draw one fourth of the object using the LINE, ARC, and FILLET commands with a grid setting of .1 and a snap of .05. Use the MIRROR command to make half of the object; then use it again to create the other half. Alternatively, you might try drawing the quarter profile as follows: Starting with a circle, draw a horizontal and a vertical line from the center; then use the OFFSET, BREAK, and FILLET commands to create one fourth of the object.

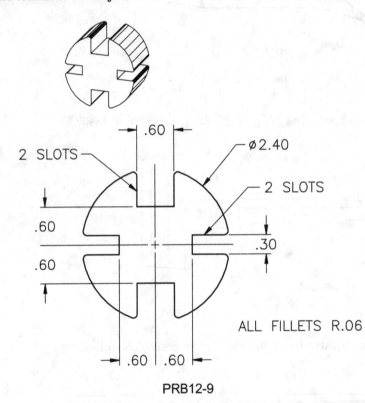

2 SLOTS

Ø2.40

2 SLOTS

.60

.60

.60

.30

.60 .60

ALL FILLETS R.06

PRB12-9

136

10. Draw the front view of this wheel.

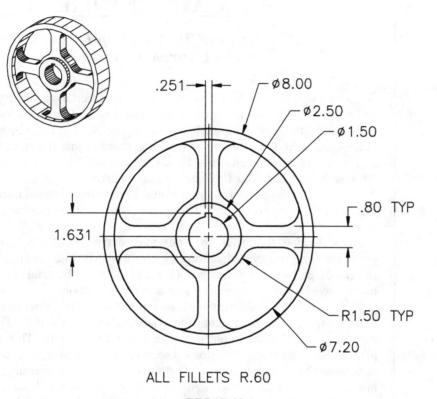

ALL FILLETS R.60

PRB12-10

Begin by constructing one fourth of the object by drawing the three arcs and four lines as shown in the first figure below. Center the arcs on a known point. (You will find the Center, Start, Angle option convenient.) Next, use the FILLET command to create the cavity shown in the second figure below. Erase the outside lines and the outside arc. Then use the MIRROR command to make three more copies of the cavity. Complete the drawing by adding arcs and circles. Try constructing the keyway by first drawing half of the shape, then using MIRROR to generate the other half. For more practice, try the drawing with one eighth of a wheel.

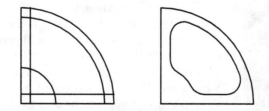

Problems 6 and 7 courtesy of Joseph K. Yabu, Ph.D., San Jose State University
Problem 8 courtesy of Dr. Kathleen P. King, Fordham University, Lincoln Center
Problems 9 and 10 courtesy of Gary J. Hordemann, Gonzaga University

AutoCAD® at Work

Get Me (and My Wedding Dress) to the Church on Time

"You'll need 15 yards," Beverly Knox told the bride-to-be. Knox had designed a wedding dress for the bride in the traditional way, using pencil and paper to plan how the dress pieces would be laid out on a length of standard-width fabric. The fabric was to be purchased in India by a friend of the bride, who happened to be traveling there at just the right time.

The friend came back with 15 yards of beautiful satin taffeta and twice-embossed lace. But the fabric was too narrow.

The pattern pieces for the dress would have to be arranged into a new layout on the narrower fabric. There was about a square yard less fabric to work with overall.

Knox decided that this was the time to try CADTERNS, a custom patternmaking program that works with AutoCAD. The alternative was to lay out pattern pieces on the 15 yards of fabric by trial and error, using a worktable no bigger than an average-size door.

With the client's measurements as input, CADTERNS generates a garment template, or sloper. The sloper is then styled in AutoCAD into a custom pattern. Plotted pattern pieces are the output. This system allowed Knox to modify the dress design on-screen and try out different arrangements of pattern pieces. "The narrower fabric left literally no margin for error," recalls Knox. "CADTERNS had to be extremely accurate to produce a layout that would work."

Wedding bells rang, and the bride was beautiful in her dress with the 5-foot train, redesigned by computer. Knox found that by modifying the train into a two-piece design she could even have a bit of fabric left over, "enough for a couple of bridesmaid purses."

Knox has used AutoCAD and CADTERNS ever since in her shop, Leather and Lace, in White Rock, British Columbia. The business advantage, she says, is that when a customer asks for changes, the original sloper can be redone almost effortlessly. Without CAD, design changes are laborious and time-consuming. With CADTERNS and AutoCAD, customer satisfaction goes up and re-design problems go down.

Lauraline M. Grosenick, president of CADTERNS Custom Clothing Inc. (White Rock, BC), believes custom-fit clothing is the way the apparel industry will progress, replacing mass production. With software like CADTERNS able to produce an individualized sloper in a fraction of the time, costs for custom fit can come down. As more and more businesses recognize this opportunity, you will be able to choose clothes that fit as though measured for you at a price that's competitive with off-the-rack clothing.

Unit 13 — The Powerful ARRAY Command

 OBJECTIVE:

To create rectangular and polar arrays

This unit uses the ARRAY command to construct two drawings: (1) a schematic diagram of computer chip sockets and (2) a bicycle wheel with spokes.

Rectangular Arrays

1 Start AutoCAD and start a new drawing from scratch.

2 In the lower left corner of the screen, draw the following and make it small (approximately 1 unit wide by 1 unit tall).

HINT: Set grid and snap. Use the COPY command to duplicate the small circle. Then use the MIRROR command to make the bottom half identical to the top half.

Modify

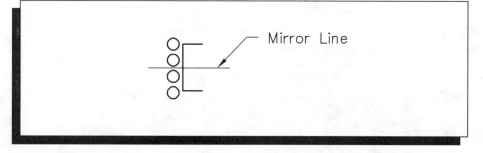

3 Mirror the object to complete the opposite side as shown below. If your drawing does not look exactly like the one here, that's okay.

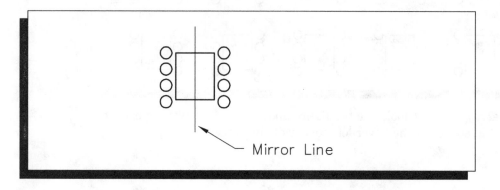

4 Save your work in a file named **chips.dwg**.

The object you have just drawn represents a computer chip socket. It resembles a schematic of the sockets found inside a computer. The sockets house the RAM chips that are currently holding the information you see on the screen.

5 Pick the **Array** icon from the Modify toolbar. (Snap should be off.)

6 In reply to the Select objects prompt, select the chip socket by placing a window around it, and press **ENTER**.

7 Enter **R** for Rectangular.

8 In reply to Number of rows, enter **3**.

9 Enter **5** at the Number of columns prompt.

10 Specify that you want **1.5** units between the rows and **1.75** units between the columns. (The distances specify the center of one object to the center of the next.)

You should now have 15 chip sockets on the screen, arranged in a 3 × 5 array as shown in the following illustration.

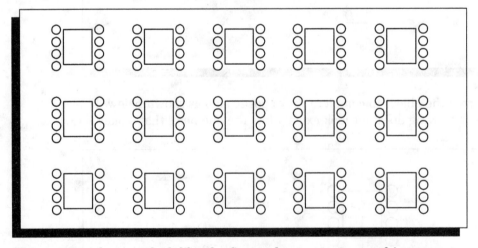

If your array does not look like the figure above, erase everything except for the chip socket in the lower left corner of the screen and try again.

NOTE:

You can produce rectangular arrays at any angle by first changing the snap rotation angle. Also, you can specify the distances between rows and columns by picking the opposite corners of a rectangle with the pointing device.

11 Save your work.

Polar (Circular) Arrays

Next, we're going to produce a bicycle wheel.

1 Begin a new drawing from scratch.

Standard

2 Draw a tire and wheel similar to the one in the following drawing. Don't worry about the exact sizes of the circles, but do make the wheel large enough to fill most of the screen.

HINT:
Use the Center object snap to make the circles concentric.

Object Snap

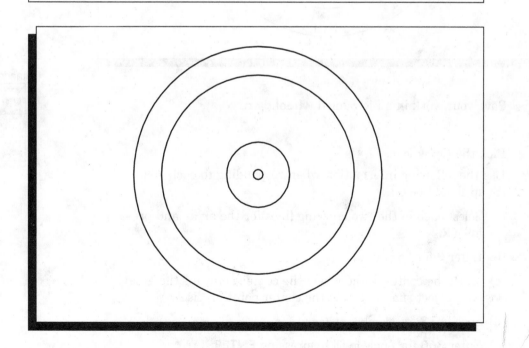

③ Draw two crossing lines similar to the ones below.

HINT: Use the Nearest object snap mode to begin and end the lines on the appropriate circles.

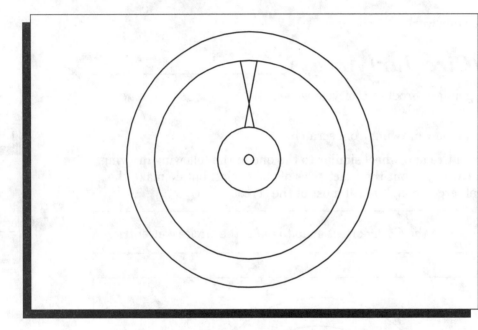

④ Save your work in a file named **wheel.dwg**.

⑤ Pick the **Array** icon.

⑥ Use the following information when responding to each step. (Snap should be off.)

(a) Select each of the two crossing lines for the array and press **ENTER**.

(b) Enter **P** for Polar array.

(c) Make the center of the wheel the center point for the array. (Use object snap to locate the center point precisely.)

(d) Specify **18** for Number of items.

(e) Enter **360** for Angle to fill by pressing **ENTER**.

(f) . . . and yes, you want to rotate the spokes as they are copied.

Standard

Modify

If you were not successful, try again. The wheel should look similar to the following.

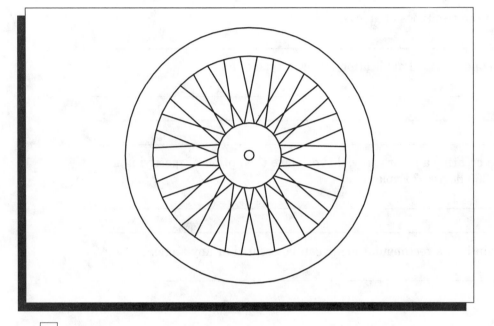

 Save your work.

8 Practice creating polar arrays using other objects. At least once, specify less than 360 degrees when replying to Angle to fill.

Modify

NOTE:

During the polar array sequence, the prompt Angle to fill (+=ccw, −=cw) appears. AutoCAD is asking for the angle (in degrees) to fill the array. +=ccw means that a positive number will produce an array in a counterclockwise direction, and −=cw means that a negative number will produce an array in a clockwise direction.

9 Save your work and exit AutoCAD.

Questions

1. Name the two types of arrays.

_____

2. State one practical application for each type of array.

_____

_____

_____

3. When creating a polar array, do you have the option of specifying less than 360 degrees? Explain.

_____

_____

4. Explain how a rectangular array can be created at any angle.

_____

_____

■ *Challenge Your Thinking: Questions for Discussion*

1. You have been asked to complete a drawing of the dialpad for a pushbutton telephone. You have wisely decided to create one of the pushbuttons and use the ARRAY command to create the rest. Each pushbutton is a 13mm × 8mm rectangle, and the buttons are arranged in the typical fashion shown in the illustration. To create a drawing that shows the true measurements, what should you answer when AutoCAD prompts you for the space between rows and space between columns? (You may assume that 1 AutoCAD unit = 1mm.)

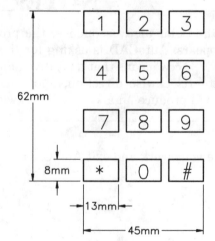

Problems

1. Develop an auditorium with rows and columns of seats. Design the room any way you'd like. Save your work as prb13-1.dwg.

2. Load the drawing named gasket.dwg. Erase each of the circles. Replace the large ones according to the locations shown below, this time using the ARRAY command. Also reproduce the small circles using the ARRAY command, but don't worry about their exact locations. The diameter of the large circles is 1.25; the small circles have a diameter of .30.

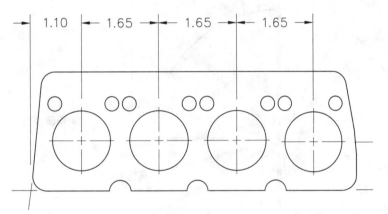

PRB13-2

3. Draw the following figure (you decide how).

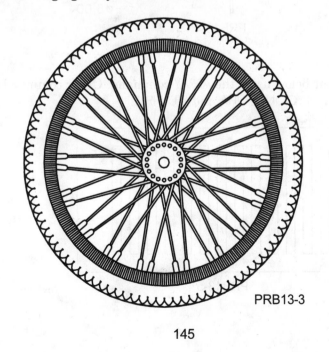

PRB13-3

145

4. Draw the top view of the power saw motor flywheel shown below. Use the ARRAY command to insert the 24 fins and arcs. You may find the OFFSET command useful in creating the first fin.

.50 TYP

3.00 TYP

.0625 TYP

15° TYP

⌀6.00

PRB13-4

5. Draw the air vent below using the ARRAY command. Other commands to consider are POLYGON, COPY, and MULTIPLE.

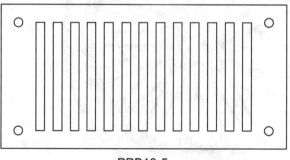

PRB13-5

6. Draw one or more of these snowflakes using the ARRAY command.

Since snowflakes are six-sided, you may want to begin by drawing a set of concentric circles. Then use the ARRAY command to insert 6, 12, or more construction lines. The example below shows the beginning of one of the snowflakes. Note that you only need to array three lines at 15° intervals for a total of 45° (or two lines if you use the MIRROR command).

PRB13-6

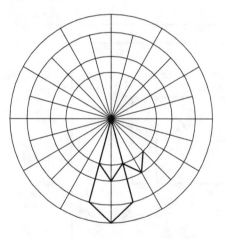

7. Create this cooling fan.

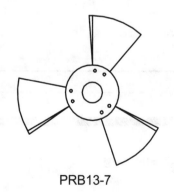

PRB13-7

Problem 3 courtesy of Bob Weiland
Problems 4 and 6 courtesy of Gary J. Hordemann, Gonzaga University
Problems 5 and 7 courtesy of Joseph K. Yabu, Ph.D., San Jose State University

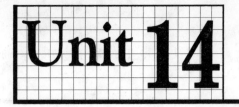

Unit 14 Modifying and Maneuvering

■ OBJECTIVE:

**To apply the STRETCH, SCALE, ROTATE, TRIM, EXTEND, and
LENGTHEN commands**

AutoCAD offers numerous editing commands for changing the appearance
of drawings. For instance, you can stretch the end of a house to make it
longer, scale it down if it is too large, rotate it to position it better on a
lot, trim sidewalk lines that extend too far, and extend driveway lines that
are too short. We will do all of this, and more, in the following steps.

1 Start AutoCAD and start a new drawing from scratch.

2 Draw the following site plan according to the dimensions shown.
With snap on, place the lower left corner of the property line at
absolute point 1,1. Omit dimensions. All points in the drawing
should fall on the snap grid.

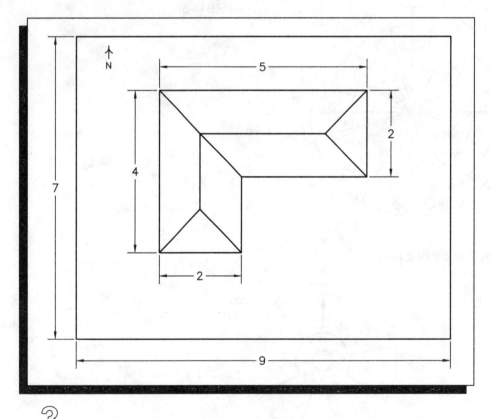

3 Save your work in a file named **site.dwg**.

STRETCH Command _____

1 Pick the **Stretch** icon from the Modify toolbar.

This enters the STRETCH command.

2 Select the east end of the house as shown in the next drawing, and press **ENTER**. Use the **Crossing** object selection procedure. (See the following hint.)

HINT: In reply to Select objects, enter C for Crossing.

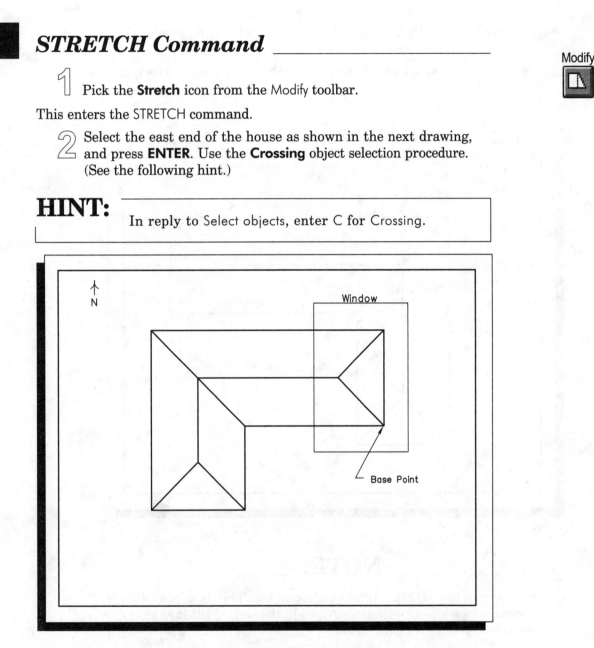

3 Pick the lower right corner of the house for the base point as shown above.

4 In reply to Second point of displacement, stretch the house 1 unit to the right, and pick a point. (Snap should be on.) The house stretches dynamically.

The house should now be longer.

5 Stretch the south portion of the house so that it looks similar to the house shown below.

Remember, if you make a mistake, use the U or UNDO command to back up.

6 Save your work.

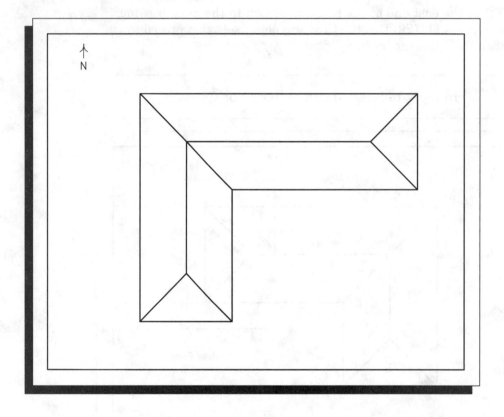

NOTE:

You can also use STRETCH to alter dimensions. This is part of the associative dimensioning feature covered in the section titled "Associative Dimensioning" in Unit 25.

SCALE Command

Since the house is now slightly too large, let's scale it down to fit the lot.

1 Pick the **Scale** icon from the Modify toolbar.

Modify

2 Select the house by placing a window around the entire house, and press **ENTER**.

3 Pick the lower left corner of the house as the base point.

At this point, the house can be dynamically dragged into place or a scale factor can be entered.

4 Turn snap off. Move the crosshairs and notice the dynamic scaling.

5 Enter **.5** in reply to Scale factor.

Suppose the house is now too small. Let's scale it up using the SCALE Reference option.

1 Enter **SCALE** (press the space bar), select the house as before, and press **ENTER**.

2 Pick the lower left corner of the house again for the base point.

3 Enter **R** for Reference.

4 For the reference length, pick the lower left corner of the house, and then pick the corner just to the right of it as the second point.

5 In response to the New length prompt, move the crosshairs about ¹/₂ unit to the right of the second point (ortho and grid should be on) and pick a point.

Did the house enlarge according to the specified length?

ROTATE Command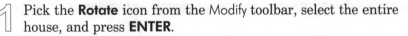

Let's rotate the house.

Modify

1 Pick the **Rotate** icon from the Modify toolbar, select the entire house, and press **ENTER**.

2 Pick the lower left corner of the house for the base point.

3 Turn off ortho and snap. As you move the crosshairs, notice that the object rotates dynamically.

NOTE:

Drag is embedded in the command sequence and therefore is entered for you automatically.

4 For the rotation angle, enter **25** (degrees).

Did the house rotate 25 degrees counterclockwise?

Let's rotate the house again, but this time with the Drag option.

Enter **ROTATE**, select the house, and press **ENTER** after you have made the selection.

Pick the lower left corner of the house for the base point.

At the <Rotation angle>/Reference prompt, drag the house in a clockwise direction a few degrees so that it is positioned similarly to the one on the next page.

Modify

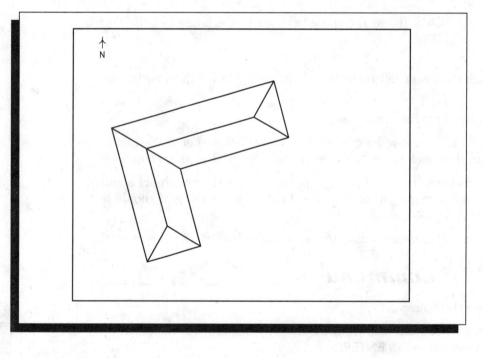

TRIM Command

With ortho on and snap off, draw two horizontal xlines near the bottom of the site plan to represent a sidewalk.

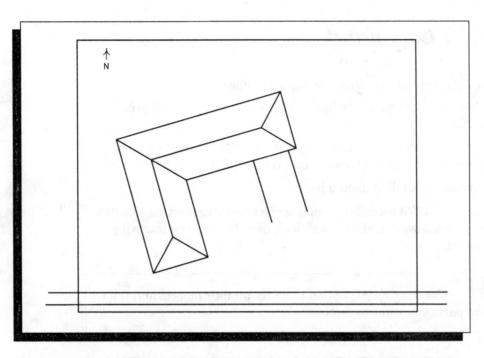

2 With ortho off, draw a partial driveway as shown in the preceding drawing.

HINT:
Use the OFFSET command and the Through option to create the second line.

Modify

Modify

3 Pick the **Trim** icon from the Modify toolbar.

4 Select the east property line as the cutting edge and press **ENTER**.

5 Select the sidewalk lines on the right side of the property line.

Does the sidewalk now end at the property line? It should.

6 Press **ENTER** to return to the Command prompt.

Modify

7 Trim the west end of the sidewalk to meet the west property line using the **TRIM** command.

8 Obtain on-line help to learn about the Project and Edge options. The UCS and View features will be covered in upcoming units.

EXTEND Command

Modify

1. Pick the **Extend** icon from the Modify toolbar.

2. Select the south property line as the boundary edge and press **ENTER**.

3. In reply to Select object to extend, pick the ends of the two lines which make up the driveway and press **ENTER**.

Did the driveway extend? It should have.

Modify

4. Using the **TRIM** command, remove the short intersecting lines so that the sidewalk and driveway look like those in the following drawing.

HINT:

Use the Crossing option to select all four objects and then select the parts you want to remove.

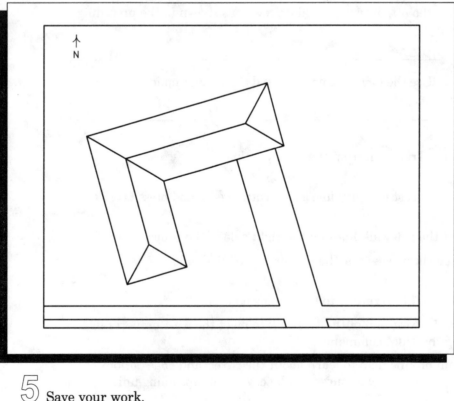

5. Save your work.

LENGTHEN Command _____

The LENGTHEN command enables you to lengthen an object dynamically or by entering a percentage. LENGTHEN is similar to the EXTEND command.

1 In the upper right corner of the site plan, create a small arc.

Modify

2 Pick the **Lengthen** icon from the Modify toolbar.

This enters the LENGTHEN command.

3 Enter **DY** for dynamic, and select the arc.

4 Moving the crosshairs, dynamically lengthen the arc and pick a point.

5 Press **ENTER** to terminate LENGTHEN and then reenter it.

6 This time, enter **P** for Percent, enter **150**, and select the arc.

The arc's length should change to 150 percent of its original size. If you receive the message Illegal length, reduce the percentage to 130 or smaller.

Standard

7 Enter **U** to undo the last change and press **ENTER** to terminate the command.

Obtain on-line help to learn about the DElta and Total options.

Modify

8 Array the arc, as shown below, to produce a tree symbol.

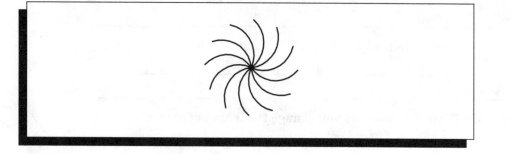

9 Save your work and exit AutoCAD.

Questions

1. Explain the purpose of the STRETCH command.

_____

_____

2. Using the SCALE command, what number would you enter to
   enlarge an object by 50 percent? _____
   enlarge it to 3 times its present size? _____
   reduce it to $1/_2$ its present size? _____

3. Explain how you would dynamically scale an object up or down.

_____

_____

4. Can you dynamically rotate an object? Explain.

_____

_____

5. How would you accurately specify a 90-degree clockwise rotation of an
   object?

_____

_____

6. Explain the purpose of the TRIM command.

_____

_____

7. Describe a situation in which the EXTEND command would be useful.

_____

_____

8. Using LENGTHEN, how would you change the length of an arc?

_____

_____

■ *Challenge Your Thinking: Questions for Discussion*

1. Refer to the site drawing you completed in this unit. If you were doing actual architectural work, you would need to place the features on the drawing much more precisely than you did in this drawing. Describe a way to place a 14-foot-wide driveway exactly 2 feet from the corner of the house. The driveway should be perpendicular to the house. (For this problem, assume that the measurements of your drawing are one tenth of the actual measurements. For example, a length of 5 equals 50 feet, a length of 2 equals 20 feet, and so on.)

2. The floor plan for a rectangular house shows that the house will be 42 feet long and 34 feet wide. The client wants to increase the length of the house to 60 feet. Using the Percent option of the LENGTHEN command, by what percentage would you increase the house length? Why might you choose a different method of lengthening the house? What method would you choose?

Problems

1. Load site.dwg, the drawing you created in this unit. Perform each of the following operations on the drawing.

   - Stretch the driveway by placing a (Crossing) window around the house and across the driveway. Stretch the driveway to the north so that the house is sitting farther to the rear of the lot.
   - Add a sidewalk parallel to the east property line. Use the TRIM and BREAK commands to clean up the sidewalk corner and the north end of the new sidewalk.
   - Reduce the entire site plan by 20 percent using the SCALE command.
   - Stretch the right side of the site plan to the east 1 unit.
   - Rotate the entire site plan 10 degrees in a counterclockwise direction.
   - Place trees and shrubs to complete the site plan drawing. Use the ARRAY command to create one tree or shrub. Duplicate and scale the tree or shrub using the COPY and SCALE commands.

   The site plan should now look similar to the one below.

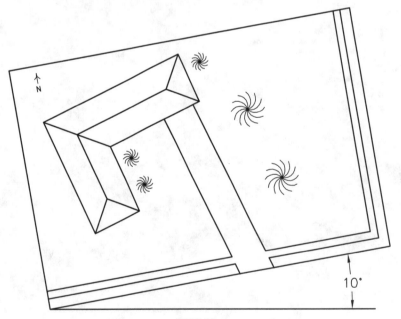

PRB14-1

2. Marketing wants to use the original design of your TV remote control device, but you must change it to meet competitive standards. Create the first drawing of the device (shown on the left). Then copy the drawing to make the following changes using the commands you learned in this unit.

- Scale the device down to 90 percent of its original size.
- Change the two buttons at the bottom into one large button.
- Space the buttons further apart.
- Reduce the overall length.

The final device should look similar to the drawing on the right. Label the two devices "Original Design" and "New Design" as shown below.

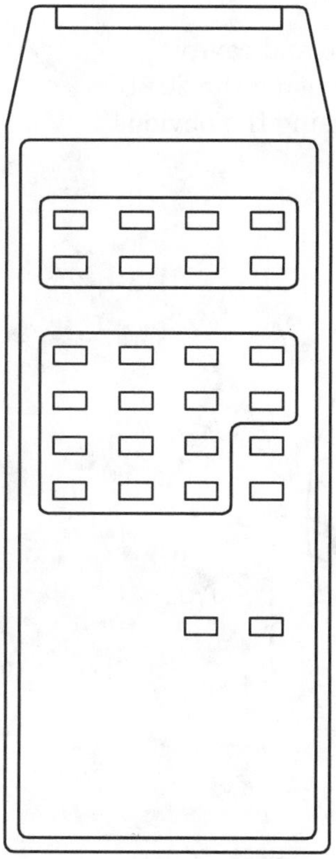

New Design

Original Design

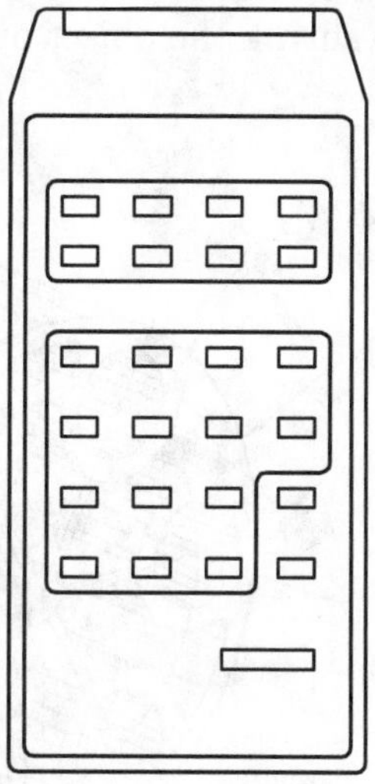

PRB14-2

3. Draw the front view of the tube bundle support shown below. Proceed as follows:

- Draw the outside circle and one of the holes.
- Offset both circles to create the width of the material.
- Draw a line connecting the center of the outside circle and the center of the hole.
- Offset the line to create the internal support structure that connects the hole with the center of the support.
- Draw another line from the center of the large circle at an angle of 30° to the first line.
- Use TRIM to create half of one cavity.
- Use MIRROR to create the other half of the cavity.
- Fillet the three corners.
- Use ARRAY to insert five more copies of the hole and cavity.

When you have completed the drawing, change it using the SCALE command to shrink the object by a factor of .25 using the obvious base point.

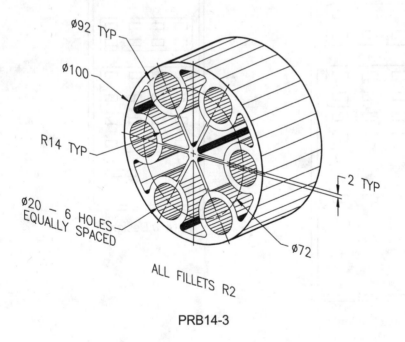

PRB14-3

Problem 2 courtesy of Joseph K. Yabu, Ph.D., San Jose State University
Problem 3 courtesy of Gary J. Hordemann, Gonzaga University

AUTOCAD® AT WORK

Quick-Start Manual Helps in One-Computer Classroom

With only one computer in the classroom, it was very difficult to have students wade through software manuals *and* learn how to design clothes. The manuals are very detailed, and class time is limited. One answer would be to ignore the computer, teach Clothing Design the way grandma used to do it. But CAD is fundamental to the modern apparel industry, and teacher Barb Ritchie of John Diefenbaker High School (Calgary) was determined that her students would learn skills they could use.

Her answer was to write a step-by-step instruction manual that enables a student who knows nothing about AutoCAD to draw simple slopers in AutoCAD and restyle them to their own taste. For Mrs. Ritchie, writing the manual was a matter of changing design language into AutoCAD commands—so that *split* and *spread* became BREAK and ROTATE. For students, it's an opportunity for hands-on experience with an industry standard CAD program. These students are in a position now to benefit more from a comprehensive introduction to AutoCAD tools and techniques.

Finding a way with limited resources is a skill needed by teachers and students everywhere. As a result of cuts in education funding, Mrs. Ritchie's one-computer classroom became a no-computer classroom for a time. Her clothing students raised $1800 toward a new computer. Meanwhile, Mrs. Ritchie asked around about plotters. A ten-year-old plotter that had been declared scrap metal—too out-of-date for industry—was donated to her program. It was great for printing patterns. The only problem was that it needed a (software) driver.

"I'm no expert," she says. "I just keep asking questions till I find the answer I need. I don't give up."

The students had raised $300 for the hookup, but a specially written driver would cost another $500. There had to be a way.

The programmer agreed to write the program as a tax write-off. Mrs. Ritchie's home economics and shop classes have enjoyed output from AutoCAD ever since.

For more information about the skirt-pattern manual or the recently completed bodice-pattern manual, contact Fran Hill, Calgary Board of Education, Alberta, Canada.

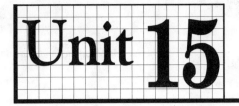 **Unit 15** Advanced Object Selection and Editing

■ OBJECTIVE:

To introduce alternative, advanced, and in some instances, faster methods of selecting and editing objects

Usually, the quickest way to edit an object is to click and drag. This is true whether you are using a $100 drawing program on a $1,000 computer or a $4,000 CAD program on a $40,000 computer. AutoCAD offers a "click 'n' drag" editing feature called *grips*. This feature permits you to pick an object and immediately stretch, move, rotate, or mirror it, *without* entering a command.

AutoCAD also offers a noun/verb selection technique. It allows you to select an object first and then enter a command.

■ *Grips*

1 Start AutoCAD and begin a new drawing from scratch.

2 Enter the **RECTANG** command and draw a rectangle of any size and at any location.

Draw

3 Select any point on the rectangle. (A command *should not* be entered at the Command prompt.)

Notice that small boxes, called grips, appear at each corner of the selection.

4 Pick any one of the four grips. (As you pick the grip, notice how the pickbox locks onto the selected grip.)

The grip turns red, showing that you've selected it. Notice also that the command line has changed, indicating that AutoCAD is in Stretch mode.

5 Press the right button on the pointing device (right-click).

This displays a pop-up menu.

6 Pick **Move**, move the rectangle to a new location, and pick a point.

7 Pick one of the four grips again and right-click.

8 Pick **Mirror** and mirror the rectangle. (Note that the original rectangle disappears.)

9 Pick one of the four grips again and right-click. Using this same method, rotate, scale, and stretch the rectangle.

Copying with Grips _____

You can also copy objects using grips.

1 Draw a circle of any size and at any location.

2 Select the circle and pick one of the four grips that lie on it.

3 Move the pointing device and notice that you can adjust the circle's radius.

4 Pick a point.

5 Pick the circle's center grip, right-click, and pick **Copy**.

6 Copy the circle three times by picking three points anywhere on the screen.

7 Press **ENTER** to terminate the copying.

_____ NOTE: _____

In this case, you must either press the ENTER key or the space bar. Right-clicking displays the Grip Mode Cursor menu.

8 Press **ESC** twice to clear the grips from the circle.

You can move a circle by clicking and dragging the circle's center grip.

1 Pick one of the circles and then pick its center grip.

2 Drag it to a new location.

You can use the grips and right-click feature to modify an object's properties.

1 Pick one of the objects on the screen, pick a grip, and right-click.

2 Pick **Properties...** from the pop-up menu.

This displays a dialog box that permits you to change an object's color, layer, linetype, and other information. You will learn about these object properties in upcoming units.

3 Pick the **Cancel** button in the dialog box.

Grips Dialog Box _____

You can make changes to the grips feature using the Grips dialog box.

1 Select **Grips...** from the **Tools** pull-down menu.

This displays the Grips dialog box, as shown in the following illustration.

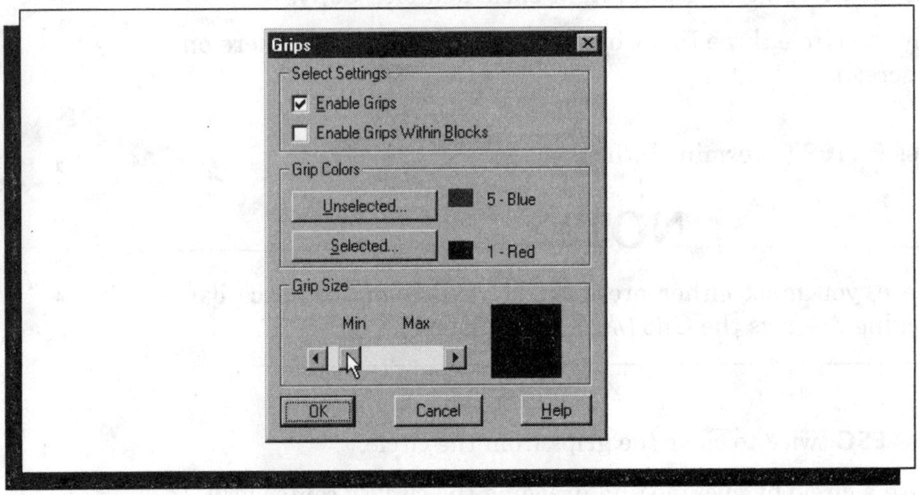

2 Move the slider bar and watch how it changes the size of the box to the right of it.

This adjusts the size of the grips box.

3 Increase the size of the grips box and pick the **OK** button.

4 Select one of the objects and notice the increased size of the grips.

5 Pick one of the grips and perform an editing operation.

6 Display the Grips dialog box again and adjust the size of the grips box, but do not close the dialog box.

The Grips dialog box also permits you to enable and disable the grips feature, as well as control the assignment of grips within blocks. You will learn about blocks in Unit 32. You can also make changes to the color of selected and unselected grips. For now, do not change these settings.

7 Pick the **Help...** button if you are interested in learning more.

8 Close the AutoCAD Help and Grips dialog boxes.

Noun/Verb Selection Technique

As mentioned at the beginning of this unit, AutoCAD lets you select one or more objects and then execute an editing command, such as ERASE, on the object(s) you selected. This is called a noun/verb technique because you first select the object, such as a *line* (the noun) and then the function, such as *erase* (the verb). You may find it convenient to use, especially if you are familiar with other graphics programs that use this technique.

1 Select one of the circles.

2 Pick the **Erase** icon or enter **E** for ERASE.

Modify

As you can see, AutoCAD erased the circle without first asking you to select objects. You have just used the noun/verb technique.

3 Pick another circle and press the **Delete** key on the keyboard.

As you can see, pressing the Delete key is the same as entering the ERASE command.

4 Pick the **Undo** icon from the Standard toolbar or enter **OOPS**.

5 Select one or more objects.

6 Enter the **ARRAY** command.

Modify

AutoCAD does not ask you to select objects because it knows you've already selected them.

7 Press **ESC** to cancel the ARRAY command.

The noun/verb technique applies to all AutoCAD commands that require object selection. Practice this technique in the future.

Object Selection Settings Dialog Box

You can change object selection modes using the Object Selection Settings dialog box.

1 Select **Selection...** from the **Tools** pull-down menu.

This enters the DDSELECT command.

```
Object Selection Settings                    [X]
┌─ Selection Modes ──────────────────────────┐
│  ☑ Noun/Verb Selection                     │
│  ☐ Use Shift to Add                        │
│  ☐ Press and Drag                          │
│  ☑ Implied Windowing                       │
│  ☑ Object Grouping                         │
│  ☐ Associative Hatch                       │
│              ┌──────────────┐              │
│              │   Default    │              │
│              └──────────────┘              │
└────────────────────────────────────────────┘
┌─ Pickbox Size ─────────────────────────────┐
│                             ┌────────────┐ │
│     Min      Max            │            │ │
│                             │     □      │ │
│   ◄│  █    │►               │            │ │
│                             └────────────┘ │
└────────────────────────────────────────────┘
        ┌──────────────────────────┐
        │   Object Sort Method...   │
        └──────────────────────────┘
   ┌────────┐  ┌────────┐  ┌────────┐
   │   OK   │  │ Cancel │  │  Help  │
   └────────┘  └────────┘  └────────┘
```

The Selection Modes area in the dialog box allows you to select from five entity selection methods. The settings, as you see them here, are adequate for most AutoCAD work and usually do not require changes.

2 Pick the **Help...** button and read the information on DDSELECT.

3 Close the AutoCAD Help and Object Selection Settings dialog boxes.

4 Exit AutoCAD. Do not save your changes.

Questions

1. When you use a graphics program, such as AutoCAD, what usually is the quickest way to edit an object?

   _____

   _____

2. Explain the primary benefit of using the grips feature.

   _____

   _____

3. What editing functions become available to you when you use grips?

   _____

   _____

4. In conjunction with the grips feature, what editing options become available when you right-click the pointing device?

   _____

   _____

5. Explain how to copy an object using grips.

   _____

   _____

6. What is the purpose of the slider bar in the Grips dialog box?

   _____

   _____

7. Explain the difference between noun/verb and verb/noun selection techniques.

   _____

   _____

▉ *Challenge Your Thinking: Questions for Discussion*

1. As you have seen, AutoCAD's grip feature is both easy and convenient to use. With that in mind, discuss possible reasons AutoCAD also includes specific commands that you must enter to perform some of the same functions you can do easily with grips.

2. Explore the use of various object snap modes with the grips feature. Is it possible to enhance the function of grips using object snap modes? Explain.

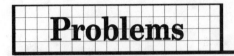

Create only one drawing file for the problems in this unit. Copy all of the objects to a new location on the screen for each problem. When you finish, you should have seven sets of objects in the drawing file. Use the noun/verb selection technique wherever possible, and use grips to perform the editing in problems 3 through 7.

1. Create the conference table and chair shown on the right. Remember to make them small enough that you can place six additional copies in the drawing as you work through the problems.

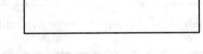

2. Use the noun/verb selection technique to array the chair to create three more chairs. Place them as shown below.

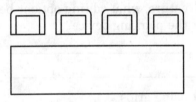

3. Mirror the four chairs to the other side of the table as shown below.

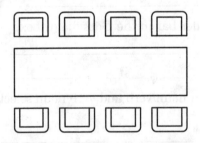

4. Stretch the table to make room for one more chair on each side.

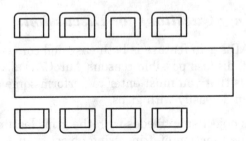

5. Copy two of the chairs twice (to create a total of four more chairs).

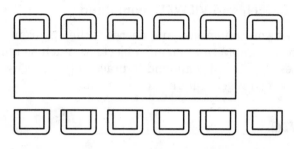

6. Move one of the extra chairs to the left end of the table. Rotate the chair to face the table. Then use a similar procedure to move the other extra chair to the other end of the table and rotate it.

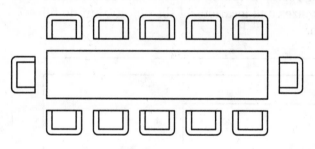

7. Scale the table and chairs to half their original size.

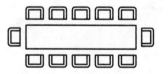

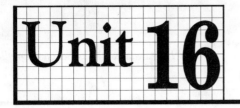 **Unit 16** AutoCAD's Magnifying Glass

■ **OBJECTIVE:**

To practice the ZOOM, REGEN, and VIEWRES commands

The ZOOM command allows you to increase and decrease the apparent size of drawings on the screen. ZOOM can magnify objects as much as ten trillion times! This unit uses the ZOOM command to practice this process, and it covers the effects of screen regenerations and redraws.

■ *ZOOM Command* _____

Let's apply the ZOOM command.

1 Start AutoCAD and start a new drawing from scratch.

2 Draw the following room, including the table and chair. Don't worry about exact sizes and locations of the objects, but do fill most of the screen.

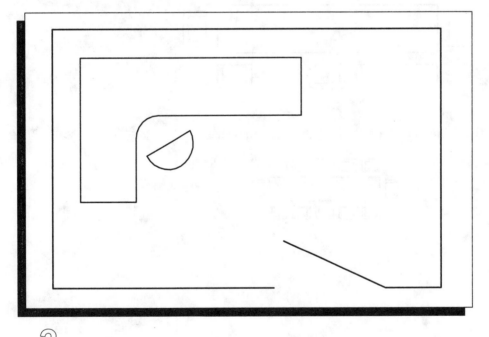

3 Save your work in a file named **zoom.dwg**.

4 Pick the **Zoom Window** icon from the Standard toolbar.

Standard

170

This enters the ZOOM command and displays the following options.

All/Center/Dynamic/Extents/Previous/Scale(X/XP)/Window/<Realtime>: _w

5 Place a window around the table and chair as shown below.

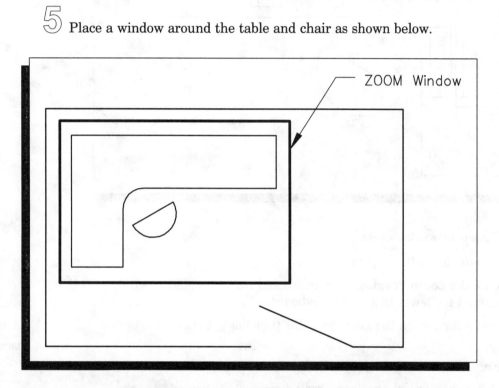

ZOOM Window

Did the table and chair magnify to fill most of the screen? They should have.

6 Next, draw schematic representations of several components that make up a CAD system as shown on the following page. Approximate their sizes, and omit the text.

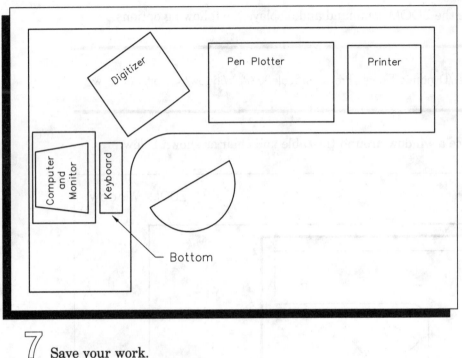

7 Save your work.

Now let's zoom in on the keyboard.

8 Pick the **Zoom Window** icon again, and this time place a window around the lower half of the keyboard.

Let's zoom in further on the keyboard, this time using the Zoom In option.

9 From the Zoom flyout, pick the **Zoom In** icon.

10 Using the vertical scrollbar, scroll down so that you can see the bottom of the keyboard. If you need help, refer to the section titled "Scrollbar Panning" in the following unit.

You will practice using the scrollbars and learn more about them in the following unit.

11 Zoom again if necessary, and adjust the vertical placement of the drawing so that the bottom of the keyboard fills most of the screen.

12 In the lower left corner of the keyboard, draw a small square to represent a key, as shown below.

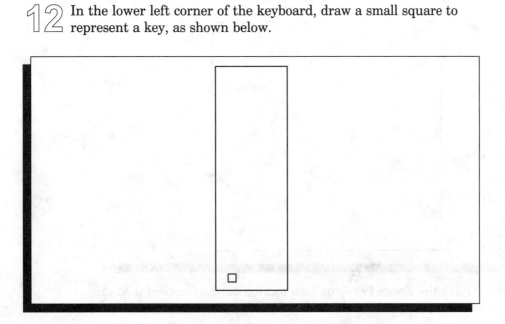

Standard

13 Zoom in on the key, this time using the **Zoom Center** icon from the Zoom flyout. Pick a center point by picking the center of the key, and then specify the height by picking points below and above the key.

NOTE:

You can also enter zoom options by entering ZOOM (or just Z) and the first letter of the option.

14 Using the **POLYGON** and **OFFSET** commands, draw a small trademark on the key as shown on the following page.

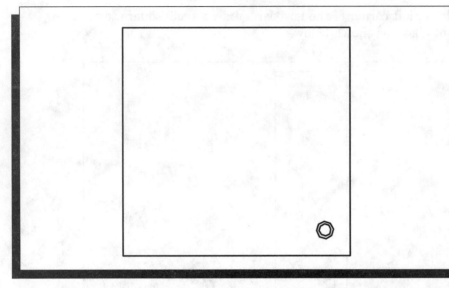

15 Next, pick the **Zoom Previous** icon located on the Standard toolbar. Watch what happens.

Standard

You should now be at the previous zoom factor. The trademark should look much smaller.

16 Array the key to create two rows and five columns as shown in the following illustration. (See the following hint.) These ten keys represent the function keys on a computer keyboard.

Modify

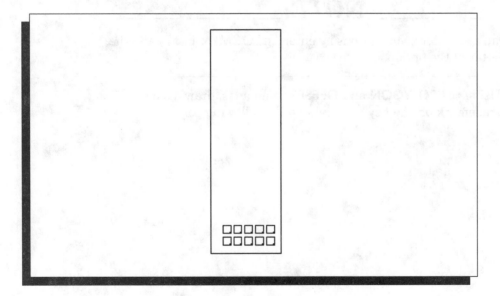

HINT: You can use the pointing device to indicate the row/column distances. The format is:

Unit cell or distance between rows (---): *Pick point 1.*
Other corner: *Pick point 2.*

The following illustrates this operation.

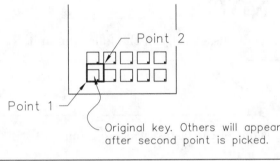

Point 2

Point 1

Original key. Others will appear
after second point is picked.

 Pick the **Zoom All** icon from the flyout, or enter **Z** and **A** at the keyboard.

AutoCAD zooms the drawing to its original size.

Pick the **Zoom Extents** icon from the flyout, or enter **Z** and **E** at the keyboard.

AutoCAD zooms the drawing as large as possible while still showing the entire drawing on the screen.

Enter **Z** (or press the space bar) and enter **2**.

This causes AutoCAD to magnify the screen by two times relative to the ZOOM All display (All = 1).

Enter **ZOOM** and **.5**.

This causes AutoCAD to shrink the screen image by one half relative to the ZOOM All display.

Pick the **Zoom In** icon from the flyout; pick it again.

Pick the **Zoom Out** icon from the same flyout.

Pick the **Zoom Extents** icon again.

Standard

Standard

Standard

Standard

Standard

Standard

Enter **ZOOM** and **.9x**.

This reduces the screen to nine tenths its current size.

 Continue to practice using the ZOOM command by zooming in on the different CAD components of the drawing and including detail on each.

NOTE:

The ZOOM Dynamic option is discussed fully in the following unit.

 Realtime Zooming

AutoCAD's realtime zooming is one of the fastest ways to zoom.

Standard

1 From the Standard toolbar, pick the **Zoom Realtime** icon.

This places you into realtime zoom mode. Notice that the pointer has changed.

2 Using the pick button, click and drag upward to zoom in on the drawing.

3 Click and drag downward to zoom out on the drawing.

You may prefer to use this method of zooming over most alternatives.

4 Press **ESC** or **ENTER** to cancel realtime zooming.

NOTE:

AutoCAD offers a time-saving shortcut for using realtime zooming and panning. At the keyboard, enter Z (for ZOOM) and then right-click the pointing device. Realtime zooming and panning become available.

 Screen Regenerations

Screen regenerations require the recalculation of each vector (line segment) in the drawing. All AutoCAD objects (including arcs, text, and other curved shapes) are made up of vectors stored in a high-precision, floating-point format. As a result, the regeneration of large, complex drawings can take considerable time.

Let's force a screen regeneration with the ZOOM command.

1 Save your work and open the drawing file named **pipes.dwg**. It is located in AutoCAD's Sample folder.

2 Zoom in on a small area repeatedly until you force a screen regeneration. You may need to toggle off the snap mode.

AutoCAD displays a warning: About to regen – proceed?

3 Pick the **OK** button.

Regenerating drawing appears at the bottom of the screen. The time it takes to regenerate depends on the size of the drawing and the speed of the computer.

HINT:

To save time when drawing with AutoCAD, you should avoid screen regenerations as much as possible. The ZOOM command will cause a regeneration when ZOOM All is issued, or if the zoom magnification goes beyond the virtual screen of 4,294,967,296 pixels in each axis. The virtual screen is the portion of the drawing that AutoCAD has previously generated and stored in memory.

Screen regenerations can also be caused by using other AutoCAD commands, including the REGEN command.

4 Enter **REGEN**.

Did the screen regenerate as it did when you entered ZOOM All? As you work with AutoCAD commands such as QTEXT and FILL, you will use the REGEN command.

VIEWRES Command

The VIEWRES command controls the Fast Zoom mode and sets the resolution for arcs, circles, and ellipses.

1 Open the drawing named **zoom.dwg**. Do not save your changes to the pipes.dwg.

2 Enter the **VIEWRES** command.

3 In reply to Do you want fast zooms? enter **Yes**.

This causes the ZOOM, PAN, and VIEW commands to avoid regenerations and perform at redraw speed whenever possible. You will apply the PAN and VIEW commands in the following unit.

4 Enter **20** for the circle zoom percent.

The circle zoom percent controls the appearance of circles, arcs, and ellipses. A high number makes circles and arcs appear smooth, but at the expense of regeneration speed.

 Construct an arc, circle, or ellipse and notice its "coarse" appearance.

As you can see, it now contains fewer vectors and therefore generates more quickly on the screen.

 Enter **VIEWRES** again; enter **Yes**; enter **150**.

Notice the smooth appearance.

Erase the arc, circle, or ellipse.

Transparent ZOOM

With AutoCAD, you can perform transparent zooms. This means you can use ZOOM while another command is in progress. To do this, enter 'ZOOM (notice the apostrophe) at any prompt that is not asking for a text string. Let's try it.

Enter the **LINE** command and pick a point on the screen.

At the To point prompt, enter **'Z**.

Notice that the list of zoom options is preceded by >>. This reminds you that you're in transparent mode.

Zoom in on any portion of the screen.

Notice the message Resuming LINE command at the bottom of the screen.

Pick an endpoint for the line and press **ENTER**.

NOTE:

Picking any of the zoom icons also permits you to perform a transparent zoom.

Transparent zooms give you great zoom magnification flexibility while you are using other commands. For instance, if the line endpoints require greater accuracy than the present display will allow, the transparent zoom provides you with a solution.

NOTE:

Some restrictions and operational hints are noted here.

• Fast Zoom mode (set by the VIEWRES command) must be ON in order for ZOOM to operate transparently.
• Transparent operations can be done only if a screen regeneration is not required.

These items also apply to the PAN and VIEW commands, which are covered in the next unit.

Standard

⑤ Erase the line and enter **ZOOM Previous**.

⑥ Exit AutoCAD. Do not save your work.

Questions

1. Explain why the ZOOM command is useful.

_____

2. Cite one example of when it would be necessary to use the ZOOM command to complete a technical drawing and explain why.

_____

_____

3. Describe each of the following ZOOM options.

All _____

Center _____

Extents _____

Previous _____

Scale _____

Window _____

4. Explain a screen regeneration.

_____

5. Why might you often choose to use the realtime zoom option instead of other zoom methods?

_____

_____

■ *Challenge Your Thinking: Questions for Discussion*

1. Both the ZOOM command and the SCALE command make objects appear larger and smaller on the screen. Write a paragraph explaining the differences between the two commands. Explain the circumstances under which each command should be used.

2. You read in this unit that AutoCAD drawings are stored in a vector format. Many other graphics programs store their drawings in a raster format. Find out the differences between the raster and vector formats. Can you convert a drawing in vector format to a raster format? Can you convert a raster drawing into a vector format? Explain.

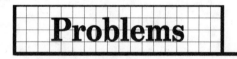

Problems

1. Create a drawing such as an elevation plan of a building, a site plan of a land development, or a view of a mechanical part. Using the ZOOM command, zoom in on the drawing and include detail. Zoom in and out on the drawing as necessary using the different ZOOM options.

2. Refer to the file named land.dwg and instructions contained on the optional *Applying AutoCAD Diskette*. The *Applying AutoCAD Diskette* is a product available from Glencoe/McGraw-Hill, 1-800-334-7344.

3. Draw the following kitchen floor plan. Then zoom in on the kitchen sink. Edit the sink to include the details shown on the following page.

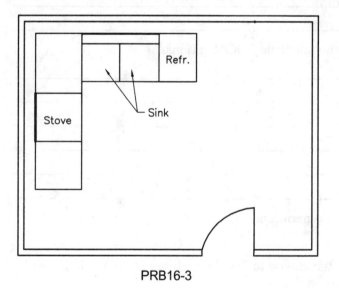

PRB16-3

180

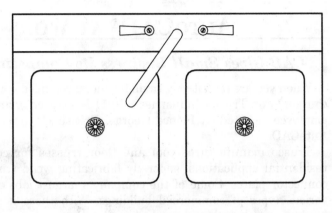

Sink Detail for PRB16-3

Problem 3 courtesy of Kathleen P. King, Fordham University, Lincoln Center

AUTOCAD® AT WORK

CAD Gives Small Business Its Competitive Edge

When you are a relatively small business, you need whatever edge you can get. For Trusco Industries (1991) Ltd., a company with twenty employees located in Prince George, British Columbia, that edge is AutoCAD.

Trusco manufactures roof and floor trusses for commercial and residential applications, generally fabricating wood trusses with steel connector plates. Some of the company's largest jobs are schools and various types of commercial buildings. With these it is necessary to produce detailed plan views and section drawings as well as accurate connection details. In AutoCAD it is possible to "zoom in" on a detail as small as a fraction of an inch within a drawing that may be hundreds of feet in size.

Layers within drawings are also important to the work. AutoCAD allows information to be isolated and worked on separately or combined (with other layers) to provide complete drawings.

Although customized software has been written especially for the truss industry, Trusco's designers find AutoCAD to be the most effective, accurate, and economical way to produce many of their drawings. Building layouts and cross sections are but two of many examples.

Accuracy is very important. Trusses are fabricated in the plant to be assembled elsewhere, and each part, or member, must fit exactly to form the entire truss system. Trusco uses AutoCAD to place all the various parts together to determine exact sizes and shapes of members.

The designer can take the output from other programs and enter it into AutoCAD in one of several formats. Then information can be added and the drawing modified to produce an accurate, professional drawing.

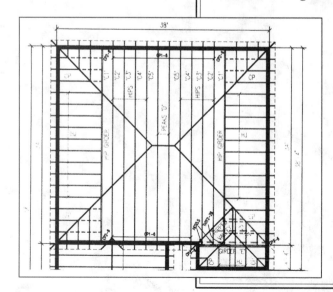

This file import capability also allows parts of a drawing that may be drawn more quickly in other programs to be added to an AutoCAD drawing.

Occasionally Trusco receives plans that have been drawn in metric, but construction subtrades and the fabrication shop still work in imperial units of measure. Converting these drawings is simple in AutoCAD. A complete drawing done in metric can be converted to imperial in a matter of minutes.

The ability to customize pull-down and digitizer menus is of paramount importance to Trusco. The ability to click a button and execute several functions, including AutoLISP routines, allows Trusco designers to produce more detailed drawings in less time—giving Trusco an important edge in a competitive business.

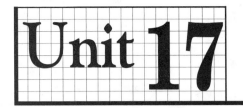

Unit 17   Getting from Here to There

■ OBJECTIVE:

To apply the PAN and VIEW commands, scrollbar panning, dynamic zooms and pans, and Aerial View

Like most CAD software, AutoCAD provides a means for moving around on large drawings so that you can examine and add detail. This unit illustrates these methods and offers suggestions for applying them in the most efficient ways possible.

Note the degree of detail in the following architectural floor plan.

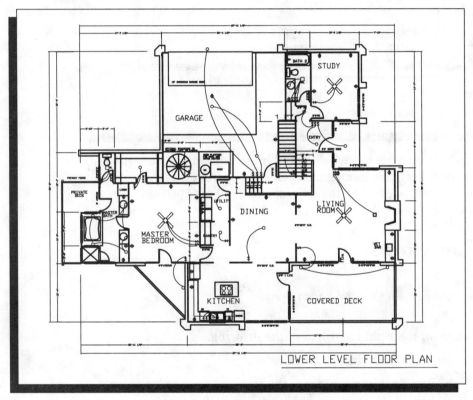

AutoCAD Drawing courtesy of Lansing Pugh, Architect

The drafter who completed this CAD drawing zoomed in on portions of the floor plan in order to include detail. For example, the drafter zoomed in on the kitchen to place cabinets and appliances.

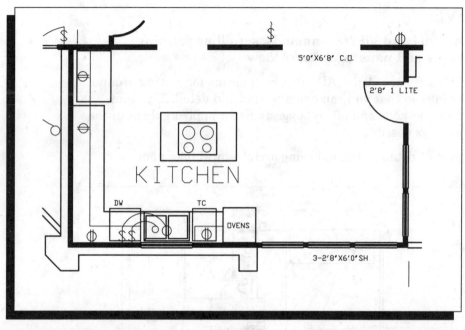

Suppose the drafter wants to include detail in an adjacent room but wants to maintain the present zoom magnification. In other words, the drafter wants to simply "move over" to the adjacent room. This operation can be accomplished using the PAN command.

PAN Command

1 Open the drawing named **zoom.dwg**.

2 Zoom in on the right one-third of the drawing.

Let's pan (move) to the left side of the drawing.

3 Pick the **Pan Realtime** icon from the Standard toolbar, or enter **P** for PAN.

Both enter the PAN command and activate realtime panning. Notice that the pointer changes to a hand.

4 In the left portion of the screen, click and drag to the right.

The drawing moves to the right.

5 Experiment further with the PAN command until you feel comfortable with it. Pan in different directions and at different zoom magnifications. Press **ESC** or **ENTER** to exit the realtime pan mode.

Scrollbar Panning

AutoCAD's scrollbars enable you to pan horizontally and vertically.

1 Zoom in on the computer and chair.

Focus your attention on the horizontal scrollbar. This is the bar that spans the bottom of the drawing area and contains left and right arrows at its ends.

2 Pick the left arrow once; pick it again.

This moves the viewing window to the left.

3 In the horizontal scrollbar, click and drag the movable scroll box a short distance to the right.

This moves the viewing window to the right. You can also move to the left or right by picking a point on either side of the movable scroll box.

Now focus on the vertical scrollbar. This is the bar that extends along the right side of the drawing area from top to bottom. It also contains arrows at its ends.

4 Pick the up arrow a couple of times.

This moves the viewing window in the upward direction.

5 Pick the down arrow; then drag the scroll box down a short distance.

As you can see, the scrollbars offer a convenient way of panning.

VIEW Command

Imagine that you are working on an architectural floor plan like the one shown previously. You've zoomed in on the kitchen to include details such as the appliances, and now you're ready to pan over to the master bedroom. Before leaving the kitchen, you foresee a need to return to the kitchen for final touches or revision. But, by the time you're ready to do this final work on the kitchen, you may be at a different zoom magnification and/or at the other end of the drawing. The VIEW command solves the problem.

Let's apply the VIEW command to the current drawing.

1 Zoom in on a small portion of the drawing, if you have not done so already.

2 Enter the **VIEW** command at the keyboard and save the present zoom window by entering **S** for Save. Give it a one-word name.

3 Now pan to a new location.

4 Enter the **VIEW** command, and restore the named view by entering **R** (for Restore) and the name of the view.

If it did not work, try again.

5 Enter the **VIEW** command once again, and this time issue the **Window** option.

6 Type another view name and press **ENTER**.

7 Define a view window by specifying two corner points.

8 Restore the new view.

9 Practice using the **VIEW** command by zooming and panning to different locations on the drawing and saving views. Define new views using the **VIEW Window** option. Then restore those named views.

After storing several views, it is possible to forget their names. Therefore AutoCAD provides a way of listing all named views.

1 Enter the **VIEW** command and then enter a question mark (**?**).

2 Press **ENTER** in reply to Views to list<*>.

AutoCAD displays a listing of all named views in the AutoCAD Text Window.

3 Close the AutoCAD Text Window.

4 Use the **VIEW Delete** option to delete one of the named views.

View Control Dialog Box _____

The View Control Dialog box offers another way of saving, restoring, reviewing, and deleting named views.

1 Select **Named Views...** from the **View** pull-down menu.

This enters the DDVIEW command and displays the View Control dialog box as shown in the following illustration.

```
┌─────────────────────────────────────────────────────┐
│  ┌──────────────────────────────────────────┐       │
│  │ View Control                         [X]  │       │
│  │ Views                                      │       │
│  │ ┌────────────────────────────────────────┐│       │
│  │ │ *CURRENT*              MSPACE          ││       │
│  │ │ COMPUTER               MSPACE          ││       │
│  │ │ DESK                   MSPACE          ││       │
│  │ │                                        ││       │
│  │ └────────────────────────────────────────┘│       │
│  │ Restore View:    *CURRENT*                 │       │
│  │ [ Restore ] [ New... ] [ Delete ] [Description...]│
│  │       [   OK   ] [ Cancel ] [ Help ]       │       │
│  └──────────────────────────────────────────┘       │
└─────────────────────────────────────────────────────┘
```

2 Using this dialog box, create two new named views on your own.

3 Using the same dialog box, restore the first named view you created in the last step and then delete it.

Transparent PAN and VIEW

The PAN and VIEW commands can be used transparently. Transparent PAN and VIEW are particularly useful for reaching line endpoints that are located off the current display.

Certain transparent pans and "view restore" operations may cause a screen regeneration, although this seldom happens. As a consequence, the transparent pan or view is not allowed. To help avoid this potential problem, use transparent pans and views only within the virtual screen boundary. For more information on the virtual screen and on screen regenerations, see the section titled "Screen Regenerations" in Unit 16.

1 Enter the **LINE** command and pick a point anywhere on the screen.

Standard

2 Pick the **Pan Realtime** icon or enter **P** at the keyboard.

3 Pan to a new location and press **ESC** or **ENTER** to exit realtime panning.

4 Pick a second point to create a line segment.

5 Erase the line.

Alternating Realtime Pans and Zooms

AutoCAD provides a fast method of alternating between realtime pans and zooms.

Standard

1　Pick the **Zoom Realtime** icon from the Standard toolbar.

2　Zoom in on the drawing.

3　Right-click the pointing device.

This displays a pop-up menu.

4　Select **Pan** from the menu and pan to a new location.

5　Right-click the pointing device to display the menu.

6　Pick one of the Zoom options.

7　When you are finished, pick **Exit** from the pop-up menu or press **ESC** or **ENTER** to exit the realtime zoom and pan mode.

Dynamic Zooms and Pans

The ZOOM Dynamic option reduces the possibility of a screen regeneration. It displays the area you should stay within during zooms and pans to avoid time-consuming screen regenerations.

Let's take a closer look.

Standard

1　Zoom in on an area of the drawing and then issue the **ZOOM Dynamic** option.

You should see something similar to the following on the screen. The dotted green box represents the current zoom magnification. The dotted blue box represents the active drawing area. This area fills the screen when you enter ZOOM All.

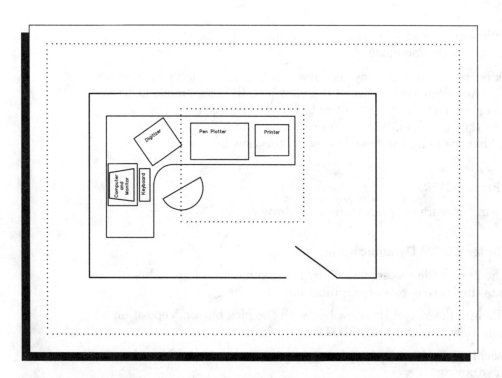

You should also see a white box on the screen with an × in the center. This is called the "view box."

2 Press the pick button on the pointing device, and move it to the left to decrease the size of the box to about one half of its original size. (Do not press ENTER. If you were to press the ENTER key at this time, AutoCAD would perform a zoom on the area defined by the view box.)

An arrow appears at the right side of the view box. This indicates Zoom mode.

NOTE:

If the view box extends outside the four corners (generated area), AutoCAD will force a regeneration of the screen. A small hourglass may appear in the lower left corner of the display to remind you that a lengthy regeneration will be required if that view is chosen.

3 Press the pick button and move the view box around the screen.

You should again see an × in the center of the view box. This means you are presently in the Pan mode.

The pick button toggles the way the view box appears. When you are able to increase and decrease the size of the view box, the Zoom mode is active. When you are able to move the entire box at a fixed size about the screen, you are in the Pan mode. When you press ENTER in either mode, the drawing changes to fill the area defined by the view box.

4 Press **ENTER**.

The drawing appears on the screen—at redraw speed.

Standard

5 Enter **ZOOM Dynamic** again.

Notice the box on the screen defined by the green dotted lines. This represents the current zoom magnification.

6 Reduce the size of the view box with the pick button, reposition the view box, and press **ENTER**.

So you see, this can be a very useful and time-effective way of moving about the drawing.

7 Practice using dynamic zoom and pan.

Standard

8 **ZOOM All**.

Aerial View

Aerial View is a fast and easy-to-use navigational tool for zooming and panning. Before you can use it, VIEWRES's Fast Zoom mode must be on.

Standard

1 Pick the **Aerial View** icon from the Standard toolbar.

This displays the Aerial View window with zoom.dwg inside, as shown in the illustration on the following page.

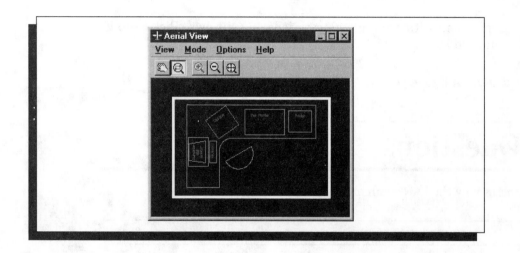

The icons in the Aerial View should look familiar. The dark rectangle represents the zoomed area.

2 Move the crosshairs into the Aerial View window.

New crosshairs appear. Notice that the Zoom icon in the Aerial View window is depressed.

3 Using the new crosshairs, click and drag a window (called a *view box*) around the computer and chair.

AutoCAD zooms in on the area defined by the view box. Meanwhile, Aerial View maintains a view of the entire zoom drawing. The view box, depicted by a white rectangle, shows the size and location of the current view in the graphics screen.

4 From the Aerial View, pick the **Pan** icon and move the pointer into the window's drawing area.

Aerial View

You now have control over the small rectangle that represents the view box you defined in Step 3.

5 Pick a new location for the rectangle.

AutoCAD immediately pans to this location. Aerial View also shows the new location of the current viewport.

6 Pick the **Zoom** icon from the Aerial View and define a new view box.

Aerial View

7 Pick the **Pan** icon from the Aerial View and pan to a new location.

Aerial View

⑧ Experiment with the remaining options on your own. Pick **Help** if you need it.

⑨ **ZOOM All**, save your work, and exit AutoCAD.

Questions

1. Explain why the PAN command is useful.

_____

_____

2. Is it possible to pan diagonally using the scrollbars? Explain.

_____

_____

3. Explain why the VIEW command is useful.

_____

_____

4. How do you list all named views in a drawing?

_____

_____

5. Explain the method of alternating between realtime pans and zooms.

_____

_____

6. Within the ZOOM Dynamic screen, what do the green and blue dotted rectangles represent?

_____

_____

7. Explain the difference between the dynamic zoom and pan modes of the ZOOM Dynamic option.

_____

_____

8. What is the primary benefit of using Aerial View?

_____

_____

■ *Challenge Your Thinking: Questions for Discussion*

1. Many zoom and pan variations were introduced in the last two units. Describe how and when you might use a combination of them, but explain also why it may be impractical to use all of them. Which methods of zoom and pan are the fastest, and which seem to be the most practical for most applications?

2. Study the three-view drawing in Problem 1 below. What does the three-dimensional object look like? Why is the relative position of each view important?

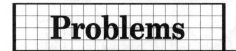

Problems

1. Draw each of the following shapes using the dimensions shown, but do not include the dimensions. Zoom in on one of the shapes and store it as a view. Then pan to each of the other shapes and store each as a view. Restore each named view and alter each shape by adding and erasing lines. Be as creative as you wish.

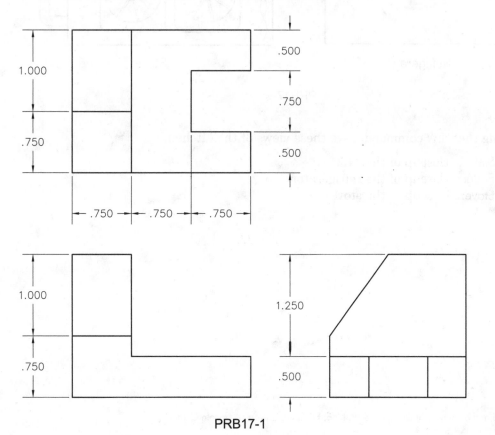

PRB17-1

2. Zoom in on one of the three views from the drawing in problem 1. Use the scrollbars to pan to the adjacent views.

3. Perform dynamic pans and zooms on the drawing in problem 1. Avoid time-consuming screen regenerations.

4. Refer to the file named land.dwg and instructions contained on the optional *Applying AutoCAD Diskette.*

5. Open the kitchen drawing you created in problem 3 of the previous unit. Add the details for the refrigerator and the stove as shown below.

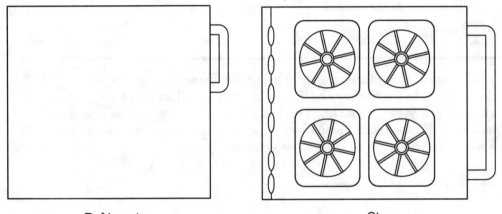

Refrigerator   Stove

6. Using the VIEW command, save these views of the kitchen.

    Sink: closeup of the sink
    Fridge: closeup of the refrigerator
    Stove: closeup of the stove

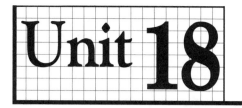

Unit 18 Viewports in Model Space

■ OBJECTIVE:

To learn how to use viewports in model space

Viewports allow you to see one or more views of a drawing at once. By default, AutoCAD begins a new drawing using a single viewport that fills the entire drawing area. You can view the drawing information in each of the viewports at any magnification, and you can draw and edit from one viewport to the next. For instance, you can begin a line in one view and complete it in another.

AutoCAD permits you to work in *model space* or *paper space*. Most AutoCAD drafting and design work is done in model space. Paper space is used to lay out, annotate, and plot views of your work.

Viewports can be applied to both model space and paper space. This unit focuses on the use of viewports in model space only. Unit 39 concentrates on the use of viewports in paper space and covers the MVIEW, PSPACE, MSPACE, and VPLAYER commands and the TILEMODE system variable.

The following illustration gives an example of applying multiple viewports to zoom.dwg. Notice that each viewport is different both in content and in magnification.

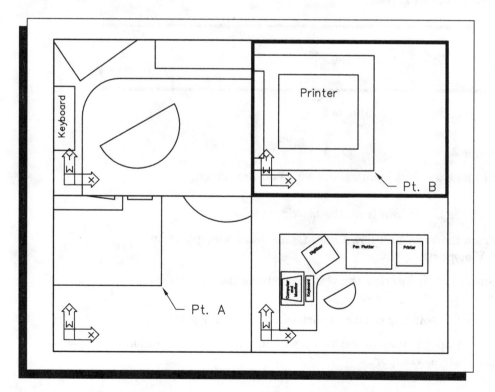

Creating Additional Viewports _____

Viewports are controlled with the VPORTS (also called VIEWPORTS) command.

1 Open the drawing file named **zoom.dwg**.

2 Enter **ZOOM All** to make the drawing fill most of the screen.

3 Enter the **VPORTS** command at the keyboard. (See the following note.)

_____ NOTE: _____

The VPORTS command is disabled when TILEMODE is set to 0 (Off). If you see a message to this effect when you enter VPORTS, enter TILEMODE and 1, and then reenter VPORTS. You will learn more about the TILEMODE system variable in Unit 39.

The following options appear.

Save/Restore/Delete/Join/SIngle/?/2/<3>/4:

As you can see, the default value is 3.

4 Enter **4**.

You should now see four identical drawings on the screen.

5 Pick the **Undo** icon from the Standard toolbar.

6 From the **View** pull-down menu, pick **Tiled Viewports** and **4 Viewports**.

This is equivalent to entering the VPORTS command and entering 4.

7 Move the pointing device to each of the four viewports.

The crosshairs appear only in the viewport with the bold border. This is the current (active) viewport.

8 Move to one of the three nonactive viewports.

An arrow appears in place of the crosshairs.

196

⑨ Press the pick button on the pointing device.

This viewport becomes the current one.

Using Viewports

Let's modify the zoom drawing using the viewports.

1 Refer to the illustration on page 195 and create four similar viewports using AutoCAD's zoom and pan features.

HINT:
Make one of the viewports current and then use the Aerial View window to zoom or pan easily to the required magnification. Repeat this process for the other three viewports. (You might have to reposition the Aerial View window to see the lower right viewport.)

2 Save this viewport configuration by entering **VPORTS** and **S.** Name it **WORKPLACE**.

3 Make the lower left viewport current.

4 Enter the **LINE** command and pick point **A**. Refer to the illustration on page 195 for point A.

HINT:
Use the object snap feature to locate the point accurately.

5 Move to the upper right viewport and make it current.

Notice that the LINE command is now active in this viewport.

6 Pick point **B** and press **ENTER**.

The line represents the edge of a hard surface for the chair.

So you see, you can easily begin an operation in one viewport and continue it in another. Also, any change you make is reflected in all viewports. This is especially useful when you are working on large drawings with lots of detail.

Let's move the printer from one viewport to another.

NOTE:

You may need to shrink the printer a small amount so that it will fit in its new location. Use the grips Scale option to scale the printer.

7 With the upper right viewport current, enter the **MOVE** command, select the printer, and press **ENTER**.

8 Pick a base point at any location on or near the printer.

9 Move to the lower left viewport and make it current. (Make sure ortho is off.)

10 Place the printer in the open area on the table by picking a second point at the appropriate location.

Did the printer location change in the other viewports? It should.

VPORTS Command Options

Let's combine two viewports into one.

1 From the **View** pull-down menu, pick **Tiled Viewports** and **Join**.

2 Choose the upper left viewport in reply to Select dominant viewport.

3 Now choose the upper right viewport in reply to Select viewport to join.

As you can see, the Join option enables you to expand—in this case, double—the size of a viewport.

4 Enter **VPORTS** and **SIngle**.

The screen changes to single viewport viewing. This single viewport is inherited from the current viewport at the time VPORTS SIngle is issued.

5 Enter the **VPORTS ?** option and press **ENTER** twice.

AutoCAD stores information about the viewports currently on the screen as well as the stored viewport configurations, such as WORKPLACE.

6 Enter the **VPORTS Restore** option and enter **WORKPLACE**.

As you can see, the viewports restored as they were stored under the name WORKPLACE. But the drawing does not restore to its earlier form. The printer remains next to the monitor, for example. Saving viewport configurations stores only the viewports themselves, not the drawing information.

7 Enter **VPORTS** and **4** or select the equivalent from the pull-down menu.

So you see, AutoCAD allows you to create additional viewports from the current viewport.

8 Make one of the four small viewports the current viewport by selecting it.

9 Enter **VPORTS** and **3**.

The following options appear.

Horizontal/Vertical/Above/Below/Left/<Right>:

---

10 Enter the **Above** option.

11 Step through the remaining options on your own. Draw and edit using the different viewport configurations.

NOTE:

The REGEN command affects only the current viewport. If you are using multiple viewports and you want to regenerate all of them, you can use the REGENALL command.

12 Save your work and exit AutoCAD.

Questions

1. How can using multiple viewports help you construct drawings?

_____

_____

_____

2. How do you make a viewport the current viewport?

_____

3. Briefly describe each of the following VPORTS options.

Save _____

Restore _____

Delete _____

Join _____

SIngle _____

? _____

2 _____

3 _____

4 _____

4. What option should you enter to obtain two viewports in the top half of the screen and one viewport in the bottom half of the screen?

_____

 Challenge Your Thinking: Questions for Discussion

1. Find out how many viewports you can have at one time in AutoCAD. Would you want to use that many? Why? Explain the advantages and disadvantages of using multiple viewports in your drawings.

Problems

1. Create each of the following viewport configurations and save each under a name of your choice. Hint: From the View pull-down menu, pick Tiled Viewports and the Layout... option.

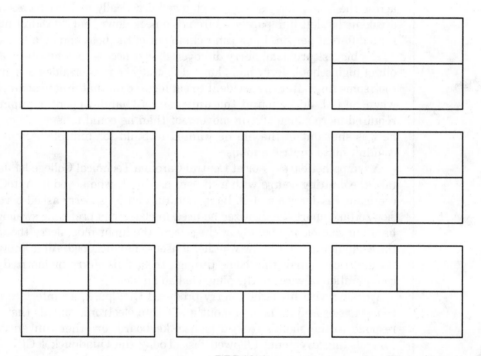

PRB18-1

2. Select one of the viewport configurations from Problem 1. Complete a drawing entirely using these viewports.

AUTOCAD® AT WORK

New Commands for AutoCAD: Sip and Puff

For fun on a weekend, says Jerry Cromer, "I'll have a buddy throw me in the speedboat, get myself situated, and take off a hundred miles an hour across the lake." Jerry's life was changed drastically in 1981 by a skiing accident that left him paralyzed from the neck down, but "it didn't make me a different person. I just can't climb out of the boat and do a flip."

His biggest problem, Jerry discovered, was access to computers. Just out of high school, Jerry had planned to study computer-aided design. It took some time after the accident to get focused on that dream again, but when he did, Jerry found the university's "handicapped" equipment required more range of arm movement than he could muster. In effect, he was shut out of the life he wanted to build for himself because he couldn't quite control a mouse.

A sympathetic professor at Central Carolina Technical College let Jerry take the drafting course with a student assistant, who keyed in AutoCAD commands as Jerry directed. Being naturally curious, Jerry asked permission to take apart the digitizer he used in class. The professor swallowed hard but said okay. Looking at the pins of the input plug, Jerry thought: the computer doesn't know whether a user is handicapped. All it knows is the electrical signals that come through these pins. Jerry memorized the configuration of wires in the plug. He had an idea.

Working with his father, Jerry produced Quadpuck, an interface that converts sips and puffs on a mouthpiece into electronic signals that correspond to the signals from a mouse, keyboard, or other conventional control. Improvements followed fast. Today the Quadpuck I CPU and Quadpuck AirStick (which emulates a mouse) allow users who cannot push buttons to work with full productivity in AutoCAD and other applications. In Windows, Quadpuck users can call up AirKeys for an onscreen keyboard, or they can use a database, word processor, or any other program that runs in Windows.

As a businessman (with a drafting firm that counts Westinghouse and Carolina Power & Light among its clients), Jerry emphasizes that Quadpuck systems can be disengaged with the flip of a switch, which means employers do not need to buy a dedicated computer for a physically challenged employee. Without elaborate setup, Quadpuck gives employers access to "some of the world's most capable and eager minds."

But Quadpuck isn't all work, any more than Jerry is. There's a Quadpuck remote control for home electronics. And with the Quadpuck AirStick I GC, a 12-function control for Super Nintendo, Jerry challenges any and all who want a test of skill—sip and puff versus handheld—in Super Mario World.

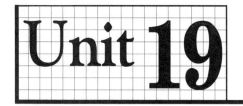 # Unit 19 AutoCAD File Maintenance

OBJECTIVE:

To practice using the Select File dialog box and the AUDIT and RECOVER commands

When working with AutoCAD, there will be times when you want to review the contents of a folder or delete, rename, copy, or move files. This unit gives you practice in using these functions, as well as commands that attempt to repair drawing files containing errors. Refer to Appendices B and C for additional information on AutoCAD file maintenance.

Select File Dialog Box

By now, you should have several drawing files. AutoCAD's Select File dialog box can help you manage these files.

1 Start AutoCAD and start a new drawing from scratch.

2 Pick the **Open** icon from the Standard toolbar.

This displays the Select File dialog box.

Standard

3 Open the **AutoCAD R14** folder (directory).

HINT:
You may need to use the Up One Level button. See the following illustration.

As shown in the illustration above, this displays all files and folders in the AutoCAD R14 folder.

4 Double-click the folder with your first name—the one you created in Unit 1.

5 Pick the **Details** button.

This provides a listing of each file, along with its size, type, and the date and time it was last modified.

6 Use the scrollbars to review the list of files.

7 Pick the **List** button to produce a basic listing of the files.

Copying Files

The Select File dialog box permits you to copy files easily.

1 Right-click and drag the **stuff.dwg** file to an empty area and release the right button on the pointing device.

A pop-up menu appears.

2 Pick **Copy Here**.

A new file named Copy of stuff.dwg appears.

Renaming Files

You can also rename files using the Select File dialog box.

1 Single-click the new file named **Copy of stuff.dwg**.

HINT:
Click on the name of the file, not the icon.

2 Single-click the file again.

You should now see a blinking cursor at the end of the file name with a box around the file name.

3 Type the name **junk.dwg** and press **ENTER**.

NOTE:

Instead of entering an entirely new name, you can also edit the name of the file by picking a new cursor location and typing new text. This method of renaming also permits you to use the backspace key and the space bar.

The new file named junk.dwg contains the same contents as stuff.dwg.

Creating New Folders

It's also fast and easy to create new folders.

 Pick the **Create New Folder** button.

A new folder named New Folder appears.

Using the renaming method described in the previous section, rename the new folder to **Junk Files**.

Moving Files

Let's move a file to a new folder.

Right-click **junk.dwg** and drag it to the Junk Files folder.

A pop-up menu appears.

Pick **Move Here**.

The file junk.dwg is now located in the Junk Files folder.

Deleting Files and Folders

It's time to delete the junk.dwg file and the Junk Files folder.

 Double-click the **Junk Files** folder.

 Select **junk.dwg** and press the **Delete** key on the keyboard.

Pick the **Yes** button in the Confirm File Delete dialog box if you are sure you have selected junk.dwg.

4 Pick the **Up One Level** button.

5 Select the **Junk Files** folder and press the **Delete** key.

6 If you are sure that you selected the Junk Files folder, press the **Yes** button in the Confirm Folder Delete dialog box to send this folder to the Recycle Bin.

Other Options

The Select File dialog box offers other useful options.

1 Pick a point in the box at the right of Look in.

This lists the current folder's parent folder, its parent, and so on. Also, this list displays other drives available to the computer. If the computer you are using is a part of a network, Network Neighborhood is available, enabling you to access other drives, folders, and files located on other computers in the network.

2 Pick a point in the **Files of type** drop-down box.

This displays a list of files that you can open. The DWG file type is AutoCAD's binary file format for drawings and models. The DXF file format is used to transfer drawing file data to and from AutoCAD and other CAD systems. The DWT format is AutoCAD's drawing template format, a file type that was not available prior to AutoCAD Release 14. You will learn more about the DXF and DWT formats in upcoming units in this book.

3 Close the Files of type listing by clicking a point in the gray area to the left or right of it.

Located in the lower right of the dialog box is the Select Initial View check box. It permits you to select a named view when you open a drawing.

If you check the Open as read-only check box, located in the lower left of the dialog box, and then open a file, you can make changes to the drawing, but you cannot save them.

4 Check the **Open as read-only** check box, open the **gasket.dwg** drawing, make a change, and try to save it.

AutoCAD displays the message Drawing file is write protected.

5 Pick the **OK** button.

AUDIT Command

The AUDIT command is available as a diagnostic tool. It examines the validity of the current drawing file and corrects errors.

1 From the **File** pull-down menu, select **Drawing Utilities** and **Audit**.

This enters the AUDIT command.

2 In reply to Fix any errors detected? press **ENTER**.

3 Press the **F2** function key to read all of the message.

Information similar to the following appears.

---

    0   Blocks audited
    Pass 1 54   objects audited
    Pass 2 54   objects audited

    Total errors found 0   fixed 0

If you had entered Yes in reply to Fix any errors detected? the report would have been the same because this file has no errors. If a drawing file contains errors and AUDIT can't fix them, use the RECOVER command explained in the following section.

The AUDIT command automatically creates an audit report file containing an adt file extension when the AUDITCTL system variable is set to 1 (On). The default setting is 0.

4 Press the **F2** function key to return to the graphics screen.

RECOVER Command

You can attempt to recover damaged files using the RECOVER command.

1 Enter **RECOVER** and pick the **No** button.

2 Select one of your drawing files, such as stuff.dwg or zoom.dwg, even though it may not be damaged. (The drawing recovery process is the same regardless.)

3 Pick the **Open** button.

AutoCAD performs an automatic audit of the drawing. If the audit is successful, AutoCAD displays an AutoCAD Message box saying that no errors were detected. An audit report file (containing an ADT file extension) is created when AUDITCTL is set to 1.

4 Pick the **OK** button.

AutoCAD loads the drawing.

NOTE:

Listed here are probable causes of damaged files.

- An AutoCAD system crash
- A power surge or disk error while AutoCAD is writing the file to disk

Listed here are typical effects of damaged files.

- AutoCAD may refuse to edit or plot a drawing file.
- A damaged file may cause an AutoCAD internal error or fatal error.

A drawing file may be damaged beyond repair. If so, the drawing recovery process will not be successful.

Each time you save an AutoCAD file, AutoCAD saves the changes to the current DWG file. AutoCAD also creates a second file with a BAK file extension. This file contains the previous version of the file—the version prior to saving. If the DWG file becomes damaged beyond repair, you can rename the BAK file to a DWG file and use it if necessary. To prevent the loss of data, save often and produce a backup copy of your DWG files frequently.

5 Exit AutoCAD without saving unless the file *was* damaged and recovered. In that case, save the changes to the file.

Questions

1. Explain the purpose of each of the four buttons located in the top middle area of the Select File dialog box.

_____

_____

_____

2. How do you copy a file in the Select File dialog box?

_____

_____

3. In the Select File dialog box, what information can you display by picking the Details button?

_____

_____

4. What is the primary purpose of the AUDIT command?

_____

_____

5. How may AutoCAD's RECOVER command be useful?

_____

_____

■ *Challenge Your Thinking: Questions for Discussion*

1. Is it possible to move and copy files from the Select File dialog box to folders on the Windows desktop? Is it possible to move and copy files from folders on the Windows desktop to the Select File dialog box? Explain why this might be useful.

Problems

Using the commands described in this unit, complete the following activities.

1. Generate a list of drawing files from one or more of AutoCAD's directories.

2. Generate a list of help files found in one or more of AutoCAD's directories.

HINT: AutoCAD help files have an extension of HLP. Therefore enter *.hlp for the search pattern.

3. Generate a list of all AutoCAD drawing BAK files.

4. Rename one of your drawing files. Then change it back to its original name.

5. Delete the BAK file for one of your drawing files.

6. While in AutoCAD, load another program such as a spreadsheet, create a file, and then return to AutoCAD.

7. Diagnose one of your drawing files and fix any errors that may be detected.

AUTOCAD® AT WORK

AutoCAD and Beyond

Ben Crow, a teacher at Casa Grande High School, entered a summer program designed to instruct and excite teachers about new technologies. The program was supported by the Arizona Department of Education, Vocational Education Division, and Arizona State University.

Along with other high school teachers, Mr. Crow underwent an intensive three weeks of study and hands-on experience with a variety of technologies. One of the areas of instruction was AutoCAD. Like most of the other teachers, Mr. Crow knew absolutely nothing about computers, much less AutoCAD, when he entered the program. In fact, the state provided a small grant that bought the computer system to get him started.

The program was an enormous help to Mr. Crow and the other teachers who attended, but the real benefits soon became evident in the classrooms around the state. The secondary students in these teachers' classrooms were the real winners.

For Ben Crow, the successes in student outcomes have been phenomenal. His students are currently designing parts and then precision-machining them. They are using AutoCAD and similar software to design and engineer products as small as mechanical linkages and as large as horse trailers! The students then market their products. Their sales help support the program.

Perhaps even more important, the students are learning about the basic skills necessary to compete in the marketplace. They are learning how economics, conservation of resources, math skills, reading and communication skills, problem solving, and teamwork can be used together to make a product or business successful.

As a result of the summer program, Ben Crow has used AutoCAD and related programs in areas that reach far beyond a substitute for mechanical drafting. His students have become more proficient in basic skills, the dropout rate has declined to almost zero, and enrollment has outstripped classroom space.

Now, Casa Grande graduates have a distinct advantage in their schooling and in the marketplace. Those who have taken Mr. Crow's classes begin their post-secondary programs with substantial knowledge of CAD and CAM. The process is ongoing. Mr. Crow now offers an in-service training class for teachers who want to learn AutoCAD, which they in turn teach to students at their schools. Exploration of what can be done with CAD programs has only begun, at least at Casa Grande.

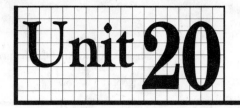

Placing Notes and Specifications

■ **OBJECTIVE:**

To practice the use of the DTEXT, STYLE, and MTEXT commands and the TEXTQLTY system variable

This unit focuses on the placement of text using the DTEXT and MTEXT commands and on the creation of new text styles using the STYLE command.

The following drawing shows the number of notes and specifications typical in many drawings. Some drawings, of course, contain more.

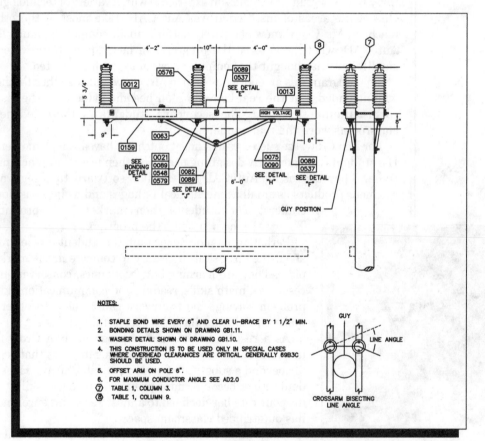

AutoCAD drawing courtesy of Russ Burns, Sacramento Municipal Utility District

As you can see, the text information is an important component in describing the drawing. With traditional drafting, the text is placed by hand, consuming hours of tedious work. With CAD, you can place the words on the screen almost as fast as you can type them.

DTEXT Command

The DTEXT command enables you to display AutoCAD text dynamically in the drawing as you type it.

1 Start AutoCAD and start a new drawing from scratch.

2 Set snap at **.25**.

3 From the **Draw** pull-down menu, select **Text** and **Single Line Text**.

This enters the DTEXT command.

NOTE:

You can also enter the DTEXT command by entering DTEXT at the keyboard.

The following options appear at the Command line.

Justify/Style/<Start point>:

4 In response to Start point, place a point on the left side of the screen. The text will be left-justified (aligned) beginning at this point.

5 Reply to the Height prompt by moving the pointing device up .25 unit from the starting point and pick a point.

6 Enter **0** (degrees) in reply to Rotation angle.

7 At the Text prompt, type your name using both upper- and lowercase letters and press **ENTER**. (See the following note.)

You should again see the Text prompt.

8 Type your P.O. box, rural route, or street address and press **ENTER**.

Where was it placed?

9 At the Text prompt, enter your city, state, and zip code.

10 Press **ENTER** again to terminate the DTEXT command.

Now let's enter the same information you entered before, but this time in a different format.

1 Enter the **DTEXT** command (press the space bar) and **J** for the Justify option.

The following options appear.

Align/Fit/Center/Middle/Right/TL/TC/TR/ML/MC/MR/BL/BC/BR:

---

2 Enter the **C** (Center) option.

3 Place the center point near the top center of the screen and set the text height by entering **.2** at the keyboard. Do not insert the text at an angle.

4 Repeat Steps 7 through 10 on the preceding page.

When you're finished, your text should be centered like the example below. If it isn't, try again.

```
           Mr. John Doe
       601 West 29th Street
      Caddsville, CA  09876
```

5 Save your work in a file named **text.dwg**.

AutoCAD has another DTEXT option very similar to Center called Middle.

1 Enter **DTEXT** again and **Justify**.

2 This time select the **Middle** option and pick a point near the center of the screen.

3 Place another string of text, such as Vance and Shirley.

How is the Middle option different from Center? If you're not sure, try the Center option again.

AutoCAD also allows for placement of text between two specified points. The Align and Fit options have similarities, but there is a difference. Let's experiment.

1. Enter **DTEXT**, **Justify**, and **Align**.

2. Pick a point at the left of the screen.

3. Pick a second point near the right of the screen. (See the following hint.)

HINT: If you want the text to appear perfectly horizontal, turn on ortho.

4. Enter the following sentence, and be sure to press **ENTER** twice.

This sentence will be aligned between two points.

Let's use the Fit option and try to determine the difference between Fit and Align.

1. Enter **DTEXT**, **Justify**, and **Fit**.

2. Pick two points approximately 4 units apart. (Turning grid on will help you estimate 4 units.)

3. Specify a height of **.25** unit.

As you may recall, "height" was *not* part of the Align option.

4. Enter the following text: **Check this out!** (Press **ENTER** a second time.)

Do you recognize the difference between Fit and Align? Fit adjusts the width of the text only, at a specified height. Align adjusts both the width and the height.

5. Experiment with each of the following DTEXT Justify options on your own.

TL   Starts the top left portion of text at a given start point
TC   Centers the top of text at a given point
TR   Ends the top of text at a given point
ML   Starts the middle of text at a given start point
MC   Centers the middle of text at a given point
MR   Ends the middle of text at a given point
BL   Starts the bottom left portion of text at a given start point
BC   Centers the bottom of text at a given point
BR   Ends the bottom of text at a given point

If you forget the meaning of the two-letter options, obtain AutoCAD help on the DTEXT command. Enter the DTEXT command and pick the Help icon. Each of these options is explained.

6 Save your work.

AutoCAD also offers the TEXT command, which is very similar to the DTEXT command. The main difference is that the TEXT command does not permit you to view the text as it is being typed, and it does not permit you to enter more than one line of text without reentering the command.

STYLE Command

It's possible to create new text styles using the STYLE command. During their creation, you can expand, condense, slant, and even draw them upside down and backwards.

1 From the **Format** pull-down menu, select **Text Style**, or enter the **STYLE** command at the keyboard.

The Text Style dialog box appears.

2 Pick the **New** button, enter **COMP1** for the new text style name, and pick the **OK** button.

3 Under Font Name, display the list of fonts by picking the down arrow.

4 Using the scrollbar, find the font file named **complex.shx** and single-click it.

Notice that the text sample in the Preview box changes to show the appearance of the complex.shx font.

5 Study the other parts of the dialog box and then pick the **Apply** and **Close** buttons.

You are ready to use the new COMP1 text style with the DTEXT command.

1 Enter the **DTEXT** command and **Justify** option.

2 Right-justify the text by entering **R** (for Right) from the list of options.

3 Place the endpoint near the right side of the screen.

4 Set the height at **.3** unit.

5 Set the rotation angle at **0**.

6 For the text, type the following three lines. Be sure to press **ENTER** twice after typing the third line.

```
         Computer-aided
    Design and Drafting
          Saves Time
```

Your text should look like the text in the illustration above.

HINT:

If you've made a spelling error, erase the object (text) and enter the correct text. You can also change the text using the CHANGE and DDEDIT commands. These commands are discussed in the following unit.

With the STYLE command, you can develop an infinite number of text styles. Try creating other styles of your own design. You can give them any name with up to thirty-one characters. The ROMANS text font is recommended for most applications.

NOTE:

The text styles you create remain within the current drawing file. They cannot be transported to another drawing unless the *entire* drawing containing the styles is inserted into the drawing.

As you create more text styles within a drawing file, you may occasionally want to check their names.

1 Enter the **STYLE** command.

2 Under Style Name, display the list of text styles by picking the down arrow.

3 Pick **STANDARD**.

Notice that the STANDARD style was developed using the TXT font file. This is the default text style.

AutoCAD provides many individual fonts. For samples of these fonts, see Appendix F. (Additional fonts can be imported as described in the *AutoCAD User's Guide.*)

4 Pick the **Close** button.

Text Fonts

AutoCAD supports TrueType fonts and AutoCAD compiled shape (SHX) fonts.

1 Reenter the **STYLE** command and pick the **New** button.

2 Enter **TT** for the new text style name and pick the **OK** button or press **ENTER**.

3 Under Font Name, use the drop-down box to find Swis721 BT.

Notice the overlapping T's located at the left of the font name. This indicates that it is a TrueType font.

4 Select **Swis721 BT**.

The Swis721 BT font displays in the Preview area. Notice also that Italic displays under Font Style.

5 Under Font Style, display the list of options, pick **Roman**, and notice how the font changes in the Preview area.

Note that 0.0000 is in the text box under Height. This 0 value indicates that the text is not fixed at a specific height, giving you the option of setting the text height when you enter the DTEXT or TEXT command.

6 Pick the **Apply** and **Close** buttons.

7 Enter the **DTEXT** command and pick a point.

8 Enter **.2** for the height and **0** for the rotation angle.

9 Type **This is TrueType.** and press **ENTER** twice.

The TrueType text should look like the text in the following illustration.

This is TrueType.

10 Enter **TEXTQLTY** at the keyboard.

This sets the resolution of text created with TrueType fonts. A value of 50 sets the resolution to 300 dots per inch (dpi), while a value of 100 sets the resolution to 600 dpi. A higher value improves the quality of the text but also increases display and plotting time. The minimum and maximum values are 0 and 100.

11 Press the **ESC** key.

12 Experiment with other TrueType fonts on your own.

Changing the Current Text Style

Next, let's set a new current text style.

1 Enter the **DTEXT** command and enter **S** for Style.

At this point, you can generate a list of styles or enter a new current style. Let's bring back the STANDARD text style.

2 Type **STANDARD** and press **ENTER**.

NOTE:

You can also enter the STYLE command and use the Text Style dialog box to change the text style.

3 Place some new text on the screen and terminate the DTEXT command.

4 Change back to the **TT** text style and terminate the DTEXT command.

5 Save your work.

MTEXT Command

The MTEXT command creates a multiple-line text object called mtext. AutoCAD uses a text editor to create mtext objects.

1 Erase the text from the upper one-half of the screen.

2 From the Draw toolbar, pick the **Multiline Text** icon.

Draw

This enters the MTEXT command and displays the text style and height at the Command line.

3 Pick a point in the upper left corner of the screen.

Notice the new list of options at the Command line.

4 Enter **W** for Width.

5 Enter **4** in reply to Specify width or, with ortho and snap on, pick a point about 4 units to the right of the first point.

AutoCAD displays the Multiline Text Editor dialog box. Notice that the font is Swis721 BT and the default text height is .2000 unit.

6 Pick the **Properties** tab.

The style is TT and the justification is Top Left. This was selected when you picked a point in the upper left area of the screen. The width is 4.0000 and the rotation is 0.

7 Type the following sentence.

I'm using AutoCAD's text editor to write these words.

8 Pick the **Character** tab.

9 Highlight the word **using** by double-clicking it and then pick the **B** (Bold) button.

10 Highlight the words **text editor** and pick the **I** (italic) button.

11 Highlight the word **words** and pick the **U** (underline) button.

The sentence should now look like the following.

12 Pick the **OK** button.

The new text appears on the screen.

13 Save your work.

Importing Text

Suppose you want to import text from a file.

1 Minimize AutoCAD.

2 From the Windows **Start** menu, pick **Programs**, **Accessories**, and **Notepad**.

3 Enter the following text.

AutoCAD's text editor permits you to import text from a file. This text will appear in AutoCAD shortly.

4 From the **File** pull-down menu, pick **Save As...**, select the folder with your name, and name the file **import.txt**.

5 Pick the **Save** button, exit Notepad, and maximize AutoCAD.

6 From the **Draw** toolbar, pick the **Multiline Text** icon.

Draw

7 Near the top center of the screen, pick the first corner.

8 Produce a rectangle that measures about 1 unit tall by 3 units wide.

9 Pick the **Import Text...** button on the right side of the Multiline Text Editor dialog box.

10 Find and double-click **import.txt**.

The text appears in the text editor.

11 Pick the **OK** button.

The text appears in the graphics screen.

12 Save your work and exit AutoCAD.

Questions

1. Describe the following options of the DTEXT command.

   Align _____

   Fit _____

   Center _____

   Middle _____

   Right _____

2. What is the purpose of the DTEXT Style option?

   _____

3. Name at least six fonts provided by AutoCAD.

   _____

   _____

4. What command do you enter to create a new text style?

   _____

5. Briefly describe how you would create a tall, thin text style.

   _____

   _____

6. How is TEXT different from the DTEXT command?

   _____

   _____

7. What might be a benefit of using the MTEXT command?

   _____

   _____

■ *Challenge Your Thinking: Questions for Discussion*

1. Obtain on-screen help and then experiment with the COMPILE command. What are the advantages and disadvantages of compiling fonts? When might you want to use this option?

2. Explain the differences between the DTEXT command and the MTEXT command. Describe a situation in which each command would clearly be a better choice.

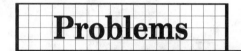

Problems

In problems 1 and 2, create new text styles using the information provided.

1. Style name: CITYBLUEPRINT
   Font file: CityBlueprint
   Height: .25 (**fixed**)
   Width factor: 1
   Oblique angle: 15

2. Style name: ITAL
   Font file: italic.shx
   Height: 0 (**not fixed**)
   Width factor: .75
   Oblique angle: 0

In problems 3 and 4, place text on the screen using the information provided. Use the DTEXT command for problem 3 and MTEXT for problem 4.

3. Use the CITYBLUEPRINT text style you created in problem 1. Right-justify the text. Do not rotate the text. The text should read:

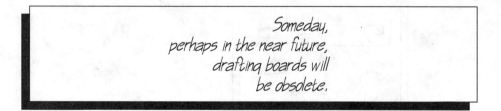

Someday,
perhaps in the near future,
drafting boards will
be obsolete.

PRB20-3

223

4. Use the ITAL text style you created in problem 2. Set the text height at
.3 unit. Rotate the text 90 degrees. Using mtext, set the width of the
text line to 4 units. The text should read:

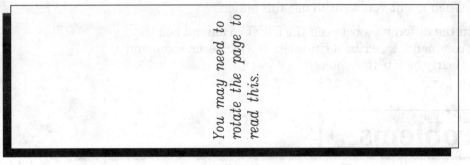

PRB20-4

5. Create this television remote with text using the STYLE and DTEXT
commands. Use all commands from Units 1-20 to your advantage.

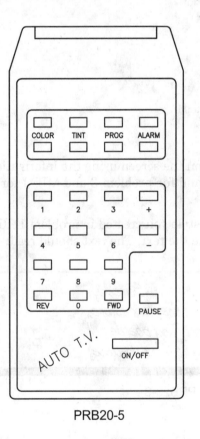

PRB20-5

Problem 5 courtesy of Joseph K. Yabu, Ph.D., San Jose State University

■ OBJECTIVE:

To edit and change text using CHANGE, DDEDIT, and the text editor and to apply several special text features, including QTEXT and the spell checker

AutoCAD offers several commands and features for editing and enhancing text and mtext entities. The CHANGE and DDEDIT commands apply to standard text. DDEDIT also works with mtext, which takes advantage of the text editor.

AutoCAD enables you to produce special characters such as degree and diameter symbols. The QTEXT command is useful to speed the display of text and mtext entities.

■ *Correcting Text with CHANGE* _____

You can make changes to text you've placed on the screen using the CHANGE command. CHANGE was introduced in Unit 12, where it was used to change endpoints of lines.

1 Start AutoCAD and open the drawing named **text.dwg**.

2 If you erased the text **Check this out!** in the previous unit, recreate it now using the STANDARD text style.

3 Enter the **CHANGE** command, select the text **Check this out!** and press **ENTER**.

4 Press **ENTER** in reply to Properties/<Change point>.

5 In reply to Enter text insertion point, move the crosshairs around and watch what happens (ortho should be off); then press **ENTER**.

6 Enter **COMP1** for the new style and move the crosshairs up and down.

7 Enter **.3** for the new height.

8 Enter **Check out this text!** in reply to New text.

DDEDIT Command

The CHANGE command is useful for changing the text height, position, or style. If only text editing is required, the DDEDIT command is faster.

1 From the **Modify** pull-down menu, select **Object** and the **Text...** item.

This enters the DDEDIT command.

2 Select **Check out this text!**

The Edit Text dialog box appears.

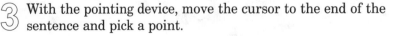

3 Move the pointer to the area containing the text and pick a point between the words out and this.

4 Press the backspace key until the words Check out are gone.

5 Type the words **Look at** and pick the **OK** button or press **ENTER**.

The line should now read Look at this text!

6 Press **ENTER** to terminate the command.

Using DDEDIT to Edit Mtext

You can also use the DDEDIT command with mtext. The command behaves quite differently, however.

1 Enter the **DDEDIT** command.

2 Select the mtext object located in the upper left corner of the screen.

The text editor appears, containing the mtext you selected.

3 With the pointing device, move the cursor to the end of the sentence and pick a point.

4 Backspace until you've deleted the word words and type the word **sentences**, but do not press ENTER. Add a period at the end.

5 Remove the underlining if it's still present by highlighting the underlined text and picking the **U** (Underline) button.

6 After the period, add a space and type the following sentence.

Mtext is particularly useful for long notes and specifications.

7 Change the style to **STANDARD**, the height to **.15**, and the width to **5**, and pick the **OK** button.

HINT:

To change the style and width, pick the Properties tab.

The mtext changes according to the changes you made in the dialog box. Note that the DDEDIT command is still entered and active.

8 Select the mtext paragraph once again.

9 Change the style to **TT**, the justification to **Top Right**, and the width to **4**.

NOTE:

Rotation specifies the rotation angle of the mtext boundary.

10 Pick the **OK** button.

AutoCAD right-justifies the text.

11 Press **ENTER** to terminate the DDEDIT command.

Special Characters

The text editor permits you to insert special characters such as the degree and plus/minus symbols.

1 Enter the **MTEXT** command.

2 In an empty area of the screen, create a rectangle measuring about 1 unit tall by 2 units wide.

3 With the help of the **Symbol** pull-down menu, write the following specification.

Drill a ∅.5 hole at a 5° angle using a tolerance of ± .010.

NOTE:

Selecting Diameter from the Symbol drop-down box causes the %%c characters to appear in the text editor. These are special codes that represent the diameter symbol. When using DTEXT or TEXT, you can enter these codes, which appear at the right in the Symbol drop-down box. When you select Degrees and Plus/Minus from the Symbol drop-down box, the actual symbols appear in the text editor.

4 Pick the **OK** button.

5 Enter the **DDEDIT** command and pick the new mtext object.

6 From the **Symbol** drop-down box, pick the **Other...** item.

This displays a character map, giving you access to many additional characters.

7 Close the character map and pick the **OK** button to close the text editor.

Find and Replace

Suppose you have lengthy paragraphs of text and you want to find and replace text.

1 With **DDEDIT** still entered, pick the mtext paragraph located in the upper left area of the screen and select the **Find/Replace** tab.

2 In the Find text box, enter **write**, and in the Replace with text box, enter **draft**.

3 Pick the **Find** button located at the right of the Find text box.

AutoCAD finds and highlights the word write.

4 Pick the **Replace** button located at the right of the Replace text box.

AutoCAD replaces the word.

If you check the Match Case check box, AutoCAD finds the text only if the case of all characters in the text match the text in the Find text box. If you check the Whole Word check box, AutoCAD matches the text in the Find text box only if it is a single word. If the text is part of another text string, it is ignored.

5 Pick the **OK** button and press **ENTER** to end the command.

Spell Checker

AutoCAD's spell checker examines text for misspelled words.

Standard

1 From the Standard toolbar, pick the **Spelling** icon.

This enters the SPELL command.

2 In reply to Select objects, pick the text **Look at this text!** and press **ENTER**.

If you spelled the words correctly, AutoCAD displays the following message.

> **AutoCAD Message** ☒
>
> Spelling check complete.
>
> [OK]

3 Pick **OK**.

NOTE:

If AutoCAD displays the Check Spelling dialog box, pick the Cancel button.

Standard

4 Reenter the **SPELL** command, select **Computer-aided**, and press **ENTER**. If Computer-aided does not exist, create it now.

AutoCAD displays the Check Spelling dialog box because it found what could be a misspelled word. "Computer-aided" is not included in the spell checker's dictionary. As you can see, it has suggested "Computer" to replace "Computer-aided."

5 Pick the **Lookup** button to review alternative suggestions and then pick the **Ignore** button to leave Computer-aided as it is.

6 Pick **OK**.

Spell Checking Mtext

The spell checker reviews an entire mtext paragraph.

First, let's misspell a couple of words in the mtext paragraph.

1 Enter **DDEDIT** and pick the mtext located in the upper left corner.

2 Using the text editor, change sentences to **sentenses** and particularly to **particulerly**.

3 Pick the **OK** button.

4 Press **ENTER** to terminate the DDEDIT command.

The mtext now contains two misspelled words.

Standard

5 Enter **SPELL**, select the mtext, and press **ENTER**.

6 If the spell checker finds AutoCAD's and suggests an alternative, pick the **Ignore** button.

The spell checker suggests the correct spelling for "sentences," as shown in the following illustration.

Check Spelling

Current dictionary: American English

Current word
sentenses.

Cancel

Help

Suggestions:
sentences

sentences.
sent.

Ignore   Ignore All

Change   Change All

Add   Lookup

Change Dictionaries...

Context
s sentence ends with the word sentenses.

NOTE:

Sometimes the spell checker finds more than one word that fits its lookup criteria, so the Suggestions area contains several choices. When this happens, you may need to pick (highlight) the correct word before you pick the Change button.

 Select **sentences.** and pick the **Change** button.

The spell checker continues searching for misspelled words.

 If it finds Mtext, pick the **Ignore** button because we do not want to change the spelling of Mtext.

It finds and suggests the correct spelling for "particularly."

Select **particularly** and pick the **Change** button; pick **OK**.

QTEXT Command

The QTEXT command saves screen redraw and regeneration time, especially when drawings contain large amounts of text.

Enter **QTEXT** and **On**.

Enter **REGEN**.

Rectangles, containing only four vectors each, replace each line of text.

Place additional text on the screen using the **DTEXT** command.

Notice that it also appears as a rectangle.

QTEXT (short for Quick TEXT) replaces text with lines that form rectangles where the text once was. The purpose of this is to speed up screen redraws and regenerations. As you may know, each text character is made up of many short lines. The greater the number of lines on the screen, the longer it takes to regenerate the screen. QTEXT temporarily reduces the total number of lines and consequently saves time, especially with heavy use of text. QTEXT is also very useful when you are using TrueType fonts, which take a relatively long time to regenerate.

Enter **QTEXT** again and enter **Off**.

5 Force another regeneration of the screen by entering **REGEN**.
The text reappears.

6 Save your work and exit AutoCAD.

Questions

1. Describe the changes you can make to text using the CHANGE command.

   _____

   _____

2. How does the DDEDIT command permit you to edit standard text?

   _____

   _____

3. How does DDEDIT permit you to edit mtext?

   _____

   _____

4. Why might the character map be useful?

   _____

   _____

5. Suppose you have paragraphs of text that contain several words or phrases that need to be replaced with new text. What is the fastest way of replacing the text?

   _____

   _____

6. How does turning on QTEXT speed screen regenerations?

   _____

   _____

Challenge Your Thinking: Questions for Discussion

1. Look again at the mtext object you edited in this unit. If you changed the word *text* to *test* and the word *write* to *rite,* what misspellings do you think the spell checker would find? Try it and see. Then write a short paragraph explaining the proper use of a spell checker.

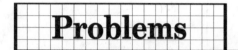

Problems

1. Shown below is a block diagram algorithm for a program that sorts numbers into ascending order by a method known as sorting by pointers. Use the LINE and OFFSET commands to draw the boxes. Then insert the text using the romans.shx font. For the words MAIN, SORT, and SWAPINT, use the romanc.shx font and make the text larger. For the symbol >, use the symath.shx font, character N.

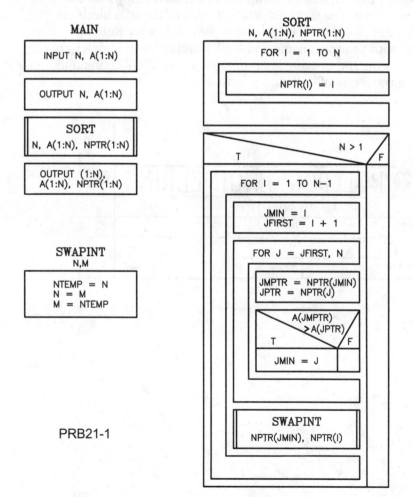

2. This border and title block is suitable for a paper size of 8.5″ × 11″. The border is drawn to allow .75″ white space on all four sides.

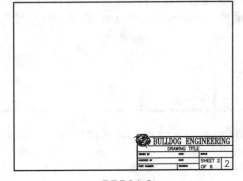

PRB21-2

Draw the border and title block using the dimensions shown in the detail drawing below. The three sizes of lettering in the title block are .0625″, .125″, and .25″. Use the romans.shx font. Replace "Bulldog Engineering" with your own name and logo. Position the text precisely and neatly by using the suitable Justify options of the DTEXT command. It will help to use a grid of .125 and a snap of .0625.

BULLDOG ENGINEERING

DRAWING TITLE			
DRAWN BY	DATE	SCALE	
CHECKED BY	DATE	SHEET 2	2
PART NUMBER	REVISION	OF 6	

.375
.25
.25
.50
.25
1.75 — .9375 — .9375 — .50

Problems 1 and 2 courtesy of Gary J. Hordemann, Gonzaga University

AUTOCAD® AT WORK

Freelancing with AutoCAD

In today's economy, many companies are reluctant to hire new employees, but they are increasingly receptive to independent contractors, or freelancers. Responding to this trend, many AutoCAD operators have gone into business for themselves to get a share of the new market for expert services.

Freelance drafting is now a very active area of enterprise, notably in facilities management. Companies that are downsizing have to reorganize office space, consolidating some departments and eliminating others. Companies experiencing rapid growth need to expand their facilities. The drawings needed for this kind of work may be little more than space plans, but to be successful the freelancer must have some ability with design and some familiarity with systems furniture.

Often, facilities drawings depict equipment with special requirements for spacing, venting, electrical supply, or drainage. Facilities drawings may include information about lighting, electrical outlets, plumbing, and HVAC. Many companies have discovered that having even a simple space plan on AutoCAD makes it much easier to accomplish asset management, energy control, motion studies, product flow studies, and so on.

Another considerable market for freelance drafters is technical illustration. Every new product needs documentation, and new products continue to arrive in a constant procession, especially peripheral devices for computers and specialized electronic and mechanical equipment.

Authorized AutoCAD dealers often need an AutoCAD expert, someone to call on for help when regular store personnel are busy. A freelance consultant can step in to help with jobs as simple as going out to hook up a

plotter. At other times, the dealer may need someone to visit a customer's site to train a new user. Perhaps the client needs a utility installed or just a quick answer to a simple question.

Developing a good relationship with AutoCAD dealers in your area can certainly pay off, even if you don't work for them directly. Dealers get calls all the time for drafters, programmers, and consultants. You may be able to stay pretty busy if they can recommend you as an AutoCAD freelancer.

Based on a story by Pete Karaiskos in *CADENCE*, November 1994, Copyright © 1994, Miller Freeman, Inc.

Unit 22 Drawing Setup

■ OBJECTIVE:

To practice drawing setup and the template development process

This unit focuses on the process of creating the foundation for new drawings. Considerations include identifying the type of drawing you are about to create, the scale, and the sheet size; determining the drawing units and limits; setting the grid and snap resolution; and checking drawing settings and parameters.

The first several steps necessary for setting up a new drawing are important and deserve special attention. Once you have set up a template file, subsequent drawing setups are fast.

■ *Creating a Template File*

An AutoCAD drawing template file contains drawing settings such as the snap and grid spacings. Some users choose to include a border and title block in their template files.

The purpose of a drawing template is to minimize the need to change settings each time you begin a new drawing. When you use a template, its contents are automatically loaded into the new drawing. The template's settings thus become the settings for the new drawing.

Template development can include the following steps. Note that the first three steps are common to the planning of manual drawings using drafting boards.

1) Determine what you are going to draw (*e.g.*, mechanical detail, house elevation, etc.).
2) Determine the drawing scale.
3) Determine the sheet size. (Steps 2 and 3 normally are done simultaneously.)
4) Set the drawing units.
5) Set the drawing area.
6) Set the grid.
7) ZOOM All. (This will zoom to the new drawing area.)
8) Set the snap resolution.
9) Enter STATUS to review the settings.
10) Determine how many layers you will need and what information will be placed in each layer; establish the layers with appropriate colors, linetypes, etc.
11) Set the linetype scale (LTSCALE).
12) Create new text styles.
13) Set DIMSCALE, dimension text size, arrow size, etc.
14) Store as a drawing template file.

This and the next three units will provide an opportunity to practice these steps in detail as well as introduce new commands and features.

1 Start AutoCAD and pick the **Use a Template** button.

AutoCAD displays a list of template files available to you.

2 Scroll down the list, and as you single-click several of them, notice that those with borders and title blocks display in the Preview area.

Notice also that the template files have a DWT file extension.

3 Find and select **Ansi_a.dwt** and pick the **OK** button.

The contents of Ansi_a.dwt display on the screen. This template is set up to fit an A-size drawing sheet according to standards established by the American National Standards Institute (ANSI).

Advanced Setup

Many AutoCAD users choose to create custom template files. We will create one also. The first step is to identify an object (mechanical part, building, etc.) on which to base the new template. For this exercise, let's create a template that we could use for stair details in drawings for homes and commercial buildings. Next, we will determine the drawing scale for the stair detail. This information will give us a basis for setting the drawing area, linetype scale, and DIMSCALE later. Let's use a scale $1/_2'' = 1'$, and let's base the template on a sheet size of $11'' \times 17''$.

1 From the Standard toolbar, pick the **New** icon.

2 Pick **Use a Wizard**.

In Unit 1, you picked Use a Wizard and stepped through the Quick Setup process.

3 Pick **Advanced Setup**, but *do not* pick OK at this time.

Read the Wizard Description. Advanced Setup uses the template file acad.dwt. So does Quick Setup.

4 Pick the **OK** button.

Advanced Setup consists of seven steps.

Unit of Measurement

Prior to Release 14, AutoCAD users were required to use either the UNITS or the DDUNITS command to set the drawing units. Both commands are still available, but now you can set the drawing units using either Quick Setup or Advanced Setup.

1. Pick the **Architectural** radio button and review the sample in the box under Sample Units.

2. At the right of Precision, pick the down arrow and scroll up and down the list.

3. Choose the **0'0–¹/₁₆"** default setting.

This means that ¹/₁₆" is the smallest fraction that AutoCAD will display.

4. Pick the **Next** button to go to the next step.

Angle Measurements

Decimal Degrees is the default setting for angle measurements. This is what we want to use for the template, so no selection is required from you. When Decimal Degrees is selected, AutoCAD displays angle measurements using decimals.

1. At the right of Precision, pick the down arrow to adjust the angle precision, and pick **0.0**.

A setting of 0.0 means that AutoCAD will carry out angle measurements to one decimal place, as shown under Sample Angle.

2. Pick the **Next** button to proceed to the next step.

This tab permits you to control the direction for angle measurements. As discussed in Unit 6, AutoCAD assumes that 0 degrees is to the right (east). Let's not change it.

3. Pick the **Next** button.

Also in Unit 6, we established that angles increase in the counter-clockwise direction.

4. Pick the **Next** button to proceed.

Setting the Drawing Area

The next step is to set the drawing area (limits). The area defines the boundaries for constructing the drawing, and it should correspond to both the drawing scale and the sheet size. The default drawing area is 12 × 9 units. Using architectural units, the drawing area is 1′ × 9″ (See Appendix D for a chart showing the relationships among sheet size, drawing scale, and drawing area.)

NOTE:

Actual scaling does not occur until you plot the drawing, but you should set the drawing area to correspond to the scale and sheet size. The limits and sheet size can be increased or decreased up to the time you plot the drawing. The plot scale can also be adjusted prior to plotting. For example, if a drawing will not fit on the sheet at $1/4''$ = 1′, you can enter a new drawing area to reflect a scale of $1/8''$ = 1′. Likewise, you can enter $1/8''$ = 1′ instead of $1/4''$ = 1′ when you plot.

As mentioned before, the drawing area should reflect the drawing scale and sheet size. Let's look at an example.

If the sheet size is 17″ × 11″, the active plotting area is approximately 15″ × 10″. This is the area on 17″ × 11″ sheets in which most plotters are able to plot. If the drawing scale is 1″ = 1′, then the drawing area should be set at 15′,10′. Why? Because 15 inches horizontally on the sheet will occupy 15 scaled feet, and 10 inches vertically on the sheet will occupy 10 scaled feet.

Let's consider another example. If the drawing scale is $1/4''$ = 1′, what should be the drawing area? Since each plotted inch on the sheet will occupy 4 scaled feet, it's a simple multiplication problem: 15 × 4 = 60 and 10 × 4 = 40. The drawing area should be set at 60′,40′ because each plotted inch will represent 4′.

Since *our* scale is $1/2''$ = 1′, what should be the drawing area?

HINT:
How many $1/2''$ units will 15″ occupy? How many $1/2''$ units will 10″ occupy?

1. Enter **30′** for the width and **20′** for the length. (See the following note.)

NOTE:

When entering 30' and 20', type it exactly as you see it here; use an apostrophe for the foot mark. If you do not use a foot mark, AutoCAD assumes that the numbers represent inches. You can specify inches using " or no mark at all. Note also that if you need to change the drawing area later, you can do so using the LIMITS command.

2 Pick the **Next** button.

Title Blocks and Drawing Layout

AutoCAD offers a selection of title blocks.

1 Under Title Block Description, pick the down arrow and scroll through the list.

2 Single-click several of the title blocks to view them in the Sample Title Block area.

Under Title Block File Name, the corresponding title block drawing file appears. The Add and Remove buttons enable you to add and remove the title block drawing files from the list.

3 Under Title Block Description, pick **No title block**.

4 Pick the **Next** button.

The final step in Advanced Setup is Layout. AutoCAD asks whether you want to use advanced paper space layout capabilities. You will learn more about paper space in Unit 39.

5 Pick the **What is paper space?** button.

6 After you've read about paper space, pick the **OK** button.

7 Pick the **No** radio button to indicate that you do not want to use paper space layout, and then pick the **Done** button.

You have completed the Advanced Setup process.

8 Save your work in a file named **tmp1.dwg**.

NOTE:

We will continue to develop the custom template file in Units 23 and 25. For now, leave it in DWG format. In Unit 23, we will save it as a template file.

 Enter the **GRID** command and set it at **1'**. (Be sure to enter the apostrophe.)

The purpose of setting the grid is to give you a visual sense of the size of the objects and drawing area. The grid fills the entire drawing area, with a distance of 1' between grid dots.

10 Enter **ZOOM All**.

This causes the drawing area to fill the screen. The grid reflects the drawing and plotting area of the sheet.

11 Enter **SNAP** and set it at **6"**. (Use the quote key for the inch mark if you wish. As stated earlier, if no mark is used, AutoCAD assumes inches.)

12 Position the crosshairs in the upper right corner of the grid and review the coordinate display.

NOTE:

It should read exactly 30'-0", 20'-0", 0'-0". The 0'-0" is the z coordinate. We will discuss the z coordinate in more detail later in this book.

Status of the Template File

1 To review your settings up to this point, select the **Tools** pulldown menu and pick **Inquiry** and **Status**.

This enters the STATUS command.

You should see a screen similar to the one on the following page. Note each of the components found in STATUS.

```
Command: '_status 35 objects in D:\r14\YourName\tmp1.dwg
Model space limits are   X:   0'-0"         Y:   0'-0"     (Off)
                         X:   30'-0"        Y:   20'-0"
Model space uses         *Nothing*
Display shows            X:   -1'-6 1/16"   Y:   0'-2 3/8"
                         X:   31'-6 1/16"   Y:   20'-2 3/8"
Insertion base is        X:   0'-0"         Y:   0'-0"      Z:   0'-0"
Snap resolution is       X:   1'-1"         Y:   1'-1"
Grid spacing is          X:   1'-1"         Y:   1'-1"

Current space:           Model space
Current layer:           0
Current color:           BYLAYER -- 7 (white)
Current linetype:        BYLAYER -- CONTINUOUS
Current elevation:       0'-0"      thickness:     0'-0"
Fill on    Grid off    Ortho off    Qtext off    Snap off    Tablet off
Object snap modes:       None
Free dwg disk (D:) space: 453.9 MBytes
Free temp disk (C:) space: 747.1 MBytes
Free physical memory: 0.5 Mbytes (out of 63.5M).
Free swap file space: 748.0 Mbytes (out of 794.6M).
```

Several of the numbers in the status information, such as Display shows and Free disk, will differ.

2 Press **ENTER** and close the AutoCAD Text Window.

Tmp1 now contains several settings specific to creating architectural drawings at a scale of $1/2'' = 1'$. It will work well for beginning the stairway detail mentioned earlier.

The next steps in creating a template file deal with establishing layers. You will create layers in the next unit and complete the final steps for creating a template file in Unit 25.

3 Save your work and exit AutoCAD.

4 Produce a backup copy of the file named tmp1.dwg.

For details on producing copies of files, refer to Unit 19.

Backup copies are important because if you accidentally lose the original (and you may, sooner or later), you will have a backup. You can produce a backup in seconds, and it can save you hours of lost work.

Questions

1. Explain the purpose and value of template files.

   _____

   _____

   _____

2. If you select architectural units, what does a precision of $0'0-1/4''$ mean?

   _____

   _____

3. What determines the drawing area?

   _____

   _____

4. If the physical drawing area of a sheet measures $22'' \times 16''$, and the scale is $10' = 1''$, what should you enter for the drawing area?

   _____

   _____

5. Describe the information displayed as a result of entering the STATUS command.

   _____

   _____

   _____

   _____

■ *Challenge Your Thinking: Questions for Discussion*

1. Discuss unit precision in your drawings. If necessary, review the precision options listed in the Units Control dialog box for units and angles. (Hint: You can display this dialog box by entering the DDUNITS command.) Then describe applications for which you might need at least three of the different settings.

2. For a drawing to be scaled at $1/8'' = 1'$ and plotted on a $17'' \times 11''$ sheet, what drawing area should you establish?

Problems

In problems 1 and 2, establish the settings for a new drawing based on the information provided. Set each of the values as indicated, and save the files as prb22-1.dwg and prb22-2.dwg.

1. Drawing type: Mechanical drawing of a machine part

   Scale: 1″ = 2″
   Sheet size: 17″ × 11″
   Units: Engineering (You choose the appropriate options.)
   Drawing area: 30″ × 20″

   Grid value: .5″
   Snap resolution: .25″

   Be sure to ZOOM All.

   Review settings with the STATUS command.

2. Drawing type: Architectural drawing of a house and site plan

   Scale: $\frac{1}{8}$″ = 1′
   Sheet size: 24″ × 18″
   Units: Architectural (You choose the appropriate options.)
   Drawing area: 176′ × 128′
       (Based on plotting area of 22″ × 16″.)

   Grid value: 4′
   Snap resolution: 2′

   Reminder: Be sure to ZOOM All.

   Review settings with the STATUS command.

In problems 3 and 4, fill in the missing data, based on the information provided. More than one answer exists in both problems.

3. Drawing type: Architectural drawing of a detached garage

   Approximate dimensions of garage: 32′ × 20′

   Other considerations: Space around the garage for dimensions, notes, specifications, border, and title block

   Scale? _____

   Paper size? _____

   Units? _____

   Drawing area? _____

   Grid value? _____

   Snap resolution? _____

4. Drawing type: Mechanical drawing of a bicycle pedal

   Approximate dimensions of pedal: 4″ × 2.75″

   Other considerations: Space around pedal for dimensions, notes, specifications, border, and title block

   Scale? _____

   Paper size? _____

   Units? _____

   Drawing area? _____

   Grid value? _____

   Snap resolution? _____

 **Layers and Linetypes**

■ OBJECTIVE:

To apply layers and linetypes and to further the template file development

This unit focuses on AutoCAD's layering and linetype capabilities. It covers creating layers; setting the current layer; assigning colors and linetypes to layers; turning layers on and off; freezing and thawing layers; and setting the linetype and linetype scale.

In this unit, you will create and set the following layers, colors, and linetypes.

Layer Name	Color	Linetype
0	white	Continuous
Border	cyan	Continuous
Center	magenta	Center
Dimensions	blue	Continuous
Hidden	green	Hidden
Notes	magenta	Continuous
Objects	red	Continuous
Phantom	yellow	Phantom

Creating New Layers _____

It may be helpful to think of layers as transparent overlays. Objects can be drawn on these layers; the layers can be visible or invisible, and specific colors and linetypes can be assigned to each layer.

For example, a house floor plan could be drawn on a layer called Floor and displayed in red. The dimensions of the floor plan could be drawn on a layer called Dimension and displayed in yellow. Furthermore, a layer called Center could contain blue center lines.

Layers and their colors are also very important when plotting because plotter colors are assigned to AutoCAD colors. You'll gain a much better feel for all of this after you have stepped through the following sequence.

1 Start AutoCAD and open the drawing named **tmp1.dwg**.

_____ NOTE: _____

Be sure to make a backup copy of tmp1.dwg if you have not already done so.

2 Pick the **Layers** icon from the docked Object Properties toolbar, or select **Layer...** from the **Format** pull-down menu.

This enters the LAYER command and displays the following Layer & Linetype Properties dialog box.

NOTE: ___

You can also enter LA for the LAYER command to display this dialog box.

First, let's create layer Objects.

3 Pick the **New** button.

This creates a new layer named Layer1.

4 Using upper- or lowercase letters, type **Objects** and press **ENTER**.

Objects replaces Layer1. If you make a mistake, you can single-click the layer name and edit it.

5 Create the layers **Border**, **Center**, **Dimensions**, **Hidden**, **Notes**, and **Phantom** on your own, as listed at the beginning of this unit. (You will set the color and linetype of the layers later in this exercise.)

Changing the Current Layer _____

Notice the 0 after the Current button. This shows that **0** is now the current layer. Let's change the current layer to Objects.

1. Pick **Objects** and pick the **Current** button.

Objects now appears after the button.

2. Pick the **OK** button.

Notice that Objects now appears in the docked Object Properties toolbar.

3. Save your work.

Assigning Colors _____

The Layer & Linetype Properties dialog box also permits you to assign screen colors to the layers.

1. Pick the **Layers** icon or enter **LA**.

Object Properties

2. Find the C... heading, which is located at the left of the Linetype heading.

3. Position the pointer between the two headings until the pointer changes to a double arrow.

4. Click and drag to the right until the names of the colors appear.

5. Adjust the column to its original size.

6. At the right of Objects, pick the empty box under the Color heading.

The Select Color dialog box appears. It contains a palette of colors available to you.

7. Pick the color red from Standard Colors and pick the **OK** button.

AutoCAD assigns the color red to Objects.

AutoCAD's colors are directly associated with plotter colors. The colors, therefore, define the relationship between the layer and the plotter colors.

8. Assign colors to the other layers as indicated in the layer listing at the beginning of this unit. (Use the Standard Colors. Cyan is light blue and magenta is purple.)

9. Pick the **OK** button.

The color red appears beside Objects in the Object Properties toolbar.

NOTE:

AutoCAD offers commands called COLOR and DDCOLOR that allow you to set the color for subsequently drawn objects, regardless of the current layer. Therefore, you can control the color of each object individually.

The ability to set the color of objects individually, using the COLOR and DDCOLOR commands, or by layer gives you a great deal of flexibility, but it can become confusing. It is recommended that you avoid use of the COLOR and DDCOLOR commands and that their settings remain at ByLayer. The ByLayer setting means that the color is specified by layer.

The Color Control drop-down box in the Object Properties toolbar allows you to set the current color. The recommended setting is ByLayer.

Drawing on Layers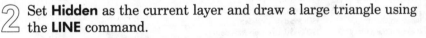

1. Draw a circle of any size on the current **Objects** layer.

It should appear in the color red.

2. Set **Hidden** as the current layer and draw a large triangle using the **LINE** command.

It should appear in the color green.

Turning Layers On and Off

1. Display the **Layer & Linetype Properties** dialog box.

Object Properties

2. At the right of Objects, click the light bulb.

This toggles off layer Objects.

3. Pick the **OK** button and notice that the circle disappears.

4 Press the space bar to display the dialog box again.

5 Toggle on layer Objects by clicking the darkened light bulb.

6 Pick the **OK** button and notice that the circle reappears.

Assigning Linetypes

Next, let's take a look at and load the different linetypes AutoCAD makes available to you.

Object Properties

1 Display the **Layer & Linetype Properties** dialog box.

2 Select the **Linetype** tab.

3 Pick the **Load...** button.

This displays a list of available linetypes, which are stored in a special linetype file named acad.lin.

```
Load or Reload Linetypes                              ? X

   File...    acad.lin

   Available Linetypes

    Linetype            Description
   Acad_iso02w100       ISO dash _ _ _ _ _ _ _ _ _ _
   Acad_iso03w100       ISO dash space _  _   _   _
   Acad_iso04w100       ISO long-dash dot ___ . ___ . ___ .
   Acad_iso05w100       ISO long-dash double-dot ___ .. ___ .. ___
   Acad_iso06w100       ISO long-dash triple-dot ___ ... ___ ... ___
   Acad_iso07w100       ISO dot . . . . . . . . . . . . . . . . . . . .
   Acad_iso08w100       ISO long-dash short-dash ___ _ ___ _ ___
   Acad_iso09w100       ISO long-dash double-short-dash ___ _ _ ___
   Acad_iso10w100       ISO dash dot

           OK          Cancel          Help
```

These linetypes conform to International Standards Organization (ISO) standards. This is important because ISO is the leading organization for the establishment of international drafting standards.

At this point, you can select individual linetypes that you want to load and use in the current drawing.

4 Using the scrollbar, review the list of linetypes.

5 Find and select each of the linetypes listed on the first page of this unit.

HINT:

Press the CTRL key when selecting them. This allows you to make multiple selections.

6 Pick the **OK** button and notice the new list of linetypes.

7 Pick the **Layer** tab.

8 At the right of layer Hidden, and under the Linetype heading, pick **Continuous**.

This displays the Select Linetype dialog box.

9 Pick **Hidden** for the linetype and pick **OK**.

10 Pick **OK** in the Layer & Linetype Properties dialog box.

Did the triangle on layer Hidden change from a continuous line to a hidden line? If not, proceed to the following section anyway.

11 Save your work.

NOTE:

You can also load linetypes using the LINETYPE command. Enter LINETYPE and pick the Load button.

LTSCALE Command

LTSCALE permits you to scale the linetypes properly.

1 Enter the **LTSCALE** command.

Let's scale the linetypes to correspond to the scale of the prototype drawing. This is done by setting the linetype scale at $1/_2$ the reciprocal of the plot scale. When you do this, broken lines, such as hidden and center lines, are plotted to the ISO standards.

HINT: As you may recall, we are creating a template drawing file based on a scale of $\frac{1}{2}'' = 1'$. Another way to express this is $1''$ = $2'$ or $1'' = 24''$. This can be written as $\frac{1}{24}$. The reciprocal of $\frac{1}{24}$ is 24, and half of 24 is 12. Therefore, in this particular case, you would set LTSCALE at 12.

2 In reply to New scale factor, enter **12**.

3 Now view the triangle.

Is it made up of hidden lines?

4 If you're not sure, zoom in on it.

Object Properties

5 Open the **Layer & Linetype Properties** dialog box.

6 Assign the **Center** linetype to layer Center and the **Phantom** linetype to layer Phantom.

7 Highlight the **Phantom** layer by single-clicking the layer name.

8 Pick the **Details** button.

This displays more information about the Phantom layer. You can change the layer name, color, linetype, and various other settings from the Details portion of the box.

9 Pick the **Details** button to remove the details display from the dialog box.

10 Now select the **Linetype** tab and pick the **Details** button.

This displays more information about the linetypes. You can change the value of LTSCALE in the Global scale factor edit box. Current object scale sets the linetype scale for newly created objects. The resulting scale is the global scale factor multiplied by the object's scale factor.

11 Pick the **Details** button and pick the **OK** button to close the dialog box.

NOTE:

AutoCAD permits you to set the linetype for subsequently drawn objects, regardless of the current layer. The Linetype Control drop-down box located in the Object Properties toolbar, permits you to change the current linetype. The default setting is ByLayer, which means that the linetype is specified by layer. This is the recommended approach—and the approach that you've used in this unit. If you change ByLayer to a specific linetype, all subsequently drawn objects will use this linetype. This gives you a great deal of flexibility, but it can become confusing.

Scaling Linetypes by Object

AutoCAD enables you to change the scale of a linetype used by an individual object.

Object Properties

1 Pick the **Properties** icon from the Object Properties toolbar.

2 In reply to Select objects, pick one of the sides of the triangle and press **ENTER**.

This causes the Modify Line dialog box to appear as shown in the following illustration.

Modify Line	✕

Properties		
Color...	BYLAYER	Handle: 53
Layer...	HIDDEN	Thickness: 0"
Linetype...	BYLAYER	Linetype Scale: 1"

From Point	To Point	Delta XYZ:
Pick Point <	Pick Point <	X: -7'
		Y: 6'
X: 29'	X: 22'	Z: 0"
Y: 9'-6"	Y: 15'-6"	Length: 9'-2 5/8"
Z: 0"	Z: 0"	Angle: 139.4

OK	Cancel	Help

Notice that 1″ is the default value for Linetype Scale.

③ Enter **2** for the new value and pick **OK**.

This increases the linetype scale by two times.

④ Pick the **Properties** icon again from the Object Properties toolbar or press the space bar.

⑤ Select the same side of the triangle and press **ENTER**.

⑥ Change the value of Linetype Scale back to **1** and pick **OK**.

Freezing and Thawing Layers _____

① Pick the **Layers** icon to display the Layer & Linetype Properties dialog box.

② Adjust the headings so that you can read the heading titled Freeze in All Viewports.

③ Adjust the column to its original size.

④ In this column, and at the right of Objects, pick the symbol representing the sun.

This changes the symbol to a snowflake and freezes the layer.

⑤ Pick the **OK** button and notice what happens to the circle.

It disappears.

⑥ Display the dialog box again, change the snowflake into the sun by picking it, and pick the **OK** button.

Did the circle reappear? As you can see, freezing and thawing layers is similar to turning them off and on. The difference is that AutoCAD regenerates a drawing faster if the unneeded layers are frozen rather than turned off. Therefore, in most cases, Freeze is recommended over the Off option. Note that you cannot freeze the current layer.

Locking Layers _____

AutoCAD permits you to lock layers as a safety mechanism. This prevents you from accidentally editing objects in complex drawings.

① Display the **Layer & Linetype Properties** dialog box.

2 At the right of Objects, pick the symbol that looks like a padlock.

The lock closes, indicating that the layer is now locked.

3 Pick the **OK** button and try to edit the circle.

AutoCAD will not permit you to edit the circle because it resides on a locked layer.

4 Reopen the dialog box and unlock layer Objects by picking the locked padlock.

Other Features

The Layer & Linetype Properties dialog box offers a few additional features.

1 Reveal the titles of the remaining headings that are hidden.

Freeze in All Viewports freezes selected layers in all viewports. Freeze in Current Viewport freezes selected layers in the current floating viewport. You will learn about floating viewports in the unit that focuses on paper space. Freeze in New Viewports freezes selected layers in the new floating viewports.

2 Change the headings and columns to their original size.

The Delete button permits you to delete selected layers. However, AutoCAD will not delete any layer that contains objects.

3 Pick the down arrow in the Show drop-down box to display a list of options.

These options determine which layers are displayed under the Name heading. You can filter layers based on whether they contain objects and on their name, state, color, and linetype. These filters are especially helpful in complex drawings that contain a large number of layers.

4 Pick the **OK** button to close the dialog box.

*Layer Control Drop-Down Box*_____

AutoCAD offers a convenient way of changing and controlling certain aspects of layers.

1 In the docked Object Properties toolbar, pick the **Layer Control** down arrow.

This displays the list of layers in the current drawing, as shown in the following illustration. By picking the symbols, you can quickly change the current status of the layers.

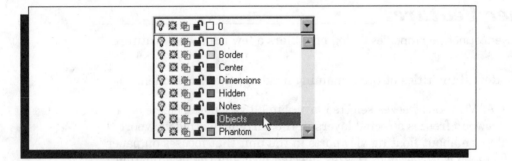

2 Pick **Objects**.

Layer Objects becomes the current layer.

3 Display the list again.

4 Beside Objects, click on the padlock.

The padlock changes to its locked position, and layer Objects becomes locked.

5 Pick it again to unlock layer Objects.

6 Beside Hidden, pick the sun.

The sun changes into a snowflake.

7 Pick a point anywhere in the drawing area.

Layer Hidden is now frozen.

8 Display the list, pick the snowflake, and pick a point anywhere in the drawing area.

This thaws layer Hidden.

⑨ Display the list again.

The third symbol in each line—the gray sun and rectangle—allows you to freeze and thaw layers in the current viewport.

⑩ Pick a point anywhere in the drawing area.

⑪ Pick the **Make Object's Layer Current** icon on the Object Properties toolbar.

Object Properties

Notice that the Command line reads Select object whose layer will become current.

⑫ Pick one of the sides of the triangle.

Hidden is now the current layer.

⑬ Using the same icon, make **Objects** the current layer.

Object Properties

Moving Objects to Another Layer

Occasionally, you may draw an object on the wrong layer. If this happens, AutoCAD offers commands that fix the problem.

Object Properties

① Pick the **Properties** icon from the Object Properties toolbar.

② Select the circle and press **ENTER**.

③ In the Modify Circle dialog box, pick the **Layer...** button.

The Select Layer dialog box appears.

④ Select layer **NOTES** and pick the **OK** button.

⑤ Pick **OK** again.

The circle changes to magenta because it now resides on layer Notes.

AutoCAD offers an even faster way of moving objects to another layer.

① Without anything entered at the Command line, select the circle.

② From the **Layer Control** drop-down box, select layer **Objects**.

This moves the circle to layer Objects.

 Press **ESC** twice to remove the selection of the circle.

Match Properties

The Match Properties icon provides a quick way to copy the properties of one object to another.

Standard

 From the Standard toolbar, pick the **Match Properties** icon.

This enters the MATCHPROP command.

 In reply to Select Source Object, pick the circle.

 Pick one side of the triangle and press **ENTER**.

This copies the properties of the first object to the second object.

4 Undo the last operation.

The line segment retains its original properties.

Standard

5 Pick the **Match Properties** icon again and pick one side of the triangle.

6 In reply to Settings/<Select Destination Object(s)>, enter **S** for Settings.

This displays the Property Settings dialog box. The checked items are copied from the source object to the destination object(s), allowing you to control the properties that are copied.

7 Pick the **Cancel** button in the dialog box and press **ENTER** to terminate the command.

8 Erase both the circle and the triangle.

9 Save your work.

Applying the (Nearly Complete) Template File

The template is nearly complete. The last few steps (12, 13, and 14 from page 236) typically involve creating new text styles and setting the dimensioning variables and DIMSCALE. All of this is covered in Units 24 and 25.

The template file concept may lack meaning to you until you have actually applied it. Therefore, let's convert tmp1.dwg to a template file and use it to begin a new drawing.

First, let's create the template file.

1 Be sure to save your work if you haven't already.

2 Select **Save As...** from the **File** pull-down menu.

3 At the right of Save as type, pick the down arrow and select **Drawing Template File (*.dwt)**.

4 Find and select the folder with your name and pick the **Save** button.

The Template Description box appears.

5 Enter **Created for stair details** and pick **OK**.

Standard

6 Pick the **New** icon in the Standard toolbar to display the Create New Drawing dialog box and pick **Use a Template**.

7 Under Select a Template, pick **More files...** and pick **OK**.

8 Find and double-click the **tmp1.dwt** template file.

AutoCAD loads the contents of tmp1.dwt into the new drawing file.

9 Review the list of layers.

Look familiar? Do you now see why drawing templates are of value?

10 Save the new drawing file as **staird.dwg** and exit AutoCAD.

NOTE:

A template file is used only to begin a new drawing. You are not locked into the template's settings. When you open staird.dwg in the future, you can add more layers and change any settings based on your needs.

Questions

1. Give at least two purposes of layers.

   _____

   _____

2. Using the Layer & Linetype Properties dialog box, how do you change the current layer?

   _____

   _____

3. Describe the purpose of the LTSCALE command and explain how to set it.

   _____

   _____

4. Why/when would you want to freeze a layer?

   _____

   _____

5. Name five of the linetypes AutoCAD makes available.

   _____

   _____

6. What is the purpose of locking layers?

   _____

   _____

7. If you accidentally draw on the wrong layer, how can you correct your mistake without erasing and redrawing?

   _____

   _____

8. Explain how to freeze a layer using the Layer Control drop-down box in the Object Properties toolbar.

   _____

   _____

■ *Challenge Your Thinking: Questions for Discussion*

1. When might you use the Layer Control drop-down box in the Object Properties toolbar instead of using the Layer & Linetype Properties dialog box? Explain.

2. As you may know, drafted documents may contain several different linetypes, as well as lines of varying thicknesses. In this unit, you discovered how to control linetypes and layers. On your own, find out how to control line thickness in AutoCAD. Discuss the implications this has for using the commands demonstrated in this unit.

3. AutoCAD allows you to use linetypes in your drawings that correspond to ISO standards. Find out more about ISO standards. When and where are they used? What is the purpose of having such standards?

Problems

1-2. Create two new drawings using the tmpl.dwt template file. Make the appropriate changes to achieve the following layers with corresponding settings. Name the drawings prb23-1.dwg and prb23-2.dwg.

Layer name	State	Color	Linetype
0	Frozen	white	Continuous
Border	On	cyan	Continuous
Center	On	yellow	Center
Dimensions	On	green	Continuous
Hidden	On	yellow	Hidden
Objects	On	red	Continuous
Phantom	On	blue	Phantom
Text	Frozen	magenta	Continuous

Current layer: Objects

PRB23-1

Layer name	State	Color	Linetype
0	On	white	Continuous
Center	Frozen	blue	Center
Dimensions	On	yellow	Continuous
Electrical	On	cyan	Continuous
Found	On	magenta	Dashed
Hidden	On	blue	Hidden
Notes	On	yellow	Continuous
Plumbing	Frozen	white	Continuous
Title	Frozen	green	Continuous
Walls	On	red	Continuous

Current layer: 0

PRB23-2

3. Refer to the layers.dwg file and instructions contained on the optional *Applying AutoCAD Diskette.* The drawing shows examples of layers, colors, and linetypes, and the instructions give suggestions for its use.

4. Create a drawing template for use with an A-size drawing sheet. Use a scale of $1/_4'' = 1'$. Use architectural units with a precision of $1/_{16}''$. Set up the following layers: Floor, Dimensions, Electrical, Plumbing, and Furniture. Call it tmp2.dwt.

5. Using the tmp2.dwt template file you created in the previous problem, create a simple floor plan for a house. Make the floor plan as creative and as detailed as you wish. Place all the objects on the correct layers.

AUTOCAD® AT WORK

A Satellite Feed into AutoCAD Mapping

Wherever on Earth you are right now, your location is within view of five of the 24 navigational satellites in the Global Positioning System (GPS). With a GPS satellite receiver, you can pick up signals from the nearest four of those satellites to determine your longitude, latitude, and altitude. With a hand-held microcomputer/GPS receiver called MC-GPS (Corvallis Microtechnology, Corvallis, OR), you can insert this position information, along with your field notes, into a map drawn in AutoCAD.

As a mapping tool, AutoCAD has applications in forest management, pipeline maintenance, archaeological siting, water quality assessment, and more. When a map needs to be updated, and someone has to go out to observe and record conditions at specific sites, a tool that can link field data and exact locations to vectors in the AutoCAD base map is invaluable.

Consider utility poles, for example. How many of them are in your neighborhood? In your city? Inspecting utility poles, especially for hazard from nearby trees, is a never-ending chore. When done with pen and paper—just think about locating each pole on a map (third pole from the corner of Maple and Madison . . .) and writing down notes—the job is laborious and makes data retrieval difficult. By contrast, it is easy to call up records embedded as a layer in an AutoCAD map—for example, all utility poles with a hazard level of 4 or above. A field inspector using the MC-GPS could identify a utility pole by coordinates and enter menu-guided notes.

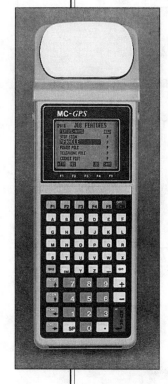

Date surveyed:	09-02-95
Condition:	good/repair/replace
Tree hazard:	none/fast growing/slow growing
Hazard level:	0 to 5 (high risk = 5)
Action required:	none/cut back/remove

To get positional information with a resolution as fine as one meter, MC-GPS and other civilian receivers currently rely on Differential GPS, a calibrating signal from a nearby ground station. This correction is necessary because GPS is a project of the U.S. Department of Defense, which reserves its Precise Positioning Service for national security purposes. However, due to advances in military security technologies that exceed the capabilities of the Precise Positioning Service, the government will begin allowing civilian access to it within the next four to ten years.

Precision defines the value of AutoCAD in mapping, the same as it does in design. Whether it's a map of fire hydrants or fossil beds, an AutoCAD drawing provides the information planners need.

Unit 24 Basic Dimensioning

■ **OBJECTIVE:**

To apply AutoCAD's basic dimensioning capabilities

One of AutoCAD's powerful features is its semiautomatic dimensioning. This unit covers the different types of dimensioning, including linear, angular, baseline, ordinate, diameter, and radial dimensioning.

Preparing to Dimension

Before we can dimension the following object, we need to prepare a few drawing settings and parameters.

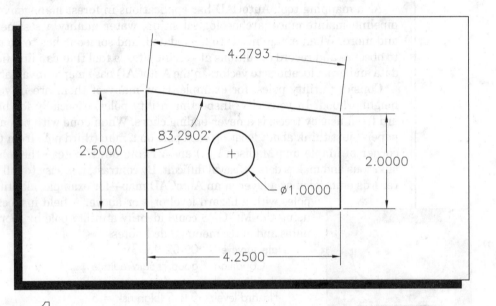

1 Start AutoCAD and start a new drawing from scratch.

2 Create two new layers. Call one of them **Objects** and the other **Dimensions**.

3 Assign the color red to Objects and the color green to Dimensions. Set layer **Objects** as the current layer.

——— NOTE: ———

When you draw the object, you will place the object lines on layer Objects and the dimensions on layer Dimensions.

4 Close the dialog box.

5 Set snap at **.25** unit and enter **ZOOM All**.

6 Draw the object shown on the previous page, and omit dimensions at this point. However, size the object according to the dimensions shown. Don't worry about the exact location of the hole. (See the following hint.)

HINT:
Begin in the upper left corner and work counterclockwise as you specify the line endpoints.

7 Save your work in a file named **dimen.dwg**.

Dimension Text Style

The DIMTXSTY system variable controls the style that AutoCAD uses for the dimension text.

1 Using the **romans.shx** font, create a new text style named **ROM**. Accept the default settings for the new style.

HINT:
Refer to Unit 20 for instructions on how to create new text styles.

2 Enter **DIMTXSTY** and **ROM**.

ROM is now the current dimension text style.

Dimensioning Horizontal Lines

1 Set layer **Dimensions** as the current layer.

2 From the **View** pull-down menu, select **Toolbars...** and check the **Dimension** check box; pick the **Close** button.

This displays the Dimension toolbar.

3 Move it to a convenient location on the screen.

4 Pick the **Linear Dimension** icon from the toolbar.

Notice that this issues the DIMLINEAR command.

Dimension

5 In reply to First extension line origin or ENTER to select, pick one of the endpoints of the object's horizontal line.

6 In reply to Second extension line origin, pick the second extension line origin (the other end of the horizontal line).

7 Move the crosshairs downward to locate the dimension line approximately 1 unit away from the object and press the pick button.

Did the dimension appear correctly? If not, undo the last operation and try it again.

Dimensioning Vertical Lines _____

Let's dimension the vertical lines in the object.

Dimension

1 Pick the **Linear Dimension** icon again from the Dimension toolbar.

2 Dimension either vertical line by selecting the first and second extension line origins as you did before.

3 Place the dimension line as you did before.

Let's dimension the other vertical line, but this time let's do it a faster way.

1 Press the space bar to reenter the **DIMLINEAR** command.

2 This time when AutoCAD asks for the first extension line origin, press **ENTER**.

The crosshairs change to a pickbox.

3 Pick any point on the vertical line.

4 Pick a point to place the dimension.

Dimensioning Inclined Lines _____

Let's dimension the inclined line by "aligning" the dimension to the line.

Dimension

1 Pick the **Aligned Dimension** icon from the Dimension toolbar.

This enters the DIMALIGNED command.

2 Proceed exactly as you did with the last dimension.

If it appears correctly on the screen, then you did it right.

Dimensioning Circles and Arcs _____

Now let's dimension the hole.

Dimension

1 Pick the **Diameter Dimension** icon from the Dimension toolbar.
This enters the DIMDIAMETER command.

2 Pick a point anywhere on the hole. (You may need to turn off snap.)

The dimension appears.

3 Move the crosshairs around the circle and watch what happens.

Notice that you have dynamic control over the dimension.

4 Pick a point down and to the right as shown in the drawing on the first page of this unit.

_____ NOTE: _____

Dimension

A Radius dimension icon is also available to dimension the radius of a circle or arc instead of its diameter.

Dimensioning Angles _____

Last, let's dimension the angle as shown in the drawing.

Dimension

1 Pick the **Angular Dimension** icon from the Dimension toolbar.
This enters the DIMANGULAR command.

2 Pick both lines that make up the angle.

3 Move the crosshairs outside the object and watch the different possibilities that AutoCAD presents.

4 Pick a location for the dimension arc inside the object.

5 If the angular dimension did not appear correctly, undo and repeat Steps 1-4.

6 Save your work.

7 Also, save your work as a template file in the folder with your name. Name it **dimen.dwt** and enter **Dimensioning Practice** for the template description.

Baseline Dimensioning _____

AutoCAD permits you to create baseline dimensions.

1 Using **dimen.dwt** as the drawing template, begin a new drawing.

2 Save the drawing and name it **base.dwg**.

3 Erase the drawing and dimensions and create the following shape on layer **Objects**. Fill the top half of the screen, and ignore all dimensions.

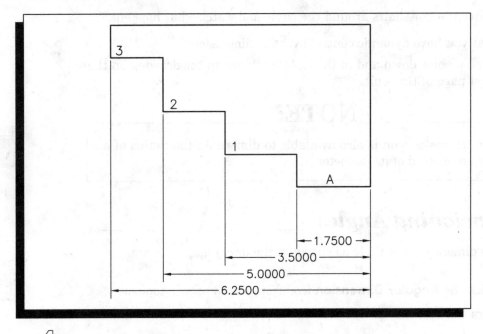

4 Make **Dimensions** the current layer.

5 Pick the **Linear Dimension** icon and dimension line A. It is important that you pick line A's right endpoint first.

6 Pick the **Baseline Dimension** icon.

This enters the DIMBASELINE command.

7 Pick Point 1.

8 Pick Point 2.

9 Pick Point 3.

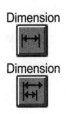

Dimension

Dimension

Dimension

14 Pick the **Ordinate Dimension** icon.

15 At the Select feature prompt, use the **Center** object snap to select the center of the hole nearest to the upper left corner.

16 Enter **X** for Xdatum and pick a point about 1 unit from the top of the object. (Ortho and snap should be on.)

The X-datum ordinate dimension appears.

17 Repeat Steps 14 through 16 to create X-datum ordinate dimensions for the remaining holes.

18 Create the Y-datum ordinate dimensions on your own.

NOTE:

True ordinate dimensioning requires that center lines be placed through the holes in the object; you will learn more about using center lines later.

19 Save your work.

Moving Dimensions _____

Using AutoCAD's grips feature, you can change the location of dimension lines and dimension text.

1 Open the drawing named **dimen.dwg**.

2 Pick one of the linear dimensions.

Five grips appear.

3 Pick one of the two grips near the arrows.

4 Move the pointing device and notice that the dimension moves accordingly.

5 Pick a new location for the dimension.

6 Pick the grip located at the center of the dimension text.

7 Move the text and pick a new location for it.

⑧ Press **ESC** twice.

⑨ Close the Dimension toolbar.

⑩ Exit AutoCAD. Do not save your changes to dimen.dwg.

Questions

1. How do you specify a text style for dimension text?

_____

2. Describe the alternative to specifying both endpoints of a line when dimensioning a line.

_____

_____

3. Which dimension icon do you use to dimension inclined lines? Angles?

_____

_____

4. Which dimension icons do you use to dimension fillets, rounds, and holes?

_____

5. Describe baseline dimensioning.

_____

_____

_____

6. Describe ordinate dimensioning.

_____

_____

_____

■ *Challenge Your Thinking: Questions for Discussion*

1. Describe the advantages of using baseline dimensions in some situations. When would you want to use baseline dimensions? Why?

2. Experiment with using grips to change dimensions. Try to change the dimension by selecting the grip at one end of one of the extension lines (closest to the object). Can you change the dimensions of the object by changing the placement of the dimension line? Explain.

Problems

1-2. Begin a new drawing for each problem. Place object lines on a layer named Objects and dimensions on a layer named Dimensions. Create a new text style using the romans.shx font. Approximate the locations of the holes. If you have the optional *Applying AutoCAD Diskette,* refer to files block1.dwg and block2.dwg as well as the instructions that apply.

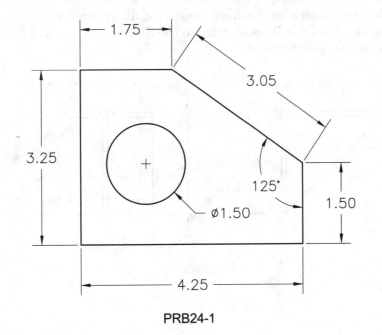

PRB24-1

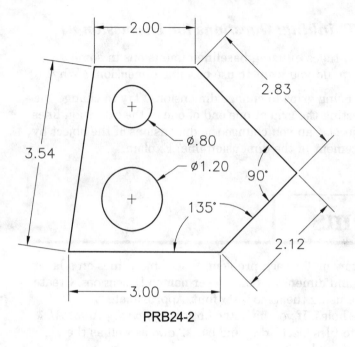

PRB24-2

3. Create the following mounting bracket drawing on layer Objects. Then dimension the bracket on a layer named Dimensions. If you have the optional *Applying AutoCAD Diskette*, refer to the file bracket1.dwg and the instructions that accompany it.

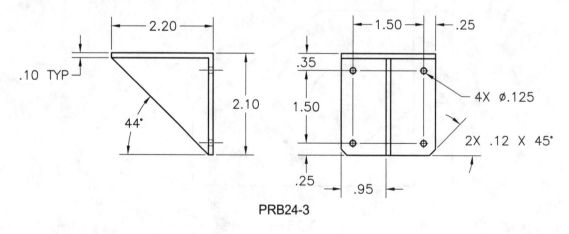

PRB24-3

4. Create a new layer named Visible and assign it the color red. Draw the front view of the shaft lock shown below, placing the visible lines on the Visible layer. Create another layer named Dimensions and assign the color green to it. Use the romans.shx font to create a new text style named Roms, and use DIMTXSTY to assign Roms as the current dimension text style. Dimension the front view, placing the dimensions on layer Dimensions. Show all pertinent dimensions except the location of the holes. Also draw and dimension the right-side view. Create a layer named Hidden and assign the color cyan and the Hidden linetype to it. Place the hidden lines for the holes on the Hidden layer. Assign the color blue and the Center linetype to another layer named Center, and draw the appropriate center lines. Change the drawing area to accommodate the dimensions of the part. If you have the optional *Applying AutoCAD Diskette,* refer to the file shaftloc.dwg and the instructions that accompany it.

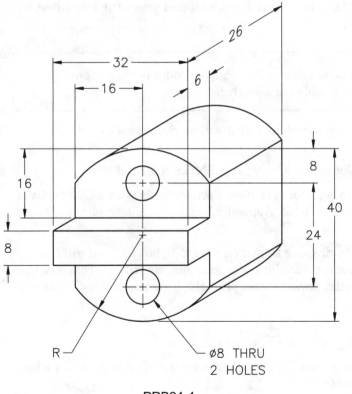

PRB24-4

Problem 3 courtesy of Joseph K. Yabu, Ph.D., San Jose State University
Problem 4 courtesy of Gary J. Hordemann, Gonzaga University

Unit 25 Advanced Dimensioning

■ OBJECTIVE:

To apply AutoCAD's advanced dimensioning capabilities, including associative dimensioning and basic dimensioning system variables

This unit introduces additional semiautomatic dimensioning techniques and dimensioning system variables and allows you to practice them. With this knowledge, you will step through the completion of the tmp1.dwt template file.

Dimensioning System Variables

AutoCAD permits you to change the appearance of dimensions with dimensioning system variables. For example, you can adjust the size of the dimension text and arrows.

 Start AutoCAD, pick **Use a Template**, and select the **tmp1.dwt** template file.

HINT:
From the Select a Template list, double-click More Files... and then find and double-click tmp1.dwt.

Tmp1.dwt is not entirely complete, but it will work for this exercise.

2 Enter **SETVAR**, enter **?**, and press **ENTER** a second time.

This produces a list of system variables. Each variable with a DIM prefix is a dimensioning system variable. Appendix L contains a description of these variables.

We will dimension the drawing shown on page 280, but first we will change some of the dimensioning variables. For example, we will use DIMSCALE to fit the dimensions to the drawing scale properly. If we don't, the dimensions will be much too small.

3 Press **ENTER** several times to view the list.

Notice the fractional inches. This is because we specified fractional inches in the drawing template file.

4 Press **ESC** and press **F2** to hide the AutoCAD Text Window.

5 Enter the dimensioning system variable **DIMASZ** (for dimension arrow size).

6 Enter $^1/_8$″.

The inch (double quote) mark is optional because AutoCAD assumes you are working in inches.

7 Enter **DIMTXT** (for the dimension text) and enter $^1/_8$″ for the new value.

This will make the dimension text $^1/_8$″ tall.

8 Enter **DIMCEN** and $-^1/_{16}$″.

DIMCEN controls the drawing of center marks and center lines at the center of arcs and circles. A positive value specifies center marks only. A negative value, such as $-^1/_{16}$″, specifies full center lines with $^1/_{16}$″ center marks.

DIMSCALE

DIMSCALE lets you enter a scale factor for all dimensions in the drawing.

1 Determine the DIMSCALE setting by calculating the reciprocal of the drawing's plot scale.

HINT: The plot scale for tmp1.dwt is $^1/_2$″ = 1′, which is the same as $^1/_{24}$, as discussed in the section on LTSCALE in Unit 23. The reciprocal, therefore, is 24.

2 Enter **DIMSCALE** and **24**.

3 Save your work in a file named **dimen2.dwg**.

Dimension Styles

A dimension style is a group of dimensioning system variable settings saved under a name. Using a saved dimension style can save time because fine-tuning a set of variables can require a considerable amount of time.

1 Create a new text style named **ROMS** using the **romans.shx** font. Accept each of the default settings.

ROMS is now the current text style.

2 Display the Dimension toolbar.

③ Pick the **Dimension Style** icon from the toolbar.

This enters the DDIM command and displays the Dimension Styles dialog box, as shown in the following illustration.

NOTE:

You can also display this dialog box by selecting Dimension Style... from the Format pull-down menu.

As you can see, +STANDARD_WS is the name of the default dimension style.

④ Double-click inside the Name edit box to select the text.

⑤ Enter **STYLE1** in upper- or lowercase letters and pick the **Save** button.

STYLE1 is now the current dimension style, which consists of the current dimension settings.

NOTE:

You can create a dimension style for each type of dimension in your drawing. Suppose you want the radial dimensions to use settings that are different from the linear dimensions. You can accomplish this by picking the proper radio button in the box labeled Family prior to saving a new style. If a family member doesn't exist, AutoCAD uses the parent dimension style.

Pick the **OK** button.

Dimension Text

The Annotation dialog box permits you to change the appearance of the dimension text, giving you an alternative to using dimensioning variables to make these changes.

1 Press the space bar to reenter the DDIM command.

2 In the Dimension Styles dialog box, pick the **Annotation...** button.

The Annotation dialog box appears.

3 In the lower right area of the dialog box, pick the down arrow on the **Style** drop-down box.

A short list containing the STANDARD and ROMS text styles appears.

4 Pick **ROMS**.

NOTE:

The previous action is equivalent to changing the DIMTXSTY dimensioning system variable from STANDARD to ROMS.

5 Pick the **OK** button.

We will review other parts of the Annotation dialog box, along with the Geometry... and Format... buttons, in the following unit.

 Pick the **Save** button to save your change in STYLE1, and pick **OK**.

Creating the Drawing

1 On layer Objects, create the drawing shown below, but omit the dimensions at this point. (Read the following hint.)

HINT:

For best results, begin at the lower left corner of the object and draw the object in a counterclockwise direction. Add the arc using the FILLET command. Turn on snap and ortho and use the From object snap to locate the center of the circle.

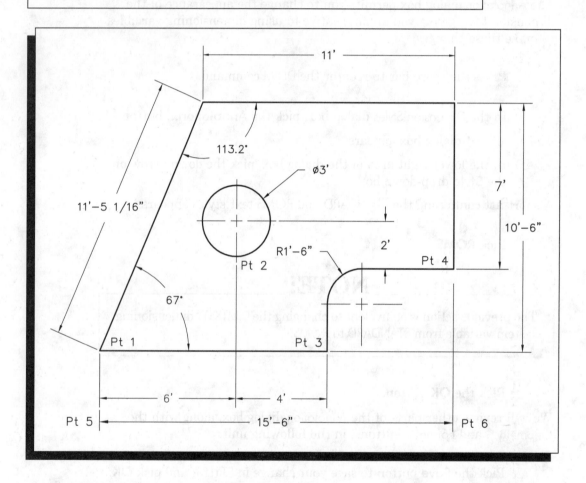

Let's dimension the hole.

2 Save your work, make **Dimensions** the current layer, and change its color to cyan.

3 Pick the **Diameter Dimension** icon from the Dimension toolbar.

Dimension

4 Dimension the hole as shown in the drawing.

280

Full center lines appear with the dimension.

Next, let's dimension the horizontal string of dimensions.

Dimension

5 Pick the **Linear Dimension** icon.

NOTE:

You can also enter many of the dimensioning commands by typing the first six letters of the command. For example, DIMLIN enters the DIMLINEAR command, and DIMRAD enters the DIMRADIUS command.

6 Pick the lower left corner of the object (point 1) for the first extension line origin.

7 Using object snap, pick the end of the vertical center line (point 2) for the second extension line origin.

8 Place the dimension line **2'** from the object.

Dimension

9 Pick the **Continue Dimension** icon.

10 In reply to Specify a second extension line origin, **pick point 3** and press **ENTER** twice.

11 Add the 15'-6" linear dimension on your own. Pick points 5 and 4. Place the dimension 1' away from the string of dimensions.

HINT:

Turn on snap or use the Endpoint object snap mode to snap to points 5 and 4.

12 Complete the remaining dimensions on your own.

13 Save your work.

14 If you have access to a printer or plotter and know how to operate it, plot the drawing using a thick pen (*e.g.,* .7 mm) for color 1 and a thin pen (*e.g.,* .3 mm) for color 3.

Associative Dimensioning

Associative dimensions are created whenever the variable DIMASO is on. The default setting for DIMASO is on (1). When DIMASO is off, the lines, arcs, arrows, and text of a dimension are drawn as separate entities.

Associative dimensioning allows dimensions to update automatically when the dimensioned object is altered using the commands SCALE, STRETCH, ROTATE, EXTEND, TRIM, MIRROR, or ARRAY.

Modify

1 Enter the **SCALE** command, enter **All** to select the entire object, and press **ENTER**.

2 Enter **0,0** for the base point.

3 In reply to Scale factor, enter **1.1**.

Notice that the dimensions change to reflect the new sizes.

Modify

4 Enter the **STRETCH** command and stretch the rightmost portion of the object a short distance to the right. (When selecting the portion to be stretched, be sure to use the Crossing option.)

The dimensions again change to reflect the new size.

5 Undo Steps 1 through 4.

LEADER Command

To create a leader followed by one or more lines of text, use the LEADER command.

1 Enter the **UNDO** command and **M** for Mark.

2 Erase the hole dimension.

Dimension

3 Pick the **Leader** icon from the Dimension toolbar.

This enters the LEADER command.

4 In reply to From point, pick a point on the hole where you'd like to place the point of the arrow.

HINT: Use the Nearest object snap mode to ensure that the point of the arrow touches the circle.

282

5 Move the crosshairs away from the circle and pick a point about 1' from the first point. (The leader should point at the center of the circle.)

6 Press **ENTER**, type the word **HOLE**, and press **ENTER** twice.

This is the fastest and simplest way of annotating a leader.

Using Mtext with LEADER

As part of the LEADER command, you can choose to create an mtext note.

Dimension

1 Undo the creation of the leader and enter the **LEADER** command by picking the **Leader** icon.

2 Repeat previous Steps 4-5.

3 Press **ENTER** three times.

The text editor appears.

4 Type **DRILL AND REAM** and pick the **OK** button.

The mtext object appears. It can be edited in the same way as any mtext object.

Use the DDEDIT command to control the width of an mtext object created with the LEADER command.

5 Enter **DDEDIT** and select the leader text (DRILL AND REAM).

The text editor appears.

6 Pick the **Properties** tab, enter a new width of **1.25** units, and pick **OK**.

7 Press **ENTER** to terminate the DDEDIT command.

Spline Leaders

The LEADER command offers a spline option.

Dimension

1 Pick the **Leader** icon.

2 Pick four points anywhere on the screen to form a complex leader. (Make the segments any length and at any angle.)

283

Enter **F** for Format.

Format controls the appearance of the leader. The Spline option draws the leader as a spline curve. Straight draws the leader using straight line segments. Arrow draws an arrow at the beginning of the leader line. None draws a leader line without an arrow.

 Enter **S** for Spline.

AutoCAD fits a spline curve through the points.

 Press **ENTER** once and then enter the text **SPLINE LEADER**. Press **ENTER** a second time to end the LEADER command.

As you can see, the LEADER command offers a lot of flexibility.

Enter **UNDO** and **B** for Back.

Close the Dimension toolbar and save your work.

The remaining LEADER command options are covered in Unit 27, "Tolerancing."

Completing the Template File

Now that we know how to perform the remaining steps (12-14 on page 236) in creating a prototype drawing, let's finish tmp1.dwt.

Open **tmp1.dwt**.

Create a new text style named **ROMS** using the **romans.shx** font, and accept the default settings.

NOTE:

If you'd like your prototype drawing to contain more text styles, create them at this time. The last style you create will be the current style in the prototype drawing. The romans.shx font is recommended for most applications.

Now let's set a few dimensioning system variables.

 Enter **DIMASZ** to review the current setting of the variable DIMASZ.

Enter a new value of $1/8''$.

3 Enter **DIMTXT** and $\frac{1}{8}$".

4 Change **DIMCEN** to $-\frac{1}{8}$".

5 Using **DIMSCALE**, scale all of the dimensioning system variables to correspond with the prototype drawing scale. Enter **24** as discussed in the hint on page 277.

Other changes in the dimensioning system variables could be made at this time, but let's stop here.

6 Save your changes in tmp1.dwt and exit AutoCAD.

The drawing template is now complete and ready for use with other new drawings.

As you continue to use this template file, as well as others you may create, feel free to modify them further to tailor them to your specific needs.

Documenting Drawing Templates

To know what is in each template, you may want to document the contents of each by printing certain information, such as the system variables. You and others will then be able to review the settings of each template prior to choosing the one you need. At the top of the printout, write the name of the template file, the directory in which it resides, and the drawing scale and paper size.

Questions

1. What is the purpose of the dimensioning system variables?

   _____

   _____

2. How do you determine the DIMSCALE setting?

   _____

   _____

   _____

3. Explain the difference between AutoCAD's center marks and center lines, and describe how to generate each.

   _____

   _____

4. Explain the use of the Continue Dimension icon (DIMCONTINUE command).

   _____

   _____

5. What is the benefit of using dimension styles?

   _____

   _____

6. Describe the purpose of each of the following dimensioning system variables.

   DIMASZ _____

   DIMTXT _____

   DIMCEN _____

   DIMSCALE _____

■ *Challenge Your Thinking: Questions for Discussion*

1. In this unit, you placed object lines on one layer and dimensions on another. Discuss the advantages of placing dimensions on a separate layer. Are there any disadvantages to placing dimensions on a separate layer? Explain.

2. Describe the changes you would need to make to tmp1.dwt to create a template for an A-size sheet at the same drawing scale.

Problems

1. Begin a new drawing and establish the following drawing settings and parameters. Store as a template file. (You could name it tmp3.dwt.)

Units:	Engineering
Scale:	1″ = 10′ (or 1″ = 120″)
Sheet size:	17″ × 11″
Drawing Area:	___?___ (You determine the drawing area based on the scale and sheet size.)
Grid:	10′
	(Reminder: Be sure to enter ZOOM All.)
Snap:	2′

Layers:	*Name*	*Color*
	Thick	Red
	Thin	Green

   Dimension text: Create a new text style using the romans.shx font. Do not make the style height fixed; leave it at 0.

   Dimensioning system variables:

DIMASZ:	.125
DIMTXT:	.125
DIMSCALE:	120
DIMCEN:	−.0625

   Create a new arrow block and enter it using the DIMBLK dimensioning system variable. (Refer to Chapter 10, in the section titled "Creating and Modifying Arrowheads," of the on-line *User's Guide* for directions on how to create and use an arrow block.)

2-3. Use the template file you created in problem 1 to create the following drawings.

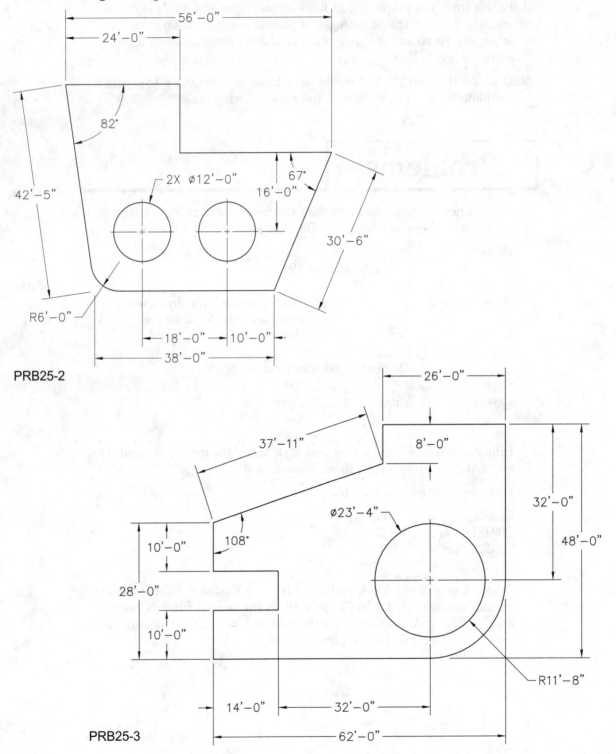

PRB25-2

PRB25-3

4. Draw and dimension the front view of the casting shown on the next page. Use the following drawing and dimension settings, and place everything on the proper layer.

Units:	Decimal with two digits to the right of the decimal point
Drawing Area:	11 × 8.5
VIEWRES:	5000
Grid:	.1
Snap:	.05

Layers:	*Name*	*Color*	*Linetype*
	Visible	red	Continuous
	Dimensions	blue	Continuous
	Text	magenta	Continuous
	Center	green	Center

LTSCALE:	Start with .5

Dimension text: Use romans.shx to create a new text style and use it for the dimension text. Leave the style at a height of 0, but insert text with a height of .125.

Dimensioning system variables:

DIMASZ:	.125	DIMCLRD:	ByLayer
DIMCLRE:	ByLayer	DIMCLRT:	magenta
DIMEXE:	.125	DIMEXO:	.0625
DIMGAP:	.0625	DIMTXT:	.125
DIMCEN:	−.06		

HINT: Use AutoCAD's help to read about the dimensioning system variables. After reading about them, find their equivalent settings in the dialog boxes made available by picking the Dimension Style icon or by entering the DDIM command.

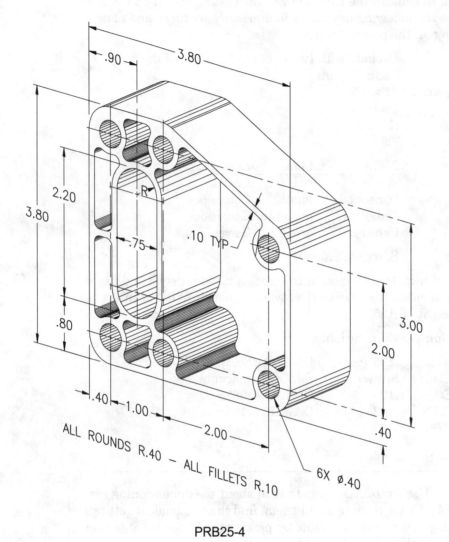

.90

3.80

2.20

3.80

R

.75

.10 TYP

.80

3.00

2.00

.40 ▸◂ 1.00

2.00

.40

ALL ROUNDS R.40 — ALL FILLETS R.10

6X ⌀.40

PRB25-4

5. Draw and dimension the front view of the plate shown below. Use the following drawing and dimension settings to set up the metric drawing, and place everything on the proper layers. Also draw and dimension the top and right side views, assuming the plate and boss thicknesses to be 10 and 5 respectively.

Units: Decimal with no digits to the right of the
                     decimal point
Drawing Area: 280 × 216
VIEWRES: 5000
Grid: 10
Snap: 5

Layers: *Name*   *Color*   *Linetype*
         Visible   red   Continuous
         Dimensions   blue   Continuous
         Text   magenta   Continuous
         Center   green   Center
         Hidden   magenta   Hidden

LTSCALE: Start with 10

Dimension text: Use romans.shx to create a new text style and use it for the dimension text. Leave the style at a height of 0, but insert text with a height of 3.

Dimensioning system variables:

    DIMASZ: 3
    DIMEXE: 1.5
    DIMEXO: 1.5
    DIMGAP: 1.5
    DIMCEN: −1.5
    DIMCLRD: ByLayer
    DIMCLRE: ByLayer
    DIMCLRT: magenta
    DIMTXT: 3

As an alternative to entering each of these system variables at the command line, find their equivalent settings in the dialog boxes made available to you by picking the Dimension Style icon.

PRB25-5

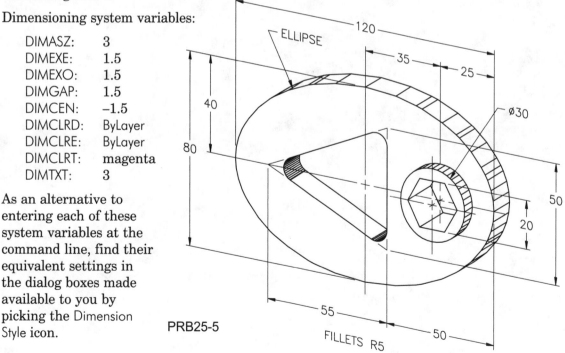

6. Draw the front view of the shaft shown below, placing the visible lines on a layer named Visible using the color green. Create a layer named Forces using the color red. Draw and label the four load and reaction forces, using leaders and a text style based on the romans.shx font. Create another layer named Dimensions using the color cyan. Dimension the front view, placing the dimensions on layer Dimensions. Use the LIMITS command to change the drawing area to accommodate the dimensions of the drawing.

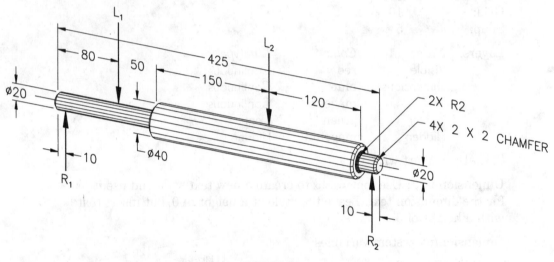

PRB25-6

Problems 4, 5, and 6 courtesy of Gary J. Hordemann, Gonzaga University

AUTOCAD® AT WORK

New Perspectives in 3D with WalkAbout™

"Looks good from the front," says the client, "but can you show me a view from inside the courtyard?" No problem. The WalkAbout utility lets AutoCAD for Windows users move around, over, or among objects in a 3D drawing. The only question is: Do you want to walk through or fly over?

To get from one point to another you press arrow keys, moving up, down, left, or right. Your progress through the drawing is flashed to the monitor at a rate of 2 to 3 frames per second, creating a sense of motion through an environment that is reminiscent of flight simulator programs. Objects appear in black and white or AutoCAD color.

In addition to the animation effects, WalkAbout works as an easy way of generating perspective views. You can take a snapshot from anywhere within a 3D drawing without going through the keyboard inputs required by the AutoCAD DVIEW command. "WalkAbout provides freedom of move-

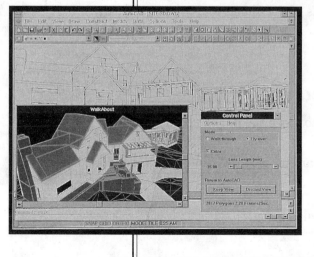

ment at a low cost," says Bob McNeel of Robert McNeel & Associates (Seattle, WA), developers of the software. "There are no assigned paths with WalkAbout."

The freedom to wander gives 3D modelers and renderers new opportunities to check design elements and evaluate how they fit together. For example, a playground designer can crawl inside a jungle gym for a kid's eye view of handholds. A facilities manager setting up for a special conference can take a quick tour to see how the layout will look to arriving guests. With WalkAbout, it's easy to gain new perspectives on your work.

Photo courtesy of Robert McNeel & Associates

293

Unit 26 — Fine-Tuning Your Dimensions

■ **OBJECTIVE:**

To alter the appearance of dimensions using dimensioning commands and dialog boxes

This unit covers special editing techniques that will help you fine-tune your dimensions. The unit steps you through the many dialog boxes that enable you to adjust the appearance of dimensions.

Editing Dimensions _____

As presented in Unit 24, you can move dimensions using grips. You can also edit dimensions using the ERASE and EXPLODE commands.

1 Start AutoCAD and open **dimen2.dwg**.

2 From the **File** pull-down menu, select **Save As...** and enter **dimen3.dwg** for the file name.

3 Attempt to erase a single element of any dimension, such as the dimension text or extension line.

AutoCAD selects the entire dimension because AutoCAD treats an associative dimension as a single object.

4 If necessary, press **ESC** to cancel the ERASE command.

Modify

5 Pick the **Explode** icon from the Modify toolbar.

This enters the EXPLODE command.

6 Select any dimension (other than the top horizontal dimension), and press **ENTER**.

The dimension is no longer an associative dimension, so its color changes. You can now edit individual elements of the dimension.

7 Erase part of the dimension.

DIMTEDIT Command _____

The DIMTEDIT command allows you to move the dimension text of an associative dimension.

Dimension

1 Display the **Dimension** toolbar and pick the **Dimension Text Edit** icon.

2 Select one of the associative dimensions.

3 Move the crosshairs and notice how the dimension and text move. (Ortho and snap affect the movement.)

4 Pick a new location for the dimension text.

5 From the Dimension toolbar, pick the **Dimension Text Edit** icon and pick the dimension you edited in Steps 1 through 3.

Dimension

6 Enter **H** for Home.

The dimension text returns to its home location.

DIMEDIT Command

The DIMEDIT command changes the appearance of dimension text and extension lines.

Dimension

1 Pick the **Dimension Edit** icon and enter **R** for the Rotate option.

2 Enter **20** (degrees) for the new text angle.

3 Select any associative dimension and press **ENTER**.

The dimension text rotates 20 degrees counterclockwise.

4 Reenter **DIMEDIT** (press the space bar or **ENTER**) and enter **N** for the New option.

The text editor appears.

5 Delete the angle brackets (<>).

6 Enter an arbitrary number, such as **25.5**, and pick **OK**.

7 Select any associative dimension and press **ENTER**.

25.5 replaces the dimension text.

8 Reenter **DIMEDIT** and enter **O** for the Oblique option.

⑨ Select the top horizontal dimension and press **ENTER**.

⑩ Enter **30** (degrees) for the obliquing angle.

The dimension's extension lines rotate 30 degrees. The oblique option is especially useful when you are dimensioning isometric drawings, such as those that follow. The drawing at the right is the result of applying oblique to the drawing at the left.

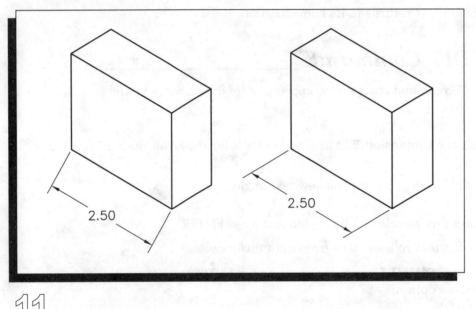

⑪ Save your work.

NOTE:

You can edit dimensions using the Windows right-click feature. Without anything entered at the Command prompt, select a dimension. Pick the grip box located at the center of the dimension so that it turns red, and then right-click. If you select Properties... from the pop-up menu, AutoCAD displays the Modify Dimension dialog box, enabling you to change the appearance of the dimension. In the Contents edit box, you can change or add characters, such as adding 2X to the beginning of the dimension for drawings that contain two holes of the same diameter.

Annotation Dialog Box _____

As you discovered in the previous unit, the Annotation **dialog box permits** you to make changes to the dimension text.

Dimension

1 Pick the **Dimension Styles** icon from the Dimension **toolbar or** enter **DDIM** at the keyboard.

2 Pick the **Annotation...** button.

Annotation	☒
Primary Units	**Alternate Units**
Units...	☐ Enable Units Units...
Prefix: [ ]	Prefix: [ ]
Suffix: [ ]	Suffix: [ ]
[1.00]	[|25.4|]
Tolerance	**Text**
Method: [None ▼]	Style: [STANDARD ▼]
Upper Value: [0.0000]	Height: [0.1800]
Lower Value: [0.0000]	Gap: [0.0900]
Justification: [Middle ▼]	Color... [■ BYBLOCK]
Height: [1.0000]	Round Off: [0.0000]
	[OK] [Cancel] [Help]

3 Pick the **Units...** button located in the upper left corner, as shown in the above illustration.

4 Change the units to **Decimal** and the dimension precision to **0.00**, and pick **OK**.

5 Pick **OK**, pick the **Save** button to save your changes to STYLE1, and pick **OK**.

All associative dimensions in the drawing change to decimal units with a precision of .00. These are the new settings you stored in STYLE1. Note that the dimension you exploded is not affected by the style change.

6 Reenter **DDIM** and pick the **Annotation...** button.

7 Pick the **Units...** button located in the upper left corner.

Notice the options located under Zero Suppression. Checking Leading or Trailing suppresses the leading or trailing zero in the dimension text. When you are using architectural units and 0 Feet is checked, AutoCAD suppresses the foot portion of dimensions that measure less than 1 foot. Otherwise, the dimension is stated as 0'-3 $\frac{1}{2}$". When 0 Inches is checked, AutoCAD drops the 0" part of a dimension that measures an even number of feet: 11'-0" becomes 11'.

The Linear item located in the Scale box specifies a global scale factor for linear dimensions. Checking Paper Space Only applies the Linear scaling factor only to dimensions created in paper space. Paper space is covered in Unit 39.

Angles, located in the upper right section of the dialog box, changes the appearance of angular dimensions. The items in the Tolerance section change the appearance of dimension tolerances.

8 Pick the **OK** button.

Alternate Units

The Alternate Units section, found in the upper right area of the Annotation dialog box, controls the display of alternate units in the dimension.

1 Pick the **Enable Units** check box and pick the **Units...** button located in the same area. (*Do not* pick Units... located in the upper left.)

The Linear scale is 25.4, which is the multiplier used to change inches to millimeters.

2 Pick the **OK** button.

The lower right section of the dialog box permits you to change the height and color of the dimension text and the gap between the text and dimension line.

3 Pick the **OK** button.

4 Pick the **Save** button to save the current settings, and pick the **OK** button.

All of the associative dimensions now contain alternate units. They change to follow the settings stored in STYLE1 because they were created using the STYLE1 dimension style. If you had not saved the changed settings to STYLE1, only new dimensions would be affected.

5 Press the space bar to reenter **DDIM**, pick **Annotation...**, and uncheck **Enable Units** in the Alternate Units area.

6 Pick **OK**, pick **Save**, and pick **OK**.

The alternate units disappear.

Format Dialog Box

The Format dialog box allows you to position the dimension text, arrowheads, and leader and dimension lines.

1 Press the space bar to enter **DDIM**, and pick the **Format...** button.

The Format dialog box appears.

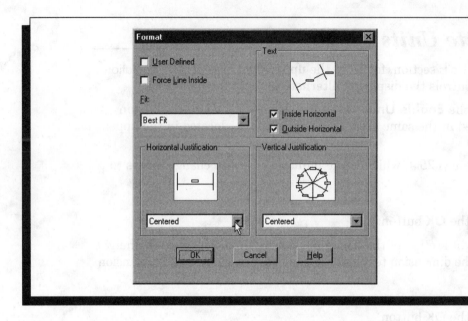

Focus your attention on the Horizontal Justification section and pay particular attention to the example dimension. Notice that the rectangle representing the text is centered on the dimension line.

2 Pick the down arrow and select **1st Extension Line**.

The rectangle in the example dimension moves close to the first extension line.

3 Try each of the remaining options in the list.

4 Focus on the Vertical Justification section, located at the right, and pick the down arrow.

5 Try each of the options in the list.

JIS conforms to the Japanese Industrial Standards.

6 Uncheck and check the check boxes located in the upper right section labeled Text, and notice the effect on the example dimensions.

7 Pick the **Help...** button to read about the remaining items in the dialog box.

8 After exiting the help screen, pick the **Cancel** button.

Geometry Dialog Box _____

The Geometry dialog box allows you to review and change certain dimensioning system variables more easily.

1 Pick the **Geometry...** button.

The Geometry dialog box appears.

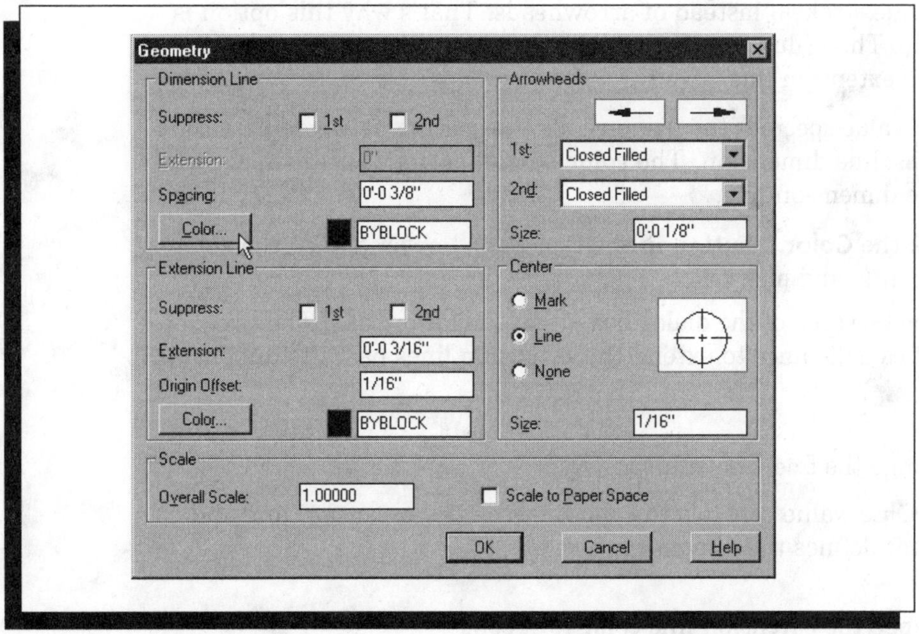

Focus on the upper left area of the dialog box. Checking the 1st and 2nd check boxes suppresses the display of the first and second dimension lines, respectively, when they are outside the dimension's extension lines. The following illustration provides an example.

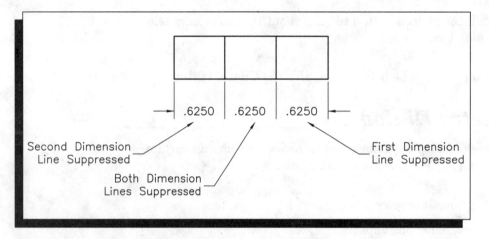

The Extension value in the Dimension Line area applies to the use of tick marks (oblique strokes) instead of arrowheads. That's why this option is not available. This value specifies the distance to extend the dimension line past the extension line.

The Spacing value specifies the distance for the spacing between dimension lines of a baseline dimension. The color button permits you to specify a color for the dimension line.

2 Pick the **Color...** button in the Dimension Line area, pick the color magenta, and pick **OK**.

The Extension Line area of the dialog box offers similar options. The Extension value specifies a distance to extend the extension lines past the dimension line.

3 Change the Extension value to $1/8$.

The Origin Offset value controls the gap between the extension line and the point that defines the dimension.

4 Change the extension line color to green.

You can see in the lower left corner that the overall scale is 24. As you may recall, you set this value using the DIMSCALE dimensioning system variable. The Scale to Paper Space check box refers to the use of paper space, which is covered in Unit 39.

The Arrowheads area of the dialog box controls the appearance of arrowheads.

5 In the Arrowheads area, pick the 1st drop-down box.

This displays a list of arrowhead types.

6 Pick **Dot Small**.

This also sets the arrowhead type for the second arrowhead. However, you can override this default by selecting a new type for the second arrowhead.

7 Pick the 2nd drop-down box next to and select **Datum Triangle**.

8 Try each of the remaining arrowhead types on your own. When you're finished, select **Dot Small** for the first and **Datum Triangle** for the second.

NOTE:

The User Arrow... selection displays the User Arrow dialog box. It expects you to specify a user-created arrowhead.

The Size value specifies the size of the arrowhead. Changing this value is equivalent to changing the value of the DIMASZ dimensioning system variable.

The Center area of the dialog box controls the appearance of center marks and lines for radial dimensions (diameter and radius).

9 In this area, pick the **Mark** radio button.

The example at the right of the radio button changes to center marks.

10 Pick the **None** radio button and notice the example; then pick the **Line** radio button.

Picking the Line radio button is equivalent to entering a negative value for DIMCEN. The Size value is equivalent to the value entered for DIMCEN.

11 Pick the **OK** button, pick **Save**, and pick **OK**.

The dimensions conform to the changes you stored in STYLE1.

12 Close the Dimension toolbar.

13 Save your work and exit AutoCAD.

Questions

1. How would you erase a piece of an associative dimension?

   _____

   _____

2. What part of an associative dimension does the DIMTEDIT command allow you to move?

   _____

3. Will the Home option of the DIMEDIT command work with a dimension that is not an associative dimension? Explain.

   _____

   _____

4. Can you rotate dimension text? Explain.

   _____

5. What is the purpose of the New option of the DIMEDIT command?

   _____

   _____

6. How do you adjust the angle of a dimension's extension lines?

   _____

7. How can you specify a global scale factor for linear dimensions?

   _____

   _____

◼ *Challenge Your Thinking: Questions for Discussion*

1. A drawing of a complex automotive part has been drawn in AutoCAD using standard imperial units. Without redimensioning the drawing, how could you change it to show metric units only?

2. Under what circumstances would you want to suppress one or both dimension lines? One or both extension lines? Explain.

Problems

1. Create a new drawing named prb26-1.dwg. Plan for a drawing scale of 1″=1″ and sheet size of 11″ × 17″. After you apply the following settings, create and dimension the following drawing. Use the dimension dialog boxes to set the dimensioning variables. In addition to saving prb26-1.dwg as a drawing file, save it as a template (DWT) file.

Drawing area:	Upper right corner 16 × 10
Snap:	.25
Grid:	1
Layers:	Create layers to accommodate multiple colors and linetypes.
Linetype scale:	.5
Font:	romans.shx
Dimension style name:	HEATHER
Dimension scale:	1
Dimension text height:	.16
Arrowhead size:	.16
Center mark size:	.08
Mark with center lines?	Yes

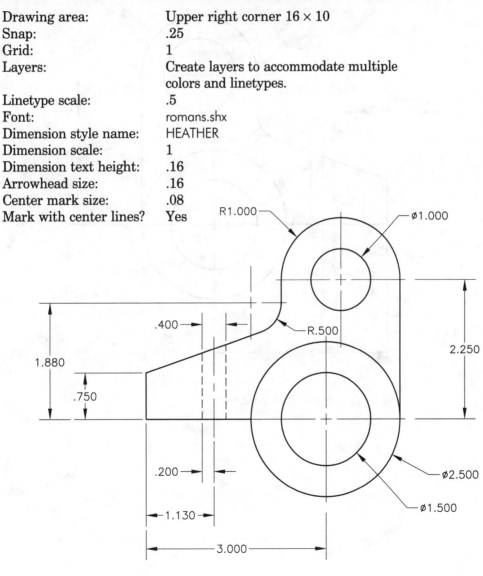

PRB26-1

2. Create a new drawing named prb26-2.dwg using the prb26-1.dwt template file. Use AutoCAD's commands, such as EXPLODE and STRETCH, and dimensioning commands, such as DIMTEDIT and DDIM, to make the following changes. Note that the 2.250 dimension is now 2.500. Stretch the top part of the object upward .250 unit and let the associative dimension text change on its own.

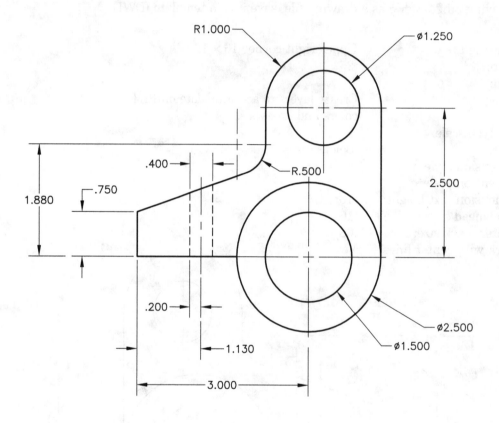

PRB26-2

3. Draw and dimension the front view of the oscillating follower shown on the next page. Use the drawing settings shown below. Draw the top and right-side views, assuming the follower to have a thickness of 8.

Units:	Decimal with no digits to right of the decimal point
Drawing area:	280 × 216
VIEWRES:	5000
Grid:	10
Snap:	5

Layers:	*Name*	*Color*	*Linetype*
	Visible	red	Continuous
	Dimensions	blue	Continuous
	Text	magenta	Continuous
	Center	green	Center
	Hidden	magenta	Hidden

Linetype scale:	Start with 10
Text style:	Use the romans.shx font to create a new text style. Leave the style at a height of 0, but insert text with a height of 3.

Create two dimension styles named TEXTIN and TEXTOUT. Use TEXTIN for those dimensions in which the text is inside the dimension lines, and TEXTOUT for the few radial dimensions in which the text should be outside the extension lines. Use the standard acad.dwt settings except for the following:

DIMASZ:	3	DIMCLRD:	ByLayer
DIMEXE:	1.5	DIMCLRE:	ByLayer
DIMEXO:	1.5	DIMCLRT:	magenta
DIMGAP:	1.5	DIMTXT:	3

Set the dimension style to TEXTIN and TEXTOUT as appropriate for each of the two styles. For the style TEXTOUT, set DIMTIX to 1 (on).

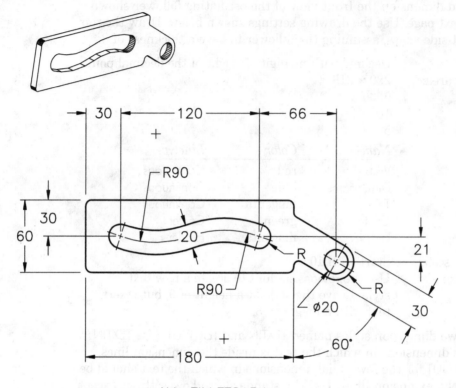

ALL FILLETS R4
ALL ROUNDS R5

PRB26-3

Unit 27 Tolerancing

■ OBJECTIVE:

To apply AutoCAD's basic and advanced tolerancing features

AutoCAD permits you to add tolerances to dimensions. Tolerances are necessary on drawings that will be used to manufacture parts because some variation is normal in the manufacturing process. The tolerances specify the largest variation allowable for a given dimension.

With AutoCAD, you can create geometric characteristic symbols and feature control frames that follow industry standard practices for geometric dimensioning and tolerancing (GD&T). Geometric characteristic symbols are used to specify form and position tolerances on drawings. Feature control frames are the frames used to hold geometric characteristic symbols and their corresponding tolerances.

■ *Basic Tolerances* _____

Using the Dimension Styles dialog box and the Annotation dialog box, you can create basic dimension tolerances.

1 Use the **dimen.dwt** template file to create a new drawing.

2 Zoom in on the drawing so that it fills most of the screen.

3 Display the Dimension toolbar.

4 Pick the **Dimension Styles** icon or enter **DDIM**, and pick the **Annotation...** button.

Dimension

The Annotation dialog box appears.

5 In the Text section, change the value of Height to **0.125**.

6 Pick the **Units...** button located in the upper left corner in the Primary Units section.

The Primary Units dialog box appears.

7 Change the precision to **0.0** in the Dimension section and **0.00** in the Tolerance section and pick **OK**.

In the Annotation dialog box, focus your attention on the Tolerance section.

8 Pick the down arrow in the **Method** drop-down box and pick **Symmetrical**.

Notice that the example dimensions change to reflect the new tolerance method. In symmetrical tolerancing (also called *symmetrical deviation*), the upper and lower values are equal. Therefore, the tolerance is written as a single value preceded by a ± symbol. Because both values are equal, you can only specify one value if you select the Symmetrical method. The lower value defaults to the negative of the higher value.

9 Change Upper Value to **0.0500** and pick the **OK** button.

10 Pick the **Save** button to save your changes in the STANDARD dimension style.

STANDARD was the dimension style you used when you created and dimensioned the dimen drawing. For this reason, you must save changes to the STANDARD dimension style to affect the existing dimensions in the dimen drawing.

11 Pick the **OK** button.

The dimensions change according to the new settings, as shown in the following illustration.

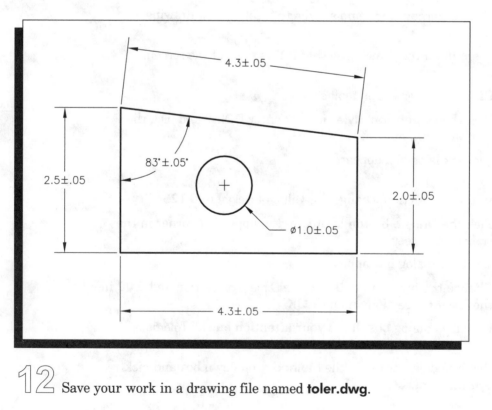

12 Save your work in a drawing file named **toler.dwg**.

Other Basic Tolerancing Options _____

AutoCAD provides other methods for basic dimension tolerances.

Dimension

1 Display the **Dimension Styles** dialog box again and pick the **Annotation...** button.

Focus once again on the Tolerance section.

2 Pick the **Method** drop-down box and select **Limits**.

The limits method of dimensioning shows only the upper and lower limits of variation for a dimension. The basic dimension is not shown.

3 Pick **OK**, pick **SAVE**, and pick **OK**.

The dimensions change according to the new settings, as shown in the following illustration.

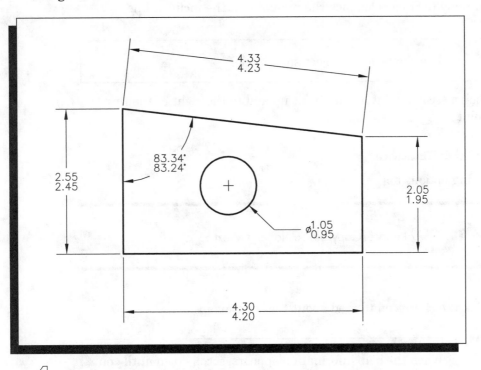

4 Press the space bar and pick the **Annotation...** button.

5 Try the Basic and Deviation tolerance methods on your own. Be sure to save the dimension style each time you change the tolerance method.

6 Change the tolerance method to **Limits** and make sure that both the upper value and the lower value are 0.0500. Pick the **Save** button to save your changes to the dimension style.

GD&T Practices

In Unit 25, you created simple and moderately complex leaders using the LEADER command. You can also use the LEADER command to create geometric characteristic symbols and feature control frames.

1 Create a new layer named **Gdt** and assign the color magenta to it. Make it the current layer, freeze layer Dimensions, and pick the **OK** button.

2 Pick the **Leader** icon to enter the LEADER command.

3 In reply to From point, pick the midpoint of the inclined line.

HINT:

Use the Midpoint object snap mode.

4 Pick a second point about 1 unit up and to the right of the first point.

5 Press **ENTER** twice.

The following options appear.

Tolerance/Copy/Block/None/<Mtext>:

6 Pick the **Help** icon to read about these options.

HINT:

When the help information appears, scroll down until you see To point (Format/Annotation/Undo) in the color green. Pick Annotation. A second help screen appears. Scroll down to read about Tolerance, Copy, and the other options.

7 After you exit the help screens, enter **T** for Tolerance.

The Symbol dialog box, which contains 14 geometric symbols, appears. These symbols are commonly used in GD&T. See Appendix E for more information about these symbols.

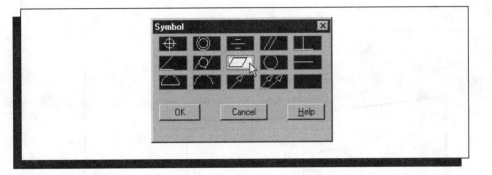

⑧ Pick the flatness symbol as shown in the illustration above, and pick the **OK** button.

This displays the Geometric Tolerance dialog box, as shown below.

This dialog box permits you to create complete feature control frames by adding tolerance values and their modifying symbols. Notice that the flatness symbol appears under Sym, which is located in the upper left corner.

Focus your attention on the Tolerance 1 area.

⑨ Under Value and to the right of the flatness symbol, enter **0.020** for the tolerance value.

10 Pick **OK**.

The feature control frame appears as a part of the leader, as shown in the following illustration.

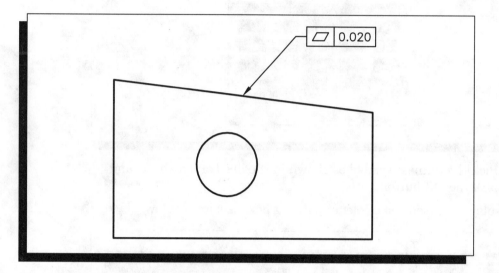

TOLERANCE Command

The TOLERANCE command also permits you to select geometric characteristic symbols and create feature control frames.

Dimension

1 Pick the **Diameter Dimension** icon from the toolbar.

This enters the DIMDIAMETER command.

2 With snap off, pick a point anywhere on the circle and pick a point to position the dimension.

3 Without anything entered at the Command prompt, select the new dimension.

4 Using the grips, reposition the parts of the dimension so that it looks like the one in the following illustration. (Ortho should be off.)

HINT:

You may need to change the location of all three grips boxes.

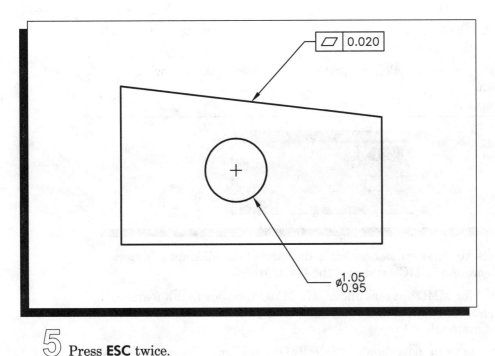

5 Press **ESC** twice.

Now we're ready to use the TOLERANCE command.

1 Pick the **Tolerance** icon from the Dimension toolbar.

This enters the TOLERANCE command and displays the Symbol dialog box.

2 Pick the true position symbol located in the upper left corner of the dialog box and pick **OK**.

The true position symbol appears under Sym.

3 In the Tolerance 1 area, pick the black box located under Dia and to the right of the true position symbol.

The diameter symbol appears.

4 Pick the box again, and again, causing it to disappear and reappear.

As you can see, this serves as a switch, toggling the diameter symbol on and off.

5 Enter **0.010** in the box located under Value and to the right of the diameter symbol.

6 Pick the black box located to the right and under MC. (MC stands for *material condition*.)

The Material Condition dialog box appears, as shown in the following illustration.

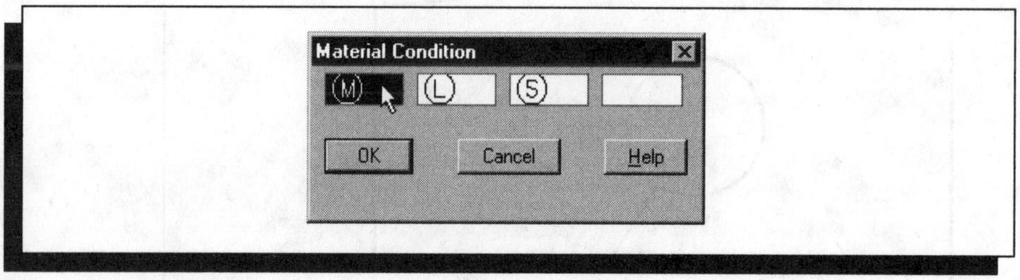

7 Pick the first symbol, which is the symbol for Maximum Material Condition (MMC), and pick the **OK** button.

The symbol for MMC appears under MC. MMC specifies that a feature, such as a hole or shaft, is at its maximum size or contains its maximum amount of material.

8 In the white boxes under Datum 1, Datum 2, and Datum 3, enter **A**, **B**, and **C**, respectively.

The black boxes located to the right of the boxes which now contain A, B, and C also permit you to select a material condition.

Focus on the second line, which enables you to create a second feature control frame.

1 Pick the bottom black box located under Sym.

The Symbol dialog box appears.

2 Pick the perpendicularity symbol located in the upper right corner and pick **OK**.

3 In the Tolerance 2 area, on the second line, pick the black box under Dia, causing the diameter symbol to appear.

4 Enter **0.006** in the box under Value.

5 Pick the black box under MC, pick the MMC symbol, and pick **OK**.

⑥ Under Datum in the Datum 1 area, enter **A**.

NOTE:

Under the Tolerance 1 area, you can enter a value for Height. It creates a projected tolerance zone in the feature control frame which controls the height of the extended portion of a perpendicular part. We will not enter a value.

You can specify projected tolerances in addition to positional tolerances to make the tolerance more specific. For example, projected tolerances control the perpendicularity tolerance zone of an embedded part.

⑦ Pick the black box located to the right of Projected Tolerance Zone.

This inserts a projected tolerance zone symbol.

⑧ Pick it again to make it disappear.

⑨ In the box located to the right of Datum Identifier, enter **-D-**.

This creates a datum-identifying symbol consisting of the letter D with a dash before and after it.

The Geometric Tolerance dialog box should now look like the one in the following illustration.

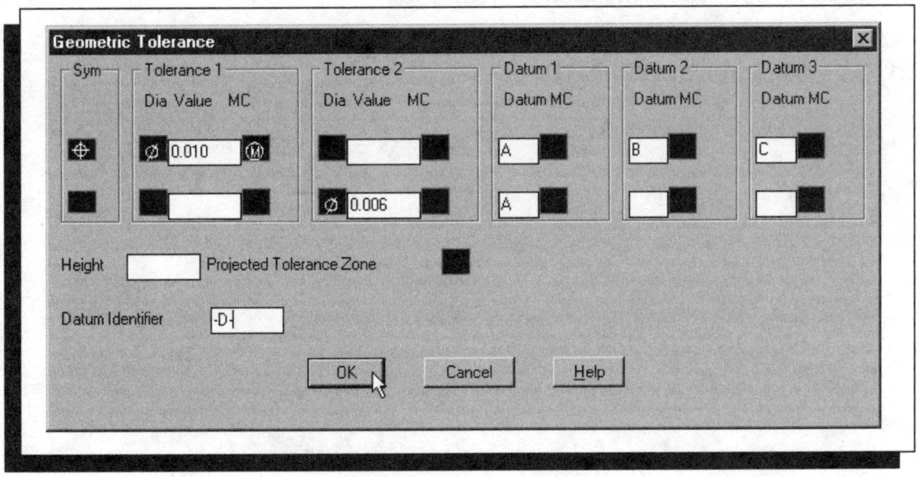

⑩ Finally, pick the **OK** button.

11 In reply to Enter tolerance location, position it as shown in the following illustration and pick a point.

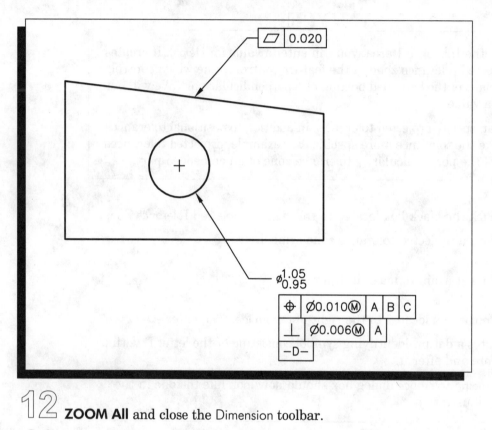

12 **ZOOM All** and close the Dimension toolbar.

13 Save your work and exit AutoCAD.

Questions

1. Give an example for each of the following four tolerancing methods made available to you in the Annotation dialog box.

   Symmetrical _____

   Deviation _____

   Limits _____

   Basic _____

2. What is the purpose of adding tolerances to a drawing?

   _____

   _____

3. How is the TOLERANCE command similar to the LEADER command?

   _____

   _____

4. Explain how you would include material condition symbols in feature control frames.

   _____

   _____

   _____

5. How would you include a second feature control frame below the first one?

   _____

   _____

■ *Challenge Your Thinking: Questions for Discussion*

1. With AutoCAD, you can quickly produce feature control frames, complete with geometric characteristic symbols. Investigate ways of editing them.

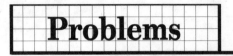

Problems

1. Create a new drawing named prb27-1.dwg using the prb26-1.dwt template file. Using AutoCAD's Dimension Styles and Annotation dialog boxes, make the following changes.

   - Change the dimension text height to 0.13.
   - Change the center marks to 0.07.
   - Change the dimension arrowheads to 0.08 dots (using Dot Small).
   - All dimensions, except the smallest hole, should show a tolerance of ±0.03. Use a leader to show that this hole does not require a tolerance. Specify 0.08 for the extension line feature offset.
   - Save the settings in the current dimension style.

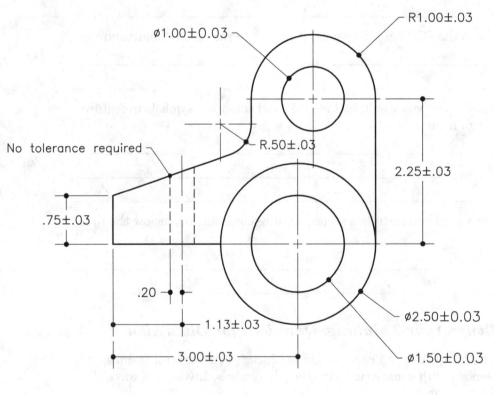

PRB27-1

2. Edit the drawing you created in problem 1 using the following criteria.

- Create a new layer named Gdt, assign the color magenta to it, and make it the current layer.
- Freeze the layer containing the dimensions.
- Change the dots to arrowheads.
- Use the LEADER and TOLERANCE commands to produce geometric characteristic symbols and feature control frames for the drawing.
- Save the settings in a new dimension style named TOLERANCE.

AUTOCAD® AT WORK

Landscaping with AutoCAD

For 150 years, a typical landscape designer's work tools—a drafting table, pen and pencil, paper, and templates—remained virtually the same. But the introduction of CAD changed the profession. Nowadays, landscape architects who don't use CAD are at a serious disadvantage. Layouts that traditionally took two days to prepare by hand can now be completed using CAD in a few hours.

Landscape architecture focuses on everything above the ground and outside a building, including vegetation, fountains, sculptures, and roads. In working with their clients, landscape architects must draw up detailed plans that can be adapted quickly and easily, and the special features of CAD programs such as AutoCAD complement the professional nicely.

The architect can create buildings, roads, boundaries, irrigation systems, and recreation areas on separate layers. Then the architect can combine any of the layers to use for presentations to clients or for construction plans for engineers and work crews.

AutoCAD's symbol library capability is especially useful. Symbols such as trees and shrubs can be used to increase drawing speed. The symbols can also contain attribute data that can later be used to generate reports.

Using AutoCAD and special telecommunications equipment, an architect in one city and a client in another can view a proposed drawing and make immediate modifications. The architect can exchange information with other professionals on the project, and field personnel—such as surveyors—can make on-site suggestions.

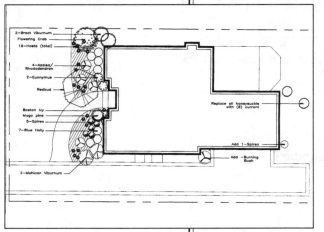

Courtesy Green View Companies, Dunlap, IL

Unit 28 — Heavy Lines and Solid Objects

■ OBJECTIVE:

To apply the TRACE, SOLID, and FILL commands

This unit focuses on thick lines and solid objects and how they are used to produce elevation drawings.

Note the heavy lines in the following drawing.

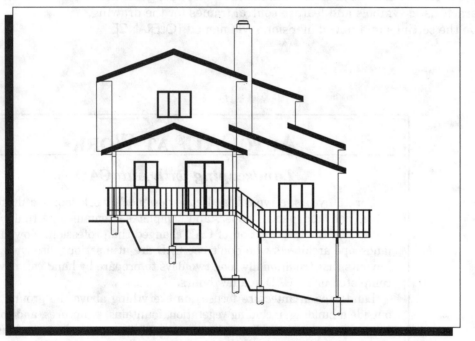

AutoCAD drawing courtesy of Tim Smith, Hyland Design

The AutoCAD SOLID and TRACE commands were used to create the thick lines and solid-filled areas. Let's draw similar lines and solid objects.

■ *TRACE Command* _____

1 Start AutoCAD and begin a new drawing using the **tmp1.dwt** template file.

2 Enter the **TRACE** command.

The TRACE command is used very much like the LINE command, except TRACE requires you to enter a trace width in units.

3 Specify a trace width of 4 inches and draw the figures shown below. Don't worry about exact sizes.

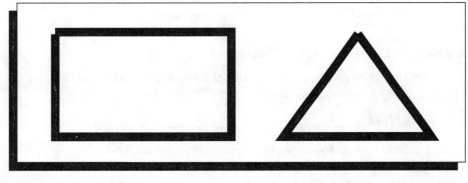

HINT:
Press ENTER to make the last segment of each object appear.

You'll notice that it is difficult to produce a perfect corner at the first and last points of a polygon. This is the nature of the TRACE command.

4 Practice using TRACE by creating several more objects.

5 Save your work in a file named **trace.dwg**.

SOLID Command

Now let's work with the SOLID command to produce solid-filled objects.

1 From the **Draw** pull-down menu, pick **Surfaces** and **2D Solid**.

This enters the SOLID command.

2 Produce a solid-filled object similar to the one below. Pick the points in the exact order shown, and press **ENTER** when you are finished.

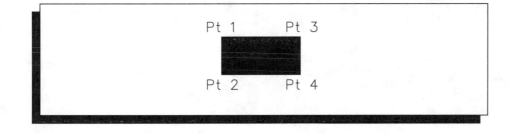

③ Pick a fifth and sixth point.

④ Experiment with the SOLID command. If you pick the points in the wrong order, AutoCAD creates an hourglass-shaped object.

⑤ Press **ENTER** to terminate the SOLID command.

Leave the objects on the screen so you can practice the FILL command.

FILL Command

The FILL command affects traces, solids, wide polylines, multilines, and solid-filled hatches. You will work with polylines in the following unit and the remaining objects in later units..

FILL is either on or off. When FILL is off, only the outline of a solid is represented. This saves time when the screen is regenerated.

① Enter the **FILL** command, and turn it off by entering **OFF**.

② Enter the **REGEN** command.

NOTE:

After you turn FILL on or off, you must regenerate the screen before the change will take place.

The objects should no longer be solid-filled.

③ Reenter the **FILL** command and turn it on.

④ Enter **REGEN** to force a screen regeneration.

⑤ Save your work and exit AutoCAD.

Questions

1. What might be a limitation of using the TRACE command?

   _____

   _____

2. How would you draw a solid-filled triangle using the SOLID command?

   _____

   _____

3. Can you draw curved objects using the SOLID command?

   _____

4. What is the purpose of FILL and how is it used?

   _____

   _____

5. What object types does FILL affect?

   _____

 Challenge Your Thinking: Questions for Discussion

1. Discuss possible uses for the TRACE, SOLID, and FILL commands. Under what circumstances might you use them? Give specific examples.

Problems

1. Construct prb28-1.dwg using the TRACE and SOLID commands. Specify a trace width of .05 unit. Don't worry about the exact size and shape of the roof.

PRB28-1

2. After you have completed prb28-1.dwg, place the solid shapes as indicated below. Don't worry about their exact sizes and locations.

PRB28-2

3. Are the trace and solid objects similar to the line, circle, and arc objects? To find out, try removing a small piece of the roof. What is your conclusion?

_____

_____

_____

4. The icon shown below is used on metric drawings that conform to the SI system and use what is called *third-angle projection*. This projection system, which places the top view above the front view, is the normal way of showing views in the United States. Draw the icon, including the outlines of the letters S and I; then use the SOLID command to fill in the letters.

PRB28-4

5. Construct the drawing shown on the first page of this unit. You may choose to use a drawing template when you begin the drawing.

Problem 4 courtesy of Gary J. Hordemann, Gonzaga University

Unit 29 — Joining Straight and Curved Objects

OBJECTIVE:

To apply polylines and spline curves using the PLINE, PEDIT, 3DPOLY, SPLINE, and SPLINEDIT commands

This unit focuses on polylines and splines. A polyline is a connected sequence of line and arc segments that is treated by AutoCAD as a single object. Polylines are often used in lieu of conventional lines and arcs because they are more versatile. With the SPLINE command, you can create non-uniform rational B-splines, known as NURBS. The examples below illustrate some uses of polylines and splines.

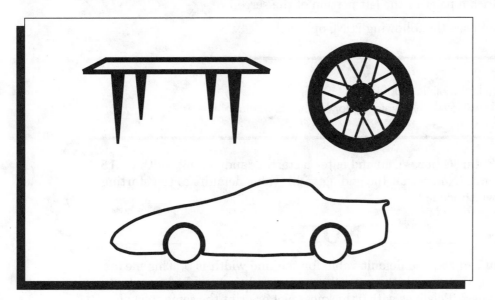

PLINE Command

Let's create the following polyline using the PLINE command.

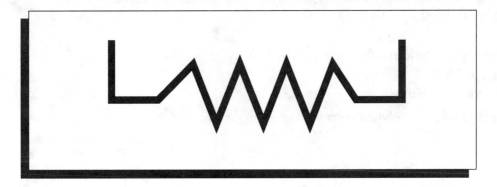

1 Start AutoCAD and start a new drawing from scratch. Do *not* use a template file.

2 Set the snap resolution at **.5**.

3 Pick the **Polyline** icon from the Draw toolbar, or enter **PL** (the PLINE command alias).

Draw

This enters the PLINE command.

4 Pick a point in the left portion of the screen.

You should see the following PLINE options.

Current line-width is 0.0000
Arc/Close/Halfwidth/Length/Undo/Width/<Endpoint of line>:

---

5 Enter **W** (for Width) and enter a starting and ending width of **.15** unit. (Notice that the ending width value defaults to the starting width value.)

NOTE:

As you can see, the default value for the line width is 0. This means that it is the same thickness as lines, arcs, and circles, which also have 0 thickness. Polylines with 0 thickness plot or print the same thickness as these other objects.

6 Draw the object by approximating the location of the endpoints. If you make a mistake, undo the segment. Press **ENTER** when you have finished the object.

7 Save your work in a file named **poly.dwg**.

8 Move the polyline a short distance.

Notice that the entire polyline is treated as a single object.

PEDIT Command

Now let's edit the polyline using PEDIT.

1 From the **View** pull-down menu, pick **Toolbars...** and open the **Modify II** toolbar.

Modify II

2 Pick the **Edit Polyline** icon from the Modify II toolbar.

This enters the PEDIT command.

3 Pick the polyline.

The polyline does not highlight as it does during a normal object selection, but the following options appear when you select a polyline.

Close/Join/Width/Edit vertex/Fit/Spline/Decurve/Ltype gen/Undo/eXit <X>:

Let's change the polyline width.

4 Enter **W** and specify a new width of **.1** unit.

As you can see, PEDIT is useful in changing the width of a series of wide lines.

Now let's close the polyline, as shown in the following illustration.

5 Enter **C** for Close.

AutoCAD closes the object.

Let's do a simple curve fitting operation.

6 Enter **F** for Fit.

Did the drawing change?

7 Enter **D** (for Decurve) to return it to its previous form.

Next, let's move one of the object's vertices as shown below.

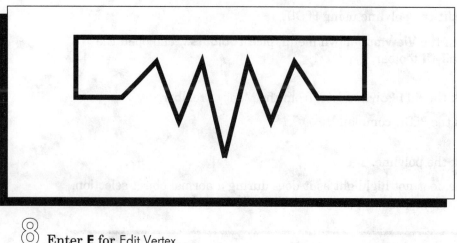

⑧ Enter **E** for Edit Vertex.

Notice that a new set of choices, shown below, becomes available. Also notice the × in one of the corners of the polyline.

Next/Previous/Break/Insert/Move/Regen/Straighten/Tangent/Width/eXit <N>:

⑨ Move the × to the vertex you want to change by pressing **ENTER** several times.

⑩ Enter **M** for Move and pick a new point for the vertex.

Try it again if it did not work.

⑪ To exit the PEDIT command, enter **X** (for eXit) twice.

Note that PEDIT contains many more editing features. Experiment with each of them on your own.

Breaking Polylines

You can remove small pieces from polylines using the BREAK command. Let's try it.

① From the Modify toolbar, pick the **Break** icon.

This enters the BREAK command.

Modify

330

2 Pick the polyline, enter **F** to enter the first breaking point, and pick two points in a counterclockwise direction.

The piece of the polyline between the two points you picked disappears.

3 Undo the break.

Exploding Polylines _____

The EXPLODE command gives you the ability to break up a polyline into individual line and arc segments.

1 Enter **UNDO** and select the **Mark** option. (We will return to this location at a later time.)

2 Pick the **EXPLODE** icon.

Modify

3 Pick the polyline and press **ENTER**.

Notice the message: Exploding this polyline has lost width information. The UNDO command will restore it.

This is the result of applying EXPLODE to a polyline that contains a width greater than 0. You now have an object that contains numerous entities for easier editing, but you have lost the line width.

4 To illustrate that the object is now made up of numerous entities, edit one of them, and then undo your change.

PLINE's Arc Option _____

In some drafting applications, there is a need to draw a series of continuous arcs to represent, for example, a river on a map. If the line requires thickness, the ARC Continue option will not work. But the PLINE Arc option can handle this task.

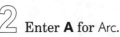
Draw

1 Enter the **PLINE** command and pick a point anywhere on the screen.

2 Enter **A** for Arc.

The following list of options appears.

Angle/CEnter/CLose/Direction/Halfwidth/Line/Radius/Second pt/Undo/Width/
<Endpoint of arc>:

3 Move the crosshairs and notice that an arc begins to develop.

4 Enter the **Width** option and enter a starting and ending width of **.1** unit.

5 Pick a point a short distance from the first point.

6 Pick a second point, and a third.

7 Press **ENTER** when you're finished.

8 Enter **UNDO** and **Back**.

Spline Curves

With AutoCAD, you can create spline curves, also referred to as B-splines.

The PEDIT Spline option uses the vertices of the selected polyline as the control points of the curve. The curve passes through the first and last control points and is "pulled" toward the other points but does not necessarily pass through them. The more control points you specify, the more "pull" they exert on the curve.

1 Pick the **Edit Polyline** icon from the Modify II toolbar and select the polyline.

Modify II

2 Enter **S** for Spline.

Do you see the difference between the Spline and Fit options?

3 Enter **D** to decurve the object.

4 Enter **X** to exit the PEDIT command.

NOTE:

In connection with PEDIT, AutoCAD offers two spline options: quadratic B-splines and cubic B-splines. An example of each is shown below.

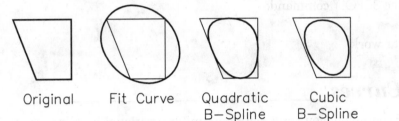

Original   Fit Curve   Quadratic   Cubic
                          B—Spline   B—Spline

The system variable SPLINETYPE controls the type of spline curve to be generated. Set the value of SPLINETYPE at 5 to generate quadratic B-splines. Set its value at 6 to generate cubic B-splines.

The PEDIT Decurve option enables you to turn a spline back into its frame, as illustrated in Step 3 above. In addition, you can view both the spline curve and its frame by setting the system variable SPLFRAME to 1.

3DPOLY Command

The 3DPOLY command enables you to create polylines consisting of *x*, *y*, and *z* vertices. 3D polylines are made up of straight line segments only, and they cannot take on a specified thickness.

You can edit 3D polylines with the PEDIT command, and you can fit a 3D B-spline curve to the vertices of a 3D polyline.

1 From the **Draw** pull-down menu, pick **3D Polyline**.

This enters the 3DPOLY command.

2 Pick a point anywhere.

Notice that fewer polyline options are given.

3 Pick a series of points and press **ENTER** when you are finished.

Modify

4 Pick the **Edit Polyline** icon and select the new polyline.

Here again, fewer options are provided.

5 Enter the **Spline curve** option and enter **X** to exit the command.

The polyline is a three-dimensional spline curve.

Refer to Units 41-49 to learn about AutoCAD's 3D environment and how you can apply the 3DPOLY command.

6 Save your work.

NURBS Curves

With the SPLINE command, you can create non-uniform rational B-splines, known as NURBS.

1 From the **File** pull-down menu, pick **Save As...**, and enter **nurbs** for the file name.

2 Erase both objects.

3 From the Draw toolbar, pick the **Spline** icon.

Draw

This enters the SPLINE command.

4 Pick a point and then a second point.

5 Slowly move the end of the spline back and forth and notice how the spline behaves.

6 Continue to pick a series of points.

7 Press **ENTER** three times to complete the spline and terminate the command.

8 Enter **SPLINE** again and pick a series of points.

9 Enter **C** for Close.

As you can see, this closes the spline.

10 Press **ENTER** to terminate the command.

NOTE:

The Fit Tolerance option changes the tolerance for fitting the spline curve to the control points. A setting of 0 causes the spline to pass through the control points. By changing the fit tolerance to 1, you allow the spline to pass through points within 1 unit of the actual control points.

You can easily convert a spline-fit polyline into a NURBS curve. Enter the SPLINE command, enter O for Object, and pick the spline-fit polyline.

Editing Spline Objects

Using grips, you can interactively edit a spline entity. The SPLINEDIT command provides many additional editing options.

1 Without anything entered at the Command prompt, select one of the spline curves.

Grip boxes appear at the control points.

2 With the crosshairs, lock onto one of the grips and move it to a new location.

This is the fastest way to edit a spline.

3 Press **ESC** twice.

Modify II

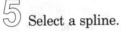

4 From the Modify II toolbar, pick the **Edit Spline** icon.

This enters the SPLINEDIT command.

5 Select a spline.

The following options appear.

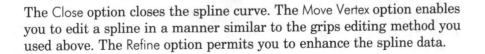

Fit Data/Close/Move Vertex/Refine/rEverse/Undo/eXit <X>:

The Close option closes the spline curve. The Move Vertex option enables you to edit a spline in a manner similar to the grips editing method you used above. The Refine option permits you to enhance the spline data.

6 Enter **R** for Refine.

The following Refine options appear.

Add control point/Elevate Order/Weight/eXit <X>:

---

7 Enter **A** for Add control point.

8 Pick several points on the spline to add new control points, and then press **ENTER**.

The Elevate Order option increases the density of control points on the spline. The Weight option adjusts the distance between the spline and a selected control point.

9 Enter **X**.

The rEverse option reverses the order of the control points in AutoCAD's database. The Undo option undoes the last edit operation. The Fit Data option permits you to edit fit data using several options.

10 Enter **X**.

11 Create a new spline of any shape.

12 Pick the **Edit Spline** icon and pick the new spline.

13 Enter **F** for Fit Data and pick the **Help** icon to read about each of the fit options.

14 After exiting the help screen, experiment with the Fit Data options on your own.

15 Close the Modify II toolbar, save your work, and exit AutoCAD.

Draw

Modify II

Questions

1. What is a polyline?

_____

2. Briefly describe each of the following PLINE options.

   Arc _____

   Close _____

   Halfwidth _____

   Length _____

   Undo _____

   Width _____

3. Briefly describe each of the following PEDIT command options.

   Close _____

   Join _____

   Width _____

   Edit vertex _____

   Fit _____

   Spline _____

   Decurve _____

   Ltype gen _____

   Undo _____

   eXit _____

4. Describe one application for the PLINE Arc option.

_____

_____

5. Of what importance is the EXPLODE command to polylines?

_____

_____

6. What effect does EXPLODE have on polylines that contain a width other than 0?

_____

7. Is it possible to remove a small piece of a polyline? Explain.

_____

_____

8. Describe the system variable SPLINETYPE.

_____

_____

9. What is a NURBS curve?

_____

_____

10. When creating a NURBS curve using the SPLINE command, what does a fit tolerance setting of 0 indicate?

_____

_____

11. What is the fastest method of editing a spline?

_____

_____

12. What editing function can you perform by entering SPLINEDIT, Refine, and Add control point?

_____

_____

 Challenge Your Thinking: Questions for Discussion

1. Explain the advantages and disadvantages of using a polyline instead of individual lines and arcs to create a complex object.

2. Discuss everyday applications of NURBS curves.

3. What is the difference between a NURBS curve and a spline curve created with the PEDIT Spline option?

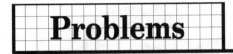

Problems

1. Create the approximate shape of the following racetrack using PLINE. Specify .4 unit for both the starting and ending widths. Select the Arc option to draw the figure.

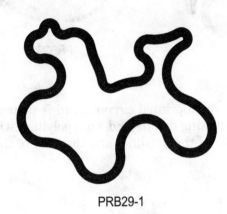

PRB29-1

2-6. Draw each of the following objects using the PLINE and PEDIT commands.

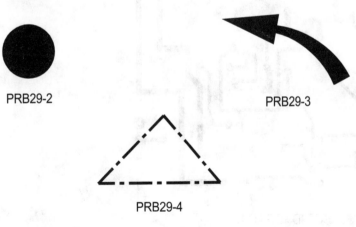

PRB29-2

PRB29-3

PRB29-4

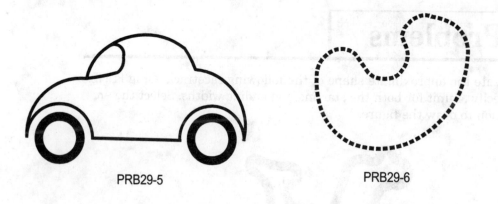

PRB29-5   PRB29-6

7. The artwork for one side of a small printed circuit board is shown
   below. Reproduce the drawing using donuts and wide polylines. Use
   the grid to estimate the widths of the polylines and the sizes of the
   donuts.

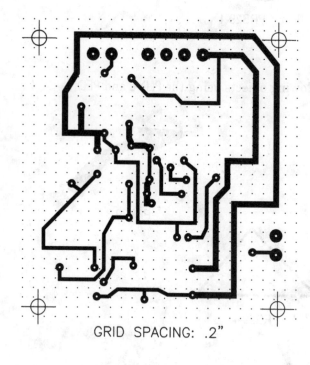

GRID SPACING: .2"

PRB29-7

8. The following graph shows the indicated and brake efficiencies as functions of horsepower for a small engine. Reproduce the graph as follows: Using a suitable scale, draw the grid and plot the given points as shown; then draw a spline through each set of points. Place the border, title border, and curves on a layer named Visible; the grid and point symbols on layer Grid; and the text on layer Text. Trim the grid around the text and arrows, and trim the curves and grid out of the symbols. Use the appropriate justify options of the DTEXT command to align the axis numbers and titles properly.

HP	INDICATED	BRAKE
1.53	3.60	2.55
2.83	5.30	3.86
4.11	5.90	5.00
5.30	7.25	4.52
6.60	8.82	5.62
9.10	8.30	5.83
10.45	6.27	5.40

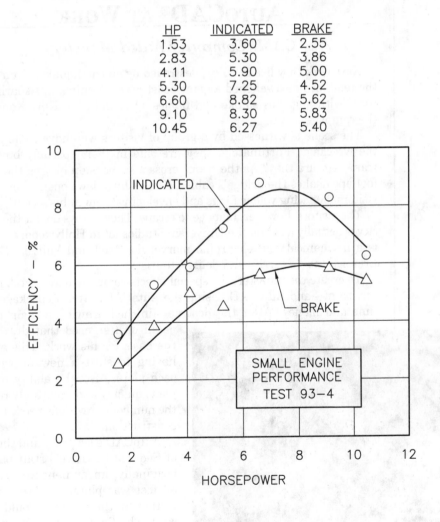

PRB29-8

Problems 7 and 8 courtesy of Gary J. Hordemann, Gonzaga University

AutoCAD® at Work

CAM (Computer-Aided Monster)

New York City lies in ruins, destroyed by an earthquake. Even worse, the temblor has awakened a giant beast. Eyes flashing, lips snarling, he roars with anger as smoke and flames surround him. King Kong is on the rampage.

The scene is witnessed by a group of visitors who have come to view the wreckage. Fortunately, they are safe in their specially built tour trams—or are they? As the trams cross a suspension bridge, the thirty-foot ape shakes the bridge's cables. Kong's huge jaws open, showing the visitors menacing yellow fangs and enveloping them in his steamy breath.

The visitors, however, manage to escape. They always do, for the ruined city is actually a set on the Universal Studios lot in Hollywood, and Kong is a mechanical and electronic marvel designed and built by Sequoia Creative Inc. to terrify and delight tourists.

To create the monster, the special-effects company used metal, plastic, fur, paint—and AutoCAD. Using the AutoCAD software package saved time and money. After the designer finished drawing a main part, a detail drafter used that drawing as the basis for his work. This avoided having to start a new drawing for each piece. Also, the ability to draw plans quickly and accurately reduced the number of revisions needed during construction.

AutoCAD's accuracy and the skills of Sequoia Creative's staff paid off. Originally, an audience distance of 80 feet was planned so that imperfections in the monster would not be seen. The final result, however, was so realistic that the tour trams now pass within 6 feet of Kong, almost close enough to shake hands with this furry celebrity!

Based on a story in CADalyst magazine, Vol. 3, No. 5

342

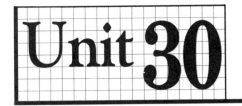

Unit 30 — A Calculating Strategy

■ OBJECTIVE:

To apply the ID, DIST, AREA, CAL, LIST, DBLIST, DIVIDE, and MEASURE commands and the PDMODE system variable

This unit focuses on commands that allow you to perform a variety of measurements and calculations on your drawings. It also covers the AutoCAD commands that reveal hidden, but important, data about specific components within the drawing.

The drawing below shows an apartment complex with parking lots, streets, and trees. It is possible to perform certain calculations on the drawing, such as determining the square footage of the parking lot or the distance between the parking stalls.

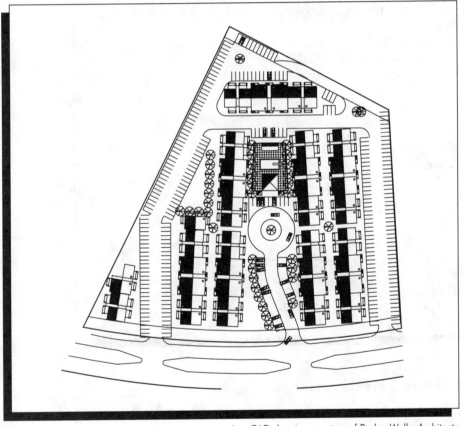

AutoCAD drawing courtesy of Buday-Wells, Architects

Let's load AutoCAD and practice these functions.

ID and DIST Commands

1. Start AutoCAD and start a new drawing from scratch.

2. Draw the following end view of a shaft with a rectangular pocket machined into it. Use the sizes shown below. Omit the numbers and dimensions, and don't worry about the exact placement of the rectangular pocket.

NOTE:

Set the snap resolution at .25 before drawing the object, and use the coordinate display as you construct it. Use a polyline to create the rectangular pocket. Create the object on a layer named Objects and use the color cyan.

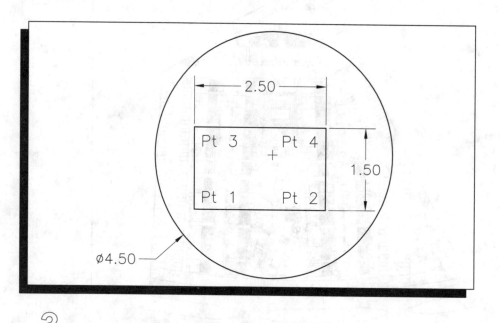

3. Save your work in a file named **calc.dwg**.

Now we're ready to perform a few simple calculations. First, let's find the absolute coordinates of point 1.

4. From the **View** pull-down menu, select **Toolbars...** and check the **Inquiry** check box.

Inquiry

5. From the Inquiry toolbar, pick the **Locate Point** icon.

This enters the ID command.

NOTE:

The items on the Inquiry toolbar are also located on a flyout in the Standard toolbar. When you want to perform only one or two operations, it may be faster to pick and hold the Distance icon and choose the appropriate icon from the flyout.

6 Pick point 1.

The absolute coordinates of point 1 appear.

7 Try it again with point 2.

The DIST command allows you to calculate the distance between two points.

Inquiry

8 From the same toolbar, select the **Distance** icon.

This enters the DIST command.

9 Pick points 1 and 4.

In addition to the distance, what other information does DIST produce?

AREA Command

Use the AREA command to calculate the area of circles, polygons, spline curves, regions, and solids. Regions and solids are covered in future units.

Inquiry

1 Select the **Area** icon from the Inquiry toolbar.

This enters the AREA command.

2 Enter **O** for Object.

3 Pick the pocket (rectangle).

The area of the pocket is 3.75 square units; the perimeter is 8 units.

The AREA command can also calculate areas with holes. Suppose we want to know the area of the end of the shaft minus the pocket.

4 Reenter **AREA** and enter **A** for Add.

345

5 Enter **O** for Object and pick the shaft.

AutoCAD displays the area (15.9043 square units) and the circumference (14.1372 units) of the shaft.

6 Press **ENTER**.

7 Enter **S** for Subtract.

8 Enter **O** for Object and pick the pocket.

AutoCAD subtracts the area of the rectangle from the area of the shaft and displays the result (12.1543 square units).

9 Press **ENTER** twice to terminate the command.

CAL Command

AutoCAD offers an on-line geometry calculator that evaluates vector, real, and integer expressions.

1 Enter the **CAL** command.

2 Type **(3*2)+(10/5)** at the keyboard and press **ENTER**.

AutoCAD calculates the answer as 8.0.

CAL also recognizes object snap modes, and you can use the command transparently.

Draw

3 Enter the **PLINE** command.

4 Enter **'CAL**. (Note the leading apostrophe.)

5 Type **(cen+end)/2** and press **ENTER**.

6 Pick any point on the shaft.

7 Pick either line near point 1.

AutoCAD calculates the midpoint between the shaft's center and point 1 and places the first point of the polyline at that location.

8 Press **ENTER** to terminate the PLINE command.

LIST and DBLIST Commands _____

These commands display database information on selected objects.

Inquiry

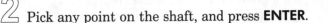

1 Pick the **List** icon from the Inquiry toolbar.

2 Pick any point on the shaft, and press **ENTER**.

The object type, layer, space, center point, radius, circumference, and area appear. You will learn more about AutoCAD's model space and paper space in Unit 56.

3 Press the **F2** function key to hide the text window.

4 Enter **LIST** again, but this time pick the pocket.

What information did you receive?

5 Enter **DBLIST** and watch what you get.

You should see information on all objects in the drawing database.

6 Close the AutoCAD Text Window.

DIVIDE Command _____

The DIVIDE command is used to divide an object into a specified number of equal parts.

1 From the **Draw** pull-down menu, pick **Point** and **Divide**.
This enters the DIVIDE command.

2 Select the shaft.

3 In reply to Number of segments, enter **20**.

It may appear as though nothing happened. Something did happen: the DIVIDE command divided the end of the shaft into 20 equal parts using 20 points; you just can't see them. Here's how to use them.

4 Enter the **LINE** command.

5 Enter the **Node** object snap. (Node is used to snap to the nearest point.)

6 Move the crosshairs along the shaft and snap to any one of the nodes.

7 Snap to the center of the shaft using the **Center** object snap. (Remember, you have to pick a point on the circle that represents the edge of the shaft.)

8 Enter the **Node** object snap again and snap to another point on the circle.

You should now have an object that looks similar to the one in the following illustration.

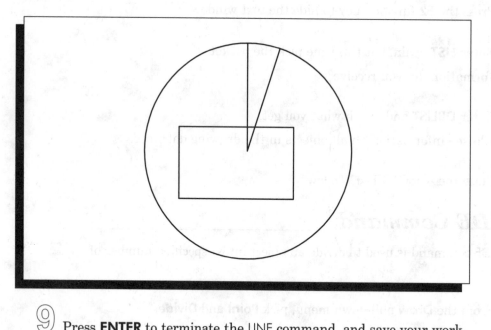

9 Press **ENTER** to terminate the LINE command, and save your work.

Making the Points Visible with PDMODE

One of AutoCAD's system variables, called PDMODE, is used to control the appearance of points. Let's use PDMODE to make the points on the circle visible.

1 Enter **PDMODE**.

2 Enter **32** for the new value, and enter the **REGEN** command.

Did 20 equally spaced circles appear on the circle as shown in the next illustration?

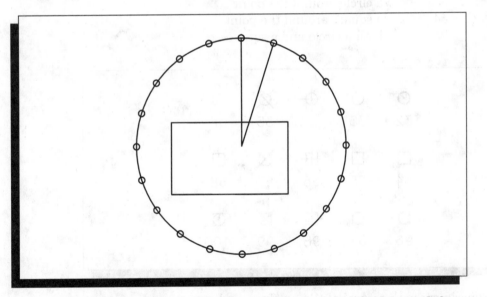

The following list and the illustration show how other values control the appearance of points.

PDMODE

Value	Draws
0	a dot at the point (default setting)
1	nothing
2	a cross through the point
3	an × through the point
4	a vertical line upward from the point

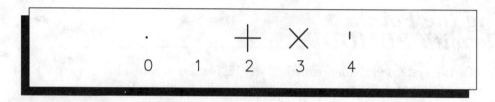

To each of the values shown above, you can add 32, 64, or 96 to select a figure to be drawn around the point in addition to the figure drawn through it, as shown in the next illustration. For example, 2 + 32 = 34, and 34 represents a cross with a circle on it.

PDMODE

Value	Draws
32	a circle around the point
64	a square around the point
96	both a circle and a square

⊙	○	⊕	⊠	◔
32	33	34	35	36
⊡	□	⊞	⊠	⊡
64	65	66	67	68
⊡	▢	⊞	⊠	⊡
96	97	98	99	100

3   Experiment with several of these values, and remember to regenerate the screen each time.

NOTE:

You can also use the DDPTYPE command to set the point style. This command presents a dialog box from which you can select the point style and specify the size of the points. Point size can be absolute or relative to the screen size.

350

MEASURE Command _____

The MEASURE command is similar to DIVIDE except that MEASURE does not divide the object into a given number of equal parts. Instead, MEASURE allows you to place markers along the object at specified intervals.

1 From the **Draw** pull-down menu, pick **Point** and **Measure**.

This enters the MEASURE command.

2 Select one of the two lines.

3 In reply to Segment length, enter **.4** unit.

AutoCAD adds points spaced .4 unit apart.

4 Further experiment with MEASURE.

5 When you're finished, close the Inquiry toolbar, save your work, and exit AutoCAD.

Questions

1. What AutoCAD command is used to find coordinate points?

_____

2. What information is produced with the AREA command?

_____

3. What information is produced with the LIST command?

_____

4. Describe the difference between LIST and DBLIST.

_____

_____

5. How do you calculate the perimeter of a polygon?

_____

6. How do you find the circumference of a circle?

_____

7. Explain how you control the appearance of points.

_____

_____

_____

_____

 Challenge Your Thinking: Questions for Discussion

1. Explain the difference between the DIVIDE and MEASURE commands. Under what conditions would you use each of these commands?

Problems

Draw the objects found in the following problems at any size, omitting all letters. Then use the appropriate commands to find the information requested above the objects. Write their values in the blanks provided.

If you have the optional *Applying AutoCAD Diskette,* refer to the files calcprb1.dwg, calcprb2.dwg, and calcprb3.dwg. These files correspond to prb30-1, prb30-2, and prb30-3, respectively.

1. Location of point A? _____

   Distance between points A and B? _____

   Area of the polygon? _____

   Perimeter of the polygon? _____

PRB30-1

2. Distance between A and B? _____

   Distance between B and C? _____

   Area of circle? _____

   Circumference of circle? _____

   Area of polygon? _____

   Perimeter of polygon? _____

   Area of polygon minus the circle? _____

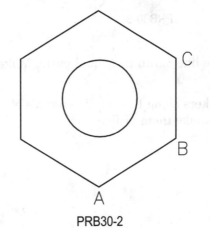

PRB30-2

353

3. What information does AutoCAD list for arc A in the figure below?

_____

_____

_____

_____

4. List the information AutoCAD provides for line B in the following figure.

_____

_____

_____

_____

5. What information does DBLIST provide for the figure below? How does this differ from the LIST information?

_____

_____

_____

_____

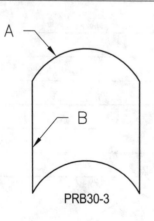

A

B

PRB30-3

6. On the above object, divide line B into five equal parts. Make the points visible.

7. On arc A above, place markers along the arc at intervals of .25 unit. If the markers are invisible, make them visible.

8. Create the floor plan shown below according to the dimensions given. Then use the appropriate AutoCAD commands to find the information requested below the floor plan. See also the optional *Applying AutoCAD Diskette.*

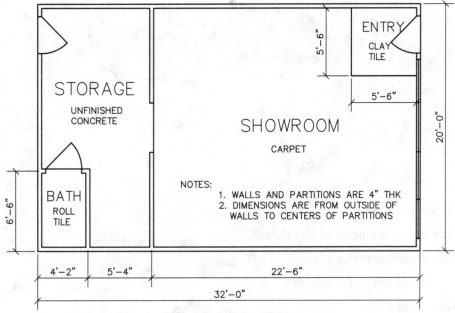

PRB30-8

Area of showroom carpet? _____

Area of entry clay tile? _____

Area of bathroom roll tile? _____

Now calculate your answers as square yards by dividing your answers above by 9 using the CAL command.

Square yards of carpet? _____

Square yards of clay tile? _____

Square yards of roll tile? _____

Calculate the distance between opposite corners of the following areas.

Showroom? _____

Entry area? _____

Bathroom? _____

9. Create the top view of a nut according to the dimensions given below. Then find the information requested below the illustration. See also the optional *Applying AutoCAD Diskette*.

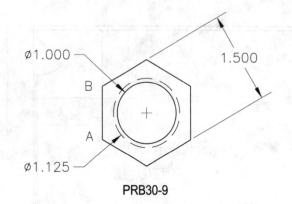

Ø1.000

B

1.500

A

Ø1.125

PRB30-9

Distance between A and B? _____

Area of the minor diameter of the thread? _____

Area of the major diameter of the thread? _____

Circumference of minor diameter? _____

Circumference of major diameter? _____

Area of top surface of the nut? _____

Area of top surface of the nut minus the minor diameter? _____

Area of top surface of the nut minus the major diameter? _____

Calculate the average of the last two answers using the CAL command.

Average: _____

Problems 8 and 9 courtesy of Mark Schwendau, Kishwaukee College

AutoCAD® AT WORK

Vashon Island Emergency Response 911

Vashon Island is a self-contained island sitting in the middle of Puget Sound, only a short ferry ride away from Seattle, Washington. The island is only about 12 miles long with about 9,000 permanent residents. In 1992, the island's street and county roads were renamed using a city street numbering method. This improvement was coupled with enhanced 911 in the hope of providing an improved response capability.

Newer enhanced 911 C.A.D. (computer-aided dispatch) systems marry three technologies: CAD drawings, GIS spatial locations, and 911 telephone systems. Until recently, only large cities could afford to implement these kinds of systems, but Vashon put together a powerful but inexpensive system using existing AutoCAD drawings, Field Notes, and Microsoft's Access.

Vashon's new system will be very simple in operation, very inexpensive, but far more powerful, using highly accurate AutoCAD base maps. This system will accompany the existing 911 system at first. As a call comes in, the dispatcher types in the phone number. Instantly, a map appears beside a form. This form shows the data about the location, name, address, access information, 911 history, and so on. Beside the form is a map, with zoom control that shows exact locations. The map and data can then be relayed to a mobile, battery-operated fax machine in any vehicle.

The company that is implementing this system for Vashon Island is Kroll Map Company. Kroll has been drafting and creating manual cartography for the Pacific Northwest region for over 85 years. This family-run company has passed down through three generations and is currently run by John Locker, who took the leap into AutoCAD with the introduction of version 10.

Kroll creates its base mapping using all available information from its map vaults. This consists of historical survey information, consultant maps, engineering construction drawings, and the weekly updated output from the County Seat. Kroll also added attributed blocks for parcel identification, tax lot acreage, government lot number, and original plat number.

The technology in this system is mostly off-the-shelf, with some database and programming front-end work. The system's utility goes far beyond 911, and property tax information, real estate listings, public utility applications, and other transportation needs. The possibilities are numerous.

Adapted from a story by David White in CADENCE, January 1995, Copyright © 1995, Miller Freeman, Inc.

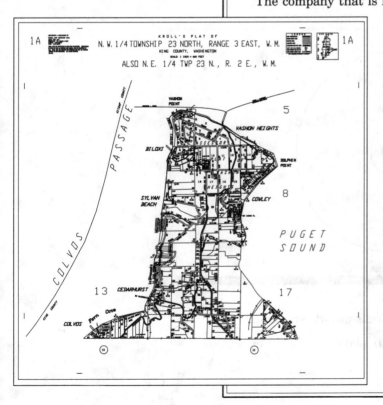

 **Unit 31** Groups

■ **OBJECTIVE:**

To create and apply groups

A *group* in AutoCAD is a named set of objects. A circle with a line through it, for instance, could be a group. Groups save time by allowing you to move, scale, erase, and perform other editing operations on several objects at one time.

 ## *GROUP Command*

The GROUP command permits you to create a named selection set of objects.

1 Start AutoCAD and open the drawing named **calc.dwg**.

2 Using **Save As...**, create a new drawing file named **groups** and erase all objects.

3 Create a layer named **Hidden** and assign the Hidden linetype and the light gray color to it. Set snap to **.25**.

4 Create the following drawing of a wheel. Place the hidden-line circle on layer Hidden and the remaining circles on layer Objects. Approximate all sizes.

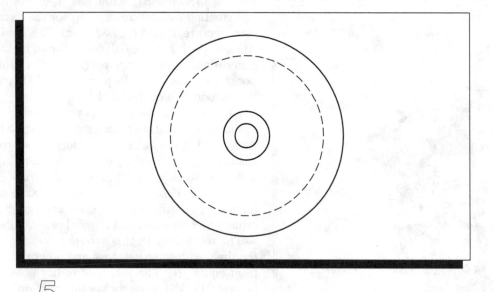

5 Pick **Object Group...** from the **Tools** pull-down menu.

This enters the GROUP command and displays the Object Grouping dialog box.

6 For Group Name, enter **WHEEL** using upper- or lowercase letters.

7 Enter **Cast aluminum wheel** for Description.

Focus your attention on the Create Group area of the dialog box. Notice that the Selectable check box is checked.

8 Pick the **New** button, select all four circles, and press **ENTER**.

When the dialog box returns, notice that WHEEL is listed under Group Name. Yes appears under Selectable because the Selectable check box was checked when you picked the New button. "Selectable" means that when you select one of the group's members (*i.e.,* one of the circles), AutoCAD selects all members in the group. There are exceptions, such as members that are on locked layers.

9 Pick the **OK** button.

10 Pick any one of the circles.

Notice that AutoCAD selected all four circles.

11 Move the group a short distance.

12 Lock layer **Hidden**.

13 Pick one of the circles again and then move the group a short distance.

The gray hidden-line circle did not move because it is on a locked layer.

14 Undo the move and unlock layer **Hidden**.

 ## Changing a Group

The Object Grouping dialog box offers several features for changing a group's definition and behavior.

1 Enter the **GROUP** command.

2 Select **WHEEL** in the Group Name list box.

3 Pick the **Highlight** button.

This highlights all members of the group. This could be particularly useful when you're editing a complex drawing containing many objects and groups.

NOTE:

> If the small Object Grouping box obscures a critical part of the drawing, you can move the box out of the way. Use the pointing device to pick the top of the box and drag it away.

4 Pick the **Continue** button.

Focus your attention on the Change Group area of the dialog box.

5 Pick the **Remove** button and pick the gray hidden-line circle.

AutoCAD removes this circle from the group.

6 Press **ENTER**.

7 Select the **Highlight** button again.

Notice that the hidden-line circle does not highlight; it is no longer part of the group.

8 Pick **Continue**.

9 Pick the **Add** button, pick the gray hidden-line circle, and press **ENTER**.

This adds the circle back to the group.

10 In the Group Name edit box, change the group name to **PULLEY**.

11 Pick the **Rename** button.

PULLEY appears in the Group Name list box near the top of the dialog box.

12 Rename the group back to **WHEEL**.

13 In the Description edit box, change the group description from Cast aluminum wheel to **Cast magnesium wheel**.

14 Pick the **Description** button (just above the OK button).

This causes the group description to change, as indicated at the bottom of the dialog box.

Remaining Options _____

The remaining buttons are Selectable, Explode, and Re-order.

1 In the Change Group area, pick the **Selectable** button.

This specifies whether a group is selectable. In the upper right corner under Selectable, notice that Yes changed to No. The Selectable button is a toggle that switches the selectable state between Yes and No.

2 Pick the **OK** button and select one of the circles in the group.

As you can see, the group is no longer selectable. Grips appear only for the specific object you select.

3 Reenter the **GROUP** command, pick **WHEEL** in the Group Name list box, and pick the **Selectable** button again.

The group is selectable once again.

4 Pick the **Re-order...** button.

This causes the Order Group dialog box to appear.

This dialog box enables you to change the numerical sequence of objects within a group. AutoCAD numbers the objects in the order in which you select them when you create the group. Reordering can be useful when you are creating tool paths for computer numerical control (CNC) machining.

5 Pick the **OK** button.

The Explode button deletes the group definition. *Do not* pick it.

6 Pick the **OK** button.

PICKSTYLE System Variable

The PICKSTYLE system variable controls the selection of groups and associative hatching. (You will learn about hatching in Unit 36.) The default value of PICKSTYLE is 3. The possible values are 0 through 3:

0	No group selection or associative hatch selection
1	Group selection
2	Associative hatch selection
3	Group selection and associative hatch selection

1 Enter **PICKSTYLE** and **0** or **2**.

2 Select one of the circles in the group.

Notice that you did not select the entire group.

3 Enter **PICKSTYLE** and **1** or **3**.

4 Select one of the circles in the group.

Group selection is enabled once again.

5 Save your work and exit AutoCAD.

Questions

1. What is a group?

_____

2. How can groups help AutoCAD users save time?

_____

_____

3. When creating a new group, what is the purpose of entering the group description?

_____

_____

4. Is it possible to add and delete objects from a group? Explain.

_____

_____

5. In the Object Grouping dialog box, is it possible to specify whether a group is selectable? Explain.

_____

_____

6. What is the purpose of the PICKSTYLE system variable?

_____

_____

■ *Challenge Your Thinking: Questions for Discussion*

1. Experiment with object selection using a group name. Notice that you can enter the name of the group at any Select objects prompt to select it for a move, copy, scale, or other operation. When might this feature be useful? Explain.

Problem

1. a.   Load the calc.dwg drawing from the previous unit. If the points are visible (if the small circles are present), set PDMODE to 1, and enter REGEN.

   b.   Create a group named WIDGET. Enter New product design for the description. The group should consist of all four objects.

   c.   Change the name of the group to WATCH and the group description to Stop watch design. Remove the two lines from the group.

   d.   Move the group to a new location, away from the two lines. Set the text height to .4 and add 12:30 inside the rectangle. Add this text to the group.

Unit 32 Building Blocks

■ OBJECTIVE:

To apply the BLOCK, INSERT, MINSERT, EXPLODE, RENAME, DDRENAME, PURGE, DDINSERT, and WBLOCK commands

If CAD systems are managed properly, their users should never have to draw the same object twice. This is a primary reason why CAD is beneficial. Success, however, depends on the techniques by which the drawings are created, stored, documented, and retrieved.

This unit focuses on commands that enable you to create, store, and reuse the symbols, drawings, and details that you need to use repeatedly.

BLOCK Command _____

The BLOCK command allows you to combine several objects into one, store it, and retrieve it at a later time. Let's work with the BLOCK command.

1 Start AutoCAD and start a new drawing from scratch.

2 Display help on the BLOCK command.

Now that you know what a block is, let's create one.

3 On layer 0, draw the following object using the **LINE** and **CIRCLE** commands. Approximate its size, but make it small.

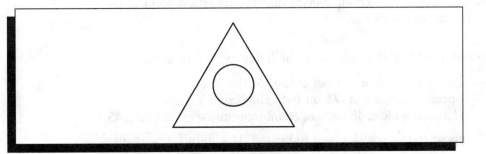

4 Save your work in a file named **blks.dwg**.

5 Enter the **BLOCK** command.

6 Create the block based on the following information.

— Name the block **MENTAL**.
— Specify the lower left corner of the object as the insertion base point.
— Select all four objects.
— Press **ENTER**.

When you press ENTER, the object disappears. Your MENTAL block is now stored in the current drawing file for subsequent insertion.

NOTE:

You can enter OOPS to make the individual objects reappear. OOPS does not affect the new block, however. The block remains stored in the drawing file.

INSERT Command

Let's insert the block.

1 Enter the **INSERT** command and enter the block name **MENTAL**.

The block appears on the screen. Notice that the insertion point of the block is attached to the cursor.

2 Insert the MENTAL block using the following information.

— Insert near the lower left corner of the screen.
— Specify the scale at **.75** on both the X and Y axes.
— Rotate the block 45 degrees counterclockwise by entering **45**.

It should appear in the position you indicated. If it didn't, try again.

3 Attempt to erase the circle from the block.

The entire object, including the circle, is treated as a single element called a block, so you cannot erase or edit any single element within it. As you can see, a block is similar to a group.

4 If you completed the ERASE command, enter **OOPS** to recover the object.

In the future, you may want to edit a block.

1. Enter **INSERT** once again.

2. This time, when typing the block name, place an asterisk (*) before it: ***MENTAL**.

3. Step through the entire INSERT command and enter whatever scale and rotation factor you wish.

4. After the block appears on the screen, try to erase the circle from the block.

Did it work? What can you conclude about the * option?

MINSERT Command

MINSERT, short for Multiple INSERT, allows you to insert a rectangular array of blocks. The MINSERT command is a combination of the INSERT and ARRAY commands. Let's apply it.

1. Enter the **MINSERT** command at the keyboard.

2. Enter **MENTAL**.

3. For the insertion point, pick an open space near the lower left area of the screen.

4. Enter **.25** for both the X and Y scale factors and **0** for the rotation angle.

5. In reply to Number of rows, enter **3**, and enter **5** in reply to Number of columns.

6. Enter **1** for the distance between rows and **1** for the distance between columns.

Fifteen MENTAL blocks appear on the screen.

Block Definition Dialog Box 2·3·99 - Skip for now.

The Block Definition dialog box offers another way of defining blocks.

1. Create an electrical symbol for a duplex outlet, as shown in the following illustration.

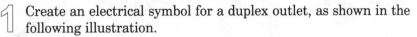

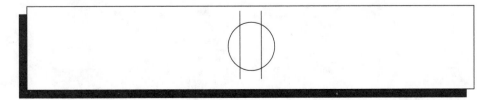

- Several WKSURFACES IN ONE DRAWING
WBLOCK for each.

2 From the Draw toolbar, pick the **Make Block** icon.

This displays the Block Definition dialog box.

3 Pick the **List Block Name...** button to see a list of block names, and then pick the **OK** button.

4 In the Block Name edit box, enter a descriptive name such as **duplex** for the symbol you created.

5 Pick the **Select Objects** button, select the objects that make up the symbol, and press **ENTER**.

The dialog box reappears.

6 Pick the **Select Point** button and pick an insertion base point for the block.

The dialog box reappears. Under the Select Point button, you can enter x, y, and z coordinates for the insertion base point instead of picking one. When the Retain Objects check box is checked, AutoCAD does not erase the objects that you selected for the block.

7 Pick the **OK** button.

8 Insert the new block.

368

Exploding Blocks _____

Often, you will insert blocks without the * option because you want to manipulate those blocks as a single element. Examples are doors and windows in a house elevation drawing or components on an electrical schematic.

1 Insert the **MENTAL** block without the * option.

Modify

2 From the Modify toolbar, pick the **Explode** icon. Pick any point on the block and press **ENTER**.

The EXPLODE command reverses the effect of the BLOCK command. The object is no longer a block entity.

3 Select the circle or one of the three sides of the triangle.

4 Press **ESC** twice.

RENAME Command _____

The RENAME command lets you rename previously created blocks.

1 Enter the **RENAME** command.

The following options appear on the screen.

_____

Block/Dimstyle/LAyer/LType/Style/Ucs/Vlew/VPort:

_____

____ NOTE: ____

As you can see by the list, RENAME can be used to rename not only blocks, but also dimension styles, layers, linetypes, text styles, user coordinate systems, views, and viewports.

2 Enter **B** for Block.

3 In reply to Old block name, enter **MENTAL**.

4 Enter **SQUARE** for the new block name.

5 Enter **INSERT** and **?** to obtain a listing of blocks.

SQUARE appears as a defined block.

Rename Dialog Box

Let's rename the block back to MENTAL.

1 Select **Rename...** from the **Format** pull-down menu.

The Rename dialog box appears, similar to the one in the following illustration.

2 Pick **Block** in the Named Objects list box and pick **SQUARE** in the Items list box.

SQUARE appears in the Old Name edit box.

3 Pick the edit box to the right of the Rename To button and type **MENTAL**.

4 Pick the **Rename To** button, as shown in the following illustration, to perform the rename.

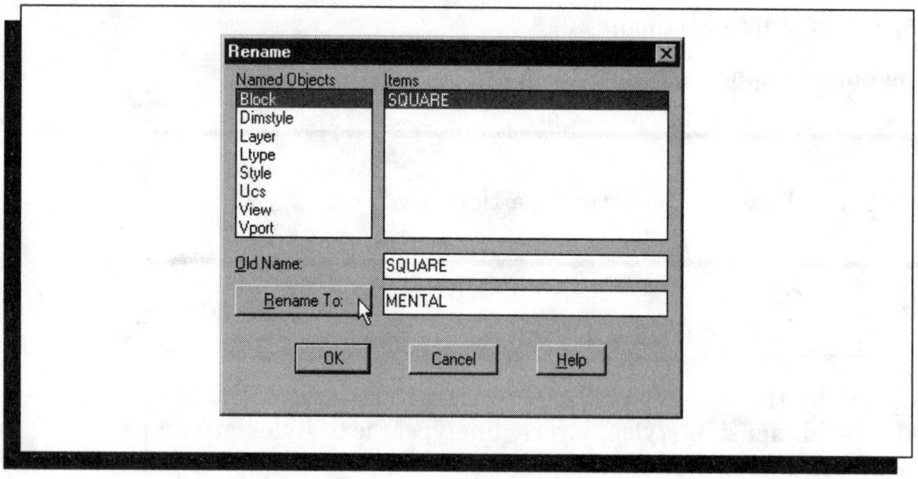

AutoCAD renames SQUARE to MENTAL in the Items list box.

5 Pick the **OK** button.

PURGE Command

The PURGE command enables you to selectively delete any unused, named objects, including blocks.

1 From the **File** pull-down menu, pick **Drawing Utilities** and **Purge**.

AutoCAD lists the object types that can be purged from the drawing.

2 Pick one type of object to purge or **All** to purge all named object types.

If you have any unused objects of the specified type, AutoCAD prompts you with the name of each object and asks whether you want to purge it.

Inserting Drawing Files

You can also insert drawing files into your existing drawing using the INSERT and DDINSERT commands.

1 Enter **INSERT**.

In addition to inserting blocks at this point, AutoCAD permits you to enter a drawing file name, if you know the name of the file and its location. File and directory names are often difficult to remember, and they're cumbersome to enter. Therefore, AutoCAD offers an alternative.

2 Press **ESC** to cancel the INSERT command.

3 From the Draw toolbar, pick the **Insert Block** icon.

Draw

This enters the DDINSERT command, which displays the Insert dialog box as shown in the following illustration.

4 Pick the **Block...** button.

This selection causes a dialog box to appear with a list of the blocks available for insertion in the current drawing. Let's not insert a block at this time.

5 Pick the **Cancel** button.

6 Pick the **File...** button and locate the **calc** drawing to insert.

7 After selecting the drawing file, pick **OK**.

Notice that the Specify Parameters on Screen check box is checked. If you were to leave it checked, AutoCAD would require you to enter the insertion point and the scale and rotation values at the Command prompt.

8 Uncheck the **Specify Parameters on Screen** check box.

9 Change the *x, y,* and *z* scale values to **.5** and the rotation value to **10**.

10 Pick the **Explode** check box and pick **OK**.

Checking Explode has the same effect as entering an asterisk (*) before the block name on the Command line.

So you see, any drawing file available to AutoCAD can be inserted into the current drawing. This enables you to combine any of your drawings and create highly sophisticated drawings in a short time.

NOTE:

The BASE command may be useful for establishing a drawing insertion base point other than 0,0 (which is the default). For instance, if the current drawing is of a part that you expect to insert into other drawings, you can specify the base point for such insertions using the BASE command.

 Undo the insertion and save your work.

WBLOCK Command

You understand that all drawing files are accessible for insertion in any drawing. But what about blocks? Is there a method of making blocks available to other drawings?

1 Enter **WBLOCK**.

The Create Drawing File dialog box appears.

2 Specify the appropriate directory, enter **ment.dwg** for a new file name, and pick **Save**.

3 Enter **MENTAL** for the block name.

Note the light or sound of the computer's disk drive as you complete the command. The computer created a new file named ment.dwg with the contents of MENTAL.

Let's review the ment.dwg file.

4 Pick the **Open** icon, pick **Yes** to save your changes, and open the new **ment.dwg** drawing file.

Now that the MENTAL block is in a drawing file format, you can insert it into any other drawing file. If you keep track of it, you'll never need to draw it again.

5 Reopen the **blks.dwg** drawing file.

6 For practice, create another block and store it as a drawing file using WBLOCK.

NOTE:

The * option, when used with the WBLOCK command, creates a new drawing file of the current drawing, similarly to the SAVE command. However, any unreferenced block definitions, layers, linetypes, text styles, and dimension styles are not written. Consider the following.

Command: WBLOCK

File Name: BK

Block Name: *

Copy and Paste

AutoCAD's copy and paste feature is an alternative to using the WBLOCK and INSERT commands. When copying one or more objects from one drawing to another, copy and paste is often a faster and simpler approach.

1 From the Standard toolbar, pick the **Copy to Clipboard** icon.

This enters the COPYCLIP command.

Standard

2 Pick a couple of objects and press **ENTER**.

This copies the objects to the Windows Clipboard.

3 Save your work.

4 Create a new drawing using the **Start from Scratch** method.

5 From the Standard toolbar, pick the **Paste from Clipboard** icon.

This enters the PASTECLIP command.

Standard

6 Move the crosshairs and notice that the objects are attached to them.

7 Pick an insertion point and accept the default values for the remaining options.

As you can see, this is a very simple way of copying objects from drawing to drawing.

Questions

1. Briefly describe the purpose of blocks.

   IT ALLOWS YOU TO STORE AND SAVE
   SEVERAL OBJECTS INTO ONE & RETIEVE IT LATER

2. Explain how the INSERT command is used.

   TYPE "INSERT" AND TYPE IN BLOCK NAME
   BRING IN LOWER PART OF, SPECIFY SCALE & ENTER
   ROTATION

3. How can you list all defined blocks contained within a drawing file?

   FROM "DRAW" TOOLBAR, PICK MAKE BLOCK ICON
   THIS DISPLAYS BLOCK DEFINITION DIALOG BLOCK

4. A block can be inserted with or without an asterisk preceding the
   name. Describe the difference between the two.

   WHEN USING ASTERISK AFTER THE BLOCK NAME
   THE LAYERS/LINE TYPES, TEXT STYLES IN NEWLY
   CREATED FILES WON'T BE STAVED, SEP. ELEMENT

5. Explain how WBLOCK works.

   BY USING THIS COMMAND IT MAKES THE
   BLOCK ACCESSABLE TO OTHER DRAWINGS

6. When would WBLOCK be useful?

   ONCE YOU DRAW IT AND KEEP TRACK
   OF THE NAMED BLOCK YOU WILL NEVER HAVE TO DRAW
   IT AGAIN

7. What is "exploding a block," and what is its purpose?

   EXPLODING A BLOCK YOU MAY USE IT AS A
   SEPARATE ELEMENT. SO YOU CAN change it REVISE.

8. How can you rename blocks?

   BY USING THE "RENAME" COMMAND

9. Explain the function of the MINSERT command.

   ALLOWS YOU TO INSERT MULTIPLE BLOCKS
   IN AT ONCE IN A RECT. ARRAY.

10. When copying an object from one drawing to another, why might
    users prefer the copy and paste method over the WBLOCK and INSERT
    approach?

    IN CASE THE BLOCK NAME ISN'T REMEMBER
    AND IT CUTS OUT THE BLOCK DIALOG BOX
    COMMANDS - ELEMINATES STEPS.

1. An electrical contractor using AutoCAD needs many electrical symbols in his drawings. He has decided to create blocks of the symbols to save time. Describe at least two ways the contractor can make the blocks easily available for all his AutoCAD drawings. Which method would you use? Why?

2. Describe the use of the PURGE command, and discuss situations in which you might need to use this command.

3. If you were to copy a block from one drawing and paste it into another, would AutoCAD recognize the pasted object as a block? Explain.

Problems

1. Begin a new drawing named livroom.dwg. Draw the furniture representations and store each as a separate block. Then draw the living room outline. Don't worry about exact sizes or locations, and omit the text. Insert each piece of furniture into the living room at the appropriate size and rotation angle. Feel free to create additional furniture and to use each piece of furniture more than once.

If you have the optional *Applying AutoCAD Diskette,* refer to livroom.dwg.

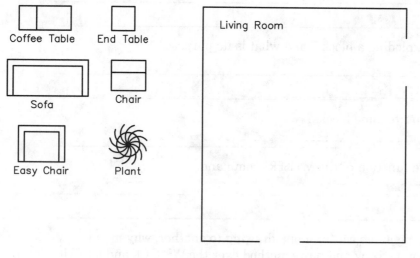

PRB32-1

2. After creating the blocks in the first problem, write two of them (of your choice) to disk using WBLOCK.

3. Copy the block of the easy chair and paste it into another drawing.

4. Explode the PLANT block and erase every fourth arc contained in it. Then store the plant again as a block.

5. Rename two of the furniture blocks.

6. Purge unused objects.

7. Using MINSERT, create a lecture room full of chairs arranged in a rectangular pattern.

8. Create a new drawing named revplate.dwg. Create the border, title block, and revisions box according to the dimensions shown below. (Do not include the dimensions.) Block the revisions box using the insertion point indicated. Then insert the block in the upper right corner of the drawing.

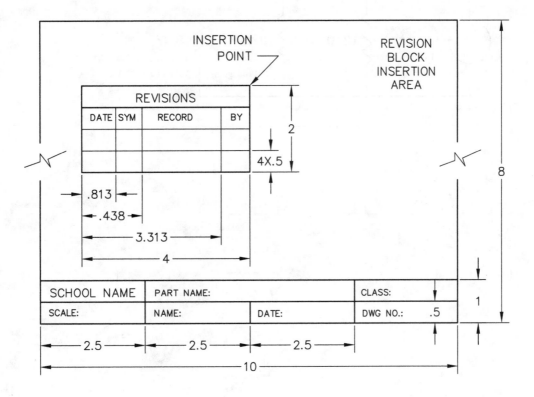

PRB32-8

9. Draw the electric circuit shown below as follows: Draw the resistor using the mesh shown; then save it as a block. Draw the circuit, inserting the blocks where appropriate. Grid and snap are handy commands for drawing the resistor and circuit. Finish the circuit by inserting small donuts at the connection points. Add the text. The letter omega (Ω), which is used to represent the resistance in ohms, can be found under the text style GREEKC (character W).

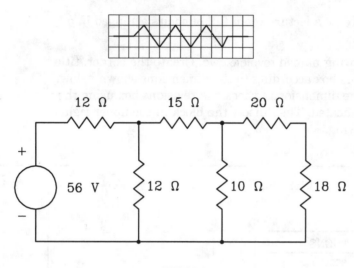

PRB32-9

Problem 8 courtesy of Mark Schwendau, Kishwaukee College
Problem 9 courtesy of Gary J. Hordemann, Gonzaga University

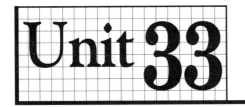

Unit 33   Symbol Library Creation

■ OBJECTIVE:

To create and use a library of symbols and details

The purpose of this unit is to create a group of symbols and details and to store them in a library. The library will then be applied to a new drawing.

The following is a collection of electrical substation schematic symbols in an AutoCAD drawing file. Each of the symbols was stored as a single block and given a block name. (In this particular case, numbers were used for block names rather than words.) The crosses, which show the blocks' insertion base points, and the numbers were drawn on a separate layer and frozen when the blocks were created. They are not part of the blocks; they are used for reference and retrieval only.

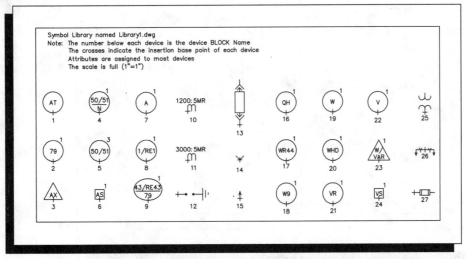

Courtesy of City of Fort Collins, Light & Power Utility

After the symbols were developed and stored in a drawing file, the file was inserted into a new drawing for creation of the electrical schematic shown on the following page.

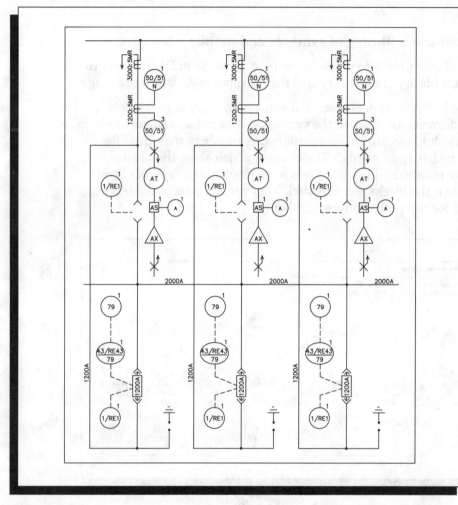

When a file contains blocks, inserting the file into a new drawing causes the blocks to be inserted into the drawing database. In this example, the blocks were then inserted into their proper locations, and lines were used to connect them. As a result, about 80 percent of the work was complete before the drawing was started. This is the primary advantage of grouping blocks in symbol libraries.

Creating a Library

Let's step through a simple version of the procedures just described.

1 Start AutoCAD and start a new drawing from scratch.

NOTE:

As described in earlier units, you should use a drawing template to save time when creating a new drawing. But for the purpose of this exercise, do not specify a template, since you have not created one specifically for this application.

2 Create the following simplified representations of tools. Set snap at **.1** and grid at **.5**. Construct each relatively small on layer 0, and omit the text.

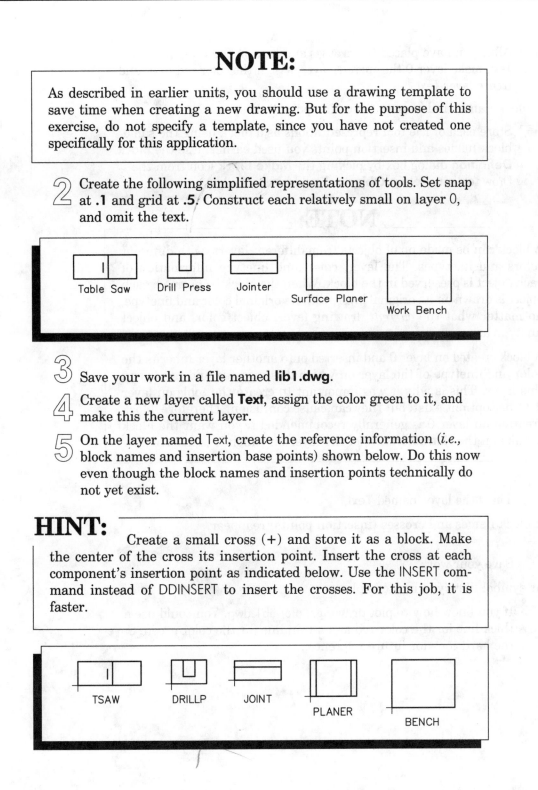

Table Saw   Drill Press   Jointer   Surface Planer   Work Bench

3 Save your work in a file named **lib1.dwg**.

4 Create a new layer called **Text**, assign the color green to it, and make this the current layer.

5 On the layer named Text, create the reference information (*i.e.,* block names and insertion base points) shown below. Do this now even though the block names and insertion points technically do not yet exist.

HINT:

Create a small cross (+) and store it as a block. Make the center of the cross its insertion point. Insert the cross at each component's insertion point as indicated below. Use the INSERT command instead of DDINSERT to insert the crosses. For this job, it is faster.

TSAW   DRILLP   JOINT   PLANER   BENCH

6 After you have placed the crosses and text on the layer named Text, make layer **0** the current layer, assign the color red to it, and freeze layer Text.

The block names and crosses (insertion points) disappear.

7 Store each of the tool representations as a block using the same block names and insertion points you used earlier. Use the **Block Definition** dialog box by picking the **Make Block** icon from the Draw toolbar. **Retain Objects** should be checked.

Draw

NOTE:

A block can be made up of objects from different layers, with different colors and linetypes. The layer, color, and linetype information of each object is preserved in the block. When the block is inserted, each object is drawn on its original layer, with its original color and linetype, no matter what the *current* drawing layer, object color, and object linetype are.

A block created on layer 0 and inserted onto another layer inherits the color and linetype of the layer on which it is inserted and resides on this layer. This is why it was important to create the tools on layer 0. Other options exist, but they can cause confusion. Therefore, block creation on layer 0 is generally recommended if you want the block to take on the characteristics of the layer on which it is inserted.

8 Thaw the layer named **Text**.

The block names and crosses (insertion points) reappear.

9 Save your work.

Your symbol library is now complete.

10 If you know how to plot drawings, plot lib1.dwg. You could use a thick line for the color red and a thin line for the color green. Save the hard copy for future reference.

Using the Library

We're going to use the new lib1.dwg symbol library to create the workshop drawing shown below.

WORKSHOP

Standard

1 Begin a new drawing using the **tmp1.dwt** template file.

2 Using **PLINE**, create the outline of the workshop as shown above. Make the starting and ending width **4″**.

Let's load and use the symbol library named lib1.dwg.

3 Save your work in a file named **workshop.dwg**.

Draw

4 Pick the **Insert Block** icon from the Draw toolbar.

5 Pick the **File...** button.

6 Find and double-click the drawing file named **lib1.dwg**.

7 In the Insert dialog box, pick **OK**, but then stop.

8 *Important:* At the Insertion point prompt, cancel (press **ESC**). Continue, and you'll see why.

⑨ Display the same dialog box (press the space bar or **ENTER**) and pick the **Block...** button to review the list of blocks.

All the blocks from lib1.dwg should be present.

So now you see why we inserted lib1.dwg and why we canceled the insertion before lib1.dwg was drawn on the screen. What we wanted from lib1.dwg were the block definitions it contained, not the graphics themselves. Now that the block definitions are present in the current drawing (workshop.dwg), we can insert each block as we wish.

⑩ Insert each of the symbols in an arrangement similar to the one in the drawing shown on the preceding page. Enter 24 for the x and y scale factors because the scale of the drawing template is $1/_2'' = 1'$, which is the same as $1'' = 24''$. Rotate each as necessary.

Because you had access to a previously created symbol library, you have just created a drawing in a fraction of the time it would otherwise have taken. Now that you know how to do it, the next time will be even faster.

HINT:

It is good practice to add to the library file continuously by storing new symbols, shapes, and details in it. Plot the library file and place it in a notebook or on the wall near the CAD system. Eventually, you will want to create new libraries for other specialized applications.

⑪ Save your work and exit AutoCAD.

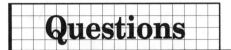

Questions

1. What is the primary purpose of creating a library of symbols and details?

   SO YOU CAN LOOK THE BLOCKS UP AND NOT REDRAW SAME OBJECT OVER & OVER

2. When you create a library, on what layer should you create and store the blocks? Why?

   ON "O" LAYERS THEN YOU CAN CHANGE THE BLOCK TO ANY LAYER YOU WANT WHEN INSERTING IT

3. In the symbol libraries discussed in this unit, block names and insertion points are stored on another layer. Why is this information important, and why should you store it on a separate layer?

   SO IT MAY TAKE ON THE CHARACTERISTICS OF THE LAYER IN WHICH IT IS INSERTED.

4. When inserting a symbol library into a drawing file, at what point do you cancel and why?

   AT THE INSERT DIALOG BOX ENTER OK THEN STOP. YOU DO THIS TO GET BLOCK DEFINITIONS IT CONTAINED NOT THE GRAPHICS THEM SELVES

Challenge Your Thinking: Questions for Discussion

1. Identify an application for creating and using a library of symbols and details. Identify the symbols the library should include. Discuss your idea with others and make changes and additions according to their suggestions.

2. Describe the major differences between groups and blocks. Why are blocks, rather than groups, used in symbol libraries? Can you think of any applications for which you could use groups as well as blocks in a symbol library? Explain.

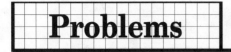

Problems

1. Based on steps described in this unit, create an entirely new symbol library specific to your area of interest. For example, if you practice architectural drawing, create a library of doors and windows. First create and/or specify a drawing template (such as the template outlined in Units 22 through 25, if it's appropriate). If you completed the first "Challenge Your Thinking" question, you may choose to create the symbol library you planned.

2. After you have completed the library symbols and details, begin a new drawing and insert the new library file as you did before. Then create a drawing using the symbols and details.

3. The logic circuit for an adder is shown below. Draw the circuit by first constructing the inverter and AND gate as blocks. Use the DONUT command for the circuit connections.

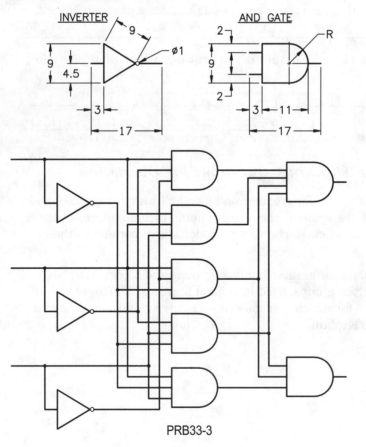

PRB33-3

Problem 3 courtesy of Gary J. Hordemann, Gonzaga University

AUTOCAD® AT WORK

Computer-Aided Drafting Plays in Peoria

For 30 years, the Central Illinois Light Company (CILCO) of Peoria, Illinois, struggled to keep its electric construction standards book up to date . . . and lost. This book, containing 550 pages of important drawings and specifications for use by design engineers and field construction people, was drafted by hand in the 1950s and has been maintained by hand ever since. It had become obsolete, but according to Senior Electric Engineer Craig D. Frommelt, "It would take so many man-hours to redo that it simply never was redone."

Frommelt and others at CILCO proposed redrawing the standards book with AutoCAD. Now CILCO is experiencing a productivity gain of about 2:1 in creating new drawings for the standards book and gains of 4:1 or 5:1 when it comes to maintaining the book.

CILCO achieved these gains by using AutoCAD to create custom libraries of symbols for commonly used parts such as transformers. Using these libraries, an engineer can rapidly create a new drawing by selecting the proper symbols and indicating where on the new drawing they should be placed.

The power to customize features of the CAD program, such as the symbol libraries, also pays off in another area—the drafting and design of electric utility substations. CAD is a great timesaver on this project because CILCO must keep many drawings for each substation: site layouts, architectural drawings of buildings, electrical schematics, and mechanical drawings used to verify clearances when replacing machines.

CILCO chose to begin the project with electrical schematics because those drawings use the same symbols repeatedly. Starting with the symbols used in the standards book, CILCO's engineers created a custom CAD library of close to 1000 schematic symbols. They then created custom menus showing those symbols, in order to make the cut-and-paste process as simple as possible. "We are able to get anywhere from a 5:1 to a 10:1 productivity advantage in making schematic diagrams," says Frommelt.

CILCO is considering other applications for its newfound CAD power. One is facilities mapping, which involves first digitizing the map of a service area (converting it into a computer-readable form) and then superimposing symbols for poles, transformers, and other equipment to make a complete geographic inventory. Using the CAD system, a person could assign attributes, such as model numbers and descriptions, to each symbol on a map. Then if it appeared that a particular model of transformer, for example, was likely to fail after ten years, an engineer could have the program identify all transformers of that model and age and highlight them in red. Those transformers could then be replaced quickly for better scheduled maintenance.

Courtesy of CILCO

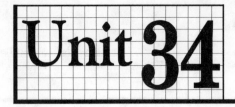 **Remarkable Attributes**

■ OBJECTIVE:

To create and display attributes with the ATTDEF, DDATTDEF, ATTDISP, ATTEDIT, and DDATTE commands and the ATTDIA system variable

The purpose of this unit is to experiment with AutoCAD's attribute feature.

Attributes are text information stored within blocks. The information describes certain characteristics of a block, such as size, material, model number, cost, etc., depending upon the nature of the block. The attribute information can be made visible, but in most cases, you do not want the information to appear on the drawing. Therefore, it usually remains invisible, even when plotting. Later, the attribute information can be extracted to form a report such as a bill of materials.

The following electrical schematic contains attribute information, even though you cannot see it. It's invisible. (The numbers you see in the components are not the attributes.)

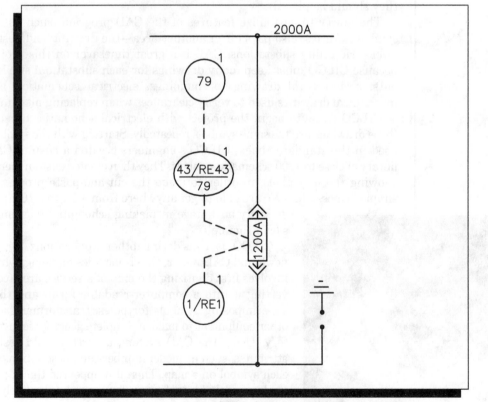

Courtesy of City of Fort Collins, Light & Power Utility

The following example shows a zoomed view of one of the components. Notice that in this example, the attribute information is displayed near the top of the component.

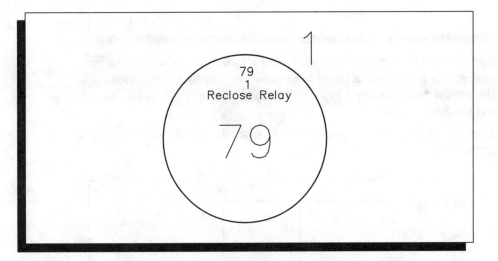

All of the attributes contained in this schematic were compiled into a file and placed into a program for report generation. The following report (bill of materials) was generated directly from the electrical schematic drawing.

DESCRIPTION	DEVICE	QUANTITY/UNIT
Recloser Cut-out Switch	43/RE43/79	1
Reclose Relay	79	1
Lightning Arrestor	--	3
Breaker Control Switch	1/RE1	1
1200 Amp Circuit Breaker	52	1

Creating Attributes _____

Attributes can be extracted, and reports produced, from any type of drawing, not just electrical drawings.

1 Start AutoCAD and open the library drawing called **lib1.dwg**. It should look somewhat like the one on the next page.

____ **NOTE:** ____

If you do not have lib1.dwg on file, create it using the steps outlined in the previous unit.

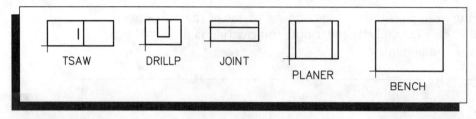

Let's assign attribute information to each of the tools so that we can later insert them and then generate a bill of materials. We'll design the attributes so that the report will contain a brief description of the component, its model, and the cost.

2 Zoom in on the first component (table saw). It should fill most of the screen.

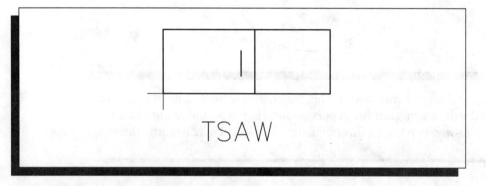

Now you're ready to assign attributes to the table saw.

3 Enter the **ATTDEF** (short for "attribute definition") command.

4 Set the attribute modes as follows, and press **ENTER** when you're finished setting them. See the following hint.

- Invisible: **Y**
- Constant: **Y**
- Verify: **Y**
- Preset: **N**

HINT:

The modes can be changed from Yes to No or from No to Yes by simply typing the first letter of each mode and pressing ENTER. For example, if you want to change Invisible to Yes, type the letter I and press ENTER.

5 Type the word **DESCRIPTION** (in upper- or lowercase letters) for the attribute tag and press **ENTER**.

6 Type **Table Saw** (exactly as you see it here) for the attribute value and press **ENTER**.

7 Place the information inside the tool, near the top. Be sure to make it small. When placing the information, use the same technique you used with the DTEXT command.

The word DESCRIPTION appears. If it extends outside the table saw representation, that's okay.

8 Press the space bar to repeat the ATTDEF command.

9 The attribute modes should remain the same, so press **ENTER**.

10 This time, enter **MODEL** for the attribute tag and **1A2B** for the attribute value.

11 Press **ENTER** in reply to the Justify/Style/<Start point> prompt.

The word MODEL now appears on the screen below DESCRIPTION.

12 Repeat Steps 9 through 12, but enter **COST** for the tag and **$625.00** for the value.

13 Save your work.

You are now finished entering the table saw attributes.

Storing Attributes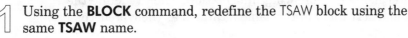

Now let's store the attributes in the block.

Draw

1 Using the **BLOCK** command, redefine the TSAW block using the same **TSAW** name.

HINT: AutoCAD displays the message Block TSAW already exists. Redefine it? <N>. Enter Y for Yes, because you want to redefine it. When selecting the block, be sure to select the attributes also, but *do not* select the cross and text reference information.

The attribute information should now be stored within the TSAW block.

 Insert the block in the same location where the table saw was before. The attribute tags should not appear.

Displaying Attributes

Let's display the attribute values using the ATTDISP (short for "attribute display") command.

1 Enter **ATTDISP** and specify **On**.

You should see the attribute values, similar to those in the drawing below.

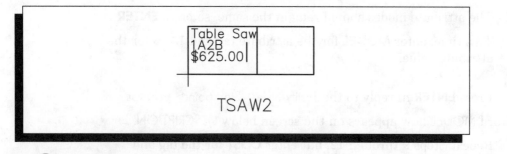

TSAW2

2 Reenter **ATTDISP** and enter **N** for Normal.

The attribute values should again be invisible.

Applying What You've Learned

Now, let's assign attributes to the rest of the power tools and components.

1 Using the **ATTDEF** command, assign attributes to the drill press and jointer using the following information. Then redefine the blocks and insert them in the same locations as their predecessors.

Description	Model	Cost PRICE
Drill Press	7C-234	$590.00
Jointer	902-42A	$750,00

HINT: To move around more easily in the drawing, display and use the Pan Realtime feature.

Standard

2 Save your work.

Attribute Definition Dialog Box _____

Let's use the Attribute Definition dialog box to assign attributes to the surface planer and work bench.

1 From the **Draw** pull-down menu, pick the **Block** and **Define Attributes...** items.

This enters the DDATTDEF command, which displays the Attribute Definition dialog box.

2 Enter **DESCRIPTION** (in upper- or lowercase letters) for the tag and **Surface Planer** for the value.

Notice that the remaining options default to the options you selected when you last used the ATTDEF command.

3 Select the **Pick Point** button, pick a start point for the attribute text, and pick the **OK** button.

DESCRIPTION appears.

4 Press the space bar to redisplay the dialog box.

5 Check the **Align below previous attribute** check box found in the lower left corner of the dialog box.

This causes AutoCAD to place the attribute tag below the previously defined attribute.

6 Insert the remaining attributes for the surface planer and the work bench. Be sure to check the **Align below previous attribute** check box for each entry.

Description	Model	Cost
Surface Planer	789453	$2070.00
Work Bench	31-1982	$825.00

7 Redefine the PLANER and BENCH blocks and insert them.

8 When you're finished assigning attributes, redefining the blocks, and inserting them, display the attribute values (using ATTDISP) to make sure they are complete.

9 Save your work.

Your symbol library, lib1.dwg, now contains attributes. When lib1.dwg is inserted into a new drawing and tools are inserted, the attributes will be contained in the blocks.

The following unit explains the processes of extracting attributes and generating a bill of materials.

Variable Attributes

Thus far, you have experienced the use of fixed attribute values. With variable attributes, you have the freedom of changing the attribute values as you insert the block. Let's step through the process.

1 Using the **tmp1.dwt** template file, begin a new drawing.

2 Zoom in on the lower $1/4$ of the display, make layer **0** the current layer, and set the snap resolution to **2"**.

3 Draw the following architectural window symbol and approximate the dimensions that are not given. Do not place dimensions on the drawing. (The symbol represents a double-hung window for use in architectural floor plans.)

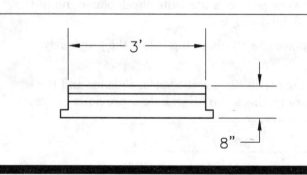

Per teacher IGNORE this

Ignore

4 Save your work in a file named **window.dwg**.

5 From the **Draw** pull-down menu, pick **Block** and **Define Attributes...**, and set the following modes.

Invisible: **Y**
Constant: **N**
Verify: **N**
Preset: **N**

Notice that, unlike before, the Constant and Verify modes are set to No (not checked). Attribute values that are not constant, called *variable* attributes, can be edited. We will practice editing variable attributes later in this unit. But for now, let's continue.

6 Type the word **TYPE** for the attribute tag.

Under Tag, notice Prompt. As you may recall, Prompt was grayed out and not available when the Constant mode was set to Yes (checked).

7 In the Prompt edit box, type **What type of window?**

You will see what this is for when we insert the block.

8 Type **Double Hung** for the attribute value.

Focus your attention on the Text Options area of the dialog box.

9 Change the text justification to **Center** and the text height to **3″**.

10 Select the **Pick Point** button.

11 Pick a center point over the top of the window. Leave space for two more attributes.

12 When the dialog box reappears, pick the **OK** button.

The word TYPE appears on the screen.

13 Press the space bar to redisplay the dialog box, and leave the attribute modes as they are.

14 Type **SIZE** for the attribute tag, **What size?** for the attribute prompt, and **3′ × 4′** for the attribute value.

15 Pick the **Align below previous attribute** check box, and pick **OK**.

The word SIZE appears on the screen below the word TYPE.

16 Repeat Steps 13, 14, and 15 using the following information.

Attribute tag: **MANUFACTURER**
Attribute prompt: **What manufacturer?**
Default attribute value: **Andersen**

17 Store the window symbol and attributes as a block. Name it **DH** (short for "double hung") and pick the lower left corner for the insertion base point.

18 Save your work.

Inserting Variable Attributes

Draw

1 Insert the block **DH**. Accept the default insertion settings.

What was different about this block insertion?

2 Press **ENTER** to use the default manufacturer, Andersen.

3 Press **ENTER** in reply to What type of window? to accept the Double Hung default value.

4 Enter **3′ × 5′** for the size.

The window appears.

5 Enter **ATTDISP** and specify **On**.

Did the correct attribute values appear?

NOTE:

You also have the option of using the attribute mode Preset. It allows you to create attributes that are variable but not requested during block insertion. In other words, when you insert a block containing a preset attribute, the attribute value is not requested but rather is set automatically to its default value. The primary purpose of the Preset mode is to limit the number of prompts to which you must respond.

Enter Attributes Dialog Box

AutoCAD offers a special dialog box for entering variable attributes.

1 Enter the **ATTDIA** system variable and enter a value of **1**.

2 Insert the block named **DH**. Accept the default insertion settings.

The Enter Attributes dialog box appears, as shown in the following illustration.

Enter Attributes

Block Name: DH

What manufacturer?   [Andersen]

What size?   3' x 4'

What type of window?   Double Hung

OK   Cancel   Previous   Next   Help

③ Change one of the attribute values or leave the values as they are.

④ Pick the **OK** button when you are finished.

⑤ Undo the insertion of the block.

Attribute Editing

One very simple way to edit attributes is to insert the block using the *
option. As you know, this inserts the object as individual pieces and not as
a block. The attributes contained in the object are also individual objects,
and any of them can be erased or modified. Using ATTDEF, new attributes
can be added, and using the BLOCK command, the object can be redefined
as a block.

ATTEDIT, short for "attribute edit," is a more powerful, but more involved,
command for editing attributes. It allows you either to edit attributes one
at a time, changing any or all of their properties, or to do a global edit on
a selected set of attributes, changing only their value strings.

You should currently have something similar to the following on the screen.

```
          Double Hung
             3'x5'
          Anderson
```

Make sure ATTDISP is On.

1 Enter the **ATTEDIT** command.

2 Accept each of the following default settings. Stop at the Select Attributes **prompt.**

Edit attributes one at a time? <Y>
Block name specification <*>:
Attribute tag specification <*>:
Attribute value specification <*>:

3 Pick the attribute value **Anderson** and press **ENTER**.

The following options appear on the screen. Also notice the × at the word Anderson.

Value/Position/Height/Angle/Style/Layer/Color/Next <N>:

4 Enter **V** for Value.

5 Enter the **Replace** option.

6 In reply to New attribute value, enter **Pella** and press **ENTER** a second time to terminate the command.

Now Pella is contained in the DH block definition in place of Andersen.

7 Practice editing other attribute values contained in the window symbol.

Edit Attributes Dialog Box _____

The DDATTE command provides dialog-oriented attribute editing.

1 Enter the **DDATTE** command.

2 Pick the block.

The dialog box pictured in the previous illustration appears. Its name, however, will be different.

3 Edit one of the attribute values and pick **OK**.

4 When you're finished, save your work and exit AutoCAD.

Questions

1. Explain the purpose of creating and storing attributes.

   ATTRIBUTES IS TEXT(INFO) STORED WITHIN BLOCKS.
   TO BE USED FOR REPORTS AS IN BILL OF MATERIALS

2. Briefly define each of the following commands.

   ATTDEF _ATTRIBUTE DEFINITION._

   DDATTDEF _EDIT ATTRIBUTE DEFINITION DIALOG BOX._

   ATTDISP _ATTRIBUTE DISPLAY._

   DDATTE _EDIT. ATTRIBUTE THROUGH DIALOG BOX._

   ATTEDIT _ATTRIBUTE EDIT._

3. What are attribute tags?

   INFO. IN BLOCKS GIVING ITEM DESCRIPTION
   DEVICE OR MODEL # QUANTY- PRICE

4. What are attribute values?

   THE PRICE OR DESCRIPTION OR MODEL
   CAN EDIT ONE OR ALL ITEMS

399

5. Explain the attribute modes Invisible, Constant, Verify, and Preset.

HINT:
Use AutoCAD's on-line help to find out more about each attribute mode.

_____

_____

_____

_____

Challenge Your Thinking: Questions for Discussion

1. Discuss the advantages and disadvantages of using variable and fixed attribute values. Why might you sometimes prefer one over the other?

2. Brainstorm a list of applications for blocks with attributes. Do not limit your thinking to the applications described in this unit. Compare your list with lists created by others in your group or class. Then make a master list that includes all the ideas.

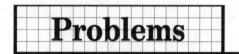

Problems

1. Load the drawing containing the furniture representations you created in Unit 32. If this file is not available, create a similar drawing. Outline a simple plan for assigning attributes to each of the components in the drawing. Create the attributes and redefine each of the blocks so the attributes are stored within the blocks.

2. Refer to schem.dwg and schem2.dwg and instructions contained on the optional *Applying AutoCAD Diskette*. Display and edit the attributes contained in the drawing files.

AUTOCAD® AT WORK

Using CAD to Design a Computer Center

The Center for Computer-Integrated Engineering and Manufacturing (CCIEM) at the University of Tennessee's College of Engineering provides students, faculty, and staff with access to several state-of-the-art computer-aided design/computer-aided engineering (CAD/CAE) systems. In addition, the center provides local industries with technical assistance in CAD/CAE, CIM (computer-integrated manufacturing), robotics, and artificial intelligence research and development.

When the process of specifying the design and layout of this multi-vendor computational facility was initiated, the proposed facility was to occupy space previously designated for classrooms. It consisted of about 2,800 sq. ft. The design had to maximize the space available and provide for future growth of the facilities. Steven R. Foster, manager and coordinator of CCIEM, recognized that the task could best be handled on a CAD system, and AutoCAD was selected for its features and moderate cost.

The design of the CCIEM facility included a number of steps, each step providing information to the next to build up the database. Information required to design the facility included specifications for hardware, a raised floor, workstations, air conditioning, lighting, power, fire protection, and security. As bid data were prepared, the requirements for the equipment were captured from the specifications and entered into the graphics database.

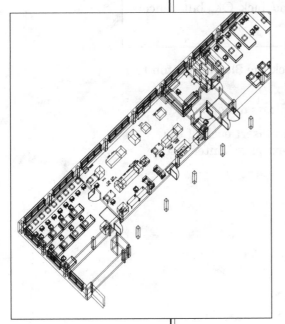

The first step in the renovation process was to make a 3D drawing of the existing layout to serve as a background drawing. Using the background drawing as a base, a composite drawing was created containing the raised floor system and the location of access doors and ramps.

Symbols were developed for each item of computer equipment to be used in the facility. Additional information for power requirements and device description was included in the attribute data assigned to each symbol. It was a simple task to insert the equipment symbols into the background drawing. Symbols were also used for the furniture, simplifying layout.

Now completed, the CCIEM represents one of the most advanced computer facilities on the University of Tennessee campus. It offers a pleasant, comfortable environment in which to conduct research and development. Additionally, as the center grows, it will be possible to utilize an up-to-date set of plans maintained in the CCIEM centralized database to change the configuration of the center.

Based on a story in CADENCE magazine, Vol. 2, No. 1.

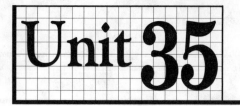

■ OBJECTIVE:

To practice report generation using the DDATTEXT command and special files that read, format, and display attributes

After finishing the attribute assignment process (Unit 34), you are ready to create a report such as a bill of materials. The first step in this process involves extracting the attribute information and storing it in a file that can be read by a computer program.

NOTE:

This unit requires the optional *Applying AutoCAD Diskette* from Glencoe/McGraw-Hill, phone 1-800-334-7344.

■ *Attribute Extraction*

1 Start AutoCAD and begin a new drawing using the **tmp1.dwt** template file.

2 Insert the latest version of the lib1.dwg library containing the attributes. (See the following hint.)

HINT:

From the Insert pull-down menu, pick Block... to issue the DDINSERT command. Enter the file name and pick OK, but cancel (press ESC) at the Insertion point prompt.

3 Reenter the **DDINSERT** command and pick the **Block...** button to list the blocks in the drawing.

Block names of the tools appear in the Defined Blocks dialog box.

4 Create the four walls of the workshop and insert each of the blocks as shown in the drawing on the next page. Enter a scale factor of **24** for each block.

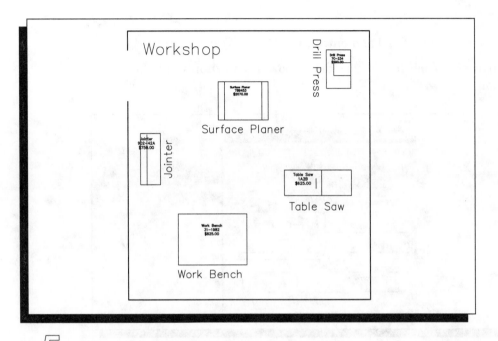

5 Save your work in a file named **extract.dwg**.

6 Enter **ATTDISP** and **On**.

Each of the tools contains attributes, as shown in the drawing.

Extracting the Attributes

The DDATTEXT command enables you to extract the attributes from the extract.dwg file. First, we'll need to copy files from the optional *Applying AutoCAD Diskette*.

1 Minimize AutoCAD.

2 From the directory created when the *Applying AutoCAD Diskette* was installed, copy **attext.txt**, **extract.dcl**, and **extract.lsp** into the folder with your name.

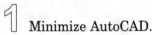

NOTE:

It is important that these three files and extract.dwg (the one currently open in AutoCAD) are in the same folder. You will learn about these three files as you use them.

③ Maximize AutoCAD and enter the **DDATTEXT** command.

This displays the Attribute Extraction dialog box, as shown in the following illustration.

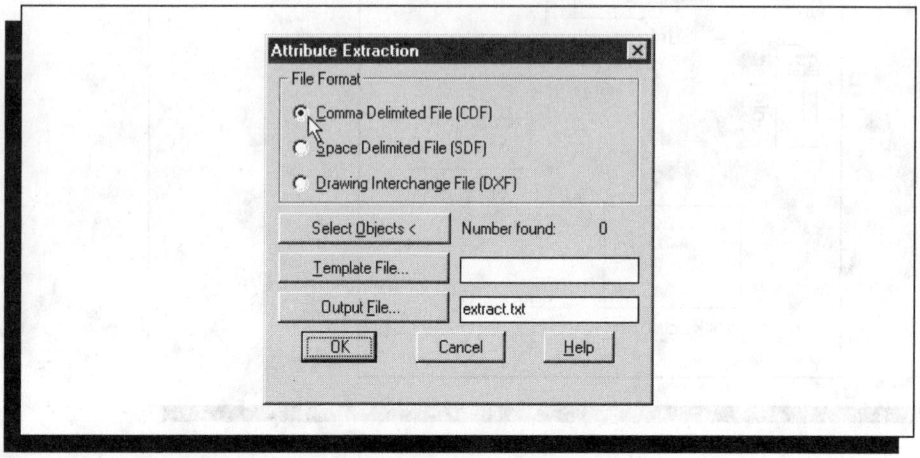

Comma delimited file (CDF) and space delimited file (SDF) are similar formats that allow you to write attributes to an ASCII text file.

The drawing interchange file (DXF) option is a variant of AutoCAD's DXF format used to import and export AutoCAD's drawing files from one CAD system to another. The variant creates a DXX file extension to distinguish it from normal DXF files.

④ Pick the **Comma Delimited File (CDF)** radio button, unless it is already selected.

⑤ Pick the **Select Objects** button, enter **All**, and press **ENTER** a second time.

The dialog box reappears.

In order to use CDF, SDF, or DXX files to display attributes in a text form that is easy to read, you must use a template extraction file. In this case, the template extraction file has been created for you.

⑥ Pick the **Template File...** button and double-click the **attext.txt** file located in the folder with your name.

Attext.txt is the template extraction file.

The file name extract.txt appears in the edit box located to the right of the Output File... button.

7 Pick the **Output File...** button, double-click the folder with your name, and pick the **Save** button.

This saves the extract.txt file in the folder with your name.

8 Pick the **OK** button.

AutoCAD creates the extract.txt file consisting of the attributes. Notice that AutoCAD displays 5 records in extract file on the Command line.

Attribute Reporting _____

You can format and display the contents of CDF files such as extract.txt using special programs such as extract.lsp.

1 Select **Load Application...** from the **Tools** pull-down menu.

This displays the Load AutoLISP, ADS, and ARX Files dialog box, as shown in the following illustration.

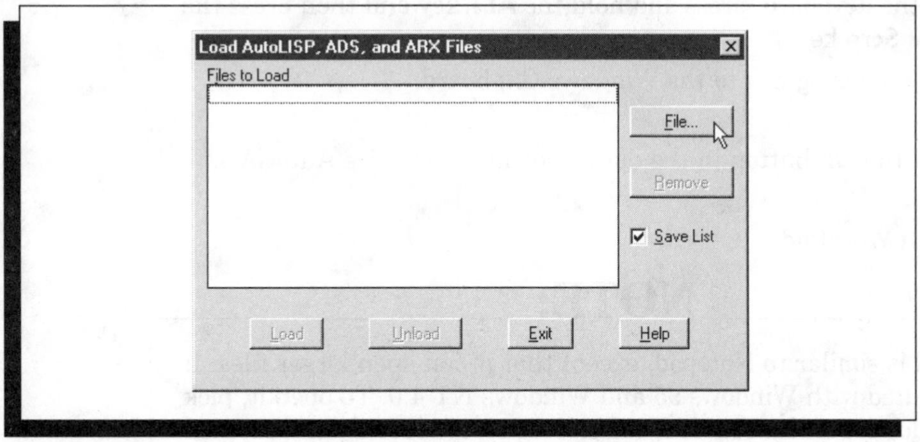

You will learn more about this dialog box in Unit 63.

2 Pick the **File...** button and locate and double-click the **extract.lsp** file.

You will learn more about LSP files and AutoLISP in Units 63-65.

3 Pick the **Load** button.

This displays the attributes in a dialog box, as shown in the illustration on the following page.

```
┌──────────────────────────────────────────────────────────────┐
│ Results of Attribute Extract [C:\Program Files\AutoCAD R14\YourNam...  [X] │
├──────────────────────────────────────────────────────────────┤
│ DESCRIPTION        MODEL        COST                          │
│ Table Saw          1A2B         $625.00                       │
│ Drill Press        7C-234       $590.00                       │
│ Surface Planer     789453       $2070.00                      │
│ Jointer            902-42A      $750.00                       │
│ Work Bench         31-1982      $825.00                       │
│                                                              │
│                         [   OK   ]                            │
└──────────────────────────────────────────────────────────────┘
```

Dialog control language (DCL) files define the appearance of dialog boxes. In this case, the extract.dcl defined this dialog box.

4 On the keyboard, press and hold the **ALT** key and then press the **Print Scrn** key.

This copies the dialog box to the Windows Clipboard.

5 Pick the **OK** button in the dialog box and minimize AutoCAD.

6 Open WordPad.

NOTE:

WordPad is similar to Notepad, except that it can open larger files. It is distributed with Windows 95 and Windows NT 4.0. To open it, pick the Windows Start button and select Programs, Accessories, and then WordPad.

7 Paste the contents of the Clipboard into it, and print the document.

NOTE:

You can use this procedure to copy and print any dialog box, as long as it is active on the screen.

8 Exit WordPad. Do not save your work.

9 Maximize AutoCAD, save your work, and exit.

Reviewing the Files

Let's review the contents of the files we used in this unit.

1 Open the folder with your name and double-click the file named **extract.txt**.

This opens the comma delimited file named extract.txt, as shown in the following illustration.

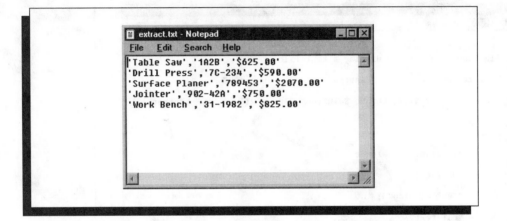

2 Review the information in the file, and then close it.

3 Double-click the file named **attext.txt**.

This displays the contents of the template extraction file as shown in the illustration on the following page. This is the file that you specified earlier in the Attribute Extraction dialog box.

4 Close the file.

5 Double-click the **extract.dcl** file.

This displays the contents of the dialog control language file, as shown in the following illustration.

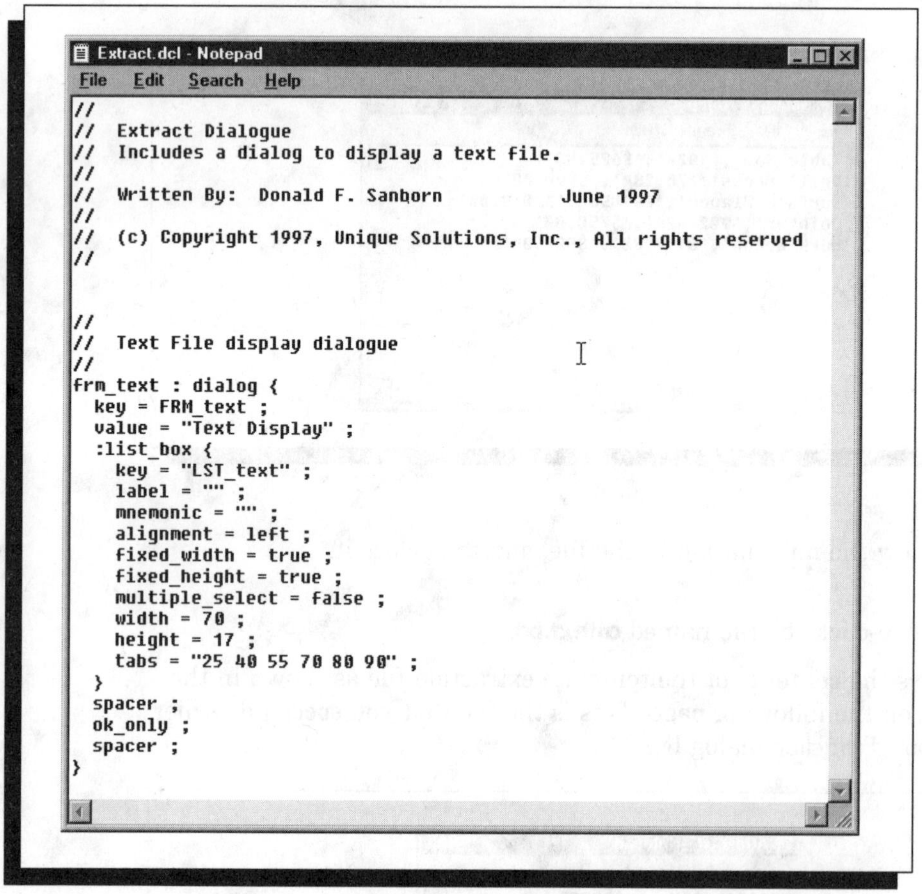

```
Extract.dcl - Notepad                                    _ □ X
File   Edit   Search   Help
//
//   Extract Dialogue
//   Includes a dialog to display a text file.
//
//   Written By:  Donald F. Sanborn           June 1997
//
//   (c) Copyright 1997, Unique Solutions, Inc., All rights reserved
//


//
//   Text File display dialogue                      I
//
frm_text : dialog {
  key = FRM_text ;
  value = "Text Display" ;
  :list_box {
    key = "LST_text" ;
    label = "" ;
    mnemonic = "" ;
    alignment = left ;
    fixed_width = true ;
    fixed_height = true ;
    multiple_select = false ;
    width = 70 ;
    height = 17 ;
    tabs = "25 40 55 70 80 90" ;
  }
  spacer ;
  ok_only ;
  spacer ;
}
```

6 Close the file.

7 Double-click the **extract.lsp** file.

This displays the contents of the AutoLISP file used to format and display the attributes. As mentioned earlier, you will learn more about LSP files later in this work-text.

8 Close the file.

408

Questions
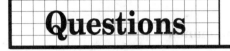

1. Describe the purpose of AutoCAD's DDATTEXT command.

   ENABLES YOU TO EXTRACT (TAKE OUT) ATTRIBUTES FROM EXTRACT.dwg file

2. What type of AutoCAD files can you read and manipulate with other computer programs?

   attext.txt, extract.dcl and extract.lsp.

3. What is the purpose of the extract.dcl file? [lsp]

   you can format & DISPLAY the CONTENTS OF CDF FILES such as extract.txt

4. What is the purpose of the extract.lsp file? [dcl]

   DCL FILES DEFINE THE APPEARANCE OF DIALOG BOXES.

■ *Challenge Your Thinking: Questions for Discussion*

1. Find out more about SDF and DXF/DXX files. What must you do first if you want to use either of these formats for attribute extract files? After you have created the files, how can you display them on the screen or print them?

Problems
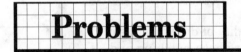

1. Load the drawing that contains furniture representations. Using the steps outlined in this unit, create a bill of materials.

2. Refer to schem.dwg and schem2.dwg and instructions on the optional *Applying AutoCAD Diskette*. Using the DDATTEXT command, CDF option, and the files you used in this unit, create bills of materials from the two drawing files.

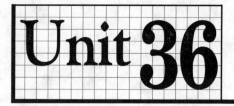

Unit 36 Dressing Your Drawings

■ **OBJECTIVE:**

To apply the HATCH, BHATCH, and SKETCH commands

This unit covers the application of hatching using the many patterns made available by AutoCAD. Sketching is also practiced using the SKETCH command.

Some drawings make use of hatching and sketching to communicate accurately and precisely the intent of the design, as illustrated in the drawing below.

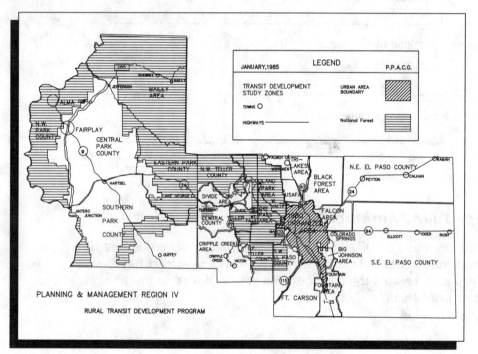

Courtesy of David Salamon, Pikes Peak Area Council of Governments

Both hatching and sketching can enhance the quality of drawings greatly, but both can consume lots of disk space. Therefore both hatching and sketching should be used only when necessary.

■ *HATCH Command* _____

Let's see what AutoCAD's HATCH command can do.

1 Start AutoCAD and begin a new drawing from scratch.

2 Create new layers named **Objects** and **Hatch**. Assign color red to Objects and green to Hatch. Make **Objects** the current layer.

3 Create the drawing shown below on layer Objects. Use the dimensions to create it accurately, but do not include the dimensions in the drawing. Include the text (Center Support) on layer Hatch.

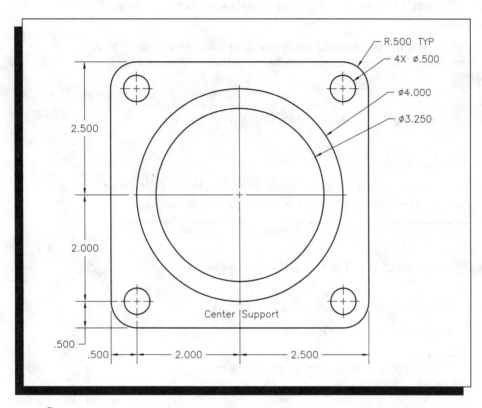

4 Make **Hatch** the current layer.

5 Save your work in a file named **hatch.dwg**.

6 Enter the **HATCH** command.

The following appears at the bottom of the screen.

Enter pattern name or [?/Solid/User defined] <ANSI31>:

---

7 Enter **?** to create a listing of all hatch patterns. Press **ENTER** a second time.

AutoCAD presents the list one screen at a time.

⑧ Continue pressing **ENTER** as prompted by AutoCAD to see the entire list.

As you can see, AutoCAD provides many hatch patterns. Let's use one of them in our drawing.

⑨ Enter the **HATCH** command again, and this time accept the hatch pattern ANSI31 (the standard crosshatch pattern) by pressing **ENTER**.

⑩ Specify a scale of **1**. (This should be the default value.)

NOTE:

The Scale for pattern should be set at the reciprocal of the plot scale so that the hatch pattern size corresponds to the drawing scale.

⑪ Specify an angle of **0**. (This should be the default value.)

⑫ Select the entire drawing and press **ENTER**.

Does your drawing look like the one below? If not, try again.

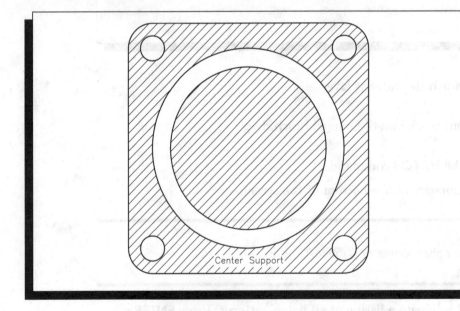

Center Support

Note the areas that received hatching.

13 Erase the hatching by picking any one of the hatch lines and pressing **ENTER**.

HINT:

The entire hatch pattern is treated as a single object. If you want the freedom to edit small pieces from the hatch pattern, precede the hatch pattern name with an asterisk (*). This is similar to inserting blocks with an asterisk (see Unit 32).

14 Hatch the outermost area only using the **O** (Outer) option. (When you enter the name of a pattern, follow it with a comma and O, as in EARTH,O.)

15 Undo the last operation and then use the **I** (Ignore) option (as in HONEY,I) to ignore the internal structures.

The hatch pattern covers the internal areas.

16 Erase the hatching.

BHATCH Command

BHATCH, an expanded version of the HATCH command, takes advantage of dialog boxes and creates associative hatches.

Draw

1 From the docked Draw toolbar, pick the Hatch icon.

This enters the BHATCH command, which displays the Boundary Hatch dialog box, as shown in the following illustration.

2 Pick the **Advanced...** button.

The Advanced Options dialog box appears, as shown in the following illustration.

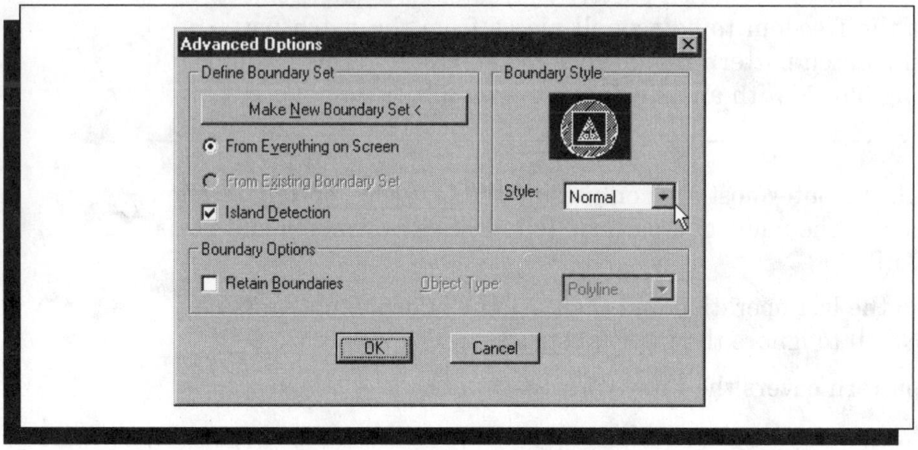

3 Pick the down arrow (located to the right of Style) and select **Normal** from the list.

Normal is equivalent to omitting Outer and Ignore with the HATCH command.

4 Pick the **OK** button.

Notice the Exploded check box. Checking it is equivalent to preceding the hatch pattern name with an asterisk (*) when using the HATCH command.

5 Pick the **Pattern...** button located in the Pattern Type area.

A list of hatch patterns appears, along with a sample of each.

6 Pick several of them and view the sample provided.

7 Pick **SOLID** and pick **OK**.

SOLID appears as the current pattern type.

8 Pick the **Select Objects** button, select the entire drawing, and press **ENTER**.

The Boundary Hatch dialog box reappears.

9 Pick the **Preview Hatch** button to preview the hatching.

The hatched areas should be the same as those shown on page 412.

10 Pick the **Continue** button.

11 Pick the **Apply** button to create the associative hatch block.

Changing an Associative Hatch

By default, hatches created with the BHATCH command are associative. This means that the hatched areas update as you change their boundaries.

1 Using the **STRETCH** command, stretch the top part of the drawing upward **.5** unit. Use the Crossing option to place a window around the top one-third of the drawing, and turn on ortho and snap.

Modify

AutoCAD updates the hatch pattern according to the new hatch boundary.

2 Make layer **Objects** the current layer and freeze layer Hatch.

3 From the **View** pull-down menu, select **Toolbars...**, open the **Modify II** toolbar, and pick the **Close** button.

Modify II

4 Pick the **Draworder** icon from the Modify II toolbar, select the entire drawing, and press **ENTER**.

5 Enter **F** for Front.

The red object lines now appear in front of other objects, permitting us to see and select them when the hatch pattern is present.

6 Thaw layer **Hatch** and make it the current layer.

Now you can select the object lines with the solid hatch present, whereas before, you could not.

7 Using the grips, change the size of one of the four small holes.

Once again, AutoCAD updates the hatch pattern according to the new hatch boundary.

8 Undo your last operation.

HATCHEDIT Command _____

This command permits you to change an associative hatch.

1. From the Modify II toolbar, pick the **Edit Hatch** icon.

This enters the HATCHEDIT command.

2. Pick the hatch.

This causes the Hatchedit dialog box to display. As you can see, it is very similar to the Boundary Hatch dialog box.

3. Make the following changes.

    Pattern: **ESCHER**
    Scale: **1.3**
    Angle: **30**

4. Pick the **Apply** button.

AutoCAD applies the changes to the drawing.

5. Experiment further with creating and changing associative hatch objects on your own.

6. Using the **Edit Hatch** icon, change the hatch pattern to **ANSI31**, the scale to **1**, the angle to **0**, and pick **Apply**.

7. Close the Modify II toolbar and save your work.

SKETCH Command _____

Now let's try some freehand sketching.

1. Using **Save As...**, create a new drawing named **sketch.dwg**, and erase the entire drawing.

2. Enter the **SKETCH** command.

3. Specify **.1** unit for the Record increment. (This should be the default value.)

The following options appear at the bottom of the screen.

Sketch. Pen eXit Quit Record Erase Connect .

The following is a brief description of each of the SKETCH options.

Pen	Raise/lower pen (or toggle with pick button)
eXit	Record all temporary lines and exit
Quit	Discard all temporary lines and exit
Record	Record all temporary lines
Erase	Selectively erase temporary lines
Connect	Connect to a line endpoint
. (period)	Line to point

4 To begin sketching, pick a point where you'd like the sketch to begin. The pick specifies (toggles) pen down.

5 Move the pointing device to sketch a short line.

6 Pick a second point (to toggle the pen up), and enter **X** to exit.

7 Move to a clear location on the screen and sketch the lake shown on the following page. Set the **SKETCH Record** increment at **.02**.

NOTE:

If you make a mistake or need to back up, toggle pen up and enter E for Erase. Then reverse the direction of the crosshairs until you have erased that which needed to be removed; press the pick button to terminate the Erase mode; press it again to toggle pen down.

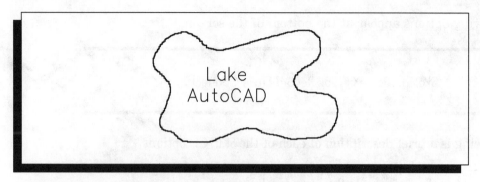

It's okay if your sketch doesn't look exactly like the one above.

⑧ When you're finished sketching the lake, type **R** for Record or **X** for eXit.

NOTE:

If you select Record, you have the option of sketching additional lines. When you're finished recording sketch lines, press ENTER.

⑨ Practice sketching by using the remaining SKETCH options. Draw anything you'd like.

⑩ When you're finished, save and exit.

Questions

1. Explain why hatch patterns are useful.

TO ENHANCE DRAWINGS, TO FILL IN (SHADE) DETAILED AREAS FOR DEFINING

2. How is the scale of a hatch pattern determined?

PREDEFINED ON DEFAULT. BUT CAN BE CHANGED IN DIALOG BOX OF BOUNDARY HATCH

3. Briefly describe the following HATCH style options.

Outer *TO HATCH THE OUTER MOST AREAS-PATTERN w/ 90*

Ignore *TO UNDO THE LAST OPERATION THEN USE 'I' FOR IGNORE TO IGNORE THE INTERNAL STRUCTURES*

4. Briefly describe the purpose of each of the following SKETCH options.

Pen *RAISE/LOWER PEN (OR toggle w/ PICK BUTTON*

eXit *RECORD ALL TEMP LINES & EXIT.*

Quit *DISCARD ALL TEMP LINES and exit.*

Record *RECORD ALL TEMP LINES.*

Erase *SELECTIVELY ERASE TEMP. LINES.*

Connect *CONNECT TO A LINE ENDPOINT.*

. (period) *LINE TO POINT.*

5. SKETCH requires a Record increment. What does it determine?

TO KEEP AND SAVE A LINE DRAWN.

HINT:
Specify a coarse increment such as .5 or 1 and notice the appearance of the sketch lines.

■ ***Challenge Your Thinking: Questions for Discussion***

1. Find out what the Inherit Properties button on the Boundary Hatch dialog box does. Explain how it can save you time when you create a complex drawing that has several hatched areas.

Problems

1-4. Construct each of the following drawings. Use the HATCH, BHATCH, and SKETCH commands where appropriate, and don't worry about specific sizes. The hatch patterns to be used are indicated below each of the drawings.

In problems 2 and 3, use the SKETCH and LINE commands to define temporary boundaries for the hatch patterns. Place the boundaries on a separate layer and freeze that layer after you are finished hatching.

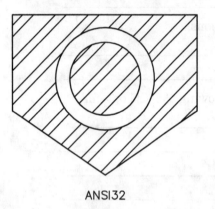

ANSI32

PRB36-1

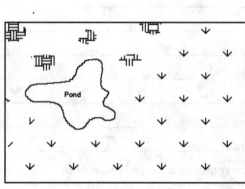

EARTH AND GRASS

PRB36-2

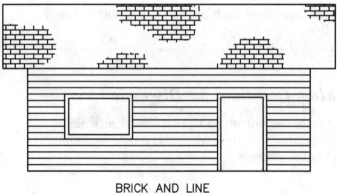

BRICK AND LINE

PRB36-3

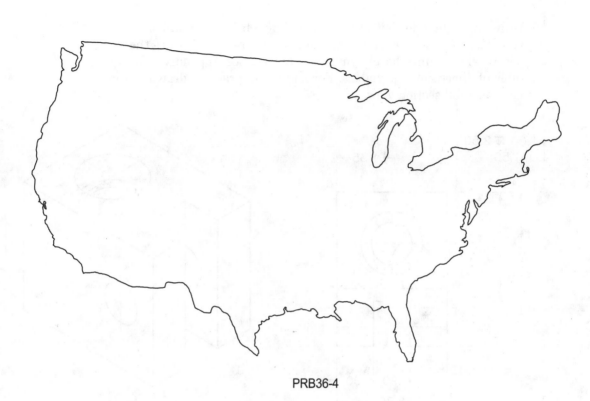

PRB36-4

5. Use the SKETCH command to draw a logo like the one shown below. Before sketching the lines, be sure to set SKPOLY to 1. SKPOLY is a system variable that controls whether the SKETCH command creates lines (0) or polylines (1). Setting SKPOLY to 1 allows you to smooth the sketched curves with the PEDIT command. Use the HATCH or BHATCH Solid pattern to create the filled areas.

PRB36-5

6. Draw the right side view of the slider shown below as a full section view. Assume the material to be aluminum. Create and store the "deep" and "counterbore" symbols as blocks. See Appendix E for a table of dimensioning symbols. For additional practice, draw the top view as a full section view.

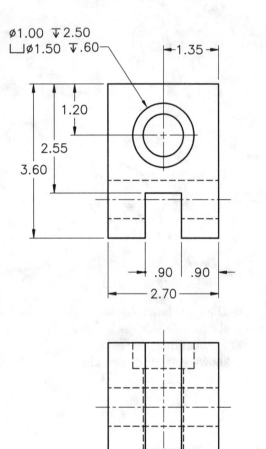

ø1.00 ⊽2.50
⊔ø1.50 ⊽.60

1.35

1.20

2.55

3.60

.90 .90

2.70

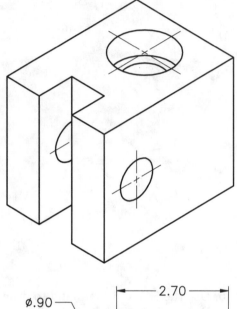

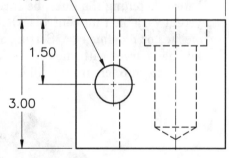

ø.90

1.50

3.00

2.70

PRB36-6

AUTOCAD® AT WORK

A Technical Illustrator's Story

Dick Clark is an Engineering Specialist at NEC Technologies, Inc. He creates illustrations for use in maintenance/repair manuals and user manuals. While he started his career using pencils, templates, and Leroy pens, today his main drawing tool is AutoCAD.

Dick works closely with the hardware technical writers during the writing of maintenance guides for both field and depot level. His main task is creating the Illustrated Parts Breakdown (IPB) for these guides. Since usable master drawings are seldom available, this process often means dismantling a complex piece of computer equipment or peripheral hardware down to the component and base assembly level to obtain the parts to be drawn and also to understand the assembly/reassembly process that is to be shown. Complex electromechanical assemblies and subassemblies are reduced to the Smallest Replaceable Unit (SRU) level and technically illustrated for the generation of the IPB. Each object is measured prior to drawing, and the final drawing is appropriately scaled to match those dimensions.

To create advertisements, data sheets, and brochures, Dick works with the marketing staff to provide technically correct drawings.

Dick does most of his drawing in isometric and 3D views on AutoCAD. For prints, he first plots the drawing, then has the plot photographically reduced to an $8^1/_2'' \times 11''$ size. When incorporating these drawings into manuals, the photo-reduced plot is digitized back into an Interleaf publishing system, using a scanner. This technique provides a very high-resolution image in that proper photo reduction improves apparent (but not actual) resolution, allowing the scanner to digitize the image to its maximum potential. Interleaf then incorporates the image with documentation to produce brochures or booklets.

The technical illustrator can wear many hats, from company space planner to advertisement production. One of Dick's special projects is the design of NEC's show booth. In addition to playing with booth layout, he provides a detailed presentation image of it. "My day is not always filled up entirely with production drawings," says Dick, "but if you're doing something new and challenging, and you love drawing, it's a joy. 'Can it be done?' is my favorite professional question."

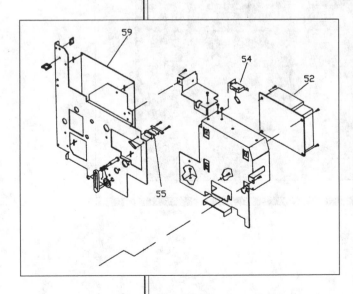

Based on a story in *CADENCE* magazine, Vol. 3, No. 8.

 **From Display to Paper**

■ OBJECTIVE:

To practice plotting and printing AutoCAD drawings using the PLOT and PRINT commands

This unit steps you through the plotting and printing process. Plotting typically refers to the use of a plotter, such as a pen or electrostatic plotter. AutoCAD printing refers to the same procedure, even though the output device may be a laser, ink jet, or thermal printer. Because AutoCAD treats plotting and printing similarly, the two terms are often used interchangeably in the software.

Plot Previewing

AutoCAD allows you to preview a plot.

1 Start AutoCAD and open the drawing named **hatch.dwg**.

2 From the Standard toolbar, pick the **Print** icon, or enter the **PLOT** or **PRINT** command at the keyboard.

Standard

The Print/Plot Configuration dialog box appears.

In the illustration, notice that DMP-61 appears in the upper left corner of the dialog box. This means a device named "DMP-61" was configured previously. In this case, a Houston Instrument DMP-61 pen plotter was chosen. The device you see in this area will probably be different. AutoCAD stores plotter settings for each configured printer, so other settings may also differ, depending on how they were last set.

NOTE:

If no printer or plotter names (such as "DMP-61") appear in the upper left corner, you must configure one before you proceed with the following steps. If Default System Printer appears, AutoCAD is configured to print using this device. Refer to the following unit for information on how to configure plotter devices.

Note the Plot Preview area located in the lower right portion of the dialog box.

3 If the Partial radio button is not selected, as shown in the previous illustration, pick it.

4 Pick the **Preview...** button.

This feature gives you a quick preview of the effective plotting area on the sheet, as shown in the following illustration.

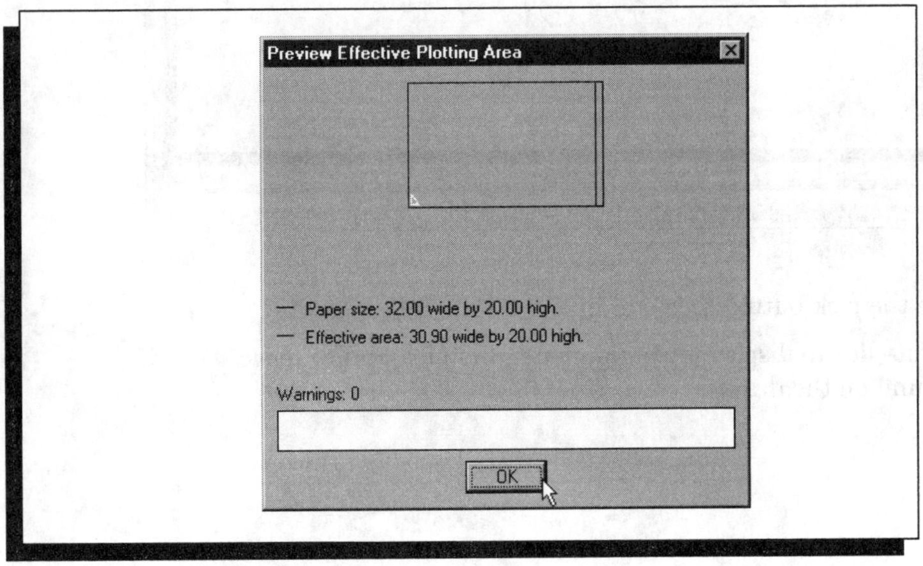

The effective plotting area, represented by the smaller rectangle (in blue) is the area the configured plotter will plot. The larger rectangle (in red) represents the sheet.

5 Pick the **OK** button.

6 Pick the **Full** radio button and pick the **Preview...** button again.

AutoCAD displays the drawing as it will appear on the sheet (shown in the following illustration).

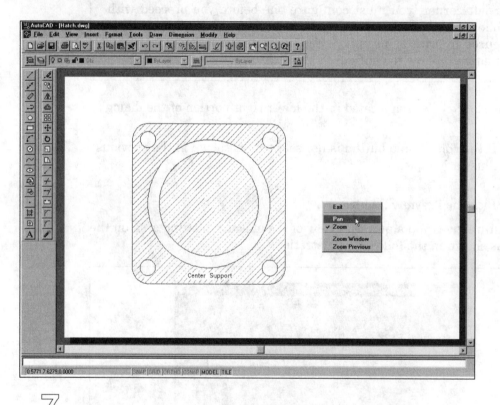

7 Use the pick button to zoom up and down on the drawing.

8 Right-click to display a pop-up menu, and pick **Pan** to move around on the drawing.

426

This enables you to examine the drawing closely before you plot it. Large drawings can take considerable time to plot, especially on pen plotters, so spotting errors before you plot can save time as well as plotting supplies.

⑨ Right-click and pick **Exit** to make the dialog box reappear.

NOTE:

> If your computer is connected to the configured plotter or printer and it is turned on and ready, continue with Step 10. If it is not, prepare the device. If you do not have access to an output device, pick the Cancel button and skip to the next section, titled "Tailoring the Plot."

⑩ Pick **OK**.

When AutoCAD has finished processing the drawing data, it displays the message Plot complete. Since we did not make changes to the plot settings before plotting, the plot may be incorrect.

Tailoring the Plot

The following steps consider sheet size, drawing scale, and other plotter settings and parameters.

① Open the drawing named **dimen2.dwg** (created in Unit 25) and review the drawing limits and layers stored in this drawing.

As you may recall, tmp1.dwt was used to create this drawing, and you considered the following criteria when you created it.

Scale: $1/_2'' = 1'$
Sheet size: $17'' \times 11''$
Effective plotting area: $15'' \times 10''$
Drawing area: $30' \times 20'$

Standard

② Pick the **Print** icon.

③ In the upper right corner of the dialog box, pick the **Inches** radio button if it has not already been picked.

Specifying Sheet Size

① Pick the **Size** button to display the Paper Size dialog box. If the Size button is not available, skip to the following section titled "Rotation and Origin."

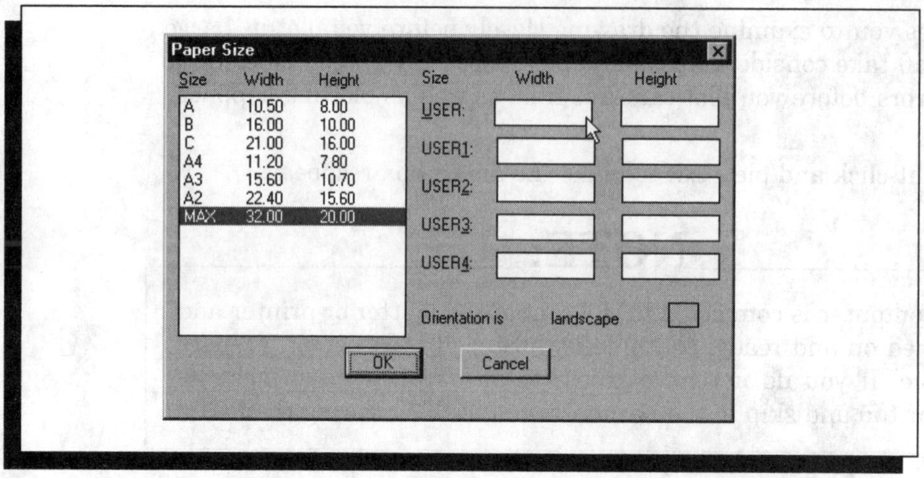

2 Review the entire list of sizes available for the configured plotter.

AutoCAD may not offer a "17 width × 11 height" B-size sheet for the configured plotter. While a "16 × 10" may be available and would be acceptable for this exercise, let's define a sheet size.

3 Enter **17** in the edit box provided in the Width column to the right of USER.

AutoCAD knows that you mean inches because the Inches radio button was selected.

4 Enter **11** in the Height column.

This new user-defined size appears in the list box, allowing you to select it in the future.

5 Pick the **OK** button.

USER appears to the right of the Size... button.

Rotation and Origin

1 Pick the **Rotation and Origin...** button.

This permits you to rotate the drawing and specify the plot origin.

2 If the **0** radio button is not selected, pick it.

Both X and Y origins should be set at 0.00. If not, . . .

3 . . . change the X and Y origins to **0.00**.

4 Pick the **OK** button.

Additional Parameters

1 Select the **Limits** radio button located in the Additional Parameters area.

We selected Limits because we want to plot the entire drawing area as defined by the 30′ × 20′ drawing limits. You may choose to select one of the other options in the future. If so, here's what they mean.

Display	Plots the current view
Extents	Similar to ZOOM Extents; plots the portion of the drawing that contains objects
Limits	Plots the entire drawing area
View	Plots a saved view
Window	Plots a window whose corners you specify; use the Window... button to specify the window to plot
Hide Lines	Plots 3D objects with hidden lines removed
Adjust Area Fill	Specifies pen width; uses width information when plotting solid-filled objects
Plot To File	Sends plot output to a file rather than to a device

NOTE:

In the future, you may choose to create a plot (PLT) file. Many third-party software products accept this file type.

429

If Hide Lines and Plot To File are checked, uncheck them.

Drawing Scale

Focus your attention on the Scale, Rotation, and Origin area of the dialog box. Notice the values contained in the edit boxes under Plotted Inches = Drawing Units. These values represent the drawing scale. When Scaled to Fit is checked, these values reflect the actual scale used to fit the drawing on the sheet.

1 If Scaled to Fit is checked, uncheck it.

2 Enter $1/2$ or **.5** in the first edit box and **1'** or **12** in the second edit box.

3 Pick the **Rotation and Origin...** button, pick **90**, and pick **OK**.

Plotting the Drawing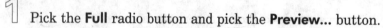

Let's preview and plot the drawing.

1 Pick the **Full** radio button and pick the **Preview...** button.

2 Right-click and pick **Exit**.

3 If your output device accepts B-size (17″ × 11″) sheets, prepare the device for plotting.

NOTE:

If your device does not accept B-size (17″ × 11″) media, pick the Size... button if it is available, select or define a size the device will accept, and pick OK. Preview the plot to make certain it fits properly on the sheet.

4 Check the connection between the computer and output device and pick the **OK** button to initiate plotting.

After plotting is complete, examine the output carefully. The dimensions on the drawing should measure correctly using a $1/2″ = 1'$ scale. For instance, the 6′ dimension should measure 3″. Text and dimension sizes, such as dimension text, arrows, and center marks, should measure $1/8″$ on the sheet.

NOTE:

Pen assignments, optimization, and device and default selection are covered in the following unit.

5 Exit AutoCAD. Do not save any changes.

Questions

1. Explain the difference between the Partial and Full plot preview options.

   PARTIAL = QUICK PREVIEW AREA LOCATED ON THE RESPECTIVE SHEET.

   FULL = DISPLAY DRAWING AS IT WILL APPEAR ON THE SHEET WHEN PRINTED OR PLOTTED

2. Is it possible to specify a unique sheet size? Explain.

   YES, PICK THE SIZE BUTTON IN THE PLOT DIALOG BOX. SELECT OR DEFINE A SIZE THE DEVICE WILL EXCEPT.

3. Briefly describe each of the following plot options.

   Display *PLOTS THE CURRENT VIEW*

   Extents *PLOTS PORTION OF DRAW THAT CONTAINS OBJECT.*

   Limits *PLOTS THE ENTIRE DRAWING AREA.*

   View *PLOTS A SAVED VIEW*

   Window *PLOTS WINDOWS WHOSE CORNERS YOU SPECIFY.*

   Hide Lines *PLOTS 3D OBJECTS W/ HIDDEN LINES REMOVED*

   Adjust Area Fill *SPECIFIES PEN WIDTH when plotting SOLID FILL OBJECTS*

   Plot To File *SENDS PLOTS to a file rather than a DEVICE*

4. The drawing plot scale for a particular drawing is 1 = 4″. What does the 1 represent and what does the 4″ represent?

   ONE UNIT EQUAL 4″ OR one segment EQUALS 4″ - When the segment is measured on the plotted drawing it should measure 4″.

431

■ Challenge Your Thinking: Questions for Discussion

1. Explain why it is important that you draw using colors if you intend to plot a drawing. What should govern which colors you choose?

2. Determine how you would add a new printer or plotter device. Hint: Select Preferences... from the Tools pull-down menu.

3. Explore ways of sending PLT files to a printer or plotter.

Problems

In problems 1 and 2, prepare to plot a drawing using the information provided. Choose any drawing to plot.

1. Unit of measure: Inches
   Sheet size: 24″ × 18″
   Rotate plot 90 degrees
   Plot origin: 0,0 on both X and Y
   Plot the limits (drawing area)
   Do not hide lines
   Adjust area fill
   Do not plot to file
   Scale: 1″ = 10″
   Perform partial and full previews

2. Unit of measure: Millimeters
   Sheet size: 285 × 198
   Do not rotate the plot
   Plot origin: 0,0 on both X and Y
   Plot the display
   Hide lines
   Do not adjust area fill
   Plot to file (using the drawing file name)
   Scale to fit
   Perform a full preview

3. Choose and plot a drawing that you created in an earlier unit. Consider the scale and drawing area so that dimensions measure correctly on the plotted sheet. Text, linetypes, and dimension sizes (such as text and arrows) should also measure correctly on the sheet. For example, $1/8$″ tall text should measure $1/8$″ in height. Ignore colors and pen assignments. They are covered in the following unit.

Unit 38 Advanced Plotting

■ OBJECTIVE:

To configure multiple output devices, adjust pen parameters, and store plot settings

Whether you are using a pen plotter or a raster output device, you will need to change the pen settings occasionally. (Laser, ink jet, thermal, and electrostatic printers are examples of raster devices.) In the future, you may want to experiment with AutoCAD's pen optimization feature. When set properly, it can reduce plot time.

As you make changes to these plot settings, you may want to store them in a file so that you don't have to enter them over and over again. AutoCAD's support of multiple output devices allows you to configure more than one device.

Configuring Plotter Devices _____

Multiple device configuration conveniently permits you to select one of them from the Device and Default Selection dialog box.

1 Start AutoCAD and open the drawing named **dimen2.dwg**.

2 Enter the **CONFIG** command or select **Preferences...** from the **Tools** pull-down menu.

AutoCAD displays the Preferences dialog box, as shown in the following illustration.

Preferences	? X
Files | Performance | Compatibility | General | Display | Pointer | Printer | Profiles	
Current printer: 600 dpi	
Default System Printer HPGL ADI 4.2 600 dpi	Set Current Modify... New... Open... Save As...
PostScript device ADI 4.3 - by Autodesk, Inc Model: 600 dpi Port: LPT1 Version: 14.0-1	Remove
OK Cancel Apply Help	

Browse the list of buttons.

③ Pick the **New...** button.

This produces a list of plotter and printer device options.

④ Select **Houston Instrument ADI 4.2 - by Autodesk, Inc**.

The selection causes the AutoCAD Text Window to appear.

⑤ Enter **8** for DMP-61MP.

⑥ Step through the remaining options by accepting the defaults.

⑦ Pick the **OK** button to close the Preferences dialog box.

⑧ Pick the **Print** icon.

⑨ Pick the **Device and Default Selection...** button located in the upper left area of the dialog box.

Standard

The DMP-61MP device you added and named should be included in the list of devices. The actual devices listed will probably be different from those shown here.

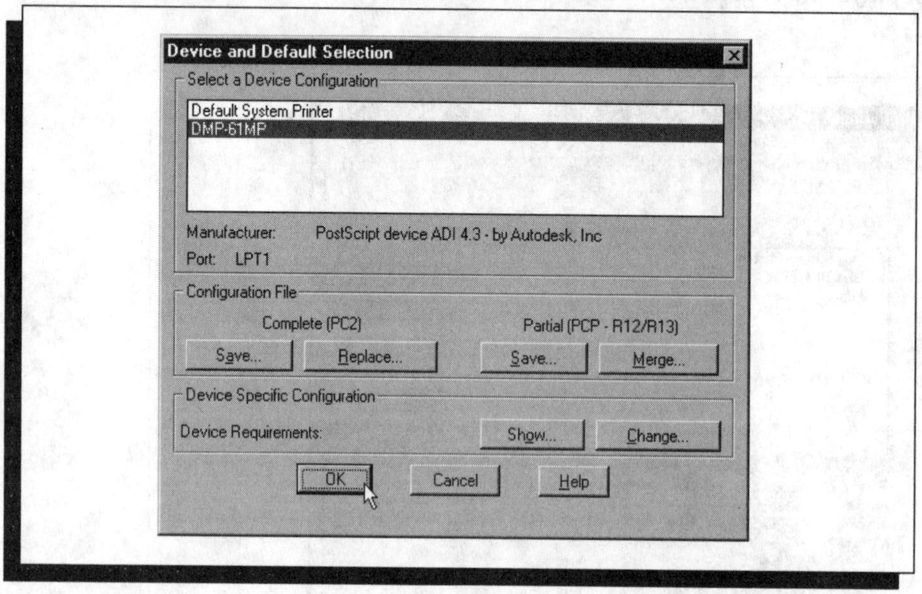

⑩ Select the **DMP-61MP** device and pick **OK**.

11 If this device is connected and ready, pick **OK** to initiate plotting. If not, pick the **Cancel** button.

Pen Parameters

AutoCAD allows you to control the color and thickness of plotted lines using the Pen Assignments dialog box. This applies to both pen plotters and raster devices.

Standard

1 Redisplay the Print/Plot Configuration dialog box.

2 Pick the **Device and Default Selection...** button and select **DMP-61MP** unless it is already selected.

3 Pick the **OK** button.

4 In the Pen Parameters area, pick the **Pen Assignments...** button.

Pen Assignments						Modify Values
Color	Pen No.	Linetype	Speed	Pen Width		
1	1	0	32	0.010		Color:
2	2	0	32	0.010		1 (red)
3	3	0	32	0.010		
4	4	0	32	0.010		Pen: 1
5	5	0	32	0.010		
6	6	0	32	0.010		Ltype: 0
7	1	0	32	0.010		
8	2	0	32	0.010		Speed: 32
9	3	0	32	0.010		
10	4	0	32	0.010		Width: 0.010
11	5	0	32	0.010		

Feature Legend...   Pen Width:

OK   Cancel

5 Pick color **1**, as shown in the previous illustration.

NOTE:

Color 1 refers to the red color 1 in the AutoCAD graphics screen, not the color produced by the output device.

AutoCAD shows the values associated with color 1 in the Modify Values area. We want pen number 1 assigned to color 1, so do not change this value. As you may recall, linetypes are set to ByLayer, so don't change the Ltype value either. Certain output devices will not permit you to make changes to values such as linetype and speed.

The Speed and Width values may require changes for each of the colors used in the dimen2.dwg drawing. The values you enter vary depending on the requirements of the output device. Consider the following two types of output devices.

Laser Printers. Suppose you are using a laser printer. For the dimen2.dwg drawing, you only need to change the Width value for color 1. For example, setting a width of .03 causes AutoCAD to plot relatively thick object lines. If you assign a width of .01 to all remaining colors, the laser printer will plot the rest of the drawing, including text and dimensions, using a thin .01 line. AutoCAD does not permit you to change the Speed value because this refers to the speed of the pens, and laser devices do not use pens.

Pen Plotters. If you're using a pen plotter, pay attention to the Pen and Speed values for each of the colors. You can ignore the Width value because the size of the pen tip controls the width of the plotted line. In the future, you will need to enter the width of each pen if you pick the Adjust Area Fill check box. You may choose to change the Speed value to a lower number, such as 12. Liquid ink pens may require a value as low as 4.

The Speed value refers to inches per second. Therefore a value of 12 means 12 inches/second. If you select MM (for millimeters) instead of Inches, the Speed value refers to centimeters/second.

⑥ Pick the **Feature Legend...** button and review the contents of the Feature Legend dialog box.

The Feature Legend dialog box may show a legend of linetypes and pen speeds. However, the legend varies from device to device and is not applicable to some devices.

7 Pick the **OK** button.

If you are using a multi-pen plotter, you can assign pens to the colors used in dimen2.dwg. Normally, you would use a thick, dark pen for color 1 so that object lines stand out. Use pens that produce thin lines for the other colors. Pen colors are personal choice, yet it is a good practice to standardize the pens you use for pen 1, pen 2, pen 3, etc. Likewise, it is best to standardize the pens you assign to color 1, color 2, color 3, etc.

8 Using the previous discussion as a guide, change the values in the Modify Values area for each color used in the dimen2.dwg drawing. Do not change Ltype values.

9 Pick the **OK** button.

10 Prepare the output device, if one is available, and pick the **OK** button. If no output device is available, pick **Cancel** and skip to the following section, "Pen Optimization."

11 Compare the plotted drawing with the values you set in the Pen Assignments dialog box.

12 If dimen2.dwg did not plot correctly, make adjustments in the Pen Assignments dialog box and plot the drawing again.

Pen Optimization

Pen optimization can minimize pen motion and can reduce plot time.

Standard

1 Display the Print/Plot Configuration dialog box.

2 Pick the **Optimization...** button.

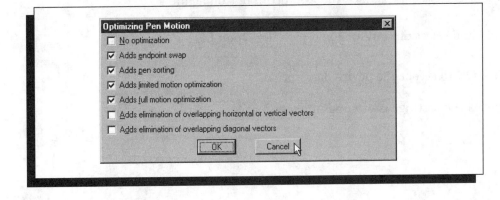

The Optimizing Pen Motion dialog box permits you to change and potentially optimize the motion of the plotter pens. The content provided in the dialog box varies from one configured device to another.

 Pick the **Cancel** button—*do not* make any changes to these settings at this time.

Device and Default Selection

The Device and Default Selection dialog box enables you to review many previously configured output devices.

 Pick the **Device and Default Selection...** button to display the Device and Default Selection dialog box.

 In the Configuration File area, under Complete (PC2), pick the **Replace...** button.

A file dialog box appears, enabling you to select a PC2 plot configuration file if one is available. PC2 files contain default settings, such as pen assignments, plot scale, and all of the other options made available in the Plot Configuration dialog box.

Pick the **Cancel** button.

Under Complete (PC2), pick the **Save...** button.

A standard file dialog box reappears. AutoCAD expects you to enter the PC2 file name.

Select a directory and pick the **Save** button to accept the dimen2.pc2 file name.

This saves the current plot settings in a file named dimen2.pc2.

Under Complete (PC2), pick the **Replace...** button again.

Find and select **dimen2.pc2** and pick the **Open** button.

Pick **OK** to accept the device description.

This retrieves the settings contained in dimen2.pc2.

The Save... and Merge... buttons are located under Partial (PCP - R12/R13), in the right half of the Configuration File area. The Save... button permits you to save plot configuration parameters to a PCP file, a format designed for AutoCAD Releases 12 and 13. The Merge... button retrieves default settings from a PCP file. Located under them are the Show... and Change... buttons, which allow you to display and change configuration information for the current device. The information varies from device to device.

9 Pick **OK**.

10 Pick the **Cancel** button.

Viewing the Contents of PC2 Files

PC2 files store plot information using standard ASCII text.

1 Minimize AutoCAD.

2 Start the Notepad text editor.

3 Select **Open...** from the **File** pull-down menu.

4 Find and select the **dimen2.pc2** file.

HINT: Next to Files of type, select All Files (*.*).

5 After you find **dimen2.pc2**, open it, if you haven't already.

As you can see, the file contains a long list of plot settings.

6 After reviewing the contents of the file, select **Exit** from the **File** menu.

7 Maximize AutoCAD.

Most companies that use AutoCAD exchange drawing files frequently. If you give a drawing file to another person, you may want to include a PC2 file with it so they don't have to guess what plot settings you intended to use.

 Exit AutoCAD. Do not save your changes.

Multiple Viewport Plotting

Until now, you have probably been plotting a single viewport only. In the future, you may want to plot multiple viewports. Multiple viewport plotting is especially useful when you create AutoCAD drawings in three dimensions.

Multiple viewport plotting is covered in Unit 39, "Viewports in Paper Space." Unit 39 focuses on the TILEMODE system variable and the MVIEW, PSPACE, MSPACE, and VPLAYER commands. You should be familiar with these commands and features before you attempt to plot multiple viewports.

Questions

1. The Pen Assignments dialog box allows you to make plotter changes related to the colors used in the AutoCAD graphics screen. Relative to these screen colors, describe the purpose of entering values for each of the following.

   Pen number THE PEN IN THIS SLOT IN THE PLOTTER OR RASTER ALLOWS THE PEN COLOR TO BE DETERMINED

   Linetype A FEATURE LEGEND DIALOG BOX COMES UP TO SELECT FROM VARIETY OF LINETYPES TO SELECT.

   Speed MAY BE REQUIRED TO CHANGE W/ EACH COLORS USED IN THE SPECIFIC PLOT.

   Pen width THE WIDTH MAY ALSO MAY NEED TO BE DETERMINED DEPENDING ON TWO TYPES OF OUTPUT DEVICES.

2. Suppose you are using a raster device that does not use pens. Are any of the previous settings applicable? Explain.

   YES, USING THESE ATTRIBUTES CAN REDUCED PLOT TIME.

3. Is it recommended that you change the Linetype values in the Pen Assignments dialog box? Explain.

   FOR THE PLOT OUTCOME, THE DRAWING MAY HAVE HIDDEN LINES, CENTER LINE ETC. IF YOU DON'T CHANGE THE LINE TYPE THE DRAW. WONT BE PRINTED OR PLOTTED W/ SUCH.

4. What is the purpose of creating and using PC2 files?

   TO REMEMBER THAT THE PLOTTING SETTINGS ARE SAVED W/IN THE FILE ALREADY & WHICH ONES SO YOU DON'T HAVE TO GUESS, ESPECIALLY IF YOU COPY IT TO A DISC.

5. Is it possible to add more than one output device? Explain.

   NO, A DEVICE IS SETUP CONFIGURED AND all you need to do IS SELECT IT.

6. How do you select an added output device?

   BY TYPING IN HOUSTON INSTRUMENT ADI- 4.2.- BY AUTO DEC & IUC. THE AUTO CAD TEXT WINDOW APPEARS- go through the steps on p. 434 to SELECT THE DEFAULTS-

1. Explain the purpose of pen optimization.

2. You are creating an architectural floor plan for a client. The client will need to plot the drawing on 17″ × 11″ paper on a CalComp electrostatic plotter, but you don't own this device. You have a Hewlett-Packard LaserJet 4 (which cannot print on paper larger than 11″ × 14″). Give a detailed explanation of how you would set up AutoCAD to handle this job, and what, specifically, you would provide to the client.

Problems

1. Open a drawing (any drawing will work) and set the following values. Certain output devices will not allow you to enter values (for linetype and speed, for instance) in the Modify Values area. If this applies to you, select another device. If another device is not available, add one.

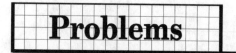

Color	Pen No.	Linetype	Speed	Pen Width
1	1	0	8	0.010
2	2	0	8	0.010
3	3	0	8	0.010
4	4	0	12	0.010
5	5	0	12	0.010
6	6	0	12	0.010
7	1	0	32	0.010
8	2	0	32	0.010
9	3	0	8	0.010
10	4	0	8	0.010
11	5	0	8	0.010

Modify Values
Color: 11
Pen: 5
Ltype: 0
Speed: 8
Width: 0.010

2. Retrieve the dimen2.pc2 file created in this unit and review the plot settings contained in it. Change the sheet size and a few pen assignments and create a new file named dimen3.pc2.

442

3. Add each of the following output devices. Unless a specific setting is given below, accept the default setting suggested by AutoCAD.

PostScript device ADI 4.3
300 dpi
Port: Parallel

Houston Instrument ADI 4.2
Model: DMP-52
Port: Serial (COM1)
Default scale: 1 = 1

AUTOCAD® AT WORK

Ropak Packs the Roe Boat

You have probably eaten or used something that came in a Ropak container. Ropak Corporation is a leading international manufacturer of plastic packaging for foods, coatings, and manufacturing materials. As a complete packaging service, Ropak designs, manufactures, and decorates containers to suit their customers' various products. The key is to develop the right package for the customer's product (which could be anything from cooking oil to caviar) at a reasonable cost. To keep costs down, Ropak adapts existing package designs using AutoCAD and AutoCAD Designer.

For shipping fish roe (eggs), for example, Ropak modified one of its basic plastic packages in several ways:

- putting on a domed lid to allow overpacking
- reshaping the bottom so the domed packages could be stacked
- adding a brine channel, which allows excess salt water to run off when the lid is pressed down on the tightly packed container.

These changes to the existing package design would have taken three weeks if the drawing all had to be done by hand. Designing a new package from scratch would take even longer, and cost much more. Using AutoCAD and AutoCAD Designer, Ropak gets a job like this done in two days.

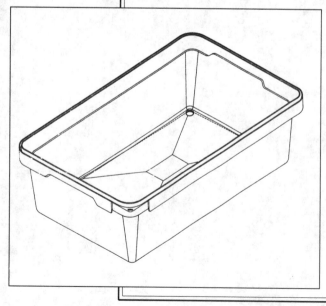

AutoCAD Designer, a parametric solid modeling program that works with AutoCAD, has proved especially useful for manufacturers like Ropak, whose product line involves a large number of variations on a basic family of parts. The software greatly reduces time spent on the extensive and repet-itive drawings required even for small variations in design specifications.

Ropak considers the AutoCAD and AutoCAD Designer software to be vital parts of its effort to compete and grow in the world's marketplace. As a design tool, this combination increas-es their productivity and their ability to serve the customer.

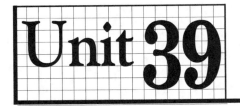

Unit 39 — Viewports in Paper Space

■ OBJECTIVE:

To apply viewports in paper space using the TILEMODE system variable and the MVIEW, PSPACE, MSPACE, and VPLAYER commands

AutoCAD allows you to work in model space and in *paper space*. Most AutoCAD drafting and design work is done in model space. Paper space is used to lay out, annotate, and plot two or more views of a drawing.

Viewports can be applied to both model space and paper space. This unit focuses on the application of viewports in paper space and the relationship of paper space to model space. Unit 18 concentrates solely on the use of viewports in model space and covers the VPORTS command.

TILEMODE System Variable

TILEMODE is a system variable that you must turn off in order to use paper space. To demonstrate the effect of TILEMODE, we will first prepare to create a simple 3D mechanical object.

1 Start AutoCAD.

2 Pick **Use a Wizard**, **Quick Setup**, and **OK**.

3 Pick the **Next** button to accept Decimal units.

4 Define a drawing area based on a scale of 1″ = 1″ and a sheet size of 11″ × 8.5″. Pick the **Done** button when you are finished.

HINT: The 11″ side should align with the X axis and 8.5″ should align with the Y axis. Assume an active plotting/printing area of 10″ × 7.5″.

5 Enter **ZOOM All**.

6 Establish the following settings.

Grid:	**Off**
Snap:	**.5** unit
Layers:	**Objects** (set color to red)
	Border (set color to cyan)
	Vports (set color to magenta)
	Make **Vports** the current layer.
Text Style:	Make a new text style named **ROMANS** using the **romans.shx** font. Use the default text style settings.

7 Save your work in a file named **pspace.dwg**.

8 Enter **TILEMODE** and **0** (off).

NOTE:

You can toggle TILEMODE on and off by double-clicking the word TILE in the status bar. Try it.

The following paper space icon replaces the standard coordinate system icon. The paper space icon is present whenever paper space is the current space. The coordinate system icon is present whenever model space is the current space.

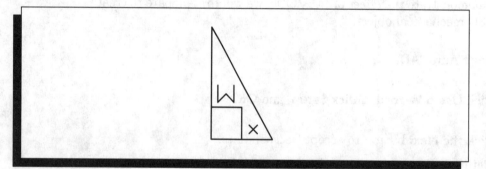

In the status bar, notice that PAPER replaced MODEL, indicating that you are now in paper space. MODEL appears when you are in model space.

9 Enter the **LIMITS** command to set the drawing area in paper space.

10 Press **ENTER** to accept the default value of 0.0000,0.0000 for the lower left corner of the drawing area.

11 Enter **10,7.5**—the same values you entered for the drawing area model space.

12 Enter **ZOOM All**.

The limits you have set in paper space are now the same as the limits set in model space.

MVIEW Command

The MVIEW command is used to establish and control viewports in paper space.

1 Enter the **MVIEW** command.

The following options appear.

ON/OFF/Hideplot/Fit/2/3/4/Restore/<First Point>:

---

2 Enter **4** to create four viewports.

3 Enter **F** for Fit.

AutoCAD fits four viewports in the available graphics area. The viewport objects are contained on layer VPORTS.

NOTE:

As an alternative to Steps 1 and 2, select the View pull-down menu and pick Floating Viewports and 4 Viewports.

MSPACE Command

The MSPACE command permits you to switch to model space.

1 Enter **MSPACE**. (See the following hint.)

HINT: The command alias MS is defined for MSPACE. Therefore, just enter MS. An alternative to this is to double-click the word PAPER in the status bar.

Model space is now the current space. Coordinate system icons appear in each of the four viewports.

NOTE: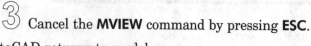

If the MVIEW command is issued in model space, AutoCAD switches to paper space for the duration of the MVIEW command.

2 Enter **MVIEW**.

Notice that AutoCAD switches to paper space.

3 Cancel the **MVIEW** command by pressing **ESC**.

AutoCAD returns to model space.

4 Make **Objects** the current layer.

5 Set the value of the **THICKNESS** system variable to **1**.

NOTE:

The THICKNESS system variable enables you to specify the thickness of an object in the z direction, resulting in a three-dimensional object. You will learn more about 3D objects in Units 41–57.

6 Make the upper left viewport the current viewport by picking a point inside it.

7 In the upper left viewport, draw the following object. Approximate its size and shape.

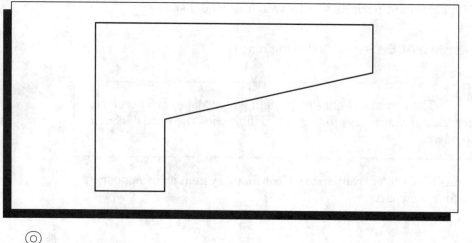

8 Save your work.

Four Individual Views _____

Taking advantage of AutoCAD's viewports, let's create four different views of the solid object.

1 Make current the lower left viewport by picking a point inside the viewport.

2 Display the Viewpoint toolbar.

Viewpoint

3 View the object from the front by picking the **Front View** icon from the toolbar.

4 Enter **ZOOM** and **1**.

5 Make current the lower right viewport.

Viewpoint

6 View the object from the right side by picking the **Right View** icon.

7 Enter **ZOOM** and **1**.

Viewpoint

8 Make current the upper right viewport and view the object from above, in front, and to the right by picking the **SE Isometric view** icon.

9 Enter **ZOOM** and **1**.

The screen should look similar to the following.

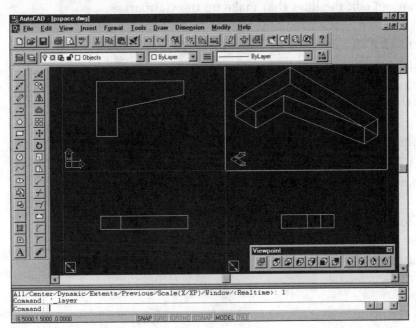

449

PSPACE Command _____

The PSPACE command is similar to the MSPACE command. Its purpose is to switch from model space to paper space.

1. Enter **PS** (the PSPACE command alias), or double-click **MODEL** in the status bar.

Little appears to change except for the coordinate system icons. However, when in paper space, you cannot edit entities created in model space. Likewise, when you are in model space, you cannot edit entities created in paper space.

2. Attempt to select the object in any one of the four viewports.

As you can see, you cannot select it because it was created in model space.

3. Draw a short line in one of the four viewports, switch to model space, and attempt to erase the line.

4. Switch back to paper space and erase the line.

5. Save your work.

Editing Viewports in Paper Space _____

Viewports in paper space are treated much like other AutoCAD objects. For example, you can edit the lines that make up the viewports.

1. Enter the **UNDO** command and the **Mark** option.

2. Enter the **MOVE** command, select one of the lines that make up one of the four viewports, and move the viewport a short distance toward the center of the screen.

3. Erase one of the viewports.

4. Enter the **SCALE** command, select one of the viewports, select a base point (anywhere), and enter **.75**.

5. Switch to model space.

6. Attempt to move, erase, or scale a viewport.

Viewports in paper space can be moved and even erased, but not in model space. Only the views themselves can be edited in model space. Paper space is used to arrange views and embellish them for plotting (you will do this in the following section), while model space is used to construct and modify objects that make up the 3D model or drawing.

7 Enter **UNDO** and the **Back** option.

The drawing returns to the point at which you entered UNDO Mark.

Plotting Multiple Viewports _____

One of the benefits of paper space is multiple viewport plotting.

1 Freeze layer **Vports** and make **Border** the current layer.

The lines which make up the viewports should now be invisible.

2 Switch to model space.

The viewport lines are invisible in model space also, but notice that the current viewport is outlined.

3 Pick each of the four viewports.

4 Switch to paper space and draw a border and basic title block similar to those in the following illustration.

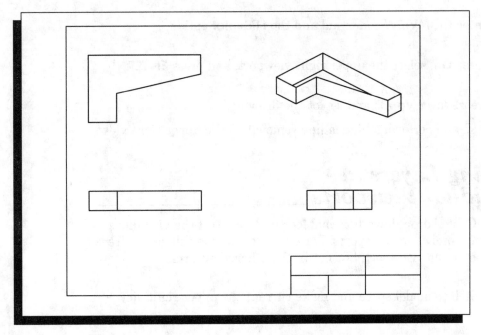

5 Plot the drawing at a scale of **1 = 1**.

HINT: Select the Limits radio button after entering the PLOT command. Refer to Units 37 and 38 for help with other plotter settings.

It is quite possible that the current position of the border and views did not plot perfectly. Regardless, make adjustments to the location of the border, title block, and views by following these steps.

1 Thaw layer **Vports**.

2 Move the individual viewports to position them better in the drawing. It is normal for them to overlap.

3 Edit the size and location of the border and title block if necessary.

4 Freeze layer **Vports**, save your work, and replot the drawing.

MVIEW Hideplot Option

The MVIEW Hideplot option performs hidden line removal on selected views.

1 First, thaw layer **Vports**.

2 Enter the **MVIEW** command and the **Hideplot** option.

3 Enter **On**, select the upper right viewport, and press **ENTER**.

4 Freeze layer **Vports** and replot the drawing.

AutoCAD should perform a hidden line removal on the upper right view.

Freezing Layers in Individual Viewports

The Layer Control drop-down box enables you to control the visibility of layers within individual viewports. For example, you may choose to freeze layer Objects in one viewport, but not in the other viewports.

1 Switch to model space and make current the upper right viewport.

2 In the docked Object Properties toolbar, pick the Layer Control **down arrow.**

Notice that the third item (the sun and rectangle) is no longer grayed out. This item allows you to freeze and thaw individual viewports in paper space.

3 Pick the sun/rectangle item next to Objects.

The sun becomes a snowflake.

4 Click anywhere in the drawing area to remove the Layer Control drop-down box.

The 3D view in the upper right viewport disappears because it is now frozen.

5 Switch to paper space and plot your work.

Everything plots except for the object on layer Objects in the upper right viewport.

6 Switch to model space and make the upper right viewport active if it is not already.

7 Display the Layer Control drop-down box, pick the snowflake/rectangle next to Objects, and click in the drawing area to remove the drop-down box from the screen.

The 3D view in the upper right viewport becomes visible.

NOTE:

You can freeze and thaw individual viewports from the Layer & Linetype Properties dialog box. You can also enter the VPLAYER command at the keyboard. Like other commands meant for use in paper space, VPLAYER is available only when TILEMODE is set to 0.

8 Close the Viewpoint toolbar, save your work, and exit AutoCAD.

Questions

1. Describe the difference between TILEMODE Off (0) and TILEMODE On (1).

   "TILEMODE ON" THE VPORTS - CAN BE MOVED AROUND monitor, OFF (LIKE THEY ARE GROUTED TOGETHER & CAN'T BE MOVED OUT

2. Explain the difference between model space and paper space.

   DESIGN & DRAFTING IS DONE IN MODEL SPACE YOU SWITCH TO PAPER SPACE TO PRINT OR PLOT ONE OR MORE VIEWS OF THE OBJECT

3. Give the primary purpose of the MVIEW command.

   TO MAKE VIEWPORTS OF THE SAME OBJECT OF DIFFERENT VIEWS IN SEPARTE SPACES.

4. The MSPACE and PSPACE commands allow you to switch to and from model space and paper space. Explain an easier and faster way to switch between these spaces.

   MS OR PS.

5. Explain why you may want to edit viewports in paper space.

   THEY CAN BE MOVED OR ERASED BUT NOT IN MODEL SPACE

6. What does the MVIEW Hideplot option permit you to do?

   HIDE LINE OF THE ISOMETRIC VIEW

7. Explain how to freeze a layer named Detail in the upper right viewport only.

   BE IN MODEL SPACE & CURRENT UPPER RIGHT VIEWPORT. IN OBJECT PROPERTIES PICK LAYER CONTROL THE SUN & RECT ARENT GRAYED OUT to pick + thaw & FREEZE LAYERS.

8. What is the main benefit of plotting in paper space?

   SO you CAN PLOT MULTIPLE VIEWS OF THE DRAWN OBJECT.

■ *Challenge Your Thinking: Questions for Discussion*

1. Experiment with viewports created in model space and paper space. Is it possible to create more than one viewport in model space, then import the model space viewports into a paper space viewport? Explain.

Problems

1. Using the commands and features related to paper space, create multiple views of the following solid object. Dimension the 2D views in paper space. Create a border and title block, and plot the multiple views on a single sheet.

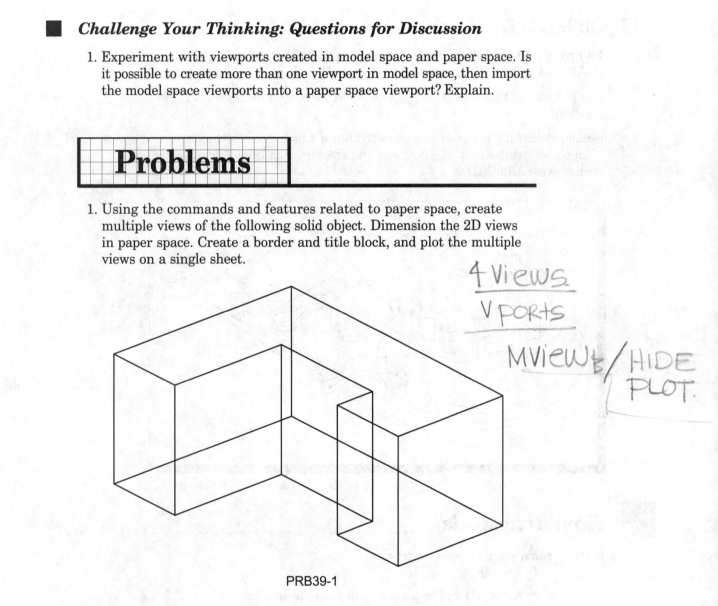

4 Views
Vports

MViews/HIDE
PLOT.

PRB39-1

2. Create a left-side profile view of the above object. The view should contain hidden lines.

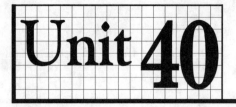

Unit 40 — Isometrics: Creating Objects from a New Angle

■ OBJECTIVE:

To apply AutoCAD's isometric drawing capabilities using the SNAP and ISOPLANE commands and the Drawing Aids dialog box

The purpose of this unit is to practice the construction of isometric drawings.

Isometric drawing is one of two ways to obtain a pictorial representation of an object using AutoCAD. Below is an example of an isometric drawing created with AutoCAD.

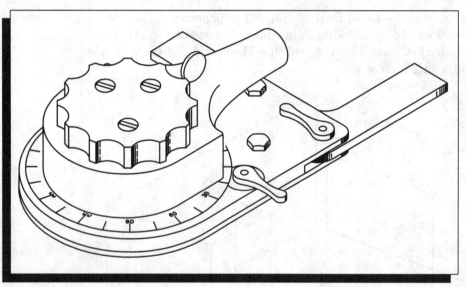

AutoCAD drawing courtesy of CAD Northwest, Inc.

Isometric Mode

Let's do some simple isometric drawing.

1 Start AutoCAD and begin a new drawing from scratch.

2 Set the grid at **1** unit and the snap resolution at **.5** unit.

AutoCAD isometric drawing is accomplished by changing the SNAP style to Isometric mode.

3 Enter the **SNAP** command.

4 Enter **S** for Style.

5 Enter **I** for Isometric.

6 Enter **.5** for the vertical spacing. (This should be the default value.)

You should now be in Isometric drawing mode, with the crosshairs shifted to one of the three (left, top, or right) isometric planes.

7 Move the crosshairs and notice that they run parallel to the isometric grid.

Drawing Aids Dialog Box

You can also use the Drawing Aids dialog box to change to the Isometric drawing mode.

1 Pick **Drawing Aids...** from the **Tools** pull-down menu.

As you can see, Isometric Snap/Grid is turned on. We did this in Steps 3 through 6. In the future, you can check the On check box instead.

2 Pick the **Top** radio button.

This shifts the crosshairs to the top isometric plane.

3 Pick **OK**.

ISOPLANE Command

The ISOPLANE command also allows you to change from one isometric plane to another.

1. Enter **ISOPLANE**.

2. Enter **L**, **T**, or **R** to change to the left, top, or right plane. (Experiment with each.)

3. Enter **ISOPLANE** and press **ENTER**.

This toggles you to the next isometric plane.

Another method of toggling the crosshairs is to enter CTRL E. You may find this the fastest method.

4. Enter **CTRL E**; enter **CTRL E** again.

5. Experiment further with each of the methods of changing the crosshairs.

Isometric Drawing

Let's construct a simple isometric drawing.

1. Create a layer named **Objects**, assign a color of your choice to it, and make it the current layer.

2. Enter the **LINE** command and draw the following box. Make it $3 \times 3 \times 5$ units in size.

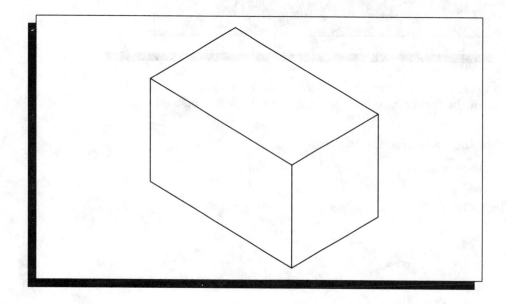

3 Save your work in a file named **iso.dwg**.

4 Alter the box so that it looks similar to the one on the next page, using the **LINE**, **BREAK**, and **ERASE** commands.

HINT: You can use all of the AutoCAD commands to construct isometric drawings.

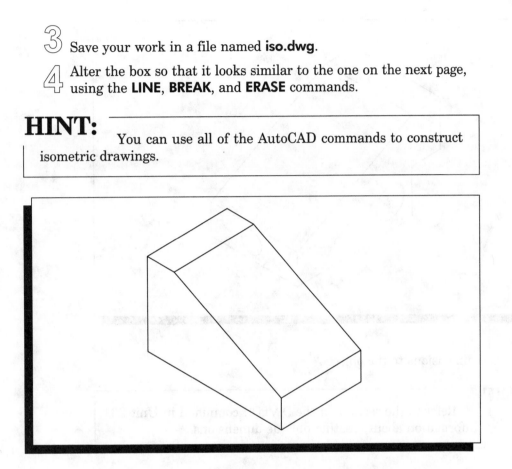

5 Further alter the object so that it looks similar to the one on the next page. Use the **ELLIPSE** command to create the holes.

HINT: Since the ellipses are to be drawn on the three isometric planes, toggle (change) the crosshairs to the correct plane. Then choose the ELLIPSE command's Isocircle option.

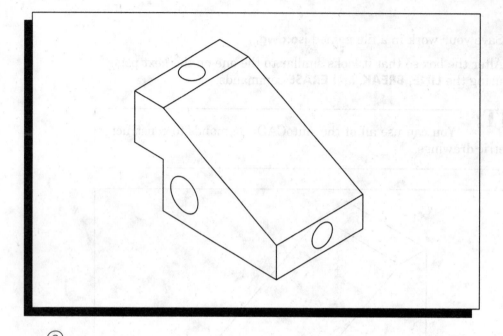

⑥ Add dimensions to the objects.

HINT: Refer to the section on the DIMEDIT command in Unit 26 for more information about creating oblique dimensions.

⑦ Further experiment with AutoCAD's isometric drawing capability by constructing other isometric objects.

NOTE:

To change back to AutoCAD's standard drawing format, enter the SNAP, Style, Standard sequence, or use the Drawing Aids **dialog box**.

⑧ Save your work and exit AutoCAD.

Questions

1. Describe two methods of changing from AutoCAD's standard drawing format to isometric drawing.

   BY GOING INTO "SNAP" COMMAND OR BY ENTERING ISOPLANE COMMAND.

2. Describe the purpose of the ISOPLANE command.

   IT ALLOWS YOU TO CHANGE FROM ONE PLANE VIEW TO ANOTHER.

3. Describe two methods of changing the isometric crosshairs from one plane to another.

   ENTER L, T, R TO CHANGE THE CROSSHAIRS TO LEFT, TOP OR RIGHT. OR CTRL E. THIS ONE IS SAID TO BE FASTER.

4. Explain how to create accurate isometric circles.

   PICK THE CORRECT PLANE IN ISOPLANE & THE ELLISPE COMMAND

5. How do you change from isometric drawing to AutoCAD's standard drawing format?

   SNAP, 'STYLE, STANDARD SEQUENCE

Challenge Your Thinking: Questions for Discussion

1. Discuss the correct way to dimension an isometric object. What factors must you take into consideration? Explain. *MUST BE IN ISOPLANE*

2. Discuss the purpose(s) of isometric drawings. When and why are they used? *TO SEE ALL SIDES*

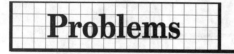
1-4. Create each object using AutoCAD's isometric capability. Approximate their sizes.

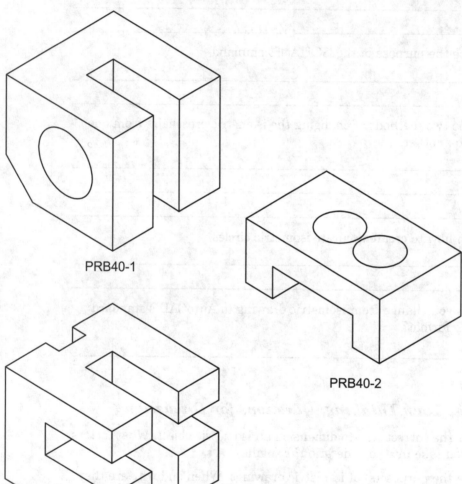

PRB40-1

PRB40-2

PRB40-3

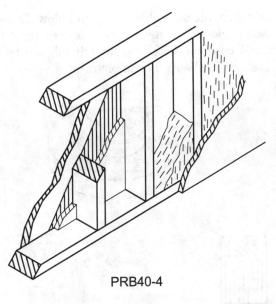

PRB40-4

5. An oblique drawing of a guide block is shown below. Draw a full-scale isometric view using AutoCAD's SNAP and ELLIPSE commands. An appropriately spaced grid and snap will facilitate the drawing.

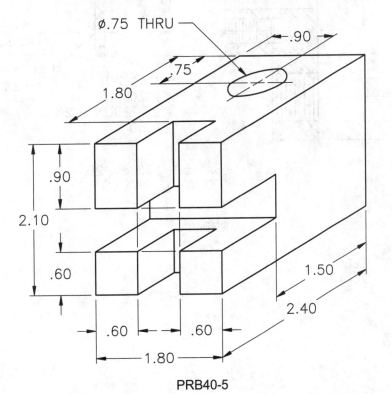

PRB40-5

6. An exploded, sectioned orthographic assembly is shown below. Draw the assembly as a full scale exploded isometric assembly, using AutoCAD's SNAP and ELLIPSE commands. Use isometric ellipses to represent the threads as shown in the isometric pictorial.

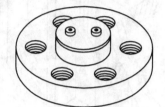

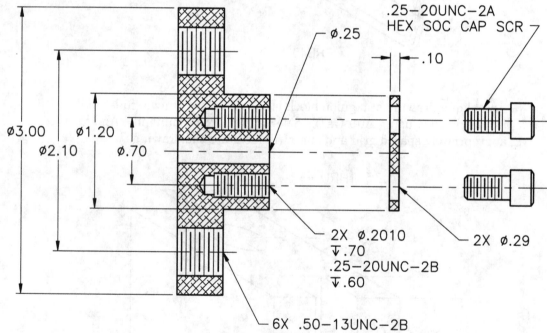

.25−20UNC−2A
HEX SOC CAP SCR

Ø.25

⌀.10

Ø3.00
Ø2.10
Ø1.20
Ø.70

2X Ø.2010
⫧.70
.25−20UNC−2B
⫧.60

2X Ø.29

6X .50−13UNC−2B
EQUALLY SPACED

PRB40-6

464

7. Accurately draw an isometric representation of the orthographic views shown below. Draw and dimension the isometric according to the dimensions provided.

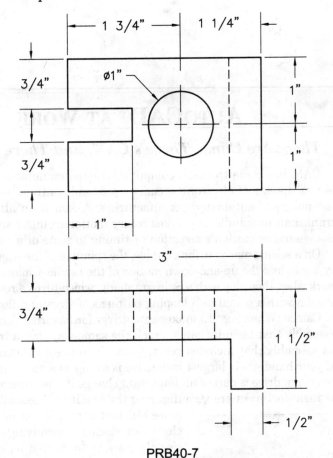

PRB40-7

AutoCAD drawing in Problem 4 courtesy of Autodesk, Inc.
Problems 5 and 6 courtesy of Gary J. Hordemann, Gonzaga University

AUTOCAD® AT WORK

There Are Cams, There's CAM, and There's Camco

CAM has come to mean computer-assisted manufacturing, but long before the age of electronics, cams were (and continue to be) a central technology of automated manufacturing. A cam is an attachment to a crankshaft or spindle that takes rotary motion as input and transmits a change in the motion's direction or timing to some other device.

On a sewing machine, for example, the spinning of the wheel is converted by a cam into the up-and-down motion of the needle. Children see a cam at work when they play with a Spirograph toy, which turns a round-and-round motion with a pencil into looping patterns of stars and flowers.

Camco (Wheeling, IL) makes cam drives for factories. A cam drive is an assembly that includes cams, much the same way that a transmission is an assembly that includes gears. Camco's smallest cams fit in the palm of your hand; their largest cam drive is as big as a merry-go-round. This large cam drive is part of an important change in the automobile industry, as manufacturers are reconfiguring the traditional assembly line into a

more efficient arrangement of assembly circles. Each circle is a revolving table that carries the cars from station to station, where robots perform assembly and welding tasks. The circular shape of the cam-driven tables allows assembly operations to be laid out in a smaller floorspace, which reduces overhead costs such as heating, taxes, and maintenance.

Cam geometry, which may be curved or straight-line, begins with a drawing. Camco engineers design in AutoCAD and save the drawings as DXF files, which can be imported by programs that generate numeric control (NC) or computer numerical control (CNC) code. The NC or CNC code guides the machine that grinds or mills the cam. Camco also uses AutoCAD to design shafts, hosings, and wheels, as well as precision-link conveyors, to perform with its cams.

466

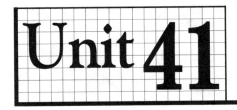

Unit 41 — The Third Dimension

■ OBJECTIVE:

To apply AutoCAD's 3D modeling capability using the ELEV, VPOINT, DDVPOINT, and HIDE commands and the THICKNESS system variable

This unit introduces AutoCAD's three-dimensional (3D) capability with four easy-to-use commands. They permit you to create simple 3D models (boxes, cylinders, etc.) and view them from any point in space. An example of an AutoCAD-generated 3D model is shown below, left. The model on the right is the same object viewed from the top.

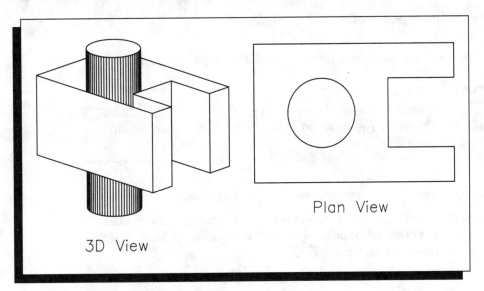

3D View

Plan View

Let's draw a simple 3D model like the one above.

■ *ELEV Command*

1 Start AutoCAD and begin a new drawing from scratch.

2 Create a new layer named **Objects**, assign any color to it, and make it the current layer.

3 Enter the **ELEV** command and set the new current elevation at **1** and the new current thickness at **3**.

NOTE:

An elevation of 1 means the base of the object will be located 1 unit above a baseplane of 0 on the Z axis. The thickness of 3 means the object will have a thickness on the Z axis of 3 units upward from the elevation plane. This is called the *extrusion thickness*. The THICKNESS system variable also permits you to specify the extrusion thickness.

467

4 Draw the top (plan) view of the object using the **LINE** command. Construct the object as shown on the right in the previous drawing, but omit the circle (cylinder) at this time. The width and length of the object is 4 × 6 units, respectively. Approximate the remaining dimensions. Use snap, ortho, and grid as necessary.

5 Save your work in a file named **three-d.dwg**.

VPOINT and HIDE Commands

[handwritten: TO UN Hide type (Regen)]

Now let's view the object in 3D.

1 From the **View** pull-down menu, pick **3D Viewpoint** and **Tripod**.

HINT:
Instead, you may choose to type the VPOINT command and press ENTER twice. You may find this to be faster than picking items from the pull-down menu.

2 Move the pointing device and watch what happens.

3 Place the small crosshairs inside the globe representation, also referred to as the *compass,* as shown in the following illustration and pick that approximate point.

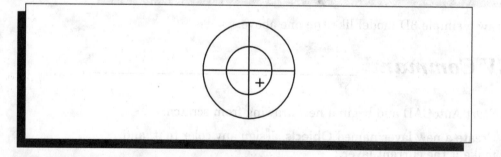

A 3D model of the object appears on the screen.

Study the following globe representation carefully. The placement of the crosshairs on the globe indicates the exact position of the viewpoint. Placing the crosshairs inside the inner ring (called the equator) results in viewing the object from above. Placing the crosshairs outside the inner ring results in a viewpoint below the object.

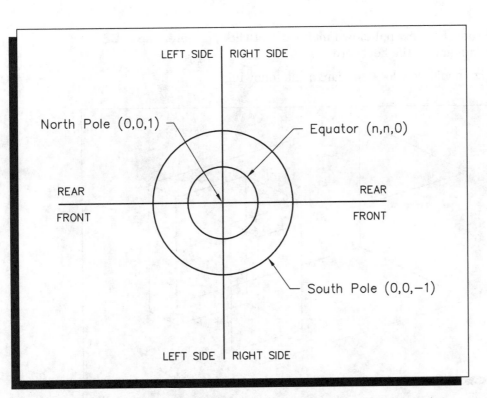

If the crosshairs are on the right side of the vertical line, the viewpoint will be on the right side of the object. Similarly, if the crosshairs are in front of the horizontal line, the viewpoint will be in front of the object.

4 Enter **Z** (for ZOOM) and **.9x**.

The object should look somewhat like the one below.

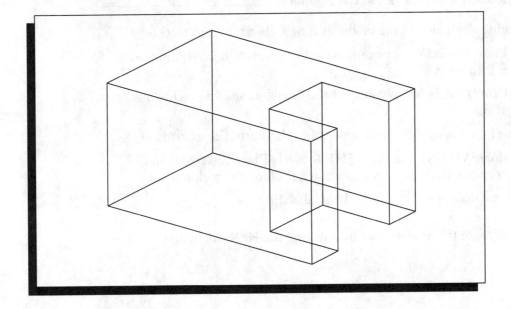

5 From the **View** pull-down menu, select **Hide**, or enter the **HIDE** command at the keyboard.

The object should now look similar to the following.

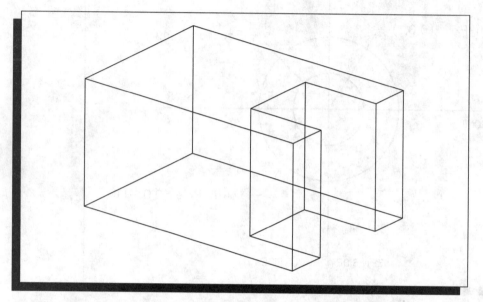

6 Return to the plan view of the object by selecting the **View** pull-down menu and then selecting **3D Viewpoint** and **Top**.

7 **ZOOM All** to obtain the original plan view of the model.

Objects of Different Elevation and Thickness

Let's add the cylinder to the model at a new elevation and thickness.

1 Enter the **ELEV** command and set the elevation at **–1** and the thickness at **6**.

2 Draw a circle in the center of the model as shown on the first page of this unit.

Visualize how the cylinder will appear in relation to the existing object.

3 Enter **VPOINT** and press **ENTER** twice. Place the crosshairs in approximately the same location as before and pick a point.

Does the model appear as you had visualized it?

4 Remove the hidden lines by entering the **HIDE** command.

Does the model now look similar to the one shown at the beginning of this unit? It should.

 Experiment with VPOINT to obtain viewpoints from different points in space, and create a 3D view of the model as shown in the following illustration.

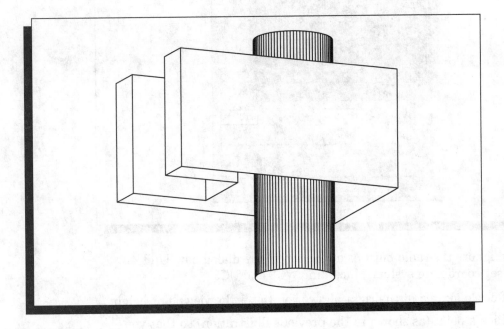

HINT: Type VPOINT and press ENTER twice. Select a point in the globe. Press the space bar (or ENTER) to reissue the VPOINT command, and press ENTER to display the globe. Repeating this is a quick way to view the object from several different viewpoints.

Viewpoint Presets Dialog Box

The Viewpoint Presets dialog box offers another way to choose viewpoints.

1 From the **View** pull-down menu, select **3D Viewpoint** and the **Select...** option.

This enters the DDVPOINT command, which displays the Viewpoint Presets dialog box.

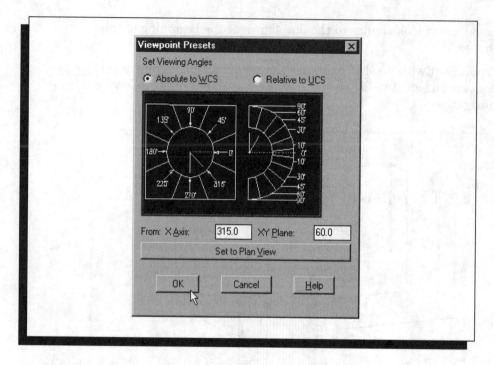

For now, ignore the radio buttons at the top of the dialog box. Unit 43 covers user coordinate systems, including WCS and UCS.

The half-circle located at the right allows you to set the viewpoint height.

2 Pick a point (as shown in the previous illustration) so that you view the object from above at a 60-degree angle and pick the **OK** button.

You should now be viewing the object from above at a 60-degree angle.

3 Display the dialog box again by pressing the space bar or **ENTER**.

The full circle (located in the left half of the dialog box) allows you to set the viewpoint rotation.

4 Pick a point so that you will view the object at a 135-degree angle and pick **OK**.

5 Enter **HIDE**.

6 Display the dialog box again.

The two edit boxes enable you to enter values for the rotation and height.

7 Pick the **Set to Plan View** button and pick **OK**.

8 **ZOOM All**.

9 Experiment further with the Viewpoint Presets dialog box.

Selecting Viewpoints from the Standard Toolbar

AutoCAD offers several predefined viewpoint icons that are available on the Standard toolbar.

1 From the Standard toolbar, pick and hold the **Named Views** icon.

Standard

A series of icons appear on a flyout toolbar.

2 Move the pointer slowly down the list so that the tooltip appears for each icon.

3 Select each of the icons. Enter **HIDE** after each selection.

As you can see, this is a fast way of producing known orientations in space.

4 Save your work and exit AutoCAD.

Questions

1. Describe the purpose of the VPOINT command.

   IT IS A FASTER COMMAND TO DO 'VIEW' 3D Viewpt. tripod.

2. The extrusion thickness of an object is specified with which command? System variable?

   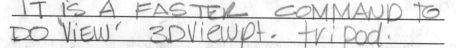
   ELEVATION - GIVES thickness FROM ELEV. PLANE ALSO KNOW AS system variable

3. Briefly explain the process of creating objects (within the same model) at different elevations and thicknesses.

   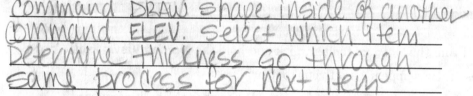
   Command DRAW shape inside of another command ELEV. select which Item Determine thickness go through same process for next Item

4. When the small crosshairs are in the exact center of the globe, what is the location of the viewpoint in relation to the object?

   THE PLACEMENT OF crosshair indicates VIEW OF OBJECT.

5. With what command are hidden lines removed from 3D objects?

   HIDE COMMAND

6. Indicate on the globe below where you must position the small crosshairs to view an object from the rear and underneath.

■ *Challenge Your Thinking: Questions for Discussion*

1. Match the following globe representations with the objects. The first one has been completed to give you a starting point.

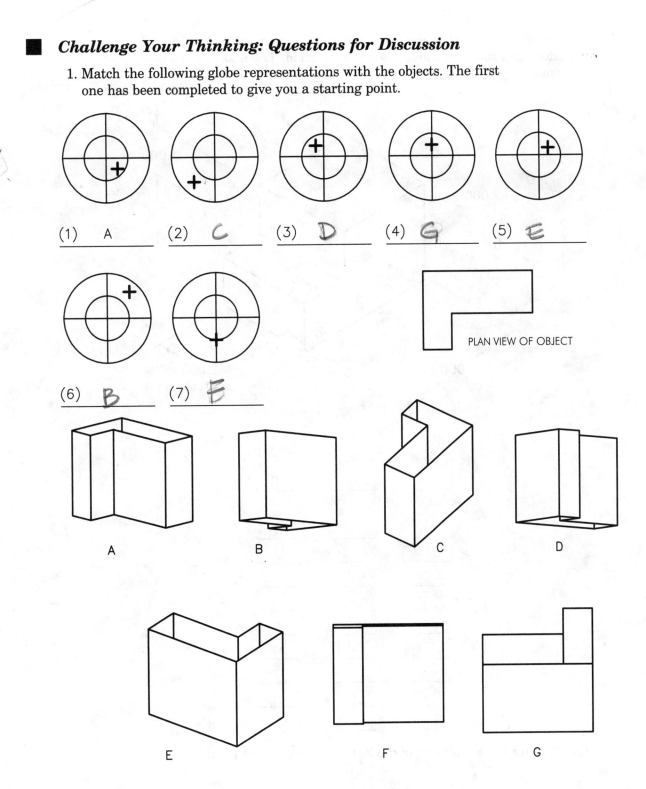

(1) A

(2) C

(3) D

(4) G

(5) E

PLAN VIEW OF OBJECT

(6) B

(7) F

A

B

C

D

E

F

G

2. Enter the x, y, and z coordinate values for the respective alphabetically identified points in the following illustration. The first one has been done for you.

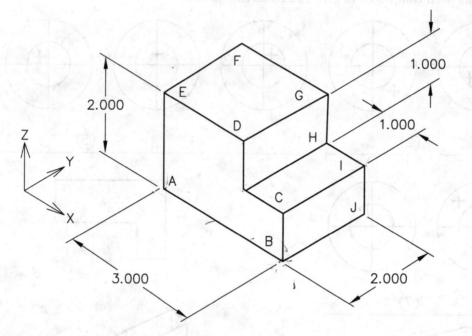

	x	y	z
A	0	0	0
B	3.00	3.000	1.0.
C	1.00	2.000	.0
D	0	2.00	-1.00
E	2.00	2.00	0
F	2.000	0.00	0
G	0	0	0
H	1.000	-2.00.	1.000
I	-1.000	0.	0.
J	0	0	1.00
?	0.	-2.00	0

(For ?, enter the coordinates of the last remaining corner point.)

476

Problems

Draw the following objects and generate a 3D model of each.

1. For problem 1, set the elevation at 1″ to create a 3D model of the drawing.

2. For problem 2, set the elevation for the inner cylinder at 0. Set the snap resolution prior to picking the center point of the first circle. Do not use object snap to select the center point of the second circle.

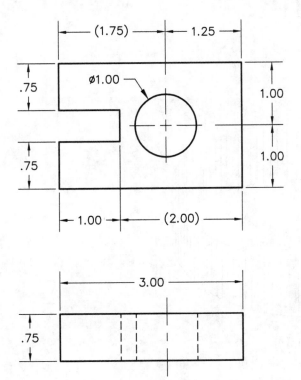

PRB41-1

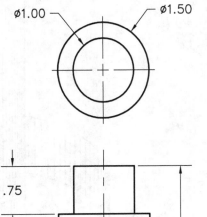

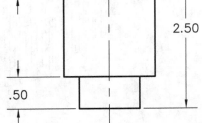

PRB41-2

3. In this problem, try your best to draw an object similar to the one shown below.

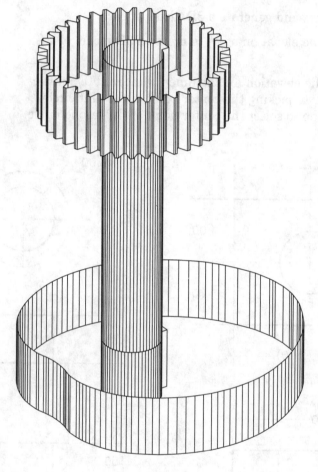

PRB41-3

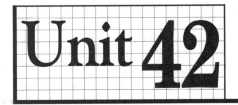

Unit 42 — X/Y/Z Point Filters

■ OBJECTIVE:

To apply AutoCAD's 3D X/Y/Z point filters using the 3DFACE command

This unit continues with AutoCAD's 3D wireframe and surface modeling capabilities. X/Y/Z filters enable you to enter coordinates using both the pointing device and the keyboard. With the LINE and 3DFACE commands, you can create lines and surfaces in three-dimensional space using x, y, and z coordinates. For instance, you can create inclined and oblique surfaces, such as a roof on a building. AutoCAD is also capable of creating 3D cones, domes, spheres, and other 3D objects with commands presented in the following units.

■ *3DFACE Command*

The 3DFACE command creates a three-dimensional object similar in many respects to a two-dimensional solid object. The 3DFACE prompt sequence is identical to that of the SOLID command. However, unlike the SOLID command, the 3DFACE command allows you to enter points in a natural clockwise or counterclockwise order to create a normal 3D face. Z coordinates are specified for the corner points of a 3D face, forming a section of a plane in space.

1 Start AutoCAD and open the drawing named **three-d.dwg.**

2 Using **Save As...**, create a new drawing named **three-d2.dwg.**

3 If you don't already have the plan view on the screen, obtain it by entering **VPOINT** and **0,0,1**. Then **ZOOM All.**

4 Erase the cylinder (circle) from the 3D model, and set the snap at **.25** unit and grid at **.5** unit.

5 Display the Surfaces toolbar.

6 From this toolbar, pick the **3D Face** icon.

This enters the 3DFACE command.

Surfaces

7 Type **.xy** and press **ENTER**.

8 In reply to of, approximate the x,y position of point 1 in the following illustration and pick that point. (Point 1 is about .5 unit from the corner of the object.)

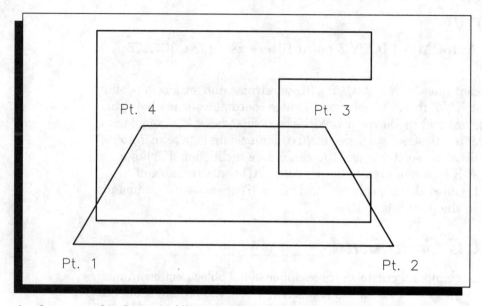

As shown on the Command line, the first point also requires a *z* coordinate.

9 Enter **3.5** for the *z* coordinate.

This method of entering 3D points is referred to as X/Y/Z filtering.

10 In reply to Second point, enter **.xy** and pick point 2, as shown in the illustration.

11 Enter **3.5** for the *z* coordinate.

12 Repeat these steps for points 3 and 4, but enter **7** for the *z* coordinate of these points. (Be sure to enter **.xy** *before* picking the point.)

13 Press **ENTER** to terminate the **3DFACE** command.

Do you know what you've created? If not, try to visualize its position in relation to the object. It's an inclined surface resembling a portion of a roof. Let's make the opposite portion.

1 Enter the **MIRROR** command, select the 3D face object you just created, and press **ENTER**.

2 Place the mirror line so that the two sections of the roof meet at the middle of the object. Do not delete the "old object."

Now let's view the object in 3D.

Modify

480

③ Enter the **VPOINT** command and obtain the globe representation.

④ Pick a point on the globe so that you view the object from above, front, and right side.

HINT:
See pages 468 and 469 for assistance.

⑤ Enter the **HIDE** command.

Does your model look similar to the one here? It should.

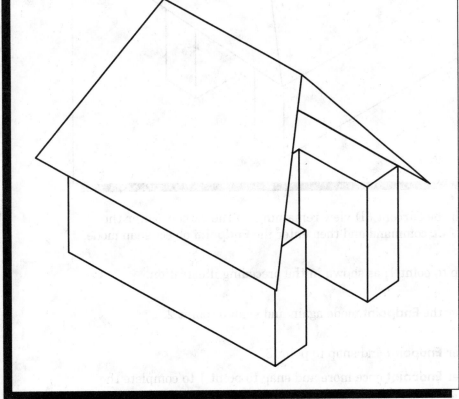

Let's finish the roof so that it looks similar to the following.

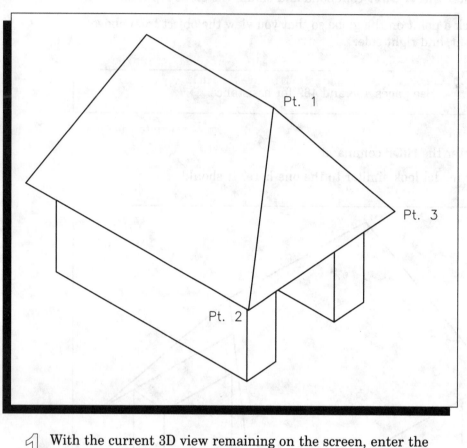

Surfaces

1 With the current 3D view remaining on the screen, enter the **3DFACE** command and then enter the **Endpoint** object snap mode.

2 Snap to point 1, as shown in the preceding illustration.

3 Enter the **Endpoint** mode again and snap to point 2.

4 Enter **Endpoint** and snap to point 3.

5 Enter **Endpoint** once more and snap to point 1 to complete the surface boundary.

6 Press **ENTER** to terminate the 3DFACE command.

7 Enter the **HIDE** command.

8 View the object from above, front, and left side using either **VPOINT** or **DDVPOINT**, and enter **HIDE**.

9 Create the remaining portion of the roof using the **3DFACE** command and object snap, and then enter **HIDE**.

HINT: To complete Step 9, you may choose to return to the plan view and use the MIRROR command to complete the roof.

10 Save your work.

11 Generate a view similar to the one below.

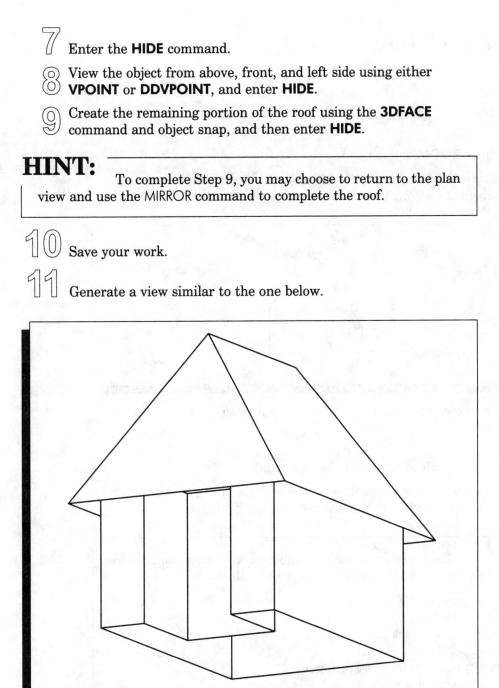

Invisible Option

3D faces can contain visible and invisible edges.

1 Generate the plan view.

2 Enter **ZOOM** and **.6x**.

3 Using **3DFACE**, create the following pentagon anywhere on the screen and at any size. Pick the points in the order shown.

Surfaces

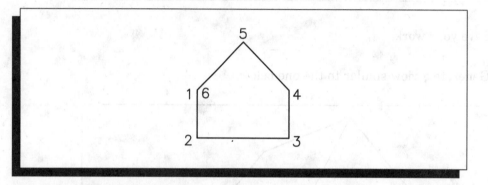

The object now contains two lines that do not appear in the previous object.

4 Press **ENTER** to terminate the command.

5 Now create another pentagon similar to the first one, but create this one according to the following hint.

HINT:

The invisible option is useful for connecting two or more 3D faces to create a 3D model. Consider the following:

```
Command: 3DFACE
First point: (pick point 1)
Second point: (pick point 2)
Third point: (pick point 3)
Fourth point: (enter i and pick point 4)
Third point: (pick point 5)
Fourth point: (enter i and pick point 6)
Third point: (press ENTER)
```

The system variable SPLFRAME controls the display of invisible edges in 3D faces. When SPLFRAME is set to a nonzero value, invisible edges are displayed. This allows you to edit them as you would a visible 3D face.

The following illustration shows the same pentagon with SPLFRAME set at a nonzero value.

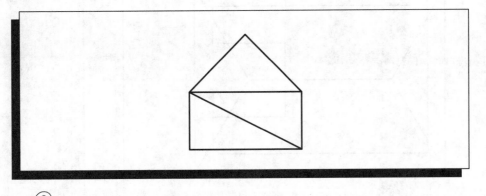

6 Enter **SPLFRAME** and enter **1**.

7 Enter **REGEN**.

The pentagon should now look similar to the one in the preceding illustration.

8 Set **SPLFRAME** back to **0** and enter **REGEN**.

9 Erase both pentagons.

Creating 3D Lines

The LINE command creates lines in 3D space. Therefore, all lines you have created with AutoCAD are three-dimensional. Their endpoints are made up of x, y, and z coordinates. Let's create a 3D property line around the building using LINE.

1 **ZOOM All**.

2 Enter the **ELEV** command and change the elevation to **1** and the thickness to **0**.

3 Enter the **LINE** command and pick one of the four corners of the property line shown in the following illustration. (Approximate its location.)

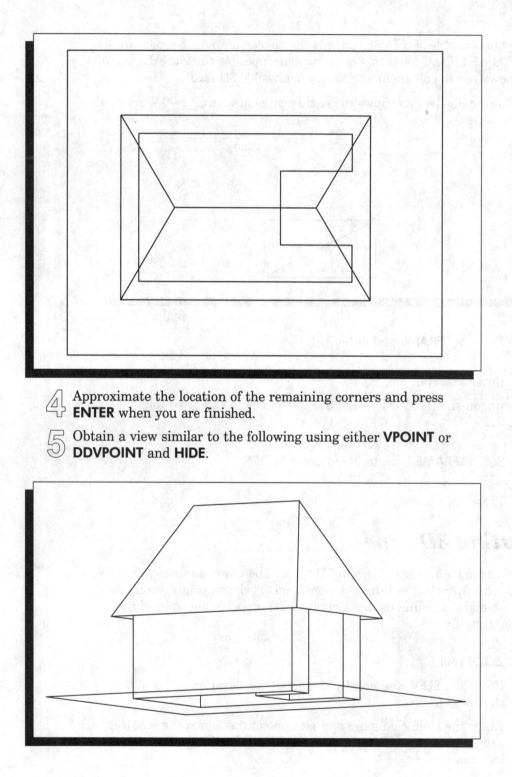

4 Approximate the location of the remaining corners and press **ENTER** when you are finished.

5 Obtain a view similar to the following using either **VPOINT** or **DDVPOINT** and **HIDE**.

6 Experiment further with the LINE and 3DFACE commands.

VPOINT Rotate Option _____

The VPOINT Rotate option lets you specify a viewpoint in terms of two angles: one with respect to the X axis (in the XY plane) and another "up" from the XY plane.

1 Enter the **VPOINT** command and **Rotate** option.

2 Enter **0** for the first angle and **0** for the second angle.

An end view (right side) of the building appears.

DDVPOINT makes it easier to take full advantage of the Rotate option.

1 From the **View** pull-down menu, select **3D Viewpoint** and the **Select...** item.

2 Select **270** degrees for the rotation and **0** degrees for the height, and pick **OK**.

A front elevation view of the building appears.

3 Remove hidden lines.

4 Create other elevation views on your own.

5 Close the Surfaces toolbar, as well as any others you may have opened.

6 Save your work and exit AutoCAD.

Questions

1. Describe a 3D face object.

ALLOWS YOU TO CREATE 3 DIM. OBJECT BUT MORE THAN 2D OBJECT. TO PLACE Pts. TO FORM SECTION OF THE PLANE.

2. What basic AutoCAD command is the 3DFACE command most like?

TWO DIMENSIONAL.

3. What can be created with the 3DFACE command that cannot be created with the LINE command? Be specific.

WITH 3DFACE THE OBJECT CAN BE GIVEN A THICKNESS UNTIL A LINE COMMAND IS GIVEN A FORM NO THICKNESS IS GIVEN

4. Describe the X/Y/Z filtering process of entering 3D points.

TO ENTER COORDINATES USING THE KEYBOARD AND THE POINTING DEVICE.

5. Describe the purpose of the VPOINT Rotate option.

TO OBTAIN A PLAN VIEW ON YOUR SCREEN YOU CAN GET IT BY USING VPOINT COMMAND DETERMINE VIEWPOINTS IN TERMS OF 2 ANGLES.

6. What is the primary benefit of using X/Y/Z point filters?

TO SELECT AND DETERMINE VIEWABLE ANGLES OF AN OBJECT.

Challenge Your Thinking: Questions for Discussion

1. Experiment with xlines and mlines in three dimensions. Can you use the X/Y/Z point filters to place these objects? In what ways do they behave differently in 3D space than ordinary AutoCAD lines?

2. Experiment with other combinations of the X/Y/Z point filters, such as .xz and .yz. In what situations might these combinations be useful?

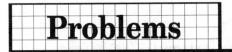

Problems

1. Using AutoCAD's 3D modeling features, add a door, window, and fireplace chimney to the object you created earlier. When finished, the model should look similar to the one below. Save the drawing as prb42-1.dwg.

HINT: Use the 3DFACE command and X/Y/Z point filters when creating the door and window.

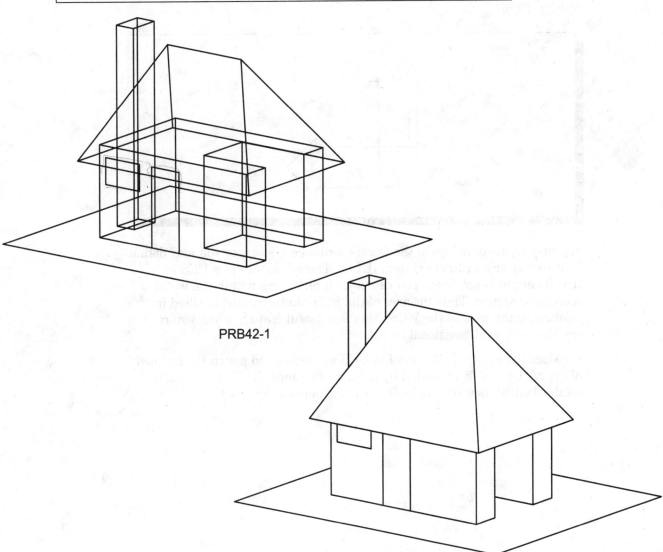

PRB42-1

2. Embellish the drawing further by adding other details to the building. Save the finished drawing as prb42-2.dwg.

User Coordinate Systems

■ OBJECTIVE:

To understand user-definable user coordinate systems (UCSs) and apply them to the construction of 3D wireframe and surface models

As you know, AutoCAD uses a coordinate system; points in a drawing are located by their *x,y,z* coordinates. AutoCAD's default coordinate system is the world coordinate system (WCS). In this system, the X axis is horizontal on the screen, the Y axis is vertical, and the Z axis is perpendicular to the XY plane (the plane defined by the X and Y axes). The WCS is indicated in the lower left corner of the screen by the following coordinate system icon.

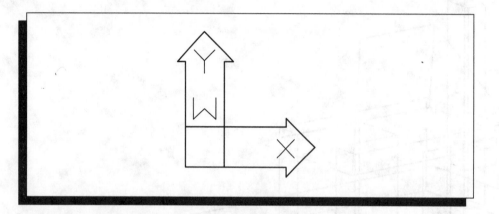

You may create drawings in the world coordinate system, or you may define your own user coordinate system (UCS). The advantage of a UCS is that its origin is not fixed. You can place it anywhere within the world coordinate system. Thus the axes of the UCS can be rotated or tilted in relation to the axes of the WCS. This is a useful feature when you're creating a three-dimensional model.

Consider the following 3D model. A UCS was defined to match the inclined plane of the roof as indicated by the arrows. Once the UCS has been established, all new objects lie in the same plane as the roof.

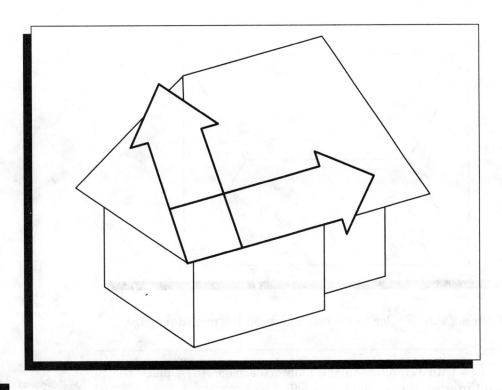

UCS Command

The UCS command enables you to create a new current UCS. Let's try it.

1 Start AutoCAD and open **three-d2.dwg**.

2 Alter the plan view so that it resembles the illustration on the next page. Remove the property line around the model.

HINT: Be sure to set the elevation at 1 and the thickness at 3 before drawing new lines.

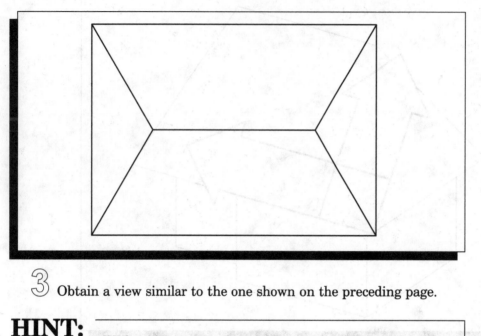

③ Obtain a view similar to the one shown on the preceding page.

HINT: With reference to the globe (see page 467), place the viewpoint in front, above, and to the left.

④ Display the UCS toolbar.

⑤ From the UCS toolbar, pick the **3 Point UCS** icon.

This enters the UCS command and displays the following list of options.

Origin/ZAxis/3point/OBject/View/X/Y/Z/Prev/Restore/Save/Del/?/<World>: _3

Because you chose the 3 Point UCS icon, AutoCAD enters the 3point option automatically. This option allows you to specify the origin, orientation, and rotation of the XY plane in a UCS. It is one of the most useful options of the UCS command.

⑥ In reply to Origin point, pick point 1 (shown on the following page) using the **Endpoint** object snap mode.

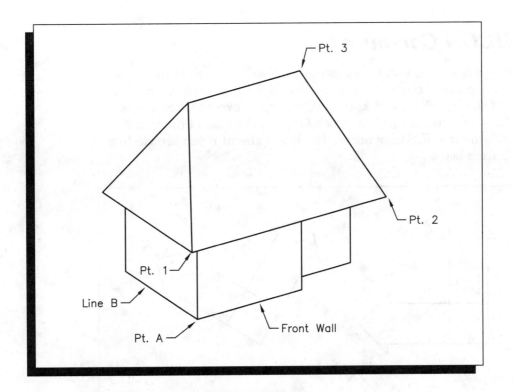

7 In reply to the next prompt, snap to point 2. (It may appear as if you did not snap to point 2.)

This defines the positive X direction from the first point.

8 In reply to the next prompt, snap to point 3. (Point 3 lies in the new XY plane and has a positive *y* coordinate.)

Note that the drawing does not change. However, the crosshairs, grid, and coordinate system icon shift to reflect the new UCS.

9 With the coordinate display on, notice the *x* and *y* values in the status bar as you move the crosshairs to each of the three points you selected.

UCS

10 From the UCS toolbar, pick the **UCS** icon.

This enters the UCS command.

11 Enter **S** for Save, and enter **FRTROOF** for the UCS name.

This names and saves the current UCS.

UCSICON Command

The coordinate system icon indicates the positive directions of the X and Y axes. W appears in the icon if the current UCS is the world coordinate system. If the icon is located at the origin of the current UCS, a + is displayed at the base of the icon. A box forms at the base of the icon if you're viewing the UCS from above. The box is absent if you are viewing the UCS from below.

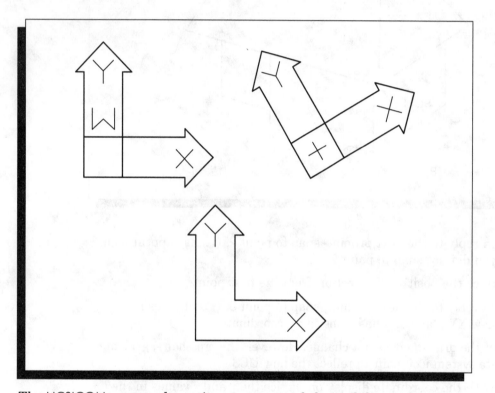

The UCSICON command permits you to control the visibility and position of the coordinate system icon.

1 Enter **UCSICON** and **OR** for ORigin.

The coordinate system icon moves to the origin of the current UCS.

2 Enter **UCSICON** and select the **Noorigin** option.

The coordinate system icon returns to its original location.

3 Enter **UCSICON** and **Off**.

The coordinate system icon disappears.

Enter **UCSICON** and **On**.

The coordinate system icon reappears.

The UCSICON All option applies changes to the coordinate system icons in all active viewports.

Using the New UCS

Suppose you want to construct the line of intersection between the roof and a chimney passing through the roof.

 Change both the elevation value and the thickness value to **0**.

With the **LINE** command, draw the line of intersection (where a chimney would pass through, as shown in the following illustration) by creating a rectangle at any location on the roof.

HINT:
Turn on ortho and snap.

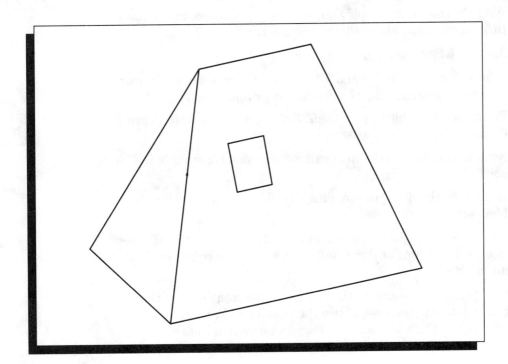

3 Pick the **World UCS** icon from the UCS toolbar to return to the WCS.

4 To prove that the rectangle lies on the same plane as the roof, view the 3D model from different points in space.

HINT: The viewpoint 4,–.1,1 illustrates it well. (Notice the position of the coordinate system icon.)

Let's create a second UCS, this time using a different UCS command option.

1 View the model from an orientation similar to the one used before (as shown on page 491).

2 Pick the **Object UCS** icon from the UCS toolbar.

This enters the OBject option of the UCS command, which allows you to align the UCS with a planar object, such as a 3D face, on the screen.

3 Select the bottom edge of the roof. (Use the same section of the roof as before.)

4 With the coordinate display on, move the crosshairs to the same three corners you chose when applying the 3point option.

Notice that the OBject option creates an identical UCS.

You are likely to use the 3point and OBject options for most applications, at least at first. However, other UCS creation options are available.

1 From the UCS toolbar, pick the **Z Axis Vector UCS** icon. (Turn off ortho.)

2 Snap to point A (see the illustration on page 491) in reply to Origin point.

3 In reply to the next prompt, pick any point on line B using the **Nearest** object snap.

This point specifies the positive Z direction of the UCS. AutoCAD then determines the directions of the X and Y axes using an arbitrary but consistent method.

The new UCS is on the same vertical plane as the front wall of the building. In relation to the wall, notice the positive X, Y, and Z directions that make up the UCS, and notice the coordinate system icon.

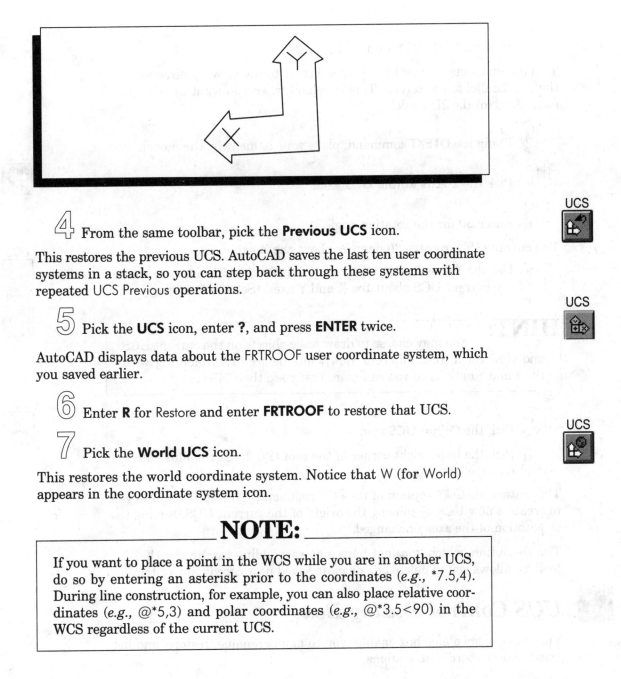

4 From the same toolbar, pick the **Previous UCS** icon.

This restores the previous UCS. AutoCAD saves the last ten user coordinate systems in a stack, so you can step back through these systems with repeated UCS Previous operations.

UCS

5 Pick the **UCS** icon, enter **?**, and press **ENTER** twice.

AutoCAD displays data about the FRTROOF user coordinate system, which you saved earlier.

6 Enter **R** for Restore and enter **FRTROOF** to restore that UCS.

UCS

7 Pick the **World UCS** icon.

This restores the world coordinate system. Notice that W (for World) appears in the coordinate system icon.

NOTE:

If you want to place a point in the WCS while you are in another UCS, do so by entering an asterisk prior to the coordinates (*e.g.*, *7.5,4). During line construction, for example, you can also place relative coordinates (*e.g.*, @*5,3) and polar coordinates (*e.g.*, @*3.5<90) in the WCS regardless of the current UCS.

8 Pick the **View UCS** icon.

This option creates a new UCS perpendicular to the viewing direction; that is, parallel to the screen. This is helpful when you want to annotate (add notes to) the 3D model.

9 Using the **DTEXT** command, place your name near the model.

10 Pick the **Z Axis Rotate UCS** icon.

11 Enter **30** for the rotation angle.

The current UCS rotates 30 degrees about the Z axis.

12 Use the **X Axis Rotate UCS** and **Y Axis Rotate UCS** icons to rotate the current UCS about the X and Y axes. (See the following hint.)

HINT:

You may choose to draw basic objects on the current UCS and view them from different viewpoints before experimenting with the X and Y options so you can more easily see their effects.

13 Pick the **Origin UCS** icon.

14 Pick the lower right corner of the roof (Pt. 2; see the illustration on page 493).

This enters the Origin option of the UCS command. This option allows you to create a new UCS by moving the origin of the current UCS, leaving the orientation of the axes unchanged.

The Del option, which does not have a corresponding icon in the UCS toolbar, allows you to delete one or more saved user coordinate systems.

UCS Control Dialog Box

The UCS Control dialog box enables you to name, rename, restore, and list existing user coordinate systems.

1 From the UCS toolbar, pick the **Named UCS** icon.

```
UCS Control                                    [X]
UCS Names
*WORLD*
*PREVIOUS*
*NO NAME*                      Current
FRTROOF
       ▶


       ┌──────────┐  ┌──────────┐  ┌──────────┐
       │ Current  │  │ Delete   │  │ List...  │
       └──────────┘  └──────────┘  └──────────┘
       ┌──────────┐  ┌──────────────────────┐
       │Rename To:│  │ FRTROOF              │
       └──────────┘  └──────────────────────┘
         ┌──────┐   ┌──────────┐  ┌──────────┐
         │  OK  │   │  Cancel  │  │  Help    │
         └──────┘   └──────────┘  └──────────┘
```

The current UCS (indicated by Current) has no name. Therefore, AutoCAD calls it *NO NAME*.

2 Select **FRTROOF** and pick the **List...** button.

This lists the points that define the FRTROOF UCS.

3 Pick **OK**.

Notice the Delete and Rename To buttons. They allow you to delete and rename existing user coordinate systems.

4 Pick the **Current** button and pick **OK**.

FRTROOF is now the current UCS.

UCS Orientation Dialog Box _____

This dialog box enables you to select predefined user coordinate systems.

UCS

1 Pick the **Preset UCS** icon from the UCS toolbar.

This enters the DDUCSP command, which displays the UCS Orientation dialog box.

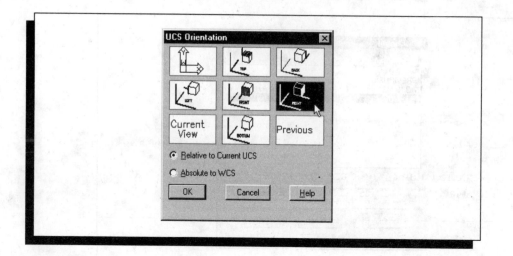

You can now select a new UCS that is at right angles to the current UCS or WCS.

2 Pick the **RIGHT** image tile, as shown in the previous illustration, and pick **OK**.

The position of the coordinate system icon reflects the new UCS.

3 Bring back the UCS Orientation dialog box and pick the **Absolute to WCS** radio button.

4 Select the image tile of the coordinate system icon, located in the upper left corner, and pick **OK**.

The WCS is now the current coordinate system.

5 Bring back the dialog box, select the **FRONT** image tile, and pick **OK**.

This produces a new UCS that is perpendicular to the WCS and parallel to the front of the building.

6 Experiment with the other options on your own.

PLAN Command

The PLAN command enables you to generate the plan view of any user coordinate system, including the WCS.

1 Enter the **PLAN** command.

2 Press **ENTER** to generate the plan view of the current UCS.

3 Enter **PLAN** and **W** for World.

A "broken pencil" icon, shown in the following illustration, replaces the coordinate system icon whenever the XY plane of the current coordinate system is perpendicular to the computer screen. This indicates that drawing and selecting objects is limited in this situation.

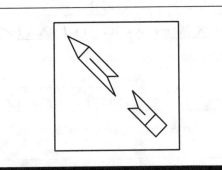

UCS

4 Pick the **World UCS** icon.

If you want to display the plan view of previously saved user coordinate systems, enter the UCS option of the PLAN command.

NOTE:

If you want AutoCAD to generate a plan view automatically whenever you change from one UCS to another, set the UCSFOLLOW system variable to 1 (on).

5 Erase your name and the chimney intersection so that only the building remains on the screen, and **ZOOM All**.

6 Close the UCS toolbar, save your work and exit AutoCAD.

Questions

1. What is the purpose and benefit of using AutoCAD's user coordinate systems (UCSs)?

   TO DRAW IN WORLD COORDINATE SYSTEM. OR TO DEFINE OUR OWN USER COORDINATE SYSTEM. THE ADVANTAGE IS NOT LIMITED. IT IS USEFUL WHEN CREATING 3-D MODEL.

2. Describe the UCS command's 3point option.

   ALLOWS YOU TO SPECIFY ORIGIN, ORIENTATION AND ROTATION OF XY PLANE IN A "UCS".

3. What purpose does the coordinate system icon serve?

   PERMITS YOU TO CONTROL THE VISIBILITY & POSITION OF COORD. SYSTEM ICON.

4. What does the W represent in the coordinate system icon?

   WORLD

5. If a box is present at the base of the coordinate system icon, what does this mean?

   IF YOU ARE VIEWING THE UCS FROM ABOVE.

6. What is the purpose of the UCS Control dialog box?

   ENABLES YOU TO NAME, RENAME, RESTORE AND LIST EXISTING USER COORDINATE SYSTEMS

7. Describe, briefly, the UCS command's View option.

   CREATES A NEW UCS PERPENDICULAR TO THE VIEWING DIRECTION. THAT IS PARALLEL TO THE SCREEN. GOOD, ESPECIALLY when you ADD NOTES TO the 3D MODEL.

Challenge Your Thinking: Questions for Discussion

1. Explain the difference between a view and a UCS, and describe how the two can be used together to create complex 3D drawings.

2. In the UCS Orientation dialog box, radio buttons allow you to choose whether a new UCS is positioned relative to the current UCS or absolute to the WCS. Explain the effect these options have on the new UCS.

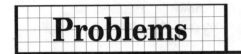

Problems

1. With the UCS 3point option, create a UCS using one of the walls that make up the building in this unit. Save the UCS using a name of your choice. With this as the current UCS, add a door and window to the building. Save this drawing as prb43-1.dwg.

2. Using VPOINT or DDVPOINT, obtain a 3D view of the building in this unit. With the UCS View option, create a UCS. Create a border and title block for the 3D model. Include your name, the date, the file name, etc., in the title block. Save your work as prb43-2.dwg.

■ OBJECTIVE:

To practice using AutoCAD's dynamic viewing capability in connection with 3D modeling

The VPOINT and DDVPOINT commands let you view 3D models from any angle in space. The DVIEW command also lets you view models from various angles in space, but in a dynamic way. It provides a range of 3D viewing features and enables you to generate 3D perspective projections—as opposed to parallel projections—of your 3D work.

Consider the two buildings shown below. Both buildings are shown from the same point in space. However, the one on the left is shown in standard parallel projection. The lines are parallel and do not converge toward one or more vanishing points. The right view is a true perspective projection. Parts of the building that are farther away appear smaller—the same as in a real-life photograph.

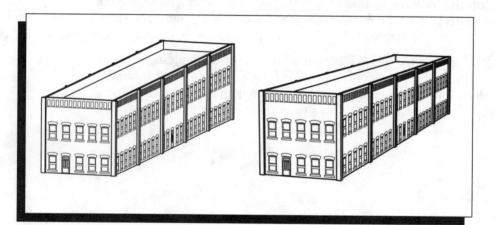

Other benefits of dynamic viewing are its exceptionally fast zooming and panning and fast hidden line removal. Plus, you can create front and back *clipping planes* of your 3D models.

■ *Perspective Projections* _____

Using the dynamic view facility, let's generate a 3D perspective projection.

1 Start AutoCAD and open **three-d2.dwg**.

2 Using either **VPOINT** or the **Viewpoint Presets** dialog box, create a view similar to the parallel projection shown in the following illustration and remove hidden lines.

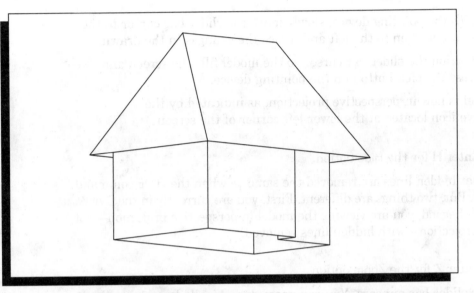

③ Enter **DV** (for DVIEW) at the keyboard or select the **View** pull-down menu and pick **3D Dynamic View**.

④ In response to Select objects, select the entire model and press **ENTER**.

NOTE:

The objects you select will be those you preview as you perform the DVIEW options. If you select too many, dragging and updating of the image may be slow. However, choosing too few may not provide an adequate preview.

The following options appear.

CAmera/TArget/Distance/POints/PAn/Zoom/TWist/CLip/Hide/Off/Undo/<eXit>:

⑤ Enter **D** for Distance.

A *slider bar* appears at the top of the screen.

The DVIEW command uses a camera and target metaphor. The slider bar is labeled from 0x to 16x, with 1x representing the current distance. Moving the slider to the right increases the distance between the camera and the target. Moving the slider to the left moves the camera closer to the target.

6　With the pointing device, slowly move the slider bar cursor to the right and then to the left and notice the changes in the drawing.

7　Position the slider bar cursor so the model fills the screen and press the pick button on the pointing device.

The model is now in perspective projection, as indicated by the perspective icon located at the lower left corner of the screen.

8　Enter **H** for the Hide option.

Notice that hidden lines are removed the same as when the HIDE command is applied. But, two things are different. First, you are currently in the DVIEW command. Second, you are viewing the model in perspective projection—not parallel projection—with hidden lines removed.

9　Enter **Z** for the Zoom option.

A similar slider bar appears. You can zoom dynamically by moving the slider back and forth.

10　Move the slider so the building fills about one third of the screen and press the pick button.

11　Enter **D** the Distance option again, and change the distance between the camera and target.

12　Enter the **Hide** option and press **ENTER** to terminate the DVIEW command.

Notice that the hidden lines reappear but the perspective view remains.

13　Enter **DVIEW**, select the model (and press **ENTER**), and select the **Off** option to turn off the perspective projection.

14　Experiment further with the DVIEW Distance, Zoom, and Hide options and press **ENTER** when you are finished.

Other DVIEW Options

Use of the camera/target metaphor helps you view the 3D model as it appears from any point in space. The line of sight (also referred to as the viewing direction) is the line between the camera and target.

1　Enter the **DVIEW** command, select the model once again, and press **ENTER**.

2　Enter the **Off** option and then the **CAmera** option.

3　Slowly move the crosshairs up and down and back and forth.

506

Notice that the movement of the crosshairs rotates the model. Technically, you are moving the camera—the point you are looking from—around the target.

4 Position the 3D model at any orientation, pick a point, and enter the **Hide** option.

5 Enter the **CAmera** option again and pick a new point.

6 Enter the **Undo** option to undo the last operation.

7 Enter the **TArget** option, move the crosshairs, and pick a point to define the new target angle.

The TArget option is similar to the CAmera option, but it rotates the target point—the point you are looking *at*—around the camera.

8 Experiment further with the DVIEW TArget option on your own.

NOTE:

The DVIEW POints option is available if you want to define new camera target points using *x*, *y*, and *z* coordinates.

As a result of using the TArget option, the model may appear partially off the screen.

9 Enter the **DVIEW PAn** option to shift the model dynamically to a new location.

The TWist option lets you rotate the model around the line of sight. It is similar to using the ROTATE command in a 2D environment.

10 Enter the **TWist** option, move the crosshairs, and pick a point at any location.

Clipping Planes

The DVIEW CLip option enables you to create front and back clipping planes. It lets you view the interior of the model in a manner similar to using conventional sectional views.

1 Using the various DVIEW options, position the model so it is similar to the one at the beginning of this unit.

2 Enter the **DVIEW CLip** option and then choose the **Front** option.

3 Slowly move the slider and pay particular attention to how the model changes.

4 Attempt to display only two-thirds of the model and pick a point.

5 Enter the **Hide** option.

The CLip Front option removes the front portion of the model—the portion between you and the clipping plane—and produces a view similar to the following.

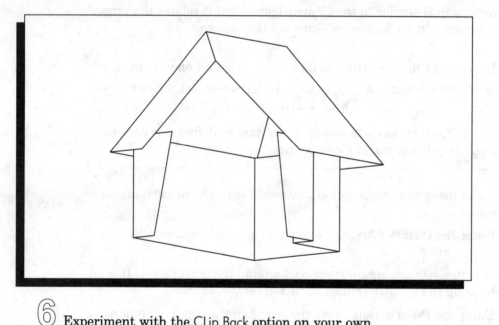

6 Experiment with the CLip Back option on your own.

7 Enter the **Hide** option to realize the effect of the CLip Back option.

Does your view look similar to the following? Notice that both front and back clipping planes are in effect.

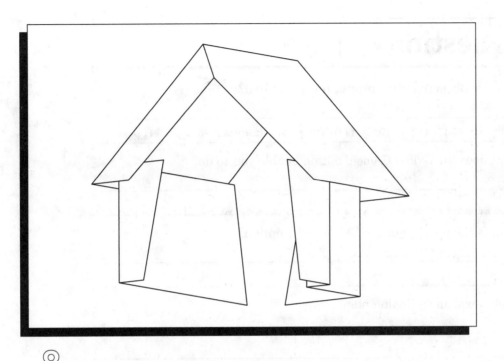

8 Enter the **Clip** option, and then **Off**.

The model returns to its previous form.

9 Terminate DVIEW by pressing **ENTER**.

10 Enter **DVIEW** and give a null response (press **ENTER**) in reply to Select objects.

AutoCAD displays a new building model.

11 Use the **DVIEW Zoom** option to resize it.

The edges of the model are aligned with the X, Y, and Z axes of the current UCS. The model is updated to reflect the changes you make while in the DVIEW command and is meant for experimentation purposes. When you exit the DVIEW command, the current drawing regenerates based on the view you selected.

12 Press **ENTER** to terminate the DVIEW command.

Notice that the drawing regenerates based on the view you selected.

13 Exit AutoCAD. Do not save any changes.

Questions

1. How are 3D perspective projections generated?

   THROUGH THE DVIEW COMMAND. IT PROVIDES
   3D VIEWING FEATURES INSTEAD OF JUST PARALLE
   PROJECTIONS

2. What does the DVIEW CAmera option enable you to do?

   HELPS YOU VIEW THE OBJECT AS IT APPEARS
   FROM ANY POINT IN SPACE. AS IF YOU WERE
   MOVING THE CAMERA
   AROUND.

3. What is the purpose of the DVIEW TWist option?

   TO ROTATE THE MODEL
   THE LINE OF SIGHT.

4. Briefly explain a clipping plane.

   ALLOWS YOU TO CLIP FRONT & BACK OF
   MODEL TO VIEW THE INSIDE OF OF MODEL
   SIMILAR TO SECTIONS

5. What is the difference between a parallel projection and a perspective projection?

   PARALLEL - LETS YOU SEE MODEL FROM ANY ANGLE
   IN SPACE. PERSPECTIVE - IS 3D MODEL ANY angle
   IN SPACE.

6. What is the purpose of the model that appears when you press ENTER at the Select objects prompt for the DVIEW command?

   IT GOES BACK TO ONE OF MORE VANISHING PTS
   IN TRUE "LIKE" PESPECTIVE.

■ *Challenge Your Thinking: Questions for Discussion*

1. Of what benefit are front and back clipping planes? Describe at least one situation in which clipping planes could save drawing time or reveal important information about the design.

Problems

1. While in the WCS plan view, create an array of lines. Create them in both the X and Y directions and place them approximately one unit apart. View the grid of lines at an arbitrary point in space using the VPOINT command. Create a perspective projection of the view using the DVIEW command. Use other dynamic view options to view the grid at various orientations and sizes in space. Save the grid as prb44-1.dwg.

2. Obtain a left view of the building using the DVIEW command. Use a front clipping plane to create a full cross section of the model. Save this drawing as prb44-2.dwg.

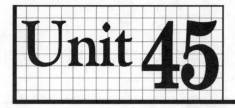

Unit 45 — 3D Revolutions

■ OBJECTIVE:

To apply the REVSURF and RULESURF commands to the construction of curved surfaces in 3D space

The REVSURF command enables you to create a surface of revolution by rotating a path curve (or *profile*) around a selected axis. The RULESURF command lets you create a polygon mesh representing a ruled surface between two curves.

The following 3D model represents a table lamp. The lamp's base was created with the REVSURF command, and the lamp shade was created with the RULESURF command.

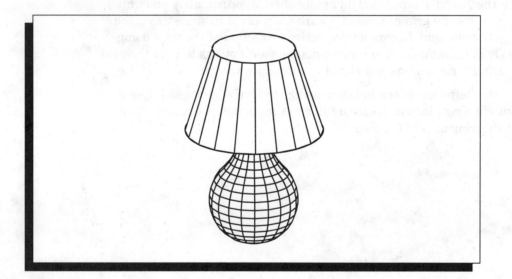

3D Drawing Template _____

In preparation for creating a lamp similar to the one shown above, let's create a new template file. The template will apply to subsequent 3D exercises in this unit and following units. Its purpose is to help you orient yourself as you create 3D models.

1 Start AutoCAD and begin a new drawing from scratch.

2 Set the grid at **1** unit and the snap at **.5** unit.

3 Create the following three layers:

Layer Name	Layer Color
Box	White
3dobject	Red
Object2	Cyan

4 Set the layer named **Box** as the current layer.

5 Enter the **ELEV** command and enter **0** for the elevation and **2** for the thickness.

6 In the center of the screen, construct a rectangular box using a polyline. Make it **4** units on the X axis by **3** units on the Y axis.

7 Specify a viewpoint in front, above, and to the right of the rectangular 3D box.

8 If necessary, use the **PAN** command to center the box on the screen.

9 Enter **ZOOM** and **.9x**.

10 Save your work in a template file named **3dtmp.dwt**. For Template Description, enter **3D drawing template**.

11 Using the **UCS** command and **3point** option, create and save each of the six user coordinate systems shown in the following illustration. Remember, the 3point option lets you specify the origin, orientation, and rotation of the XY plane.

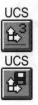

HINT:

In the following illustration, each UCS name lies on its respective UCS. Likewise, each leader points to its respective UCS origin (0,0,0). Do not include the leaders and text. You may want to display the UCS toolbar.

12 List the new user coordinate systems.

You should have six of them.

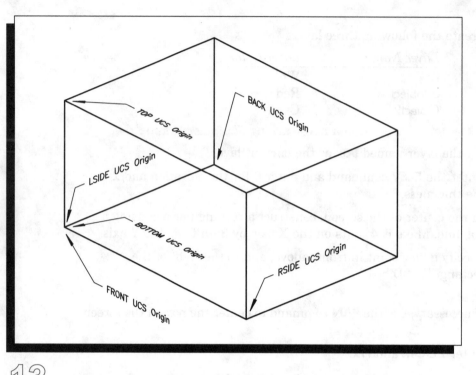

13 Make **3dobject** the current layer.

SURFTAB1 and SURFTAB2

The system variables SURFTAB1 and SURFTAB2 control the density (resolution) of 3D meshes created by 3D commands such as RULESURF and REVSURF.

1 Enter the **SURFTAB1** system variable.

2 Enter **20** for its value.

3 Enter the **SURFTAB2** system variable.

4 Enter **8** for its value.

NOTE:

See the section titled "3D Polygon Meshes," later in this unit, for more information about SURFTAB1 and SURFTAB2.

5 Enter **ELEV** and set both elevation and thickness to **0**.

6 Save your work.

REVSURF Command _____

1 Using **Save As...**, begin a new drawing and name it **lamp.dwg**.

2 Pick the **Named UCS** icon from the UCS toolbar and make **FRONT** the current UCS.

UCS

3 Enter the **HIDE** command to remove hidden lines.

4 Using the right front corner of the rectangular 3D model as the starting point, approximate the following shape with the **PLINE** command and the **Arc** option. (See the following hint.)

HINT: _____
Begin at the bottom corner. Use the Undo option if you need to redo segments of the polyline.

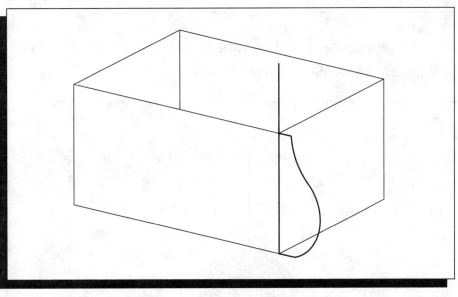

5 With the **LINE** command, draw a line that passes through the corner of the rectangular model and extends upward as indicated.

6 Display the Surfaces toolbar.

7 From this toolbar, pick the **Revolved Surface** icon.

Surfaces

This enters the REVSURF command.

8 Pick the polyline in reply to Select path curve.

9 Pick the line in reply to Select axis of revolution.

10 Enter **0** (the default) in reply to Start angle.

11 Enter **360** (the default, <Full circle>) in reply to Included Angle.

A 3D model of the lamp base appears.

12 Freeze the layer named **Box**.

13 Enter the **HIDE** command to remove hidden lines.

NOTE:

The HIDE operation may take several seconds if you are working on a relatively slow computer.

14 Thaw the layer named **Box** and enter **HIDE** again.

Your lamp base should look similar to the one shown here.

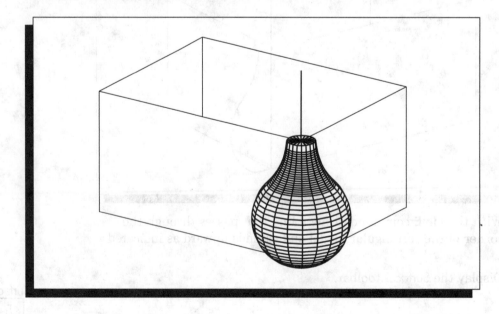

RULESURF Command _____

Now let's create the lamp shade.

1 Make **Object2** the current layer.

2 Pick the **Named UCS** icon and make **TOP** the current UCS.

UCS

3 Enter the **CIRCLE** command.

4 Pick the upper right front corner of the rectangular model for the circle's center point. (Refer to the following illustration.)

5 Enter **1.25** units for the radius of the circle.

This circle lies in the TOP UCS, and it will remain on this plane.

6 Create another circle using the same center, but enter a radius of **.75**.

This circle also lies in the TOP UCS, but you will move it upward in Step 8.

UCS

7 Make **FRONT** the current UCS.

This will allow you to move the circle upward in the next step.

8 With the ortho and snap modes on, move the smaller circle up **1.5** units.

HINT: _____

Turn on the coordinate display and watch it as you move the circle upward.

The small circle should now be 1.5 units above the TOP UCS.

Your 3D model should now look similar to the one shown below.

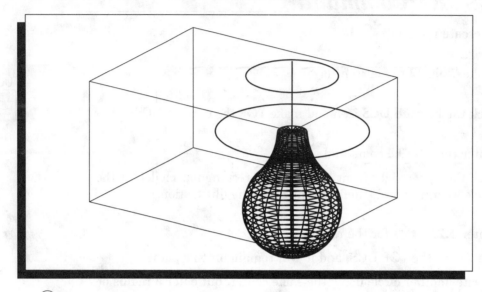

Surfaces

9 From the Surfaces toolbar, pick the **Ruled Surface** icon.

This enters the RULESURF command.

10 Pick one of the two circles in reply to Select first defining curve.

HINT: You may need to turn snap off.

11 Pick the other circle in reply to Select second defining curve.

Did the lamp shade appear as you envisioned it?

12 Freeze layer **Box**.

13 Enter the **HIDE** command.

Your 3D model should look similar to the one found at the beginning of this unit.

14 Enter the **DVIEW** command and view the lamp from various orientations in space.

15 Close the Surfaces and UCS toolbars, as well as any other toolbars you may have opened.

16 Save your work and exit AutoCAD.

HINT:

When constructing 3D models, consider using multiple viewports. The illustration below provides an example.

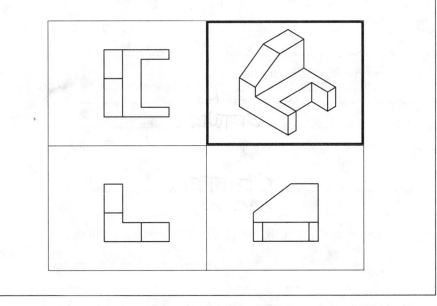

3D Polygon Meshes

A polygon mesh is defined in terms of $M \times N$ vertices. Envision the vertices as a grid consisting of columns and rows, with M and N specifying the column and row position of any given vertex.

The system variables SURFTAB1 and SURFTAB2 control the density (or resolution) of the 3D mesh. The system variable SURFTAB1 controls the N direction of a polygon mesh. An example is the resolution of the lamp shade created by the RULESURF command.

Consider the lamp base. Both SURFTAB1 and SURFTAB2 come into play because both M and N vertices are applied by the REVSURF command.

If you increase the values of these two system variables, the appearance of 3D models may improve, but the model will require more time to generate on the screen. The HIDE command will also consume more time. If you decrease the numbers excessively, the model will generate quickly on the screen, but the 3D model on the screen may not adequately represent your design.

NOTE:

You can edit 3D meshes using the PEDIT command. You can also explode meshes using the EXPLODE command.

Questions

1. What is the purpose of the REVSURF command?

   TO CREATE A SURFACE OF REVOLUTIONS
   by ROTATING A PATH CURVE.

2. Describe the use of the RULESURF command.

   TO CREATE A POLYGON MESH REPRESENTING
   A RULED SURFACE BETWEEN 2 CURVES

3. How are user coordinate systems beneficial when using the REVSURF
   and RULESURF commands?

   ALLOWS TO DRAW AT DIFFERENT VIEWS
   OF CUBE BY MAKING ONE THE CURRENT

4. Explain the purpose of the SURFTAB1 and SURFTAB2 system variables.

   CONTROLS THE DENSITY OF MESHES (FILL IN)
   OF THE 3D MODELS.

5. What are the consequences of entering high values for the SURFTAB1
   and SURFTAB2 variables?

   THE APPEARANCE OF THE MODELS WILL IMPROVE
   (closer) LINES BUT will take Longer to regenerate.

6. What are the consequences of entering low values for the SURFTAB1
   and SURFTAB2 variables?

   MAY NOT ADAQUATELY DISPLAY OR REPRESENT
   YOU DESIGN

7. In AutoCAD terms, 3D polygon meshes are made up of what?

   A GRID OF COLUMNS + ROWS TO FILL IN DESIGN

Challenge Your Thinking: Questions for Discussion

1. Could you have created the lamp in this unit without creating and
   using the six user coordinate systems? Explain.

2. It is possible to edit the lamp you created in this unit using PEDIT,
   the same command you have used previously to edit 2D polylines.
   Experiment with the use of PEDIT with objects such as the lamp base
   and lamp shade. Be sure to save the file with a new file name first.
   Then write a short essay explaining the use of the PEDIT command
   with 2D and 3D objects. How do the options differ? What do the 3D
   options do?

Problems

1-4. Using the 3D features covered in this unit, create each of the following 3D models. Approximate all sizes. Use REVSURF to create the drawings in problems 1 and 4, and use RULESURF to create the drawings in problems 2 and 3. Also, use the 3dtmp.dwt template file.

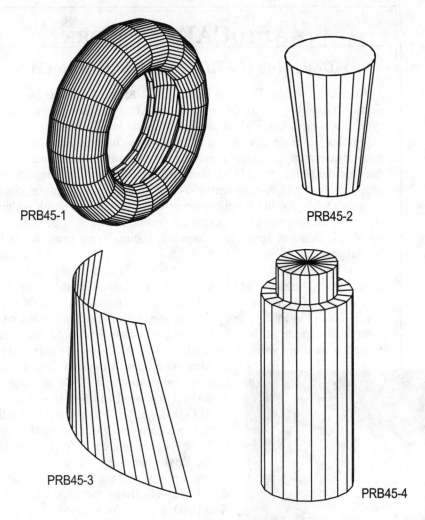

PRB45-1

PRB45-2

PRB45-3

PRB45-4

5. Refer to visual.dwg and its description contained on the optional *Applying AutoCAD Diskette*. Load visual.dwg and review its contents. Apply it to the development of a new 3D model.

6. Refer to 3dport.dwg contained on the optional *Applying AutoCAD Diskette*. Load the file and review the model orientation in each viewport. Move from viewport to viewport and change the orientation and magnification of each view.

AUTOCAD® AT WORK

Designing the Right Stuff with AutoCAD

Engineers at the Robotics branch of NASA's Goddard Space Flight Center (Greenbelt, MD) are using AutoCAD to develop a completely new type of robotic hand. Unlike previous robotic hands, the five-finger hand resembles the basic skeletal structure of the human hand and incorporates its dexterity. Instead of relying on elaborate computer architecture for this dexterity, the NASA hand mirrors the natural, intuitive hand movements of a human operator. The operator wears a special glove that records and transmits hand movements to the robotic hand, which then repeats each movement. Someday, the robotic hand may be used in space by NASA mission specialists to work remotely on projects requiring human dexterity.

The robotic hand research project is one of four that Charles Engler, a mechanical engineer at NASA's Goddard Space Flight Center (GSFC), is working on simultaneously. Because his time is so valuable, Engler says, "I can't afford to spend a lot of time on design iterations, even though design iterations are a basic part of research and development. What's nice about using AutoCAD is that I don't have to start over from scratch every time I make a design change." Engler estimates his time savings with AutoCAD at more than 50 percent.

At GSFC, quality is the watchword. Before prototypes can be built, design drawings must be created to certain Goddard specifications. Engler explains, "I can use AutoCAD to set up my format parameters so every drawing meets these specifications. By using AutoCAD, along with decreased design time, I get high-quality drawings."

Already, three space-flight hardware designers and five engineers are using AutoCAD for their projects. "Now, all I have to do is put my drawing on a floppy disk, take it to a designer, and ask him for his opinion," Engler says.

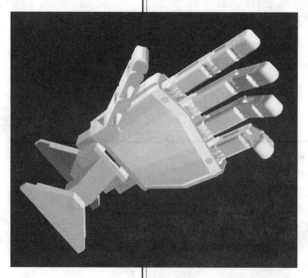

Courtesy of Autodesk, Inc.

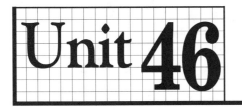

Unit 46 — Advanced 3D Modeling

■ OBJECTIVE:

To practice the 3D modeling capabilities of the TABSURF, EDGESURF, and 3DMESH commands

AutoCAD's TABSURF command enables you to create a polygon mesh representing a tabulated surface defined by a *path* (profile) and *direction vector*. The EDGESURF command lets you construct a *Coons surface patch* from four adjoining edges. The most basic of the 3D-specific commands is 3DMESH. It enables you to define a 3D polygon mesh by specifying its size (in terms of *M* and *N*) and the location of each vertex in the mesh.

The following illustration shows basic 3D models created with each of these commands.

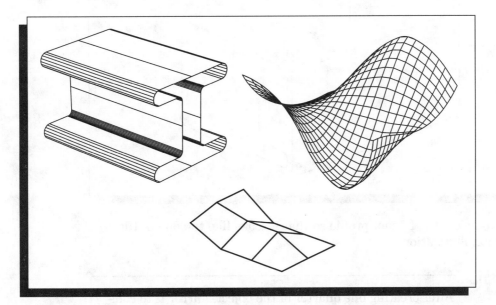

Let's create each of these 3D objects.

■ *TABSURF Command*

We'll use the TABSURF command to construct the I-beam shown above.

1 Start AutoCAD and begin a new drawing using the template file named **3dtmp.dwt**.

2 Make **RSIDE** the current UCS.

UCS

HINT:

For this unit, you may want to display the UCS and Object Snap toolbars or use the flyouts available from the standard toolbar.

③ Using the **PLINE** command's **Arc** and **Line** options, approximate the following figure. It represents one quarter of the I-beam profile.

NOTE:

It's important to use the PLINE command rather than the LINE and ARC commands so that AutoCAD will treat the profile as a single object.

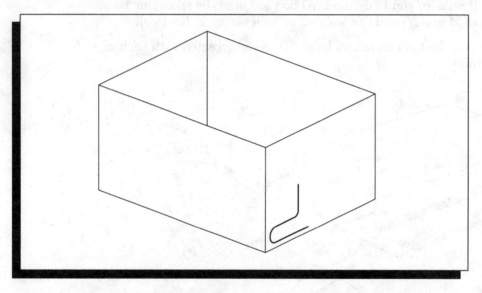

④ Complete the I-beam profile so that it looks like the one in the next illustration.

HINT:

After creating one quarter of the object, mirror it to complete half of the object. Then mirror that to complete the entire object.

⑤ Save your work in a file named **i-beam.dwg**.

⑥ Make **TOP** the current UCS and make **Object2** the current layer.

UCS

⑦ Beginning at the RSIDE UCS as shown in the following illustration, draw a line approximately 3 units long.

HINT:

Turn on the snap mode. It should already be set to .5 unit. Monitor the line's length at the coordinate display.

The line will provide the *direction vector* for use with the TABSURF command.

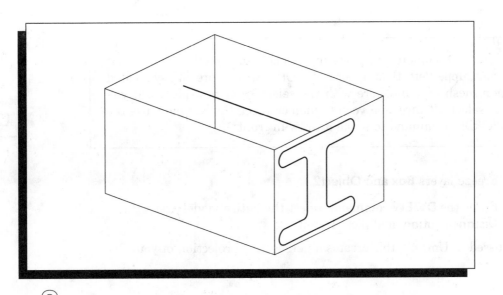

8 Make **3DOBJ** the current layer.

9 Enter the **SURFTAB1** system variable and enter **8**.

____NOTE:____

System variable SURFTAB1 controls the density of the tabulated surface.

10 Display the Surfaces toolbar.

Surfaces

11 From this toolbar, pick the **Tabulated Surface** icon.

This enters the TABSURF command.

12 In reply to Select path curve, pick a point on the lower right quadrant of the I-beam profile.

HINT:

You may need to turn off the snap mode.

13 Pick a point on the line in reply to Select direction vector. The point *must* be closer to the right endpoint of the line than the left endpoint in order for the tabulated surface to extend in the desired direction.

One quarter of the I-beam appears.

14 Complete the remaining parts of the I-beam using the **TABSURF** command. (See the following hint.)

HINT:

Complete the parts in a clockwise direction—lower left quadrant, upper left, then upper right. Otherwise, a previously created polygon mesh may interfere with the selection of new points. If you cannot select the direction vector when creating the last part, use the DRAWORDER command to make it draw in front.

15 Freeze layers **Box** and **Object2**.

16 Enter the **DVIEW** command, select the entire model, issue the **Distance** option, and pick a point.

As discussed in Unit 43, this creates a perspective projection of your 3D model.

17 Enter the **Hide** option.

Your 3D model of the I-beam should look similar to the one found at the beginning of this unit.

18 Press **ENTER** to terminate the **DVIEW** command and save your work.

EDGESURF Command

The EDGESURF command is used to construct a *Coons surface patch*. A Coons patch is a 3D surface mesh interpolated (approximated) between four adjoining edges. Coons surface patches are used to define the topology of complex, irregular surfaces such as land formations and manufactured products such as car bodies.

Let's apply the EDGESURF command to the creation of a topological figure similar to the one shown at the beginning of this unit.

1 Use the **3dtmp.dwt** template file to begin a new drawing.

2 Make **Object2** the current layer.

3 Using the **LINE** command, draw four vertical lines directly on top of the existing four vertical edges of the 3D box.

The lines will permit you to snap to the corners of the box during the construction of the Coons patch.

4 Enter **OSNAP**, pick the **Clear All** button, check **Endpoint** and **Nearest**, and pick **OK**.

5 Make **3dobject** the current layer.

UCS

6 Make **FRONT** the current UCS.

7 Using the **Arc** option of the **PLINE** command, approximate the construction of polyline A as shown in the following illustration.

__NOTE:__

When picking the first and last points of the polyline, use the Nearest object snap mode. This will snap these points onto the vertical line. If you do not do this, the polyline endpoints will not meet accurately. If they do not meet, the EDGESURF command will not work.

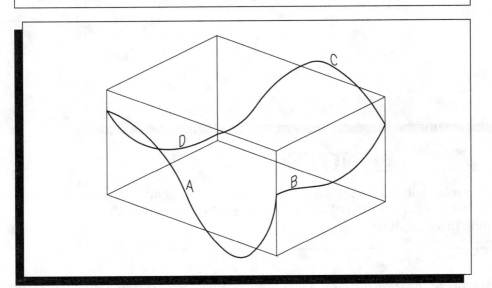

UCS

8 Make **RSIDE** the current UCS.

9 Approximate the construction of polyline B as shown in the previous illustration. Use the **Endpoint** and **Nearest** object snap modes for the first and last endpoints of the polyline.

10 Make **BACK** the current UCS and construct polyline C using the same procedure described in the preceding step.

11 Make **LSIDE** the current UCS and construct polyline D. Use the **Intersection** object snap mode to connect polyline D to polyline A.

12 Freeze layers **Box** and **Object2**.

13 Save your work in a file named **contour.dwg**.

Surfaces

14 Pick the **Edge Surface** icon from the Surfaces toolbar.

This enters the EDGESURF command.

15 In reply to Select edge 1, pick a point on polyline A near corner 1. (Refer to the following illustration.)

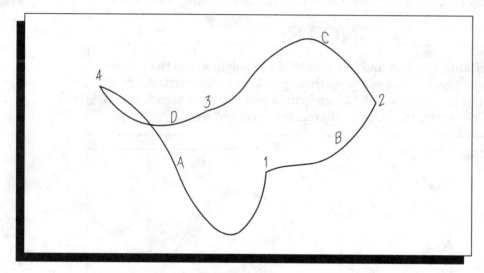

NOTE:

When you select polyline A, it is important that you pick a point on the polyline that is near corner 1. The same is true when you select the remaining three polylines.

16 In reply to Select edge 2, pick a point on polyline B near corner 2.

17 Select polylines C and D in the same fashion.

A contour similar to the one shown at the beginning of this unit appears.

18 Enter the **HIDE** command.

19 View the contour from various orientations in space.

20 Save your work.

3DMESH Command

The 3DMESH command produces a 3D polygon mesh.

1 Begin a new drawing using the **3dtmp.dwt** template file.

Surfaces

2 Pick the **3D Mesh** icon from the Surfaces toolbar.

This enters the 3DMESH command.

3 Enter **4** in reply to Mesh M size.

4 Enter **3** in reply to Mesh N size.

5 Enter the following in reply to the series of Vertex prompts. Be sure to include the decimal points.

Vertex (0, 0): **5,4,.2**
Vertex (0, 1): **5,4.5,.3**
Vertex (0, 2): **5,5,.3**
Vertex (1, 0): **5.5,4,0**
Vertex (1, 1): **5.5,4.5,.2**
Vertex (1, 2): **5.5,5,0**
Vertex (2, 0): **6,4,0**
Vertex (2, 1): **6,4.5,.2**
Vertex (2, 2): **6,5,0**
Vertex (3, 0): **6.5,4,0**
Vertex (3, 1): **6.5,4.5,0**
Vertex (3, 2): **6.5,5,0**

6 If the mesh does not appear, **ZOOM All** or **Extents**.

Your mesh should look similar to the one at the beginning of this unit.

As you can see, specifying even a small three-dimensional polygon mesh is very tedious. The 3DMESH command is meant to be used primarily with AutoLISP and not in the fashion presented above.

7 Freeze layer **Box**.

NOTE:

The PFACE command is similar to 3DMESH. PFACE produces a polygon mesh of arbitrary topology called a *polyface mesh*.

8 Save your work in a file named **3dmesh.dwg**.

Editing 3D Polygon Meshes

The vertices of a mesh can be edited with the PEDIT command in a manner similar to editing a polyline.

Modify II

1 Enter the **PEDIT** command, pick the mesh, and enter the **Edit vertex** option.

An × appears at one corner of the mesh.

2 Press **ENTER** four times.

This moves the × to four consecutive vertices.

3 Enter the **Move** option and pick a new location for the vertex.

4 Enter **eXit** twice to exit the PEDIT command.

Types of Mesh Surfaces

3D polygon meshes can be viewed in any one of these surface types.

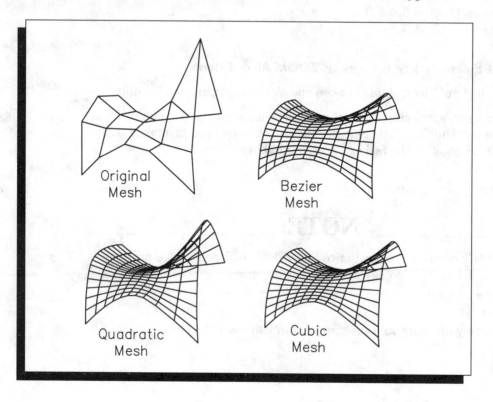

Original Mesh

Bezier Mesh

Quadratic Mesh

Cubic Mesh

The type of mesh displayed is controlled by the SURFTYPE system variable.

Value of SURFTYPE	Surface Type
5	Quadratic mesh
6	Cubic B-spline mesh
8	Bezier mesh

Let's create a quadratic B-spline from our original mesh.

1 Enter the **SURFTYPE** system variable.

2 Enter **5** to specify the quadratic B-spline surface.

Modify II

3 Enter the **PEDIT** command, pick the mesh, and enter the **Smooth surface** option.

The 3D polygon mesh changes to a quadratic B-spline surface.

NOTE:

System variables SURFU and SURFV control the accuracy of the surface approximation by changing the surface density in the *M* and *N* directions.

4 Enter the **eXit** option to exit the PEDIT command.

5 Close the Surfaces toolbar, as well as any others you may have opened.

6 Save your work and exit AutoCAD.

Questions

1. Describe the steps in using the TABSURF command.

EX. *START A DRAWING NAME 3DTMP, dWT - START ON RSIDE THE CURRENT - USE PL" COMMAND. ARC & LINE, MAKE A 1/4 OF DESIGN /mirror/ Then mirror that ~~SAVE AS~~ I. BEAM.*

2. In connection with the TABSURF command, what is the purpose of the direction vector?

CONTROLS THE DENSITY OF THE TABULATED SURFACES

3. Describe the EDGESURF command.

IS USED TO CONSTRUCT A "COONS SURFACE PATCH" WHICH IS 3D SURFACE MESH INTERPOLATED ABOUT BETWEEN 4 ADJOINING EDGES

4. How are user coordinate systems used in conjunction with the EDGESURF command?

TO BE ABLE TO SCALE & PROPORTION THE DESIGN.

5. Why is use of the 3DMESH command not generally recommended for even the simplest 3D meshes?

THE COMMAND IS VERY TEDIOUS. & TIME CONSUMING

■ Challenge Your Thinking: Questions for Discussion

1. Refer again to the I-beam drawing you created in this unit. Describe a way you could have created the I-beam by using TABSURF only once.

2. Find out more about the differences among AutoCAD's original mesh, a Bezier mesh, a quadratic mesh, and a cubic mesh. What are the applications of each surface type?

Problems

1-4. Using the commands and techniques presented in this unit, construct the following 3D models. Approximate all sizes.

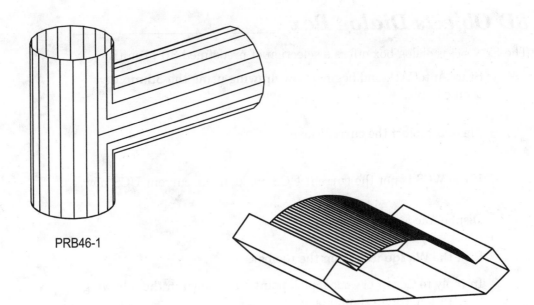

PRB46-1

PRB46-2

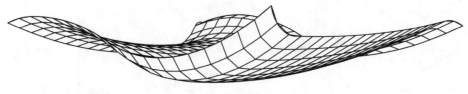

PRB46-3

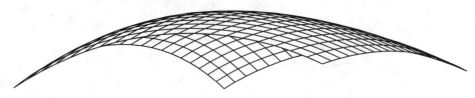

PRB46-4

Unit 47 Creating and Editing 3D Primitives

■ OBJECTIVE:

To practice creating and editing 3D primitives

With AutoCAD, you can create basic three-dimensional objects from predefined shapes called *primitives*. You can edit these and other 3D objects using the ALIGN, ROTATE3D, and MIRROR3D commands.

3D Objects Dialog Box

The 3D Objects dialog box offers a selection of predefined 3D primitives.

1 Start AutoCAD, and begin a new drawing using the **3dtmp.dwt** template file.

2 Make **3dobject** the current layer.

3 If the WCS is not the current UCS, make it the current UCS now.

4 Display the Surfaces toolbar.

Surfaces

5 Pick the **Wedge** icon from the toolbar.

6 In reply to Corner of wedge, pick point 1 as shown in the following illustration. (Snap should be on.)

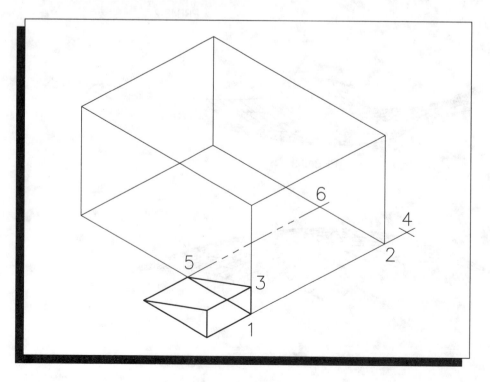

7 Enter **1.5** for the length, **1** for the width, and **.5** for the height, but do not yet enter a rotation angle.

8 Move the crosshairs and notice that you are able to rotate the wedge around the Z axis.

The Z axis is perpendicular to the current UCS.

9 Enter **180** for the rotation angle.

This rotates the wedge 180 degrees counterclockwise. The wedge appears as pictured in the previous illustration.

ALIGN Command

The ALIGN command enables you to move objects in 3D space by specifying three source points and three destination points.

1 Enter **ZOOM** and **.7x**, if necessary, to create working space around the drawing.

2 From the **Modify** pull-down menu, select **3D Operation** and **Align**.

This enters the ALIGN command.

3 Select the wedge primitive and press **ENTER**.

4 Pick point 1 (see the previous illustration) for the first source point.

5 Pick point 2 for the first destination point.

6 Pick point 3 for the second source point.

7 Pick point 4 for the second destination point. (Point 4 is located .5 unit from point 2 in the positive Y direction.)

8 Pick points 5 and 6 for the third source and destination points, respectively.

The wedge moves and rotates as shown in the illustration on the following page.

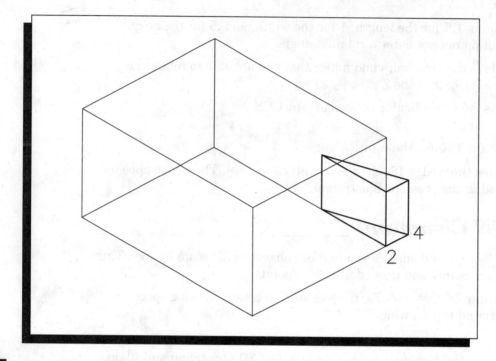

ROTATE3D Command

The ROTATE3D command permits you to rotate an object around an arbitrary 3D axis.

1 From the **Modify** pull-down menu, select **3D Operation** and **Rotate 3D**.

This enters the ROTATE3D command.

2 Select the wedge and press **ENTER**.

AutoCAD presents you with several options for defining the axis of rotation. Notice that 2points is the default setting.

3 Pick points 2 and 4 as shown in the previous illustration.

4 Enter **90** for the rotation angle.

The wedge rotates as shown in the illustration on the next page.

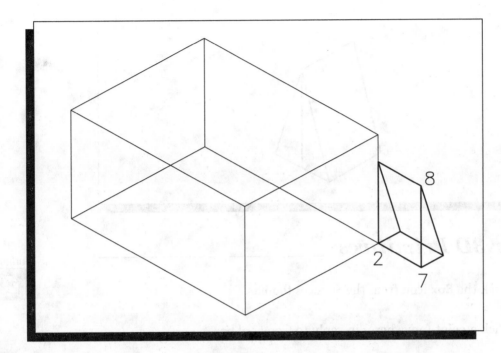

MIRROR3D Command

You can mirror 3D objects around an arbitrary plane with the MIRROR3D command.

1 From the **Modify** pull-down menu, select **3D Operation** and **Mirror 3D**.

This enters the MIRROR3D command.

2 Select the wedge and press **ENTER**.

AutoCAD presents several options for defining the plane. Notice that 3points is the default setting.

3 Pick points 2, 7, and 8 as shown in the previous illustration to define the plane, and do not delete the old object.

The mirrored object appears.

4 Freeze layer **Box** and remove hidden lines.

Your drawing should look very similar to the one in the following illustration.

5 Save your work in a file named **edit3d.dwg**.

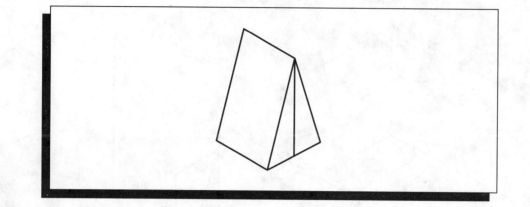

Other 3D Primitives

1 Pick the **Box** icon from the Surfaces toolbar.

2 Pick a point (anywhere) in reply to Corner of box.

3 Enter **2** for the length.

NOTE:

At this point, you could create a cube by entering C for Cube.

4 Enter **1** for the width, **.75** for the height, and **15** for the rotation angle.

5 Remove hidden lines.

The object should appear similar to the box in the following illustration.

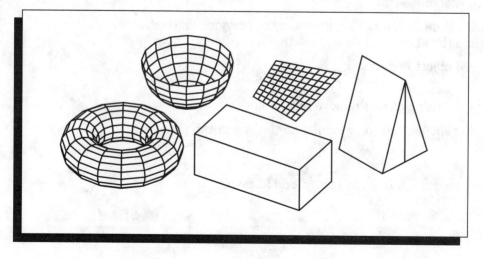

Note, however, that your 3D box may appear different because the viewpoint may be different.

Surfaces

1 Select the **Torus** icon from the Surfaces toolbar.

2 Pick a center point (anywhere) and enter **1** for the radius of the torus.

The torus radius is the distance from the center of the torus to the center of the tube.

3 Enter **.3** for the tube radius.

4 Accept the next two default values.

These values specify the density of the wireframe mesh used to create and display the torus.

5 Remove hidden lines.

The torus should appear similar to the one in the previous illustration.

Surfaces

1 Select the **Dish** icon.

2 Pick a center point (anywhere) and enter **.75** for the radius.

3 Accept the next two default values.

A dish appears.

4 Remove hidden lines.

3D Command

The 3D command displays the entire list of primitives as command options. This makes it possible to enter B for Box, C for Cone, and so on, to gain access to the 3D primitives.

1 Enter **3D**.

2 Enter **M** for Mesh.

3 Pick four points to form a polygon of any size and shape.

4 Enter **10** for both the *M* size and the *N* size.

HINT:
Refer to the end of Unit 45 for an explanation of *M* and *N* vertices.

5 Remove hidden lines.

6 Experiment with the remaining primitives on your own.

7 Practice moving, rotating, and mirroring the objects using the ALIGN, ROTATE3D, and MIRROR3D commands.

8 Close the Surfaces toolbar, as well as any others you may have opened.

9 Save your work and exit AutoCAD.

Questions

1. Name at least five 3D primitives that you can create using the Surfaces toolbar.

   ALIGN COMMAND , ROTATE3D COMMAND
   MIRROR3D COMMAND , BOX COMMAND
   CUBE COMMAND , TORUS COMMAND

2. Explain how you can move and rotate a 3D object using the ALIGN command.

   BY USING ZOOM / MODIFY , SELECT 3D OPERATION
   AND ALIGN , THEN A SERIS OF PICKING
   POINTS FOR THE SOURCE DESTINTATION

3. Briefly state the purpose of the ROTATE3D command.

   ALLOWS YOU TO ROTATE AND OBJECT
   AROUND AN ARBITRARY 3D AXIS

4. Give the purpose of the MIRROR3D command.

   YOU CAN MIRROR 3D OBJECTS AROUND
   AN ARBITRARY MIRROR 3D command

5. When creating a torus, you must specify the torus radius and tube radius values. What is the difference between these two values?

   THE VIEW POINT WILL BE DIFFERENT

6. After selecting the 3D Mesh icon from the Surfaces toolbar, what must you enter to create a 3D mesh?

   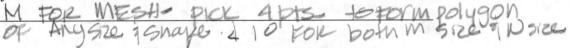
   M FOR MESH - PICK 4 pts to form polygon
   of Any Size & Shape . & 10' FOR both M size & N size

Challenge Your Thinking: Questions for Discussion

1. Discuss the use of 3D primitives in AutoCAD. Under what circumstances might these objects save drawing time? When would you use them?

2. Most 3D models require more than one primitive or a combination of primitives and other objects. Think of an everyday object that could be drawn using a combination of at least three of the primitives, with or without additional objects. Name the object and describe how you would construct it using the primitives.

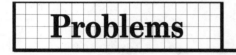
1. Create the following 3D object. Make it 2.5 units long by 1 unit wide by .5 unit high, and rotate it –90 degrees.

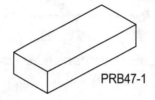

PRB47-1

2. Using either the ALIGN or the ROTATE3D command, reorient the previous object so that it looks like the one in the following illustration.

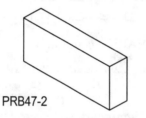

PRB47-2

3. Apply MIRROR3D to the previous object to create the following object.

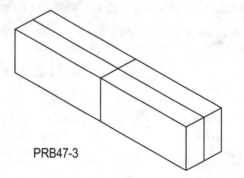

PRB47-3

AUTOCAD® AT WORK

KBJ Architects on the Fast Track

When KBJ Architects undertook the Orlando International Airport expansion, everyone knew they would have to turn out a lot of drawings in a short time. It was a fast-track project, which meant design was just months ahead of initial structural construction.

"On any fast-track job, design and construction proceed in parallel," explains KBJ Vice President Kim Goos. "It's a dynamic process, and it is not unusual to have ongoing construction impose new design constraints along the way." On the Orlando International Airport expansion, "as the design evolved, the documents had to be changed in a hurry before things got built. AutoCAD certainly helped us adjust quickly to changing conditions. That was a big plus."

The scope and complexity of the $700 million expansion demanded accurate as well as fast design work. KBJ Architects drew plans for an automated ground transportation (AGT) tram system connecting the airside terminal to the landside terminal. On top of the three-level landside terminal they built a new seven-story hotel and parking garage. Expansion of the landside terminal, which more than doubled the function of the original structure, required floor plans and ceiling plans for one million square feet of space.

Providing specifications for the AGT people-mover system was another formidable challenge. Architects had to make layouts for the entire airport, including gates and plane locations. "There are many rules and regulations governing plane spacing and which planes can park where," explains Goos. "AutoCAD was really beneficial at that stage because we were able to simulate the different types of planes and rotate jetways about their pivot points, which greatly speeded up the process."

The Orlando International Airport project was the first for which KBJ Architects used AutoCAD. Now it is standard for every project from the smallest to the major expansion of Luis Muñoz Marín International Airport in San Juan, Puerto Rico. "In today's marketplace," Goos says, "the benefits of accuracy, speed, and flexibility cannot be overstated."

Photo by Kathleen McKenzie

Unit 48 Shading and Rendering

■ OBJECTIVE:

To shade and render 3D models using the SHADE and RENDER commands

AutoCAD offers two basic options for producing shaded views of your 3D work. SHADE produces a quick shaded view of a 3D model, while RENDER produces a higher quality rendering.

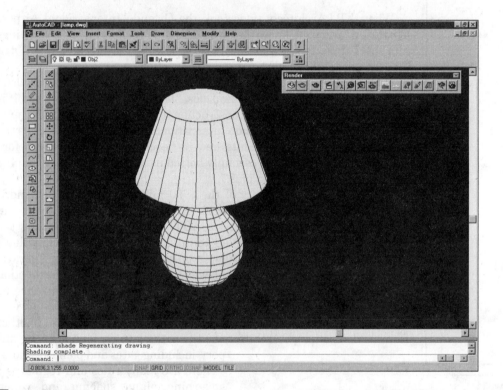

SHADE Command

You can quickly shade any 3D object in the current viewport using the SHADE command.

1 Start AutoCAD and open the drawing you created in Unit 45 named **lamp.dwg**.

2 Display the Render toolbar.

3 From the toolbar, pick the **Shade** icon.

This enters the SHADE command.

Render

AutoCAD gives the percentage complete as it performs its shading calculations. When the calculations are complete, AutoCAD displays a shaded view of the lamp, as shown in the previous illustration.

NOTE:

The SHADEDGE and SHADEDIF system variables allow you to control the shade style and lighting. However, most AutoCAD users seldom need to change their values.

 Zoom in on the lamp, filling most of the screen.

RENDER Command

The RENDER command allows you to produce higher quality shaded views of 3D models.

Render

Pick the **Render** icon from the same toolbar.

This enters the RENDER command, which initializes the Render program. The Render dialog box appears, as shown in the following illustration.

Uncheck the **Smooth Shade** check box located in the Rendering Options area.

You will learn more about this dialog box in the following unit.

Pick the **Render** button.

As AutoCAD performs its rendering calculations, it displays messages about its progress. After several seconds—depending on the speed of the computer—AutoCAD renders the lamp.

NOTE:

If AutoCAD does not render the lamp, it is possible that the computer does not contain adequate system memory.

The flat surfaces you see (called *facets*) become smaller and less noticeable as you increase the values of the SURFTAB1 and SURFTAB2 system variables prior to creating the 3D objects. In the following unit, you will have the opportunity to practice smooth shading, which smooths the facets and produces a higher quality rendering.

4 After AutoCAD renders the lamp, force a regeneration.

Statistics Dialog Box

The Statistics dialog box displays details about the last rendering. You can save this information in a file, but you cannot change it.

Render

1 Pick the **Statistics** icon from the Render toolbar.

This enters the STATS command, which displays the Statistics dialog box.

2 Review the information and pick **OK**.

Save Image Dialog Box

Using the Save Image dialog box, you can save a rendering as a BMP, TGA, or TIFF file. Certain other graphics programs are capable of reading, displaying, and printing these file types.

BMP   Device-independent bitmap format
TGA   32-bit RGBA Truevision version 2.0 format
TIFF   32-bit RGBA Tagged Image File Format

Render

1   Render the lamp drawing.

2   Enter the **SAVEIMG** command or pick **Display Image** and **Save...** from the **Tools** pull-down menu.

The Save Image dialog box appears, as shown in the following illustration.

Save Image	⊠
Format	**Portion**
⊙ BMP	Active viewport Size : 925 × 572
○ TGA	Offset Size
○ TIFF	X: [0] X: [925]
[Options...]	[Reset] Y: [0] Y: [572]
[OK] [Cancel] [Help]	

3   Pick the **TIFF** radio button and pick the **OK** button.

4   Name the file **lamp.tif**, select a directory of your choice, and pick the **Save** button.

AutoCAD creates a file named lamp.tif.

Replay Dialog Box

You can display BMP, TGA, and TIFF files using the Replay dialog box.

1 Enter the **REPLAY** command or pick **Display Image** and **View...** from the **Tools** pull-down menu.

The Replay version of the standard Files dialog box appears, as shown in the following illustration.

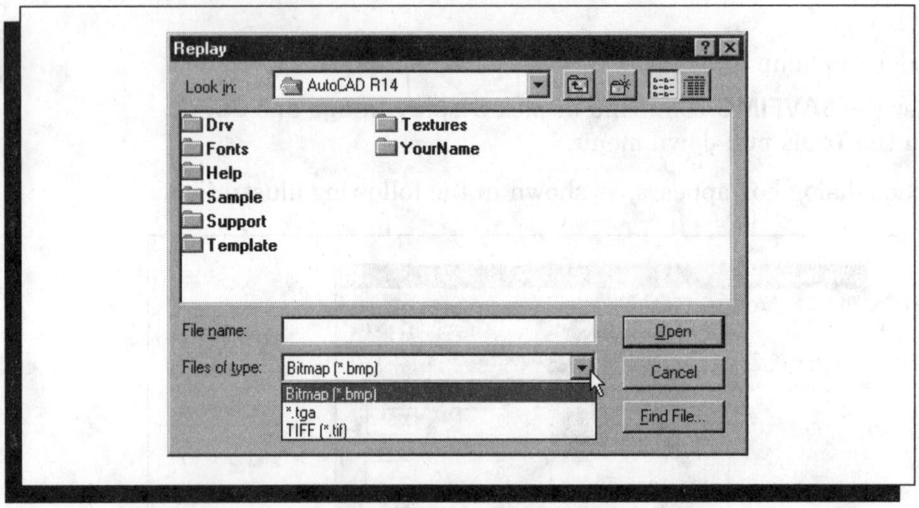

2 To the right of Files of Type, pick the down arrow.

3 Pick **TIFF (*.tif)** from the list.

4 Find and double-click **lamp.tif**.

This displays the Image Specifications dialog box.

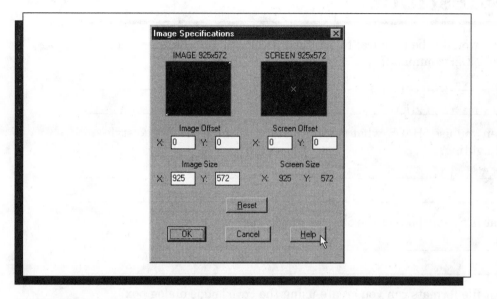

5 Pick the **Help** button if you want to read about this dialog box, or pick the **OK** button to view the file.

AutoCAD reads the lamp.tif file and displays it.

6 Close the Render toolbar, as well as any other toolbars you may have opened.

7 Exit AutoCAD without saving any changes.

Questions

1. When would it be practical to use the SHADE command rather than the RENDER command?

   FOR A QUICK SHADED VIEW/RENDER IS HIGHER Quality rendering

2. When you use SHADE, what system variables control the shade style and lighting?

   SHADEDGE & SHADEDIF-

3. What is the purpose of the Statistics dialog box?

   DISPLAYS DETAILS ABOUT THE LAST RENDERING. YOU CAN SAVE IT IN A FILE BUT CANT CHANGE IT

4. What file formats can you create using the Save Image dialog box?

   SAVES RENDERING AS A BMP, TGA OR A TIFF FILE-

5. What is the purpose of the Replay dialog box?

   TO DISPLAY THE TIFF, BMP OR TGA FILES

▪ *Challenge Your Thinking: Questions for Discussion*

1. Experiment further with the system variables that control the appearance of shaded objects. What options are available? Explain why people seldom need to change the initial values, and identify a situation in which the options might need to be changed.

2. In this unit, you saved a rendered image of the lamp.dwg file as a TIFF file. The TIFF format can be read by many graphic and illustration programs other than AutoCAD. Find out what programs available in your school or company can display a TIFF file. If a suitable program is available, load lamp.tif using that program and view the file. How does it compare to the original AutoCAD image? Explain any differences you see.

Problems

1. Shade and render three-d2.dwg, which you created in Units 42-44.

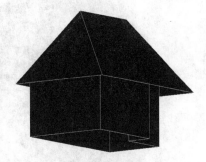

three-d2.dwg

2. Create a BMP file of the rendering of three-d2.dwg. Name it three-d2.bmp.

3. Display the three-d2.bmp rendering in AutoCAD.

4. Shade and render i-beam.dwg from Unit 46.

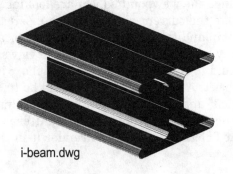

i-beam.dwg

5. Display statistics on the rendering of the I-beam. How many faces and triangles does it contain?

6. Draw and render an object similar to the electric motor cover shown here in three-dimensional and rendered views.

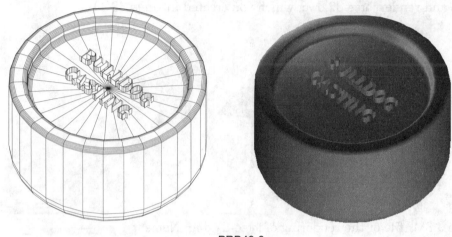

PRB48-6

Begin by drawing a polyline and center line as shown below. Use the REVSURF command with a large value for SURFTAB1 to create the cover. To add the lettering, first change the viewpoint so you are looking at the top view; change the UCS to the current view and then insert the lettering. Use the CHANGE command to give the text the appropriate elevation and thickness. It may be easier to manipulate the views if you use the ROTATE3D command to rotate the object so the plan view is truly the top view of the object. A better approach would be to set up three viewports for top, front, and isometric (viewpoint coordinates 1,–1,1) views, then draw the object in the front viewport. After you have drawn the cover, use the RENDER command to render it in the isometric view.

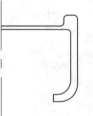

Problem 6 courtesy of Gary J. Hordemann, Gonzaga University

AUTOCAD® AT WORK

3D Lets You See

A two-dimensional overhead view, or floorplan, has always been the basic planning tool for laying out a trade show exhibit. Yet it takes a three-dimensional view to answer questions like: What will our sign look like within the exhibit space? What impression will this display make on customers as they arrive? With 3D drawings in AutoCAD, planners can combine the precision of a blueprint with the visual quality of an "artist's conception" drawing.

Being able to see the finished product before it is constructed is the heart of business at TechniCAD (Brookside, NJ), where consultant Phil Gauntt specializes in rendering "virtual environments" for customers ranging from civil engineers to interior designers. Some customers are architects who, after creating a 3D AutoCAD drawing like the one of an exhibit accompanying this story, want output that is photorealistic. Phil scans in samples of a wallpaper pattern, carpet color, and any textures and finishes the customer wants in the final image. Using 3D Studio, he applies the textures and finishes to the blocks created in AutoCAD. The result is a visually convincing image of a place that only existed, till that moment, in the mind's eye of the designer. When output to videotape, the image can come to life with animation. A customer can pore over a proposal for hours, literally, and not get as much from it as from ten seconds of video.

Helping AutoCAD users rely less often on consultants is another aspect of business at TechniCAD. For users who want 3D graphics but are leery of the training and time required to develop 3D expertise, Phil provides a custom setup that allows even a novice operator to assemble a 3D model by

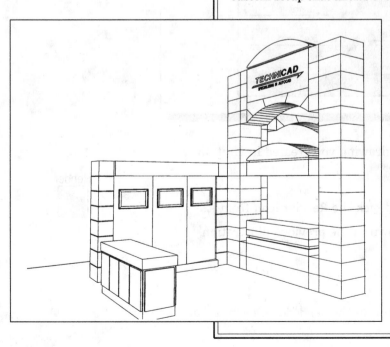

selecting and positioning objects from a customer-specific menu of pictures (furniture, fixtures, lighting, anything you can picture). With 3D graphics capability in-house, Phil's clients don't need to get in line at the consultant's door for output.

Phil got into the CAD consulting business after some years as a drafter, "pushing a pen on mylar." He enrolled in a semester-long CAD course and found new employment as an AutoCAD operator. Eventually Phil set out on his own to found TechniCAD. As an Autodesk-certified instructor, he teaches at Pratt Manhattan, an Autodesk Training Center (ATC), where he works with three kinds of students: those who want retraining for a new career, those who work with AutoCAD in their present job, and those who get into it, as he did almost 10 years ago, because of its "high coolness factor."

Unit 49 Advanced Rendering

■ OBJECTIVE:

To change rendering settings and create smooth-shaded renderings

Render options enable you to change the appearance of renderings. Instead of views with visible facets, you can create smooth shaded renderings. You can also adjust the lighting, apply materials and surface finishes, and add backgrounds.

Render Dialog Box

You can change the appearance of a rendering using the Render dialog box.

```
┌─────────────────────────────────────────────────┐
│ Render                                      [X]  │
│ Rendering Type:     [Render            ▼]        │
│  Scene to Render       ┌─Rendering Procedure─┐   │
│  *current view*        │ ☐ Query for Selections│  │
│                        │ ☐ Crop Window        │   │
│                        │ ☐ Skip Render Dialog │   │
│                                                  │
│                        Light Icon Scale:  [1  ]  │
│                        Smoothing Angle:   [45 ]  │
│  ┌─Rendering Options─┐ ┌─Destination─┐ ┌─Sub Sampling─┐│
│  │ ☑ Smooth Shade   │ │[Viewport ▼] │ │[1:1 (Best) ▼]││
│  │ ☑ Apply Materials│ │             │ │              ││
│  │ ☐ Shadows        │ │Width : 925  │ │              ││
│  │ ☐ Render Cache   │ │Height: 585  │ │ [Background...]││
│  │                  │ │Colors: 32-bits│ │            ││
│  │ [More Options...]│ │[More Options...]│[Fog/Depth Cue...]│
│                                                  │
│        [Render]      [Cancel]      [Help]        │
└─────────────────────────────────────────────────┘
```

1 Start AutoCAD and open the drawing you created in Unit 45 named **lamp.dwg**.

2 Display the Render toolbar and pick the **Render** icon from it.

The Render dialog box appears, as shown in the previous illustration.

Render

3 Review the information presented in the Render dialog box.

4 Pick the **More Options...** button in the Rendering Options area.

The AutoCAD Render Options dialog box appears. It enables you to select either Gouraud or Phong shading. Gouraud shading produces a smooth (non-faceted) rendering by interpolating the adjacent color intensities of a 3D model. Phong uses a more advanced interpolation than Gouraud to create a more realistic rendering.

Discarding back faces prevents AutoCAD from reading the back faces of a 3D solid object. The last check box controls which faces AutoCAD considers back faces in a drawing. If this is unclear to you, take a moment to read the help information.

5 Pick the **OK** button.

6 If **Smooth Shading** is not checked, check it now.

Smooth shading causes AutoCAD to blend the facets and produce a smooth shaded rendering using either the Gouraud or Phong technique. Note, however, that the smooth shading takes longer than facet shading. As the quality of the rendering increases, so does the time to produce the rendering.

7 Pick the **Help** button and read about the remaining items in this dialog box.

8 After exiting Help, pick the **Render** button.

This causes AutoCAD to render a scene, which is a combination of a named view and one or more lights. In this case, since we did not define a scene, AutoCAD uses the current view and a default *over-the-shoulder* distant light source.

9 Regenerate the screen to remove the rendering.

Materials Dialog Box

You can define and attach materials and surface finishes to objects using the Materials dialog box. For example, you can make the surface of an object appear shiny by using a glass finish.

Render

1 Pick the **Materials** icon from the Render toolbar.

This enters the RMAT command, which displays the Materials dialog box.

Objects without a material finish attached to them use the default *GLOBAL* finish.

2 Pick the **Modify...** button.

This displays the Modify Standard Material dialog box, as shown in the following illustration.

The Modify Standard Material dialog box allows you to change various attributes of the *GLOBAL* finish.

3 Pick the **Help** button and review the meaning of each of the attributes listed on the left side of the dialog box, and exit when you're finished.

4 Using the radio buttons and slider bar, increase (considerably) the values for both **Ambient** and **Reflection** and pick the **OK** button; pick **OK** again.

Render

5 Pick the **Render** icon and pick the **Render** button.

The model of the lamp should be brighter (due to the increased ambient light) with a more reflective, mirror-like surface.

NOTE:

If you are using a VGA display, changes to the surface finish will be less noticeable than if you are using a more sophisticated display subsystem.

Render

6 Redisplay the **Materials** dialog box and pick the **Help** button to review the other parts of the dialog box.

7 Exit help and pick the **Cancel** button.

Lights Dialog Box

You can control several types of lighting with the Lights dialog box.

First, let's save the current view.

Standard

1 Enter the **VIEW** command and save a view named **VIEW1**.

This will allow us to return to this view later in the exercise.

Render

2 Pick the **Lights** icon from the Render toolbar.

This enters the LIGHT command, which displays the Lights dialog box, as shown in the following illustration.

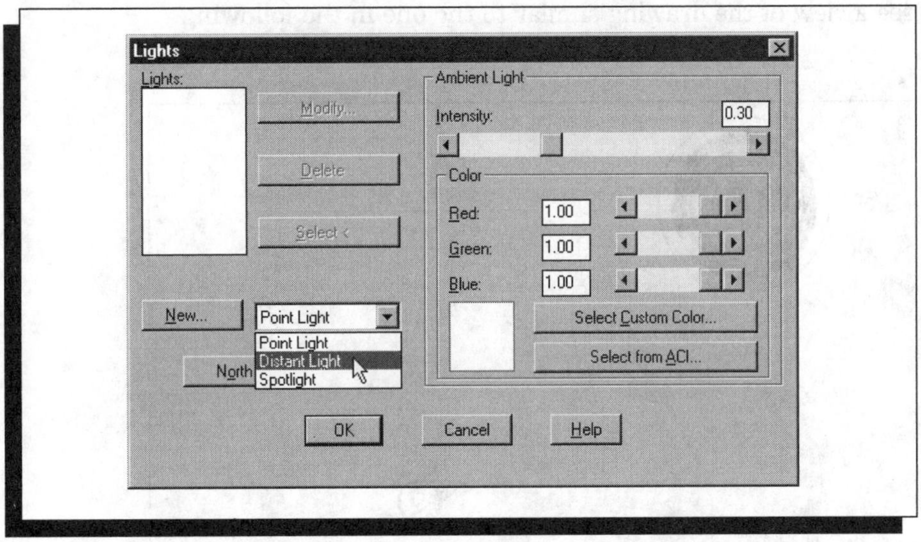

3 Pick **Distant Light** from the drop-down box found in the lower left area of the dialog box.

4 Pick the **New...** button.

AutoCAD displays the New Distant Light dialog box. AutoCAD now expects you to name the new light.

5 Enter **BRIGHT** for the name of the light and pick **OK**; pick **OK** again.

AutoCAD inserts a light block on the plan view of the WCS. Because its location is in line with the current viewpoint, it may appear as a line.

6 Enter the **PLAN** command and press **ENTER** to accept the current UCS.

This provides you with a better view of the light block location and angle.

NOTE:

You may need to move one or more toolbars on the screen to see the light block.

7 Undo the last operation.

8 Enter the **PLAN** command and, this time, opt for the world coordinate system.

You should see a view of the drawing similar to the one in the following illustration.

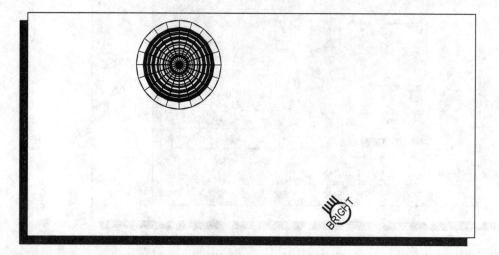

Let's move the light to a new location.

9 Pick the **Lights** icon.

Render

10 Select **BRIGHT** from the **Lights** list box (if it is not already selected), and pick the **Modify...** button.

11 Pick the **Modify** button in the Modify Distant Light **dialog box.**

The dialog box disappears temporarily and AutoCAD displays the graphics screen. At the Command line, notice that the prompt has changed to Enter light direction TO.

12 Press **ENTER** to accept the current default setting.

In reply to Enter light direction FROM, let's specify a point using the X/Y/Z point filter method.

13 Enter **.xy** and pick a point anywhere in the upper right area of the screen. (Turn off ortho if it is on.)

14 Enter **–1** for the *z* coordinate.

The Modify Distant Light dialog box reappears.

15 Pick **OK**; pick **OK** again.

16 With the **VIEW** command, restore **VIEW1**.

17 Pick the **Render** icon and pick the **Render** button.

AutoCAD considers the new light as it renders the lamp.

18 Enter the **PLAN** command and choose the current UCS.

Notice the new location of the light.

19 Undo the last step and save your work.

Standard

Render

Scenes Dialog Box

The Scenes dialog box produces a list of saved scenes. A scene is similar to a named view, but a scene can contain one or more light sources.

Render

1 Pick the **Scenes** icon from the Render toolbar.

This enters the SCENE command, which displays the Scenes dialog box, as shown in the illustration on the following page.

The Scenes dialog box.

2 Pick the **New...** button.

The New Scene dialog box expects you to name the scene.

3 Enter **SCENE1** and pick **OK**.

SCENE1 appears in the Scenes list box, indicating that you have created a new scene. AutoCAD will use the selected scene the next time you issue the RENDER command.

Render

4 Pick **OK** and pick the **Render** icon.

Notice that SCENE1 appears in the Scene to Render list box and that it is selected.

5 Pick the **Render** button to render SCENE1.

Adding Backgrounds

In the Render dialog box, you may recall seeing the Background... button. This button displays the Background dialog box.

Render

1 Pick the **Background** icon from the Render toolbar.

This also displays the Background dialog box. Notice that many options are not available in the dialog box.

2 Pick the **Gradient** radio button.

Many of the unavailable options become available, as shown in the following illustration.

3 Pick the **OK** button.

4 Pick the **Render** icon and pick the **Render** button.

AutoCAD produces a three-color gradient background.

5 Redisplay the **Render** dialog box and pick the **Background...** button.

6 Pick the **Image** radio button and pick the **Find File...** button.

7 Find the file named bpsplsh.bmp, which is located in the following directory.

\Program Files\AutoCAD R14\Sample\Activex\ebatchp

8 Double-click this file and pick the **Preview** button in the Background dialog box.

AutoCAD displays a preview of the bpsplsh.bmp image.

9 Pick the **OK** button and pick the **Render** button.

AutoCAD displays the bpsplsh.bmp in the background.

10 Redisplay the **Render** dialog box and pick the **Background...** button.

 11 Pick the **Help** button to read about the other options available in the Background dialog box.

12 When you are finished reading the help information, pick the **Solid** radio button and pick **OK**; then pick the **Render** button.

The standard background reappears.

Photorealistic Rendering

AutoCAD offers additional rendering options, such as photorealistic and raytrace rendering.

1 Display the **Render** dialog box and pick the **Rendering Type** drop-down box located at the top of the dialog box.

Render

The Photo Real and Photo Raytrace options produce more realistically shaded images of 3D models.

2 Select **Photo Real** and pick the **Render** button.

After many calculations, shown on the Command line, AutoCAD produces the image.

NOTE:

Depending on your computer's display subsystem, you may not see a significant difference in rendering quality. Factors include the number of colors available and the resolution of the subsystem.

3 Save your work and open the file named **chevy.dwg**, which is located in AutoCAD's Sample folder. (This is a large drawing file, so it may take several seconds to open.)

Render

4 Pick the **Render** icon.

5 In the Destination area of the Render dialog box, pick the down arrow and choose **Viewport**.

6 Select **Render** in the Rendering Type drop-down box and pick the **Render** button.

This could take several seconds—possibly a minute or longer—depending on the rendering speed of your computer. Note the spinning line on the Command line. This indicates that AutoCAD is processing data.

Render

7 Display the **Render** dialog box again and select **Photo Raytrace** from the drop-down box.

Notice that the Shadow check box becomes available and is checked.

8 Pick the **Render** button.

This rendering will take much longer. After AutoCAD produces the raytrace image, notice the shadow on the sidewalk near the front left fender.

Render

9 Pick the **Materials** icon from the Render toolbar and notice the list of materials in the Materials dialog box, as shown in the following illustration.

Using material types such as these, you can present objects in a more realistic way.

10 Double-click **CAR BODY**.

Notice the color given to this material.

11 Pick the **Cancel** button; pick **Cancel** again.

Other Rendering Options

With fog and depth cueing, you can provide more visual information about the distance of objects from the camera.

1 Redisplay the **Render** dialog box and pick the **Fog/Depth Cue...** button.

This displays the Fog/Depth Cue dialog box. The Fog icon in the Render toolbar also displays this dialog box.

2 Check the **Enable Fog** check box.

Most of the options in the dialog box become available.

3 Pick the **Help** button and read about the settings in this dialog box.

4 After reading the help information, pick the **Cancel** button in the Fog/Depth Cue dialog box.

5 Pick the **More Options...** button.

The Photo Raytrace Render Options dialog box appears because Photo Raytrace is the rendering option selected in the Render dialog box.

6 Pick the **Help** button and read about the options available in this dialog box.

7 After reading the help information, pick the **Cancel** button to close the Photo Raytrace Render Options dialog box.

8 Pick the **Help** button in the Render dialog box to read about the remaining options. Close it and the dialog box when you are finished.

Landscape Objects

A landscape object is an object with a bitmap image mapped onto it. Landscape objects permit you to add realistic landscape items such as trees and bushes to a drawing. A bitmapped image mapped to an object gives the object a grain, texture, or other quality that can make it look real.

1 Begin a new drawing using the **3dtmp.dwt** template file. *Do not* save any changes to the chevy.dwg file.

2 Freeze layer Box and make **3dobject** the current layer.

Render

3 From the Render toolbar, pick the **Landscape New** icon.

This displays the Landscape New dialog box, as shown in the following illustration.

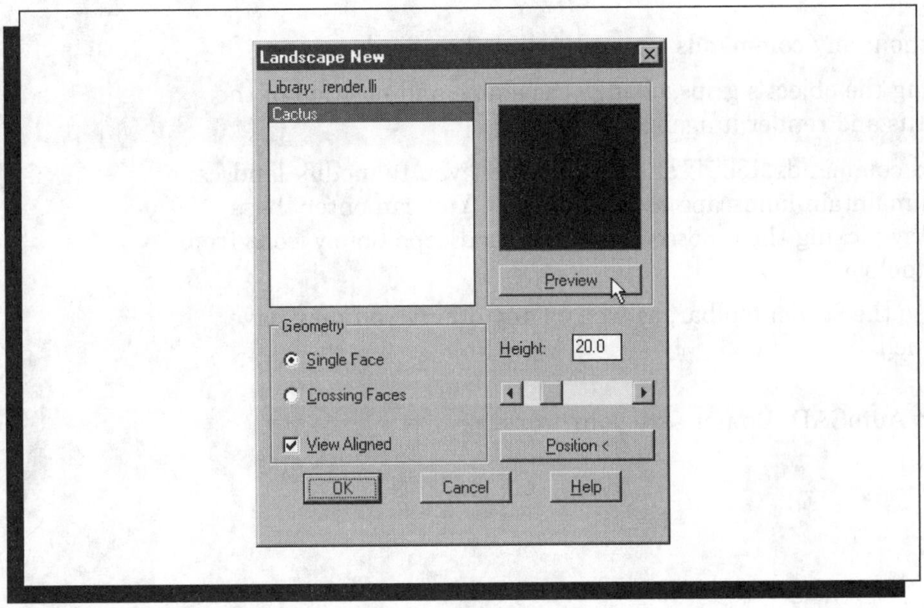

Notice that the landscape library named render.lli contains one object named Cactus.

4 Pick the **Preview** button to preview Cactus.

5 Pick the **Position** button and pick a point anywhere near the center of the screen for the base of the landscape object.

6 When the Landscape New dialog box reappears, pick the **OK** button.

7 Enter **ZOOM All** and turn off the grid.

8 Display the **Render** dialog box, select **Photo Real** for Rendering Type, and pick the **Render** button.

A landscape object of a cactus appears.

9 View the object from a new viewpoint in space and then render it again using the **Photo Real** option.

10 Enter **REGEN**.

Landscape objects, before being rendered, have grips at their base, top, and each corner. You can use the base grip to move the object, the top grip to adjust its height, and the bottom corner grip to scale it. All standard AutoCAD grip editing modes, such as stretch, scale, and rotate, work with landscape objects.

11 Without any commands entered, select the object.

12 Using the object's grips, change the shape and position of the cactus and render it again.

Two related commands, LSEDIT and LSLIB, permit you to modify landscape objects and maintain landscape object libraries. You can enter these commands by picking the Landscape Edit and Landscape Library icons from the Render toolbar.

13 Close the Render toolbar, as well as any others you may have opened.

14 Exit AutoCAD. Do not save your work.

Questions

1. Why does smooth shading require more computer processing time than faceted shading?

   _____

   _____

2. In the Modify Standard Material dialog box, what do Ambient and Reflection represent?

   _____

   _____

3. On what coordinate system does AutoCAD insert light blocks?

   _____

   _____

4. When viewing the plan view of the WCS, how can you move the light icon in the X, Y, and Z directions?

   _____

   _____

5. What does a scene contain that a saved view does not contain?

   _____

6. Explain the difference between the Render and Photo Real rendering types in the Render dialog box.

   _____

   _____

   _____

7. What is a landscape object?

   _____

   _____

Challenge Your Thinking: Questions for Discussion

1. Experiment further with the options presented in the Lights dialog box. Notice that AutoCAD allows you to assign colors to the lights. What happens when you shine a light of one color onto an object of another color? When and why might this be useful? Explain.

2. Find out more about the material finishes supplied with AutoCAD. To do this, open the Materials dialog box and pick the Materials Library button. Experiment with the various items in the Library List. What does the Preview button do? Describe the procedure for applying materials from the Materials Library to the lamp you rendered in this unit.

Problems

1. Create a smooth rendering of the edit3d.dwg drawing from Unit 47. You may choose to make each object a different color.

2. After changing several attributes, such as Color, Ambient, Reflection, and Roughness, create a smooth rendering of edit3d.dwg.

3. In the same drawing, create and insert a "distant light." After moving the light block in the X, Y, and Z directions, render the drawing. Produce a new scene using the new light.

AutoCAD® at Work

Hark! What Light Through Yonder Window "Works"?

The actor "struts and frets his hour upon the stage," as Shakespeare said. Then the crew dismantles everything, making way for the next show. Temporariness is part of the challenge for a stage lighting director. Where residential/commercial lighting is standard, permanent, and governed by codes, stage lighting gets reconfigured with every change of the marquee.

Yet a lighting director relies on schematic drawings like any electrician. With row upon row of spotlights to keep track of, suspended in a grid of pipes up near the ceiling, not to mention boom-mounted lights and footlights, the paperwork can get complicated. A typical six-foot section of footlights might have three separate circuits and the capability to accept various configurations of up to 12 lamps. That versatility makes it possible to set a misty mood for the double-double-toil-and-trouble scene in Act I and then crank everything up for the blaze of battle in Act V.

Enter Light "Works," a program that runs with AutoCAD to help lighting directors manage the versatility of stage lighting. With a built-in library of lighting symbols, Light "Works" makes it easy to draw a schematic and even easier to modify a schematic to suit an upcoming show. With only six commands (plus eight AutoCAD commands), the user can make yesterday's floods become twinks tonight.

As blocks in an AutoCAD drawing, the lights in Light "Works" have attributes (*e.g.,* instrument number, channel number) that tell where the light is, what color it is, and which circuit it's on—an important bit of information when you have to keep track of maximum load on wiring. In Light "Works" a change made on the drawing effects the same change in the database. One-step control makes the job easy for people who know lighting but don't necessarily know computers.

Light "Works" is the brainchild of Mark Weaver, who developed the software while a graduate student at the Yale School of Drama, Technical Design and Production Department. Now a staff lighting director for television's ABC News, Mark has also done lighting for the network's sports and entertainment divisions. "Monday Night Football" and "College Football Scoreboard," which are broadcast from the same studio, are examples of Mark's work with Light "Works," which he markets through TECHNICAL Art "Works" (New York).

Mark observes that while AutoCAD has applications throughout the world of theater, it tends to be used most efficiently in schools, where equipment and software are freely available to all backstage users—the set designer and costume designer as well as the lighting director. As a result, tomorrow's theater graduates will view programs such as AutoCAD and Light "Works" as standard tools of the trade.

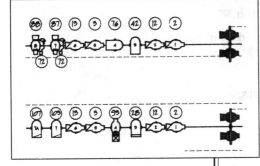

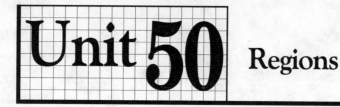

■ OBJECTIVE:

To practice AutoCAD's region commands

You can produce closed 2D areas known as *regions* using AutoCAD. Regions are the result of combining two or more 2D objects, which involves the Boolean union, subtraction, and intersection operations. Boolean operations are available in certain solid modeling systems, such as AutoCAD. The region commands are a subset of AutoCAD's 3D solid modeling commands.

■ *REGION Command*

The REGION command creates objects called *regions,* which consist of one or more closed 2D shapes called *loops.* Loops can be a combination of lines, polylines, arcs, circles, elliptical arcs, ellipses, and splines, 3D faces, traces, and solids. A loop must be a closed shape.

Let's create an object resembling a bicycle sprocket guard manufactured from sheet metal.

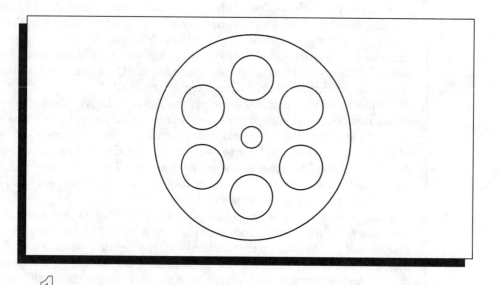

1 Start AutoCAD and begin a new drawing from scratch.

2 Create a new layer named **Objects**, assign the color red to it, and make it the current layer.

3 Set snap at **1** unit and draw a circle **7** units in diameter.

Draw

4 From the Draw toolbar, pick the **Region** icon.

This enters the REGION command.

5 Select the circle and press **ENTER**.

AutoCAD converts the circle to a region primitive.

6 Enter the **LIST** command, pick the region, and press **ENTER**.

Details on the region appear.

7 Make a note of the area of the region.

8 Close the AutoCAD Text Window.

9 Save your work in a file named **region.dwg**.

SUBTRACT Command

The SUBTRACT command creates a composite region by subtracting the area of one set of 2D objects or regions from another set. A composite region is the result of applying one or more Boolean operations, such as a subtraction.

1 Create a circle with a diameter of **1.5**. Make its center **2** units upward vertically from the center of the large circle.

HINT:
With snap on, move the crosshairs 2 units upward from the center of the large circle.

Modify

2 Array the smaller circle 360 degrees as shown in the previous illustration.

3 At the center of the large circle, create a circle with a diameter of **.75** unit.

Draw

4 Enter the **REGION** command, select the seven new circles, and press **ENTER**.

5 Enter **DBLIST**.

As you can see, AutoCAD created seven individual regions.

6 Close the AutoCAD Text Window.

7 Open the **Modify II** toolbar.

8 From the Modify II toolbar, pick the **Subtract** icon.

This enters the SUBTRACT command.

Modify II

9 Pick the large circle and press **ENTER**.

10 Pick the seven small circles and press **ENTER**.

AutoCAD subtracts the areas of the seven smaller region primitives from the large region primitive. The large circle is called the region's *outer loop*. The smaller circles are called the *inner loops*. The result is a single composite region.

11 Enter **LIST**, select the region, and press **ENTER**.

Notice the area of the region. How does it compare with the area before you subtracted the holes?

Render

12 Close the AutoCAD Text Window and enter the **SHADE** command.

AutoCAD shades the region.

13 Enter **REGEN** and save your work.

Calculating a Region's Area

Another way to find the area of a region is to use the AREA command.

1 Open the **Inquiry** toolbar.

2 Pick the **Area** icon from the toolbar and enter the **Object** option.

Inquiry

3 Select the region.

AutoCAD calculates the surface area as 27.4398 square centimeters. The perimeter of the large circle is 52.6217.

MASSPROP Command _____

The MASSPROP command calculates the engineering properties of a region.

1 From the Inquiry toolbar, pick the **Mass Properties** icon.

This enters the MASSPROP command.

2 Select the composite region and press **ENTER**.

AutoCAD displays the following information about the region.

```
---------------------------     REGIONS     ---------------------------
Area:                          27.4398
Perimeter:                     52.6217
Bounding box:        X:        2.5000    --    9.5000
                     Y:        0.5000    --    7.5000
Centroid:            X:        6.0000
                     Y:        4.000
Moments of inertia:  X:      534.1841
                     Y:     1082.9810
Product of inertia:  XY:     658.5564
Radii of gyration:   X:        4.4122
                     Y:        6.2823
Principal moments and X-Y directions about centroid:
                     I:       95.1465 along [1.0000  0.0000]
                     J:       95.1465 along [0.0000  1.0000]
```

Some of your values may differ from what you see here.

3 Enter **Yes** in reply to Write to a file?

4 When the dialog box appears, choose a directory to store the file, and pick the **Save** button.

This creates a file named region.mpr.

5 Minimize AutoCAD.

6 Open Notepad.

7 Select **Open...** from the **File** pull-down menu.

 8 Locate and open the file named **region.mpr**.

The file contains the mass properties information.

9 Pick **Exit** from the **File** pull-down menu.

10 Maximize AutoCAD.

UNION Command

Use the UNION command to create a composite region by combining the area of two or more 2D objects or regions.

1 Set snap at **.25** and draw a rectangle across the top circle using a polyline, as shown in the following illustration.

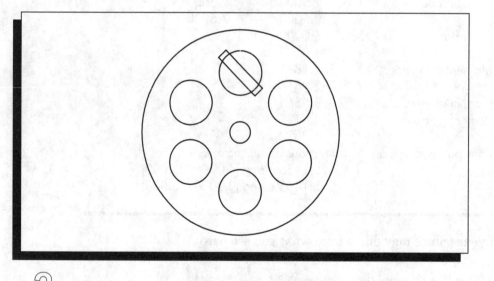

2 Create a region primitive from the polyline.

3 Array the new region primitive so that it appears in the other five circles. Rotate the objects as they are copied.

Modify II

4 From the Modify II toolbar, pick the **Union** icon.

This enters the UNION command.

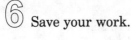 Select the composite region as well as the six new region primitives, and press **ENTER**.

AutoCAD combines the regions to create a new composite region similar to the one shown below.

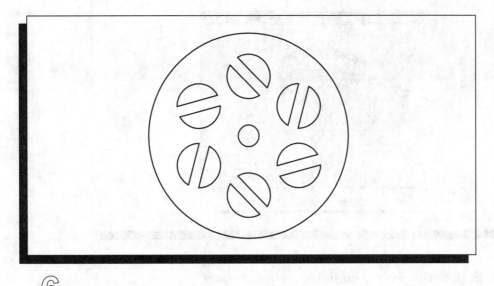

 Save your work.

BOUNDARY Command

The **BOUNDARY** command creates a region or polyline from overlapping objects. These objects must define an enclosed area.

1 Begin a new drawing from scratch.

2 Create two overlapping circles of any size, as shown in the following illustration.

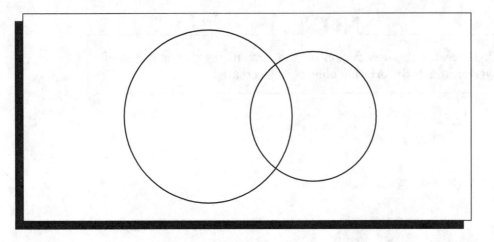

③ Pick **Boundary...** from the **Draw** pull-down menu.

This displays the Boundary Creation dialog box.

④ In the Object Type drop-down box, choose **Region**.

The Make New Boundary Set button is available in case you need to select objects for the boundary set. From Everything on Screen is checked, so AutoCAD will automatically select the two circles. Island Detection specifies whether internal objects, called *islands,* are used as boundary objects.

⑤ Pick the **Pick Points** button.

⑥ Pick a point inside the area defined by the overlapping circles and press **ENTER**.

This creates a region in the shape of the area defined by the overlapping circles.

NOTE:

AutoCAD creates the new region on the current layer, regardless of the layer on which the original objects were created.

Enter **MOVE** and **Last** and move the region to a new location on the screen.

 Close the floating toolbars and exit AutoCAD. Do not save your work.

Solid Modeling

You can apply most of the commands and Boolean concepts you've used in this unit to solid modeling. You can also produce solid models from the regions by giving them thickness. For example, you could use the EXTRUDE command to produce the following solid object.

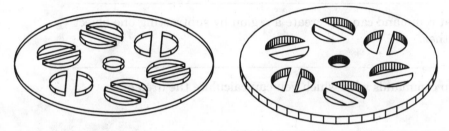

The following units cover AutoCAD's solid modeling commands and features.

Questions

1. Explain inner and outer loops.

   _____

   _____

2. Describe a composite region.

   _____

   _____

   _____

3. With what command can you create a region by subtracting one object from another?

   _____

4. Name two commands with which you can calculate the area of a region.

   _____

   _____

5. What is the purpose of the MASSPROP command?

   _____

   _____

6. Explain the purpose of the UNION command.

   _____

   _____

7. What two object types can the BOUNDARY command create?

   _____

   _____

■ *Challenge Your Thinking: Questions for Discussion*

1. Review the information provided by the MASSPROP command. If you do not know what the properties are, find out. In what types of occupations might people need to use this information? Explain.

2. Is it possible to create solid models from regions? Explain.

Problems

1. Using a circle and rectangular polyline, create the following object as a composite region.

PRB50-1

2. From the object you created in problem 1, create the following object.

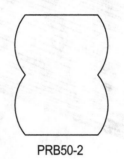

PRB50-2

3. From the object you created in problem 2, create the following object.

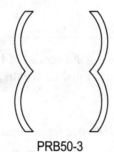

PRB50-3

4. Calculate the area (in square inches) of each of the objects you created in problems 1 through 3.

5. Display the engineering properties of the object you created in problem 3.

6. Shade the object you created in problem 3.

Unit 51

■ OBJECTIVE:

To apply basic solid modeling features using the REVOLVE and EXTRUDE commands and the ISOLINES and FACETRES system variables

The appearance of CAD solid models on the screen is similar to that of surface models, but the two are different. Solids contain volume information and physical and material properties. An example of a solid model, representing a shaft, is shown below.

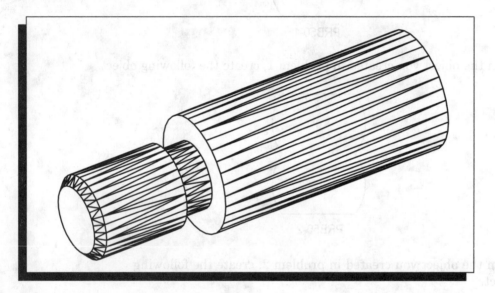

■ *REVOLVE Command* _____

We'll use the REVOLVE command to create a solid model of the shaft. REVOLVE is similar to the REVSURF command used to create a surface of revolution.

1 Start AutoCAD and begin a new drawing from scratch.

This drawing will serve as a template for the solid models you create in this and the following units.

2 Set the grid to **1** and snap to **.25**.

3 Create and make current a new layer named **Objects**, and assign the color red to it.

4 Save your work in a template file named **solid.dwt**. For the template description, enter **For solid models**, and pick **OK**.

5 Use **solid.dwt** to create a new drawing.

6 Approximate the size and shape of the following figure using the **PLINE** command. Omit the numbers and place the figure in the center of the screen. Make sure the polyline is closed.

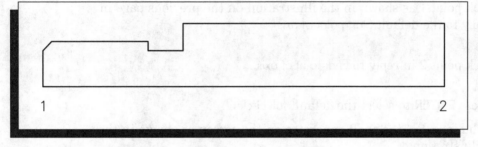

7 Type the **VPOINT** command and press **ENTER**. Press **ENTER** again to accept the View point default value.

The globe representation appears.

8 Select a point that is to the left, above, and in front of the polyline.

HINT: Unit 41, titled "The Third Dimension," shows a labeled illustration of the VPOINT globe.

You should now see a view similar to the one in the following illustration.

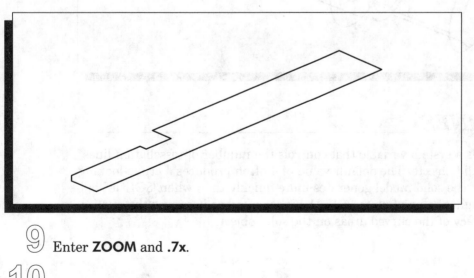

9 Enter **ZOOM** and **.7x.**

10 Save your work in a file named **shaft.dwg.**

11 Display the Solids toolbar and pick the **Revolve** icon.

This enters the REVOLVE command.

12 Select the polyline and press **ENTER**.

13 Pick point 1 as shown in the illustration on the previous page in reply to the default Start point of axis.

14 Pick point 2 in reply to End point of axis.

15 Press **ENTER** to select the default full circle.

The 3D solid model of the shaft generates on the screen as shown in the following illustration.

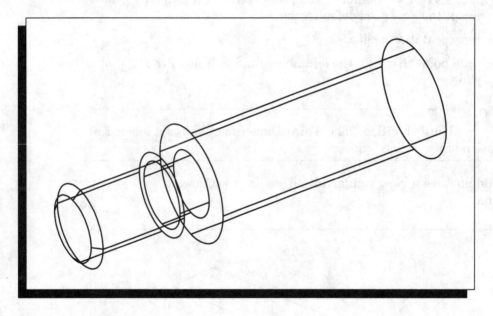

ISOLINES

ISOLINES is a system variable that controls the number of tessellation lines used in solid objects. The default value of 4 often produces a crude-looking object, yet the solid model generates more quickly than when ISOLINES is set at a higher value of up to 2047. A higher number improves the quality and accuracy of the curved areas on the solid object.

1 Enter **ISOLINES** and a new value of **20**.

2 Enter **REGEN**.

As you can see, increasing the value of ISOLINES improves the appearance of the solid model. However, it increases the time required by the subtraction, union, and intersection operations. You will have the opportunity to practice these operations in the following units.

Note that the solid model looks like a conventional 3D surface model—not a solid model. However, it is a solid model. Let's prove it.

3 Using **LIST**, list information on the solid object.

Details on the solid model appear. Notice that the object is a 3D solid.

4 Close the AutoCAD Text Window.

FACETRES

The FACETRES system variable controls the smoothness of solid objects when shading and performing hidden line removals. You can enter a value between .01 and 10, with higher values providing a smoother appearance.

Render

Render

1 Display the Render toolbar, and pick the **Hide** icon from it; then pick the **Shade** icon.

2 Enter **FACETRES** and **3**.

3 Enter **HIDE**.

The appearance of the solid object is smoother.

NOTE:

It takes considerably more time to hide and shade solid models when FACETRES is set to a higher value.

4 Enter **SHADE**.

You should see a similarly smooth shaded view.

5 Enter **FACETRES** and **.5**.

Render

6 Enter **HIDE**; then enter **SHADE**.

Both appear rougher than before.

7 Save your work.

EXTRUDE Command

The following model was created using the EXTRUDE command.

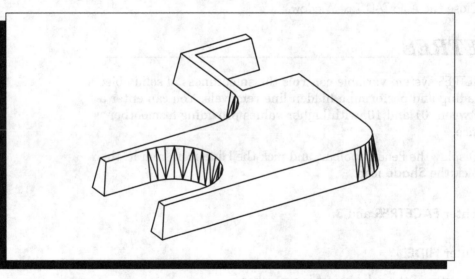

1 Begin a new drawing using the **solid.dwt** template file.

2 Using the **PLINE** command's **Arc** and **Line** options, approximate the size and shape of the following figure. Pick the points in the order shown. Do not place the numbers on the drawing.

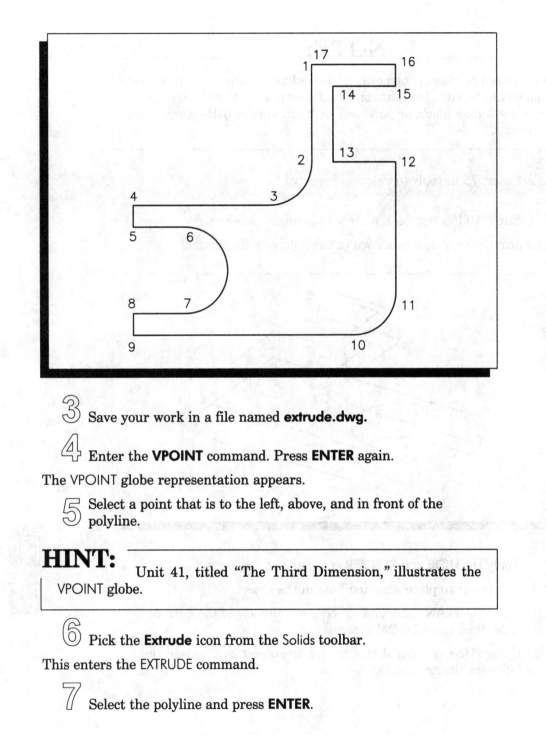

 Solids

③ Save your work in a file named **extrude.dwg**.

④ Enter the **VPOINT** command. Press **ENTER** again.

The VPOINT globe representation appears.

⑤ Select a point that is to the left, above, and in front of the polyline.

HINT: Unit 41, titled "The Third Dimension," illustrates the VPOINT globe.

⑥ Pick the **Extrude** icon from the Solids toolbar.

This enters the EXTRUDE command.

⑦ Select the polyline and press **ENTER**.

NOTE:

Only closed polylines, polygons, circles, ellipses, splines, donuts, and regions can be extruded. The command does not work with 3D objects, objects within a block, or polylines with crossing or self-intersecting segments.

⑧ Enter **.75** in reply to Height of Extrusion.

⑨ Enter **10** (for degrees) in reply to Extrusion taper angle.

A solid primitive appears, as shown in the following illustration.

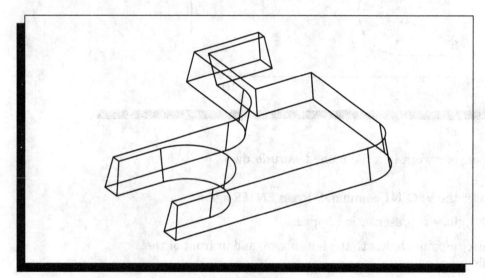

⑩ Apply the **HIDE** and **RENDER** commands to the object.

Render

Render

Now let's attempt to place a slotted hole in the object.

① Enter the **PLAN** command to see the plan view of the current UCS; then enter **ZOOM Extents**.

② Using **PLINE**, place a slotted hole in the object as shown in the following illustration.

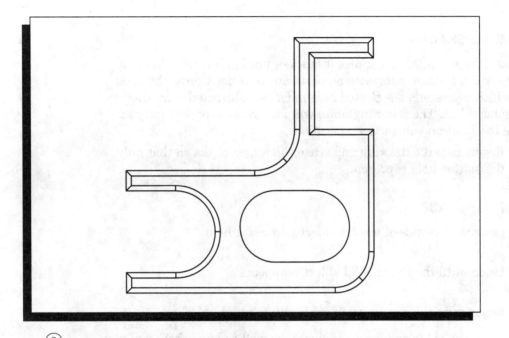

③ Next, view the object in 3D space as before.

Note that the "hole" is still two-dimensional.

Render

④ Pick the **Extrude** icon, select the new polyline, and press **ENTER**. Enter **.75** for the height and **0** for the taper angle.

The object should now look similar to the one in the following illustration.

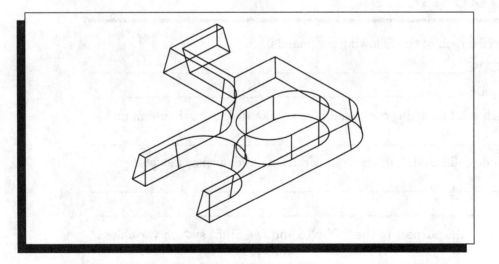

5 Enter **SHADE**.

The slotted hole is filled in because it is solid, not hollow. Creation of a hole requires a Boolean subtraction operation. In other words, the solid object which represents the slotted hole must be subtracted from the larger solid object. The following units will give you the opportunity to practice the Boolean subtraction operation.

6 Regenerate the drawing and erase the larger object so that only the slotted hole is present.

7 Enter **SHADE**.

As you can see, it is indeed a solid object and not a hole.

8 Undo until the larger solid object reappears.

9 Regenerate the drawing and erase the "slotted hole."

10 Close the Solids and Render toolbars, as well as any others you may have opened.

11 Save your work and exit AutoCAD.

Questions

1. Describe each of the following commands.

REVOLVE _____

EXTRUDE _____

2. Which solid modeling command is most like the REVSURF command?

_____

3. How do solid models differ from wireframe and surface models?

_____

_____

4. Describe the purpose of the ISOLINES and FACETRES system variables.

_____

_____

_____

■ *Challenge Your Thinking: Questions for Discussion*

1. Discuss the factors you should consider when setting the ISOLINES and FACETRES system variables. Be specific.

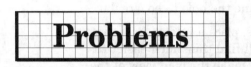

1. Construct a solid bowling pin as shown below (left).

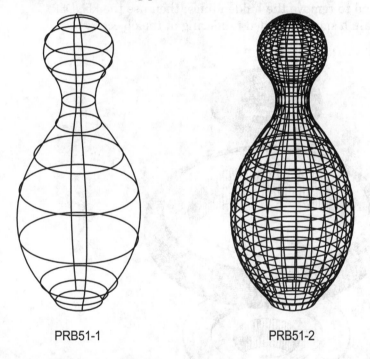

PRB51-1   PRB51-2

2. Increase the number of tessellation lines and create a second bowling pin, as shown above (right).

3. Identify a simple object in the room. Create a profile of the object and then extrude it to create a solid model. Note that only certain objects—those which can be extruded—can be used for this problem.

4. The bumper assembly for a bumper pool table is shown below. Use the dimensions of the three parts (shown on the next page), and use the REVOLVE command to model the parts as solids. Create each part as a separate drawing.

Note that the bumper support has a square hole. Later you will find out how to create such a hole. For now, assume the hole to be circular.

After you have finished each of the three parts, place it into an isometric view by using the VPOINT command. You will find this much easier to do if you first use the ROTATE3D command to make the plan view of the world coordinate system coincide with the top view of the object. Use the HIDE command to remove the hidden lines; then use the RENDER command to obtain a smooth-shaded rendering of the object.

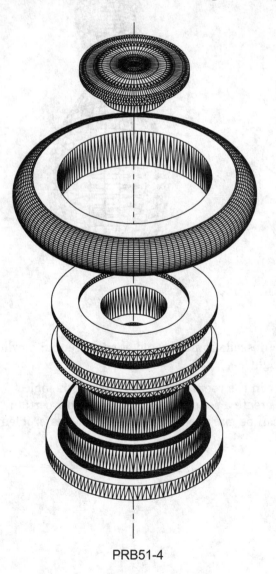

PRB51-4

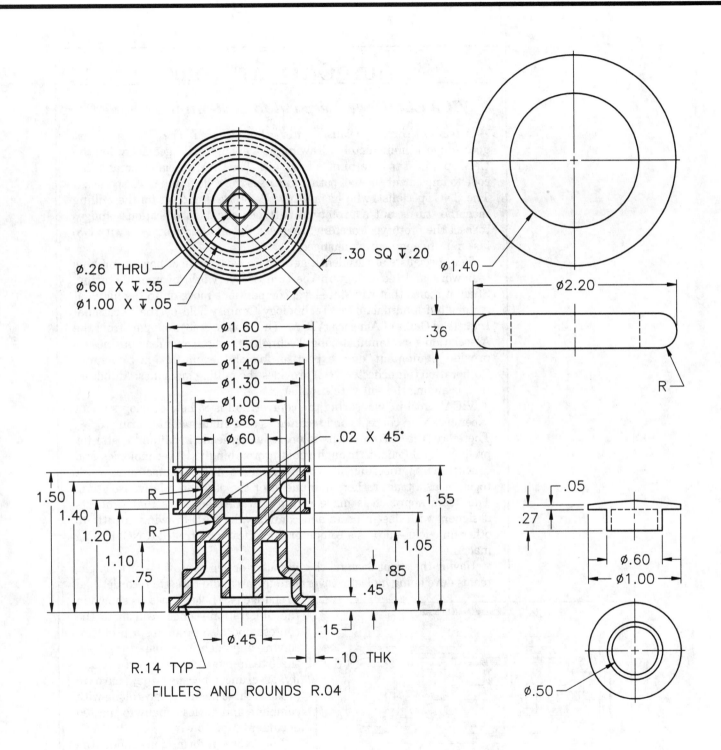

Ø.26 THRU
Ø.60 X ⊽.35
Ø1.00 X ⊽.05

.30 SQ ⊽.20

Ø1.40

Ø2.20

.36

R

Ø1.60
Ø1.50
Ø1.40
Ø1.30
Ø1.00
Ø.86
Ø.60

.02 X 45°

R

1.50
1.40
1.20
1.10
.75

R

1.55

1.05

.85

.45

.15

Ø.45

.10 THK

R.14 TYP

FILLETS AND ROUNDS R.04

.05

.27

Ø.60

Ø1.00

Ø.50

Problem 4 courtesy of Gary J. Hordemann, Gonzaga University

AUTOCAD® AT WORK

VICA Challenge: Respond to Last-Minute Change

A team of three specialists enters the test area. The CAD specialist goes to work immediately, drawing just enough part geometry for the CAM specialist to determine a tool path. The CAM specialist creates NC code to implement the tool path and passes the code to the CNC specialist. The CNC specialist, who has meanwhile stocked and set up the milling machine, turns out a prototype part per client specifications. Judges inspect the prototype carefully, then ask for another prototype with two last-minute engineering changes.

The specialists are students, getting an early taste of how to combine teamwork and technology in the competitive world of manufacturing. Student teams that win state-level competitions move on to the national Automated Manufacturing Technology Contest held by the Vocational Industrial Clubs of America (VICA). The contest enjoys support from the Association for Manufacturing Technology, whose member businesses provide equipment and expertise for the event. Light Machines Corporation (Manchester, NH) provides the milling machines. Autodesk, Inc., is among the software suppliers.

VICA's goal in sponsoring the competition is to help develop student experience with concepts and techniques that guide manufacturing today. Therefore, the current focus is on reducing lead times (and costs) for product development through rapid-prototyping (RP) technologies and concurrent engineering, in which designers and manufacturers work together as a team rather than one after the other in relative isolation. This team approach is made possible by CAD software that connects designers with design users and connects the CAD drawing directly to other functions, such as a computer numerically controlled (CNC) milling machine.

Having the client demand changes after design approval—for example, teams developing a plastic coverplate for an audio cassette recorder had to change two hole shapes, as shown in the illustration—lends realism to the contest and also promotes a problem-solving approach in a shared task. All three team members take part in quality assurance, because they have to produce three identical parts along with complete and professionally formatted documentation to win.

For more information about the VICA Automated Manufacturing Technology challenge (high school or post-secondary division), write to VICA at P.O. Box 3000, Leesburg, VA 22075 or telephone (703) 777-8810.

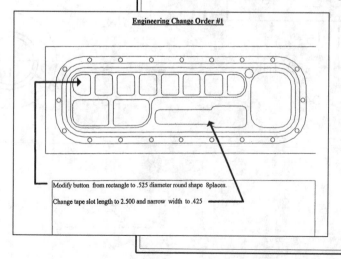

Engineering Change Order #1

Modify button from rectangle to .525 diameter round shape 8 places.

Change tape slot length to 2.500 and narrow width to .425

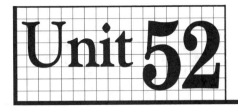 **Unit 52** Predefined Primitives

■ OBJECTIVE:

To create solid primitives using the CYLINDER, TORUS, CONE, WEDGE, BOX, and SPHERE commands

AutoCAD permits you to generate predefined solid primitives by specifying the necessary dimensions. Examples include boxes, cones, cylinders, and spheres.

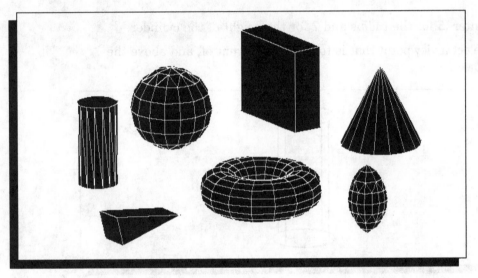

Predefined and user-defined primitives make up the basic building blocks for creating moderately complex solid models, as will be illustrated in the following units.

CYLINDER Command _____

The CYLINDER command allows you to create solid cylinders of any size.

1 Start AutoCAD and begin a new drawing using the **solid.dwt** template file.

2 Set snap to **.5**.

3 Display the Solids toolbar.

4 From this toolbar, pick the **Cylinder** icon.

This enters the CYLINDER command.

Solids

NOTE:

The Elliptical option enables you to create an elliptical cylinder using prompts similar to those used in the ELLIPSE command.

5 In reply to the center point default setting, pick a point at any location on the screen.

6 Enter **.5** for the radius and **2** for the height of the cylinder.

7 Select a viewpoint that is to the left, in front of, and above the object.

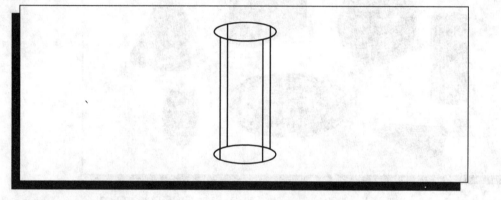

8 Display the Render toolbar and shade the cylinder.

TORUS Command

The TORUS command lets you create donut-shaped solid primitives.

1 Enter **ZOOM** and **.5x**. Pan if necessary to make space for the torus.

2 Pick the **Torus** icon from the Solids toolbar.

This enters the TORUS command.

Solids

3 Pick a center point for the torus (anywhere) and enter **1** for the radius of the torus.

The radius of the torus specifies the distance from the center of the torus to the center of the tube.

4 Enter **.5** for the radius of the tube.

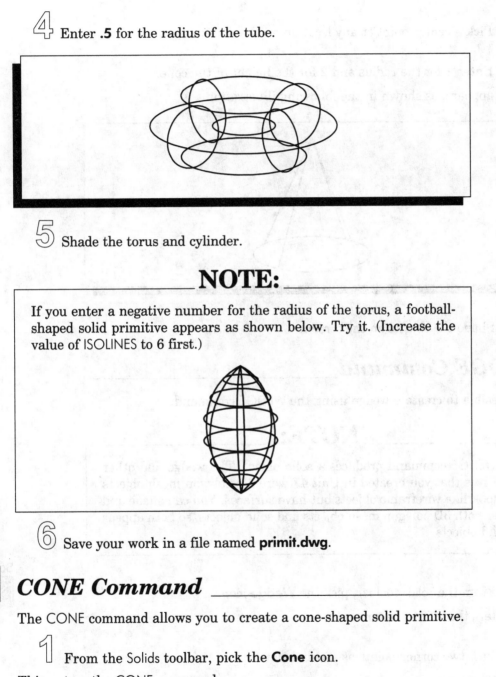

5 Shade the torus and cylinder.

NOTE:

If you enter a negative number for the radius of the torus, a football-shaped solid primitive appears as shown below. Try it. (Increase the value of ISOLINES to 6 first.)

6 Save your work in a file named **primit.dwg**.

CONE Command

The CONE command allows you to create a cone-shaped solid primitive.

Solids

1 From the Solids toolbar, pick the **Cone** icon.

This enters the CONE command.

NOTE:

The function of the Elliptical option is similar to that of the CYLINDER command.

② Pick a center point at any location on the screen.

③ Enter **1** for the radius and **2** for the height of the cone.

A cone appears, as shown in the following illustration.

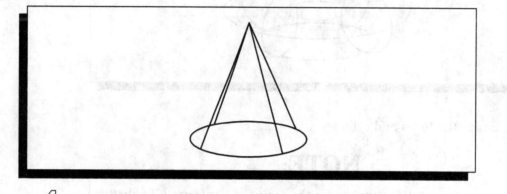

④ Shade the objects on the screen.

WEDGE Command

It is possible to create a wedge using the WEDGE command.

NOTE:

The WEDGE command produces a solid object. The wedge and other primitives that you created in Unit 47 were 3D polygon mesh objects that look like wireframe objects but have surfaces. You can shade and render both 3D polygon mesh objects and solid objects, so both appear as solid objects.

Solids

① From the Solids toolbar, pick the **Wedge** icon.

This enters the WEDGE command.

② Pick two corner points as shown below.

③ Enter **.75** for the height.

A wedge-shaped solid primitive appears.

④ Shade the objects.

■ *BOX Command*

The BOX command enables you to create a solid box.

Solids

① From the Solids toolbar, pick the **Box** icon.

This enters the BOX command.

___ NOTE: ___

The Center option permits you to create a box by specifying a center point.

② Pick a point to represent one of the corners of the box.

③ Pick the diagonally opposite corner of the base rectangle as shown in the following illustration. You may pick the two points in either order.

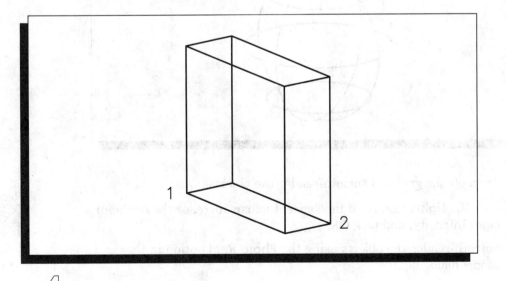

④ Enter **2** for the height.

The box that appears should be similar to the one shown above.

5 Shade the solid primitives.

NOTE:

To create a solid cube, you can select the BOX command's Cube option and enter a length for one of its sides.

SPHERE Command

The SPHERE command creates a solid sphere.

Solids

1 From the Solids toolbar, pick the **Sphere** icon.

This enters the SPHERE command.

2 Pick a point for the center of the sphere.

3 Enter **1** for the radius of the sphere.

The sphere appears as shown below.

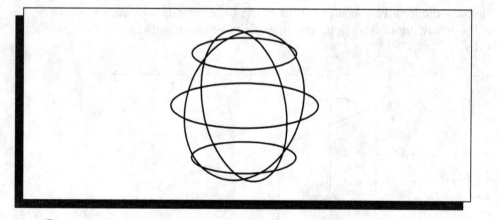

4 Turn off the grid and smooth-render the objects.

Render

5 Pick the **Lights** icon from the Render toolbar, increase the **Ambient Light** intensity, and pick **OK**.

6 Smooth-render the objects using the **Photo Real** option in the Render dialog box.

7 Close the Solids and Render toolbars, as well as any others you may have opened.

8 Save your work and exit AutoCAD.

Questions

1. Describe the function of each of the following commands.

   CYLINDER _____

   _____

   TORUS _____

   _____

   CONE _____

   _____

   WEDGE _____

   _____

   BOX _____

   _____

   SPHERE _____

   _____

2. With what command can you create a football-shaped primitive?

   _____

3. With what command can you create a solid cube?

   _____

4. What is the purpose of the Elliptical option of the CYLINDER and CONE commands?

   _____

   _____

■ *Challenge Your Thinking: Questions for Discussion*

1. Find out how to change properties of existing solid primitives. How could you change the color or layer?

Problems

1. Create a football-shaped primitive on top of a large cube.

2. Use the following three different ways to create a solid cylinder with a radius of .4 and a length of 2. Set ISOLINES to 8 before you begin.

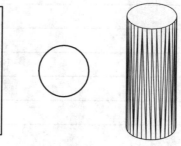

   a. Draw a rectangle measuring .2 by 2 and revolve it 360°.

   b. Draw a circle with a radius of .4 and extrude it to a height of 2.

   c. Use the CYLINDER command to create a solid cylinder with a radius of .4 and a height of 2.

PRB52-2

   Now use the AREA command to determine the surface area of all three objects. Calculate the exact answer and compare it to the area AutoCAD calculated for each of the three cylinders.

3. Change ISOLINES to 1 and create the three objects again. Use AREA to determine the surface area of the three objects again.

4. Set ISOLINES back to 8 and repeat the process for a right prism measuring .8 × .8 × 2. Use the following three ways to create the prism.

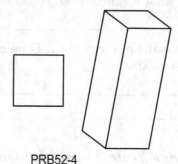

   a. Draw a rectangle measuring .8 by 2. Extrude it to a height of .8.

   b. Draw a square .8 on each side and extrude it to a height of 2.

   c. Use the BOX command to insert a box measuring .8 × .8 × 2.

PRB52-4

   Use the AREA command to find and compare the surface area of the three objects.

Problems 2, 3, and 4 courtesy of Gary J. Hordemann, Gonzaga University

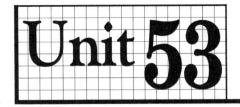

Unit 53 Boolean Operations

■ OBJECTIVE:

To create composite solids using the SUBTRACT, UNION, AMECONVERT, ACISIN, and ACISOUT commands

Much of AutoCAD solid modeling is based on the principles of Boolean mathematics. AutoCAD Boolean operations, such as union, subtraction, and intersection, are used to create composite solids. Composite solids are composed of two or more solid primitives.

Composite Solids

Using Boolean operations, let's create the following composite solid.

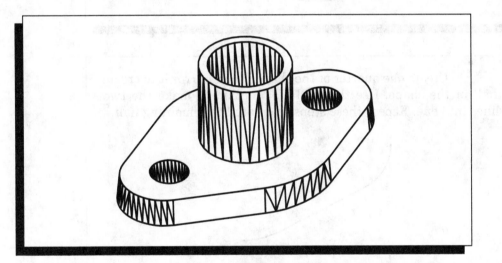

1. Start AutoCAD and begin a new drawing using the **solid.dwt** template file.

2. Using the **PLINE Line** and **Arc** options, approximate the following shape.

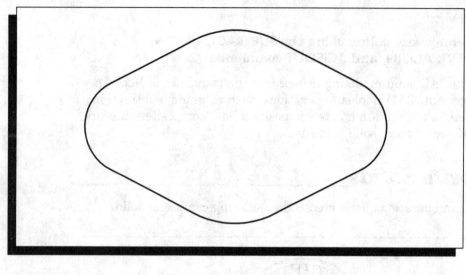

HINT:
Create one quarter of the shape. Then mirror it to create one half of the shape. Use the PEDIT Join option to make the two polylines into one. Repeat these steps to create the remaining half.

③ Select a viewpoint that is to the left, in front of, and above the object.

④ Set **ISOLINES** to **6** and save your work in a file named **compos.dwg**.

⑤ Display the Solids and Modify II toolbars.

Solids

⑥ From the Solids toolbar, pick the **Extrude** icon.

This enters the EXTRUDE command.

7 Select the polyline and press **ENTER**.

8 Enter **.5** for the height and specify a taper angle of **5** degrees.
A view similar to the following appears.

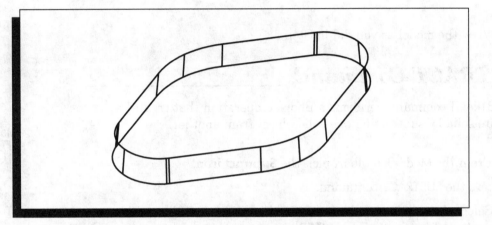

9 Enter the **PLAN** command and press **ENTER** again to accept the default setting.

The plan view allows you to see the taper, as shown in the following illustration, but without the two holes.

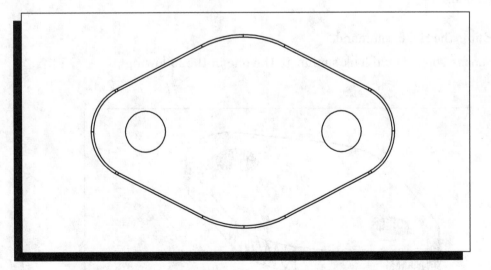

10 Pick the **Cylinder** icon from the Solids toolbar and place two cylinders as shown in the previous illustration. Make the height of the cylinders **.5** unit, and approximate their radius and location.

HINT:

Create one of them and use the COPY or MIRROR command to create the second.

You have just created two solid cylinders within the original extruded solid.

11 View the model as you did in Step 3.

SUBTRACT Command

The SUBTRACT command performs a Boolean operation that creates a composite solid by subtracting one solid object from another.

Modify II

1 From the Modify II toolbar, pick the **Subtract** icon.

This enters the SUBTRACT command.

2 Select the extruded solid (but not the cylinders) as the solid to subtract from, and press **ENTER**.

3 Select the two cylinders as the solids to subtract, and press **ENTER**.

AutoCAD subtracts the volume of the cylinders from the volume of the extruded solid.

4 Enter the **HIDE** command.

The composite solid should look similar to the one in the following illustration.

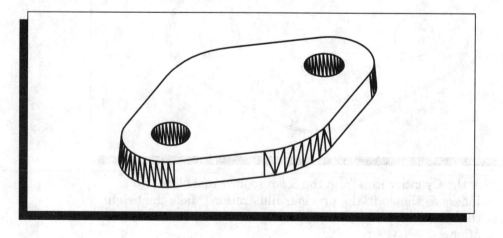

UNION Command _____

UNION permits you to join two solid objects to form a new composite solid.

 Enter the **PLAN** command and press **ENTER** to view the plan view of the current UCS.

2 Place a cylinder at the center of the model. Make the diameter **2** units, and make it **2** units in height.

HINT:
> To find the center of the model, use a horizontal xline through the centers of the smaller holes and a vertical xline through the midpoints of the top and bottom arcs of the solid.

3 Using the same center point, place a second cylinder inside the first. Specify a smaller radius and the same height as the first cylinder.

4 Specify a viewpoint in 3D space as you did before.

5 Subtract the smaller cylinder from the larger cylinder.

Modify II

HINT:
> First select the larger cylinder and press ENTER, and then select the smaller cylinder and press ENTER.

The model is made up of two separate solid objects instead of a single composite solid.

Modify II

6 From the Modify II toolbar, pick the **Union** icon.

This enters the UNION command.

7 Select the two solid objects and press **ENTER**.

The two objects join to form a new composite solid.

8 Enter **HIDE**.

Your model should look similar to the one in the following illustration.

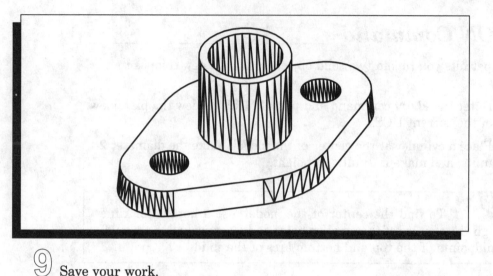

9 Save your work.

Converting AME Models

Using the AMECONVERT command, it's possible to convert solid models created with Autodesk's Advanced Modeling Extension (AME) Release 2 or 2.1 to AutoCAD Release 14 solids or regions.

1 Enter **AMECONVERT**.

2 Select the solid model you created in this unit and press **ENTER**.

NOTE:

See also the optional *Applying AutoCAD Diskette*.

AutoCAD displays the message Ignored 1 object that is neither an AME solid nor a region because the object you selected was not an AME model. If the model had been created using AME, AutoCAD would have converted it to a Release 14 solid model.

Converting ACIS Files

ACIS is the 3D geometry engine that AutoCAD uses to create solid models. CAD software products that use the ACIS engine can output ASCII files known as SAT files, which are readable directly by other ACIS-based products. SAT files are machine-independent geometry files. These files contain topological and surface information about models that were created using the ACIS engine. Using AutoCAD's ACISIN command, you can import objects stored in the SAT format. With ACISOUT, you can create SAT files from AutoCAD NURBS surfaces, regions, and solid objects.

1 Select **ACIS Solid...** from the **Insert** pull-down menu.

NOTE:

See also the optional *Applying AutoCAD Diskette.*

AutoCAD displays the Select ACIS File dialog box. Notice that the default file type is ACIS (*.sat). If SAT files were available, you could select one at this time.

2 Pick the **Cancel** button.

3 Enter **ACISOUT**, select the solid model, and press **ENTER**.

AutoCAD displays the Create ACIS File dialog box. Notice that the default file type is ACIS (*.sat). Also, notice the file name that it suggests.

4 Choose a directory and pick the **Save** button.

This creates a file named compos.sat that can be read into products that import SAT files.

5 Minimize AutoCAD and start Notepad.

6 Select **Open...** from the **File** pull-down menu, click in the **File name** text box, and replace txt with **sat**.

7 Locate and open the file named **compos.sat**.

The following shows the first few lines of the file.

```
106 514 1 0
body $1 $2 $-1 $-1 #
f_body-lwd-attrib $-1 $3 $-1 $0 #
lump $4 $-1 $5 $0 #
ref_vt-lwd-attrib $-1 $-1 $1 $0 $6 $7 #
ref_vt-lwd-attrib $-1 $-1 $-1 $2 $6 $7 #
shell $8 $-1 $-1 $9 $2 #
refinement $-1 0 0 0 0 0.023929836228489876 0 0 0 2 #
vertex_template $-1 3 0 1 8 #
ref_vt-lwd-attrib $-1 $-1 $-1 $5 $6 $7 #
face $10 $11 $12 $5 $-1 $13 reversed single #
color-adesk-attrib $-1 $14 $-1 $9 256 #
face $15 $16 $17 $5 $-1 $13 reversed single #
loop $-1 $18 $19 $9 #
plane-surface $-1 2.75 5.5 0.5 0 0 -1 -1 0 0 0 I I I I #
fface-lwd-attrib $-1 $20 $10 $9 #
color-adesk-attrib $-1 $21 $-1 $11 256 #
face $22 $23 $24 $5 $-1 $25 reversed single #
loop $-1 $-1 $26 $11 #
loop $-1 $27 $28 $9 #
coedge $29 $30 $31 $32 $33 1 $12 $-1 #
ref_vt-lwd-attrib $-1 $-1 $14 $9 $6 $-1 #
fface-lwd-attrib $-1 $34 $15 $11 #
color-adesk-attrib $-1 $35 $-1 $16 256 #
face $36 $37 $38 $5 $-1 $39 reversed single #
loop $-1 $40 $41 $16 #
cone-surface $-1 3.375 4.875 0.25 0 0 1 0.375 0 0 1 I I 0 1 0 I I I I #
coedge $42 $26 $26 $43 $44 1 $17 $-1 #
loop $-1 $45 $46 $9 #
coedge $47 $28 $28 $48 $49 0 $18 $-1 #
copar-lwd-attrib $-1 $-1 $-1 $19 #
coedge $50 $51 $19 $52 $53 1 $12 $-1 #
coedge $54 $19 $55 $56 $57 1 $12 $-1 #
```

⑧ After you have reviewed the file, pick **Exit** from the **File** pull-down menu and maximize AutoCAD.

⑨ Close the floating toolbars, save your work, and exit AutoCAD.

Questions

1. Give an example of a Boolean operation.

_____

2. What is a composite solid?

_____

_____

3. Explain the purpose of each of the following solid modeling commands.

SUBTRACT _____

_____

UNION _____

_____

AMECONVERT _____

_____

ACISIN _____

_____

ACISOUT _____

_____

▮ *Challenge Your Thinking: Questions for Discussion*

1. Describe how you would create each of the following solid models. Be specific.

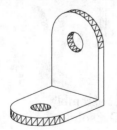 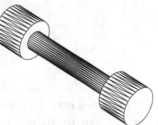

Problems

1-2. Apply the commands you learned in this unit to create the following composite solids. Approximate all sizes. Two views of prb53-2 are provided to help you visualize the entire part.

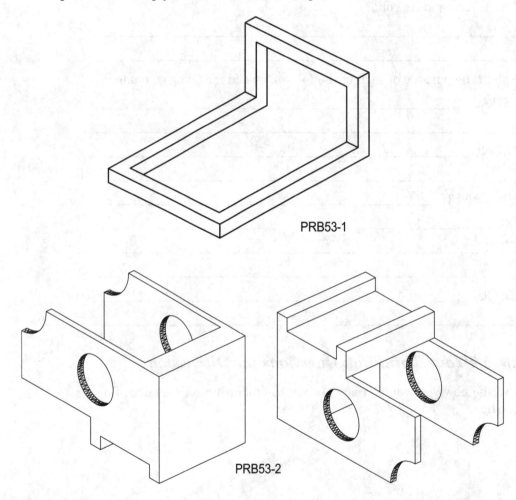

PRB53-1

PRB53-2

3. Identify a machine part, such as a gear on a shaft or a bracket assembly. Sketch the part in your mind or on paper. Using AutoCAD's solid primitives and Boolean operations, shape the part by adding and subtracting material until it is complete.

4. If you have the optional *Applying AutoCAD Diskette,* open the file named t-conn.dwg and convert the AME solid model to a Release 14 model.

5. If you have the optional *Applying AutoCAD Diskette,* start a new drawing file from scratch. Create a new layer called Objects and make it the current layer, giving it the color of your choice. Then import the SAT file named plug.sat. Save the file as prb53-5.dwg.

6. Create the fluid coupling end cap shown on the right using the REVOLVE command. Begin by creating a layer named Endcap, and set the color to cyan. Set ISOLINES to 8. The object's profile is shown below.

After you have finished the object, use the VPOINT command to place it into an isometric view. You will find this much easier to do if you first use the ROTATE3D command to make the plan view of the world coordinate system coincide with the top view of the object. Use the HIDE command to remove the hidden lines. Use the RENDER command to obtain a shaded rendering of the object.

PRB53-6

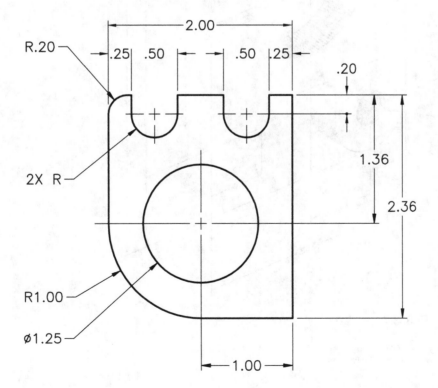

7. Create the shaft clamp shown below using the SUBTRACT and EXTRUDE commands. Create the .40 diameter hole by inserting and subtracting a cylinder.

After you have finished the object, place it into an isometric view using the VPOINT command. Use the HIDE command to remove the hidden lines and the RENDER command to obtain a shaded rendering of the object.

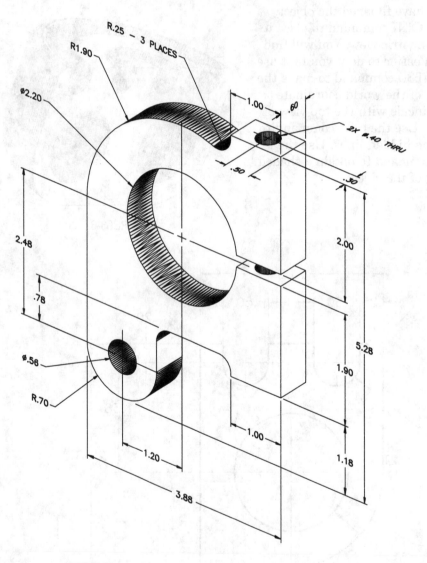

PRB53-7

Problems 6 and 7 courtesy of Gary J. Hordemann, Gonzaga University

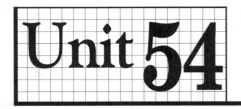

Unit 54 Tailoring Solid Models

■ OBJECTIVE:

To create and modify solid objects using the INTERSECT, INTERFERE, FILLET, and CHAMFER commands

AutoCAD provides several commands and options that permit you to modify primitives and solid composites. The above commands were used to create the following two solid composites.

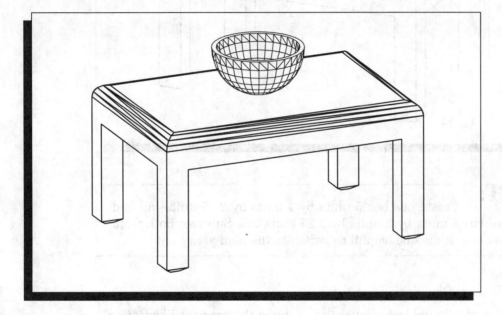

Solidifying the Basics

1 Start AutoCAD and begin a new drawing using the **solid.dwt** template file.

2 Display the **Solids** and **Modify II** toolbars.

3 Pick the **Box** icon from the Solids toolbar and create a solid box **6** units in the X direction and **4** units in the Y direction. Make it **3** units in height and place it in the center of the screen.

Solids

4 Select a viewpoint that is to the left, in front of, and above the object.

5 Enter **ZOOM** and **.9x**.

Solids

6 Using the **BOX** and **SUBTRACT** commands, create and subtract two rectangular boxes from the first one to form the following shape.

Modify II

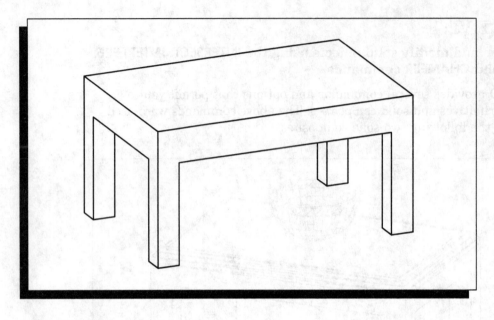

HINT:
Create one box 5 units by 4 units by 2.25 units tall and
another box 6 units by 3 units by 2.25 units tall. Subtract both from
the first box. It may be helpful to switch to the plan view.

7 Enter **HIDE**.

The solid model should look similar to the one in the previous illustration.

8 Save your work in a file named **table.dwg**.

9 Enter **REGEN**.

Beveling and Rounding

The CHAMFER and FILLET commands permit you to bevel and round parts
of a solid model.

Modify

1 Enter the **CHAMFER** command and select the bottom of one of the
four table legs.

2 If AutoCAD selects the side of the table rather than the bottom of
the leg, type **N** for Next, and press **ENTER**. When AutoCAD selects
the bottom, press **ENTER**.

3 Enter a base surface distance of **.15**.

At the Enter other surface distance prompt, enter **.1**.

 At the Loop/<Select edge> prompt, pick all four edges that make up the bottom of the leg and press **ENTER**.

AutoCAD chamfers the edges.

 Repeat Steps 1 through 5 to chamfer the remaining three legs.

The bottoms of the legs should now look similar to those in the following illustration.

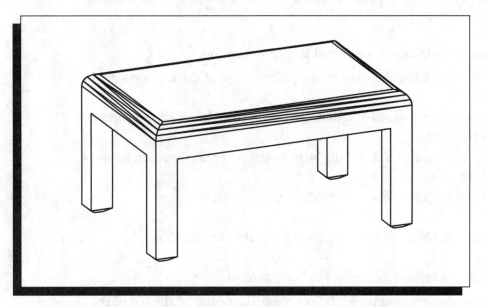

 Enter **FILLET**, select any edge of the table top, and enter **.3** for the radius.

 Enter **C** for Chain, select the remaining three edges of the table top, and press **ENTER**.

 Enter **HIDE** and save your work.

The model should look very similar to the one in the previous illustration.

Modify

 ## *INTERSECT Command* _____

You have learned that you can create composite solids by adding and subtracting solid objects. You can also create them by overlapping two or more solid objects and calculating the solid volume that is common (intersecting) to each of the objects. AutoCAD's INTERSECT command performs this Boolean operation.

1 Enter the **BOX** command and create a 2-unit cube at any location on the floor (WCS) under the table. (The floor and WCS are on the same plane.) Be sure snap is on.

HINT: You may want to return to the plan view to perform this step. After picking the first corner of the box, select the Cube option and enter 2 in reply to Length.

2 View the objects in 3D as you did before. If necessary, move the box away from the table or select a viewpoint so the two objects do not overlap.

3 Enter the **UCS** command and the **3point** option.

4 Using the **Endpoint** snap mode, pick any one of the top corners of the cube.

5 To specify the positive X direction, snap to an adjacent top corner of the cube.

6 For the positive Y direction, snap to the other adjacent top corner of the cube.

The UCS and UCS icon change to reflect the new UCS.

7 Enter **PLAN** to review the plan view of the current UCS.

Solids

8 Pick the **Sphere** icon from the Solids toolbar.

9 Place the center point of the sphere at the center of the top of the cube and enter **1** for the radius.

A solid sphere appears.

10 Move the box and sphere so they are centered directly under the table.

11 Select a viewpoint that is similar to the previous viewpoint.

You should see a view that is similar to the one in the following illustration.

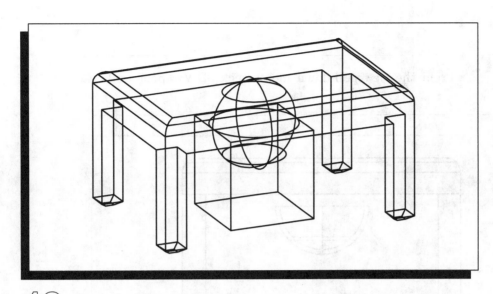

Modify II

12 From the Modify II toolbar, pick the **Intersection** icon.

13 Select the cube and sphere and press **ENTER**.

A half-sphere (hemisphere) appears.

AutoCAD has calculated the solid volume that is common (intersecting) with each of the primitives.

Suppose we want to make the hemisphere into a dish. We must subtract a smaller hemisphere from the one on the screen.

1 Return to the plan view of the current UCS.

2 Copy the hemisphere to a new location so the two are next to one another. Be sure snap is on.

Modify

3 Enter the **SCALE** command, select the second hemisphere, and press **ENTER**.

4 Select a point at the center of the hemisphere in reply to Base point and enter **.9** for the scale factor.

A slightly smaller hemisphere appears.

5 Move the smaller hemisphere inside the larger one.

The smaller hemisphere may (or may not) be in the proper location for the subtraction operation.

6 Produce the view shown in the following illustration.

HINT: From the View pull-down menu, pick 3D Viewpoint and Front.

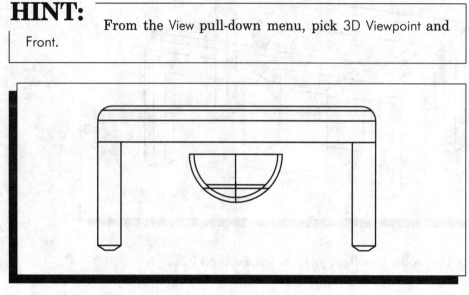

7 Zoom in closer on the hemispheres until they fill the screen and examine them closely.

8 If necessary, move the smaller hemisphere into place. (You may need to turn off ortho.)

Modify II

9 Subtract the smaller hemisphere from the larger one, and save your work.

Moving into Position

Let's move the dish to a more logical location—on top of the table.

1 Enter **ZOOM** and **.3x**.

2 Move the dish upward and position it on top of the table.

HINT: Use the UCS View option to set the UCS flat to the screen. Then, with ortho on, move the dish to the top of the table.

The dish should now be sitting on top of the table.

3 Select a viewpoint that is to the left, in front of, and above the object.

4 Move the dish near the center of the table.

5 Give the dish a new color.

6 Shade the solid objects.

7 Turn off the grid and render the solid objects.

INTERFERE Command

The INTERFERE command enables you to check for interference (overlap) between two or more solid objects. It is particularly useful when fitting together solid objects into an assembly.

1 Regenerate the drawing to remove the shading, and zoom in on the dish.

2 From the Solids toolbar, pick the **Interfere** icon.

Solids

This enters the INTERFERE command.

3 Pick either the table or the dish and press **ENTER**.

4 Pick the other composite solid and press **ENTER**.

The message Solids do not interfere should appear. If you don't receive this message, the two interfere with one another. If the solids don't interfere with each other, proceed to Step 6.

5 If they do interfere, press **ESC** to cancel.

6 Move the dish downward **.25** unit.

HINT: In reply to Second point of displacement, enter the .xy point filter and pick a point in the same location as the base point. Enter –.25 for the *z* coordinate.

Solids

7 Enter the **INTERFERE** command again.

8 Pick one of the two composite solids and press **ENTER**; pick the other and press **ENTER**.

You should receive the message Interfering pairs: 1. This is because the two interfere with one another.

When solids interfere with one another, you can create "interference solids" to show the parts of the solids that overlap.

9 Enter **Yes** in reply to Create interference solids?

The interfering portions of the dish and table appear. This is the new solid object that represents the interference.

10 Using **MOVE Last**, move the new solid away from the dish and table.

11 Enter **HIDE**.

The new solid should look similar to the one in the following illustration.

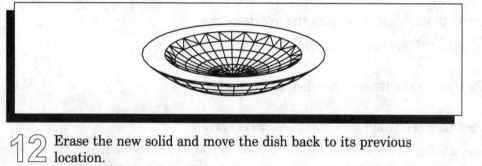

12 Erase the new solid and move the dish back to its previous location.

13 **ZOOM All** and enter **HIDE**.

14 Close the floating toolbars, save your work, and exit AutoCAD.

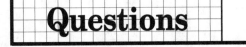
Questions

1. How do the CHAMFER and FILLET commands affect the appearance of solid objects containing outside corners?

_____

_____

2. Explain how the INTERSECT command is used to form a composite solid.

_____

_____

3. Describe the purpose of the INTERFERE command.

_____

_____

4. How might the INTERFERE command be useful when designing parts of an assembly?

_____

_____

 Challenge Your Thinking: Questions for Discussion

1. How do the CHAMFER and FILLET commands affect the appearance of solid objects containing inside corners?

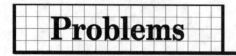

Problems

1-2. Apply the commands you learned in this unit to create the following solid composites. Approximate all sizes. Use the INTERSECT command to complete prb54-1. Modify prb54-1 to complete prb54-2. (Set FACETRES to 3 before you begin.)

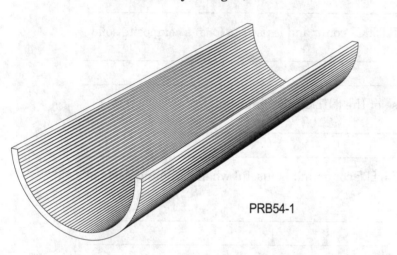

PRB54-1

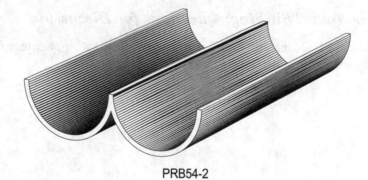

PRB54-2

3. Using the SUBTRACT and UNION commands, model the plate. The dimensions are given in the orthographic projections shown below.

Use the FILLET command to create a fillet where the boss joins the main part of the plate.

After you have finished the model, obtain an isometric view. (First rotate it using the ROTATE3D command.) Then remove hidden lines.

PRB54-3

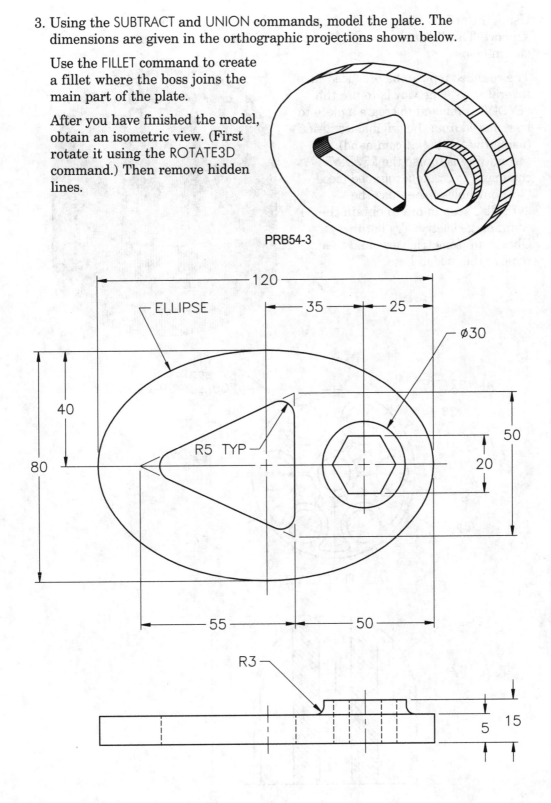

623

4. Using the SUBTRACT and EXTRUDE commands, model the tube bundle support. The dimensions are given in the orthographic projections shown below.

The rounded top can be produced in several ways. One way is to use the REVOLVE command to create a piece to be removed from the original extrusion (using the SUBTRACT command). A second way is to use the REVOLVE command to create a rounded piece (without holes); then use the INTERSECT command to obtain the common geometry. Try both ways. Obtain an isometric view, and then remove the hidden lines.

PRB54-4

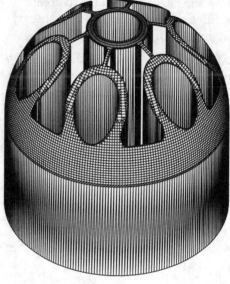

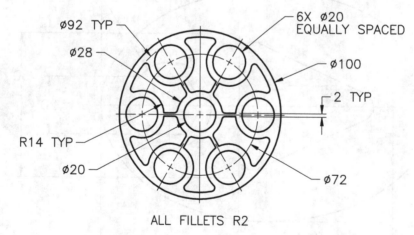

Ø92 TYP

Ø28

6X Ø20
EQUALLY SPACED

Ø100

2 TYP

R14 TYP

Ø20

Ø72

ALL FILLETS R2

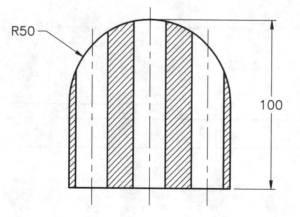

R50

100

5. Three orthographic views of a T-swivel support are shown below. Model the piece as a solid. Use the FILLET command to fillet all of the indicated edges.

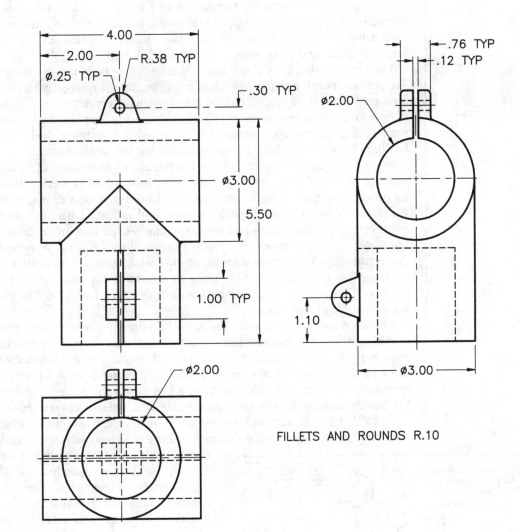

FILLETS AND ROUNDS R.10

PRB54-5

Problems 3, 4, and 5 courtesy of Gary J. Hordemann, Gonzaga University

AutoCAD® at Work

A Giant Step in the Stair Business

During a busy season, the Century Stair Co. (Haymarket, VA) cuts a thousand risers a day for straight staircases, which is the staircase type most customers want. A curved staircase, though elegant, takes much longer to build and costs more.

On a curved staircase, every step must be cut to its own specification. A machine operator might need half a day to cut out risers, treads, and stringer boards for a typical curved staircase, which gives you some idea of how much time has to be invested in calculating each dimension and rendering the drawings. Century Stair Co., using a system called TRIAD, has found it can now deliver a curved staircase five times faster.

TRIAD is an integration of "adjustable template" software and robotic machining to make the most of AutoCAD. Here is how it works: When an order for a curved staircase comes in, Century Stair Co. makes an AutoCAD drawing. The drawing is generated quickly with the aid of Synthesis, a parameter-control program that works like an adjustable template. The user specifies certain dimensions as inputs; Synthesis determines the remaining dimensions by calculation. The result is a drawing free of computational errors in a remarkably short time. For Century Stair Co., a customer-approval drawing can be ready in two hours instead of two weeks.

On the manufacturing side, the AutoCAD drawing continues to play a role, not just as documentation, but as the data source for robotic machining. A program called NC-AUTO-CODE implants into the AutoCAD drawing the numeric code to guide a CNC (Computer Numeric Control) machine. With just one CNC machine, a two-head router, Century Stair Co. has the capacity to turn out ten exquisitely curved staircases per week.

AutoCAD has its uses in-house apart from the TRIAD system, as when a builder needs to determine dimensions for a staircase with a winder (a component that connects two straight staircases, usually at a 45-degree angle, to create the effect of a curved staircase). The builder makes a stick drawing in AutoCAD and gets dimensions with accuracy "down to a couple thousandths of an inch, which is more than you generally need for a residential staircase," says Director of Planning Bob Snitzer—but assures a good fit for Century Stair customers throughout Maryland, New Jersey, eastern Pennsylvania, and northern Virginia.

Based on a story by Wayne Turner in CADENCE, June 1990, Copyright © 1990, Miller Freeman, Inc.

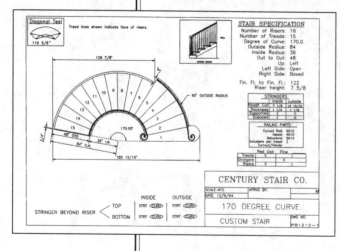

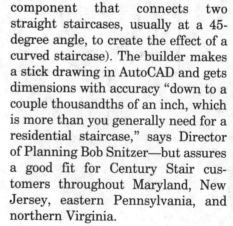

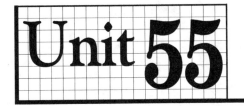

Unit 55 Downstream Benefits

■ OBJECTIVE:

To create a full section and profile view from a solid model using the SECTION and SOLPROF commands

Solid modeling provides many benefits that are "downstream" from the solid model. Examples include mass properties generation, detail drafting, finite element analysis, and the fabrication of physical parts. In this unit you will create full section and profile views of a pulley.

■ *Creating a Solid Pulley*

The following sequence steps you through the creation of a pulley.

1. Start AutoCAD and begin a new drawing using the **solid.dwt** template file.

2. Set snap to **.125**.

3. Set **ISOLINES** to **30**, and display the Solids toolbar.

4. Create the following solid model. Approximate all sizes, but *do not* make the overall height of the pulley more than 8 units.

HINT:

Create the following shape as a polyline in the top half of the screen. *Do not* make the overall height of it more than 4 units. Then create a solid by revolving the polyline using the REVOLVE command. Position the axis of revolution as shown below.

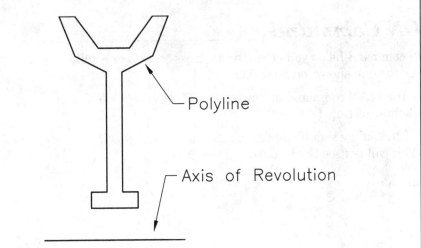

627

A solid model, similar to the view on the left in the following illustration, should appear.

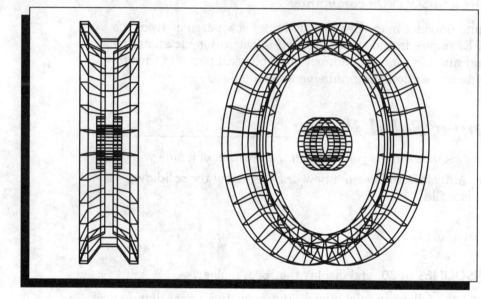

 Save your work in a file named **pulley.dwg.**

6 Select a viewpoint that is to the left and in front of the model, and enter **ZOOM .8x**.

Your view of the model should be similar to the one on the right in the previous illustration.

SECTION Command

The SECTION command helps you create a full cross section of a solid model, such as the one shown on page 632.

1 Enter the **PLAN** command and press **ENTER** to accept the Current UCS default setting.

A plan view of the pulley should appear, as shown in the following illustration. You will *not* see the box around the pulley.

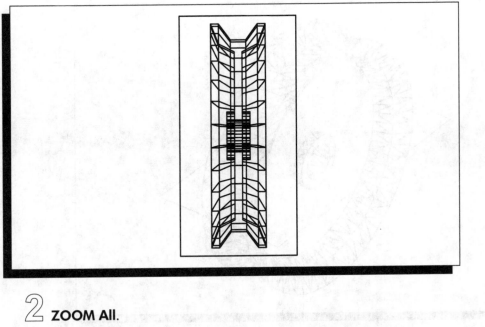

2 **ZOOM All.**

3 If necessary, move the pulley so that it is entirely within the drawing area. (The grid reflects the drawing area.)

4 Using the **PLINE** command, draw a box around the pulley as shown in the previous illustration. Approximate its size.

5 Create a view that is above, to the left, and in front of the pulley.

The polyline object passes through the center of the pulley.

6 Enter **HIDE**.

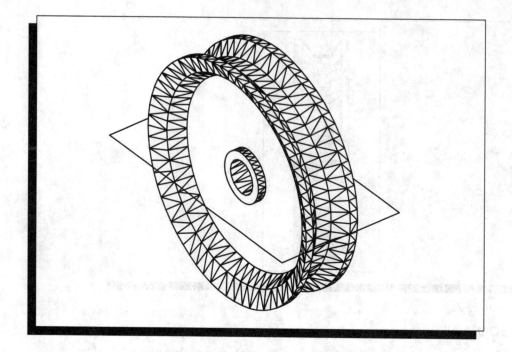

As you can see, the polyline indeed cuts through the middle of the pulley.

7 Enter **PLAN** to obtain the plan view of the current UCS and enter **ZOOM All**.

8 Make layer **0** the current layer.

9 Pick the **Section** icon from the Solids toolbar.

This enters the SECTION command.

10 Pick the pulley and press **ENTER**.

AutoCAD asks you to define a plane.

11 Enter **O** (for Object) and pick the polyline.

AutoCAD creates a cross section on the same plane as the polyline object.

12 Enter **MOVE Last** and move the cross section to the right of the pulley.

13 Erase the rectangular polyline object.

The screen should look similar to the following illustration.

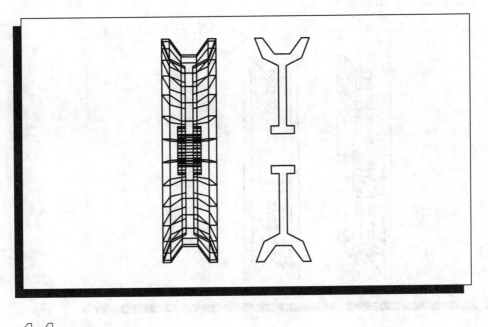

14 Save your work.

Completing the Section

The full section is not complete until you hatch it and add lines.

1 Hatch the cross section using the **ANSI31** hatch pattern.

HINT: Pick the Select Objects button, pick the object that you want to hatch, press ENTER, and then pick the Apply button.

Draw

2 Add lines to the section, as shown in the following illustration, to complete the sectional view.

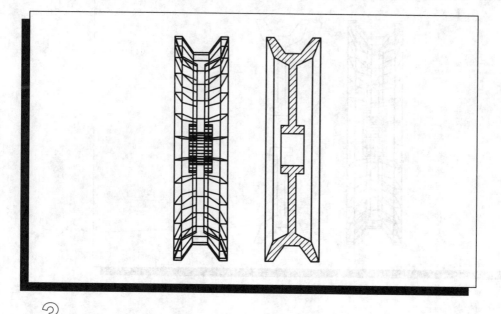

3 Save your work.

4 Make **Objects** the current layer, and freeze layer 0.

5 Produce a left view of the pulley.

SOLPROF Command

The SOLPROF command creates 2D profile objects from a solid model. SOLPROF requires knowledge of the TILEMODE system variable, the MVIEW command, and paper space. The Setup Profile icon on the Solids toolbar enters the SOLPROF command.

1 Pick the **Setup Profile** icon from the Solids toolbar.

Solids

2 Scroll up the Command line or press the **F2** function key.

AutoCAD displays the following error message.

> You must be in paperspace (TILEMODE = 0) to use SOLPROF!

③ Enter **TILEMODE** and change its value to **0** (off) or double-click **TILE** on the status bar.

The paper space icon appears in the lower left corner of the drawing area, but the pulley disappears because viewports have not yet been defined in paper space.

④ Enter the **MVIEW** command and **Fit** option.

AutoCAD creates a single viewport sized to fit the graphics screen, and the pulley reappears.

⑤ Switch from paper space to model space.

HINT: Double-click PAPER in the status bar.

The broken pencil icon reappears. Now it's possible to use the SOLPROF command.

⑥ Enter **ZOOM .5x** and move the pulley to the right half of the screen.

⑦ Pick the **Setup Profile** icon from the Solids toolbar, select the pulley, and press **ENTER**.

Solids

⑧ Enter **Yes** in reply to Display hidden profile lines on separate layer?

This means that AutoCAD will place hidden lines on a separate layer, allowing you to control the display of these lines.

⑨ Enter **Yes** in reply to Project profile lines onto a plane?

This projects the visible and hidden profile lines onto a single plane.

⑩ Enter **Yes** in reply to Delete tangential edges?

This deletes transition lines, called *tangential edges,* that occur when a curved face meets a flat face.

AutoCAD creates profile objects on top of the pulley.

⑪ With ortho on, move the pulley away and to the left of the profile objects.

12 Using the **LIST** command, select the profile objects.

AutoCAD created a block for the visible profile lines and a block for the hidden profile line. Also, AutoCAD created new layers to hold the profile objects.

13 Assign the **Hidden** linetype to the new layer beginning with Ph.

AutoCAD stores hidden profile lines on this layer and visible lines on the layer that begins with Pv.

The new profile should look similar to the one in the following illustration.

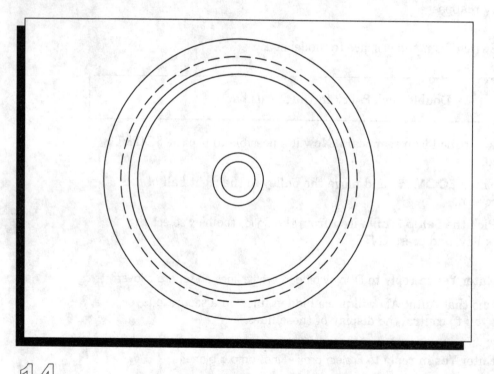

14 Close all floating toolbars, save your work, and exit AutoCAD.

Questions

1. Name two downstream benefits of solid modeling.

_____

_____

2. Explain the purpose of the SECTION command.

_____

_____

3. Explain how to define the cutting plane when using the SECTION command.

_____

_____

4. What kind of objects does the SOLPROF command produce?

_____

_____

5. What is the purpose of the layers that SOLPROF creates?

_____

_____

 Challenge Your Thinking: Questions for Discussion

1. Discuss how you would orient and print together the full section and profile views that you created in this unit.

Problems

1. Select any one of the solid models you created in any of the previous units, or create a new one on your own, and create a full section of the model.

2. The next page shows a bumper assembly for a bumper pool table. The dimensions for the bumper support, bumper ring, and cap are given in Unit 51. Model these three objects as solids.

Note that the bumper support has a square hole with a chamfer of .02 × .02. Create a box and use the SUBTRACT command to create the hole.

After you have created the three solids, put them together as they would fit if assembled. Create a full section view using the SECTION command. Use the INTERFERE command to check for interferences among the three pieces.

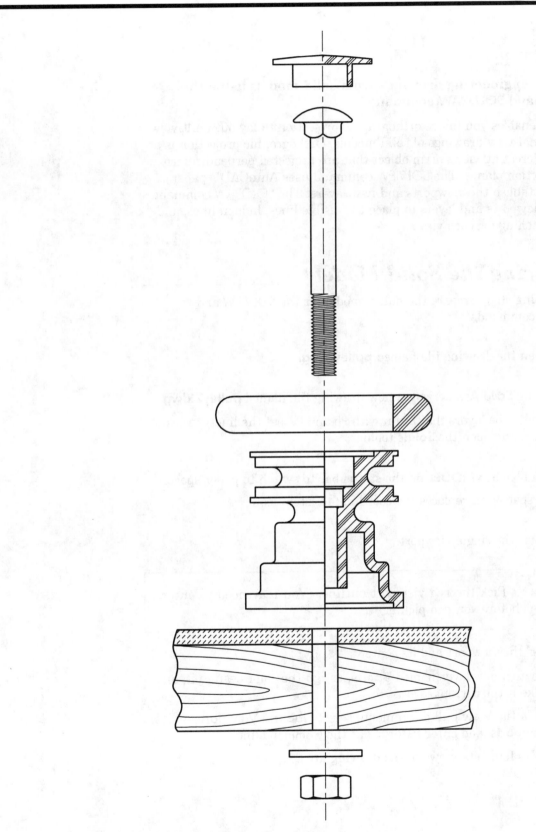

Problem 3 courtesy of Gary J. Hordemann, Gonzaga University

Unit 56 Documenting Solid Models

■ OBJECTIVE:

**To create engineering drawings from solid models using the
SOLVIEW and SOLDRAW commands**

AutoCAD enables you to use orthographic projection to lay out multiview
and sectional-view drawings of solid models. Orthographic projection is a
method of creating views of an object that are projected perpendicularly
onto projection planes. The SOLVIEW command uses AutoCAD's paper
space to establish the viewports and new layers. The SOLDRAW command
uses the viewports and layers to place the visible lines, hidden lines, and
section hatching for each view.

■ *Preparing the Solid Model*

The following steps prepare the pulley model for the SOLVIEW and
SOLDRAW commands.

1 Open the drawing file named **pulley.dwg**.

2 Using **Save As...**, create a new drawing file named **pulley2.dwg**.

3 Freeze the layers that begin with Ph and Pv, set the linetype scale
to **.5**, and open the **Solids** toolbar.

4 Double-click **MODEL** on the status bar to switch to paper space.

The paper space icon replaces the coordinate system icon.

5 Erase the single viewport.

> **HINT:**
> Pick the red viewport outline. (You may need to enter
> ZOOM All before you can pick it.)

This deletes the viewport and its contents.

6 Create a new layer named **Vports**, assign the color cyan to it, and
make it the current layer.

7 From the **View** pull-down menu, select **Floating Viewports** and **4
Viewports**, and enter **F** (for Fit) at the Command line.

AutoCAD fits four new views in the drawing area.

Your screen should look similar to the following illustration.

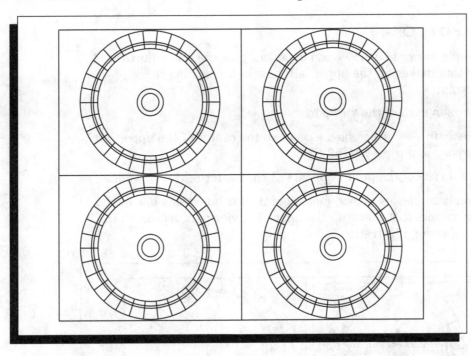

 Save your work.

SOLVIEW Command

SOLVIEW calculates orthographic, auxiliary, and sectional views. View-specific information is saved with each viewport you create and is used by the SOLDRAW command. You will use SOLDRAW to produce the final drawing view.

1 Pick the **Setup View** icon from the Solids toolbar.

Solids

This enters the SOLVIEW command, which displays four options at the Command line. The Ucs option creates a profile view relative to a user coordinate system. The Ortho option creates an orthographic view from an existing view. Auxiliary creates an auxiliary view from an existing view. An auxiliary view is one that is projected onto a plane perpendicular to one of the orthographic views and inclined in the adjacent view. The Section option creates a sectional view, complete with crosshatching.

Let's create an orthographic view in the upper left viewport.

2 Enter **O** for Ortho.

3 In reply to Pick side of viewport to project, pick the lower horizontal line that makes up the upper left viewport, snapping to its midpoint.

A grid appears in each of the viewports.

4 In reply to View center, pick a point in the center of the upper left viewport, and press **ENTER** twice.

An orthogonal view of the pulley appears at the center of the viewport.

5 In reply to Clip first corner, pick a point near but inside one of the four corners that make up the upper left viewport, as shown in the following illustration.

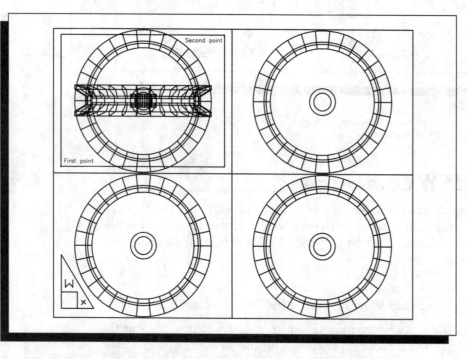

6 In reply to Clip other corner, pick the opposite corner to form a rectangle that is slightly smaller than the viewport, as shown in the previous illustration.

7 In reply to View name, enter **top** and press **ENTER** a second time to exit the command.

⑧ Display the list of layers as shown in the following illustration.

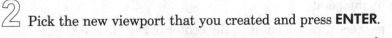

Note the three new layers that SOLVIEW created for visible lines, hidden lines, and dimensions.

⑨ Erase the old upper left viewport. *Do not* erase the new viewport containing the new orthographic view.

⑩ Save your work.

SOLDRAW Command

SOLVIEW saves view-specific information with each viewport it creates. SOLDRAW uses this information to generate the final drawing view.

Solids

① Pick the **Setup Drawing** icon from the Solids toolbar.

This enters the SOLDRAW command.

② Pick the new viewport that you created and press **ENTER**.

AutoCAD creates an orthographic view from the viewport, complete with hidden lines, as shown in the following illustration.

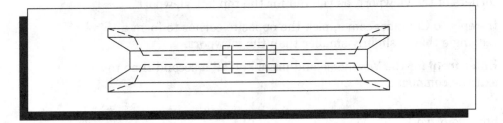

3 Switch to model space.

4 Enter **ZOOM** and **.7x**.

5 Save your work.

Creating a Second Orthographic View

Let's create a second orthographic view in the bottom left area of the screen.

1 Click inside the lower left viewport to make it the current viewport.

2 View the pulley in this viewport from the front.

HINT: From the Standard toolbar, locate the flyout that contains the Front View icon and pick it.

Standard

3 Enter **ZOOM** and **.7x**.

4 Pick the **Setup View** icon from the Solids toolbar to enter the **SOLVIEW** command, and enter the **Ortho** option.

Solids

5 Pick the left vertical line that makes up the lower left viewport, snapping to its midpoint.

6 In reply to View center, pick a point in the center of the viewport and press **ENTER**.

7 In reply to Clip first corner, pick a point inside one of the four corners of the viewport, as you did for the top left viewport.

8 In reply to Clip other corner, pick the opposite corner to form a rectangle that is slightly smaller than the viewport.

9 Enter **front** for the View name and press **ENTER** a second time to exit the command.

10 Save your work.

11 Switch to paper space and erase the old lower left viewport. *Do not* erase the new viewport containing the orthographic view.

Solids

12 Pick the **Setup Drawing** icon from the Solids toolbar to enter the SOLDRAW command.

13 Pick the newest viewport and press **ENTER**.

AutoCAD creates an orthographic view from the viewport drawing, complete with hidden lines, as shown in the following illustration.

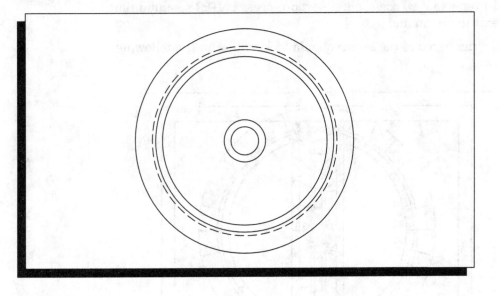

14 Save your work.

Creating a Full Section View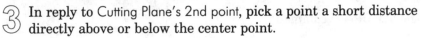

Let's use the SOLVIEW command's Section option to create a full section.

Solids

1 Enter the **SOLVIEW** command and enter **S** for Section.

2 In reply to Cutting Plane's 1st point, snap to the center of the pulley in the lower right viewport.

3 In reply to Cutting Plane's 2nd point, pick a point a short distance directly above or below the center point.

This defines the cutting plane.

4 In reply to Side to view from, pick a point on either side of the center point. (The resulting section is the same either way.)

5 In reply to Enter view scale, press **ENTER** to accept the default value.

6 In reply to View center, pick a point near the center of the screen and press **ENTER**.

7 For the first and second corners, define a viewport outline as you did before.

8 In reply to View name, enter **section**; press **ENTER** a second time to exit the command.

The lower right area of the screen should look similar to the following illustration.

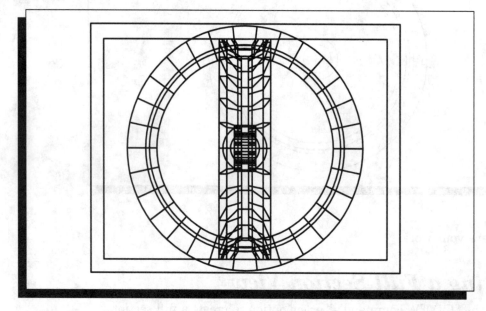

9 Review the list of layers and save your work.

10 Erase the old lower right viewport. *Do not* erase the new viewport containing the new view.

11 Enter the **SOLDRAW** command.

Solids

12 Pick the newest viewport and press **ENTER**.

AutoCAD creates a full section from the viewport drawing, complete with section lines, as shown in the following illustration.

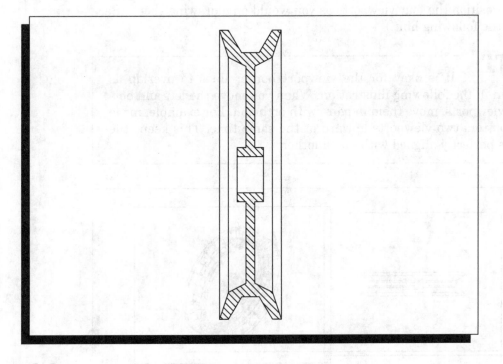

13 Switch to model space, enter **ZOOM** and **.7x**, and save your work.

Completing the Drawing Views _____

The drawing is a few steps from being complete.

1 Click the upper right viewport and enter **ZOOM** and **.7x**.

2 Produce an isometric view of the pulley in this viewport.

HINT: From the Standard toolbar, locate the flyout that contains the SW Isometric View icon and pick it.

Standard

The pulley in the upper right viewport should look similar to the one in the following illustration.

 Position the four viewports as you would on a drawing sheet. (See the following hint.)

HINT:

It is okay for the viewport border lines to overlap as shown in the following illustration. When you move the left and bottom viewports, move them in pairs with ortho on. For example, move the bottom two viewports upward at the same time. This keeps the views perfectly aligned with one another.

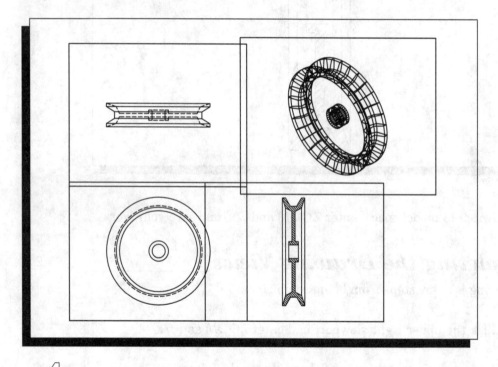

4 Enter the **MVIEW** command and the **Hideplot** option.

5 Enter **On**, select the viewport containing the isometric view, and press **ENTER**.

The MVIEW Hideplot option removes hidden lines from the viewport when you plot the drawing.

6 Make **Objects** the current layer and freeze layer Vports.

Standard

7 Switch to paper space.

8 Pick the **Print** icon from the Standard toolbar.

9 In the Additional Parameters area, pick the **Display** radio button.

10 In the Scale, Rotation, and Origin area, check the **Scaled to Fit** check box.

11 Do a full preview, make corrections if necessary, and print the drawing.

The printed sheet should look similar to the following illustration.

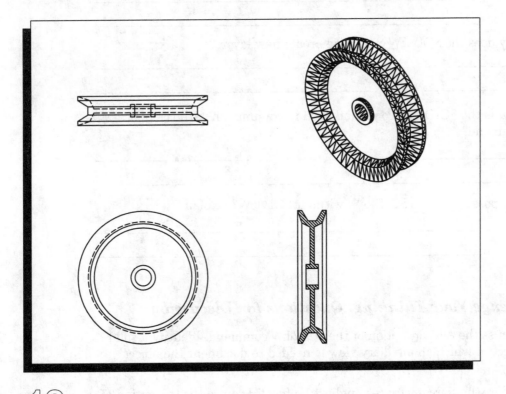

12 Save your work, close floating toolbars, and exit AutoCAD.

Questions

1. What is the purpose of the SOLVIEW command?

   _____

   _____

2. Explain the purpose of each of the following SOLVIEW options.

   UCS _____

   Ortho _____

   Auxiliary _____

   Section _____

3. Why does the SOLVIEW command create new layers?

   _____

   _____

4. How is the SOLDRAW command used in conjunction with the SOLVIEW command?

   _____

   _____

5. Is it possible to use SOLDRAW without SOLVIEW? Explain.

   _____

   _____

Challenge Your Thinking: Questions for Discussion

1. Discuss the Auxiliary option of the SOLVIEW command. Identify a solid object in which an auxiliary view is needed to document the object fully.

2. Under what circumstances might it be helpful to use the Ucs option of SOLVIEW?

Problems

1. Create a new drawing named pulley3.dwg from pulley2.dwg. Add dimensions to the views and add a border and title block.

2. Create a new drawing file named compos2.dwg from the file named compos.dwg. Using SOLVIEW and SOLDRAW, create four new views similar to the ones you produced in this unit.

3. Create a new drawing file named compos3.dwg from the file named compos2.dwg. Add dimensions, a border, and a title block.

AUTOCAD® AT WORK

Digitizing 50 Years of Data

Highland Park, Texas, is located slightly north of bustling downtown Dallas and is known for its lovely homes and affluent population. The town has a geographic area of 2.5 square miles and a population of about 9000.

The town's public works infrastructure was built primarily during the 1920s and 1930s. The original mapping system consisted of five sets of maps dating back to 1931. Each set was made up of 97 section maps for water, sanitary sewer, storm sewer, street lights, and paving detail. Town engineer James B. Dower was faced with the task of bringing these maps, which had not been changed since 1951, up to date.

He determined that computer-based mapping was the best solution, and AutoCAD was chosen as the software. A local company, Laser Data-Images, digitized the old maps for use in AutoCAD. Laser Data-Images provided a composite map made up of each of the five sets of 97 detail maps. "The resulting map was very clean and accurate and is now a complete working copy," says Dower. "We have since made one thorough update of the electronic maps to include changes and fixtures not clearly shown on the originals and have now placed these maps in service."

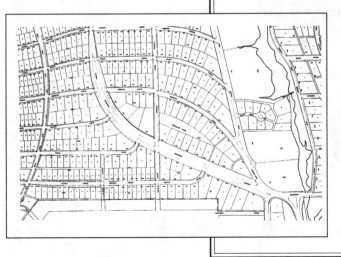

Dower is very pleased with the new system. The maps can be updated easily, and new maps can be created from the base map files. "Given the success we've had with our new electronic mapping system, I'd strongly recommend that other municipalities consider implementing such a system.

Unit 57 Physical Benefits of Solid Modeling

■ OBJECTIVE:

To apply the SLICE, MASSPROP, and STLOUT commands

This unit focuses on splitting the pulley you created in Unit 55 in half and using it to create a physical prototype. It also discusses calculating physical properties of the split pulley.

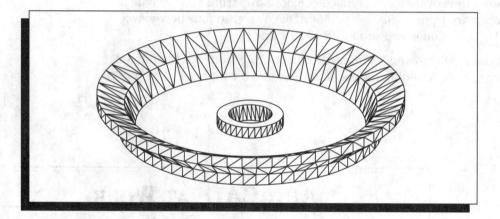

■ *SLICE Command*

The SLICE command cuts through a solid and retains either or both parts of the solid. The command is especially useful if you want to create a mold for half of a symmetrical part, such as the pulley.

 Start AutoCAD and use the **pulley.dwg** drawing file to create a new file named **half.dwg**.

Let's split the pulley and retain one half of it. We will use it (hypothetically speaking) to create a mold.

2 Switch to paper space and erase the single viewport.

The screen should now be blank.

3 Double-click **TILE** on the status bar and **ZOOM All**.

4 Freeze all layers except for layer Objects.

5 With snap on, move the pulley to the center of the screen.

Standard

6 View the pulley from the front and save your work.

The view of the pulley should be similar to the one in the illustration following Step 10, without the center line.

7 Display the **Solids** toolbar.

8 From this toolbar, pick the **Slice** icon.

This enters the SLICE command.

9 Pick the solid, press **ENTER**, and enter the **YZ** option.

AutoCAD asks you to pick a point on the YZ plane. Try to visualize the orientation of this plane, which is vertical and perpendicular to the screen. "Vertical" is the Y direction and "perpendicular to the screen" is the Z direction.

10 Pick a point anywhere on the center line as shown in the following illustration.

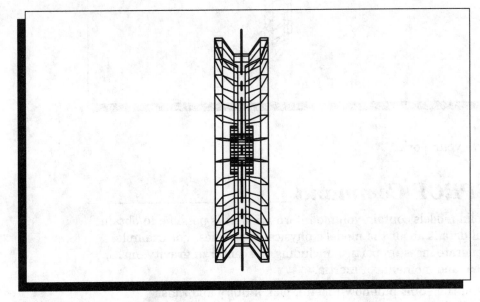

This defines the cutting plane.

11 Pick a point anywhere to the right of the cutting plane.

This tells AutoCAD that you want to keep the right side of the pulley.

AutoCAD cuts the pulley in half and leaves the right half on the screen, as shown in the following illustration. (Ignore the leaders and text.)

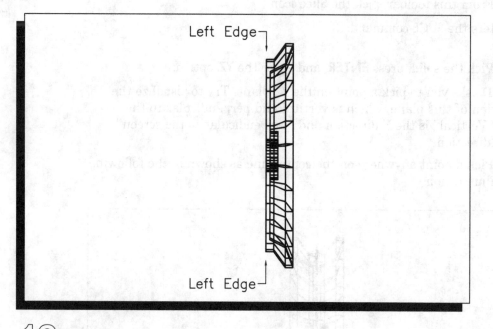

12 Save your work.

MASSPROP Command

Because solid models contain volume information, it is possible to obtain meaningful details about the model's physical properties. For example, you can generate mass properties, including the center of gravity, mass, surface area, and moments of inertia.

1 From the **Tools** pull-down menu, pick **Inquiry** and **Mass Properties**.

This enters the MASSPROP command.

2 Select the solid model and press **ENTER**.

AutoCAD generates a report similar to the one shown in the following illustration, but your numbers will differ. Properties such as mass and volume permit engineers to analyze the part for structural integrity.

3 Press **ENTER** to see the entire report.

```
-------------------------    SOLIDS    -------------------------

Mass:                    18.1501
Volume:                  18.1501
Bounding box:      X:     4.5000  --   6.2500
                   Y:     0.7500  --   7.7500
                   Z:    -3.5000  --   3.5000

Centroid:          X:     5.3750
                   Y:     4.2500
                   Z:     0.0000

Moments of inertia: X:   467.1366
                    Y:   596.7763
                    Z:   924.6118

Products of inertia: XY: 414.6154
                     YZ:   0.0000
                     ZX:   0.0000

Radii of gyration:  X:     5.0732
                    Y:     5.7341
                    Z:     7.1374

Principal moments and X-Y-Z directions about centroid:
                     I:  139.3011 along [1.0000  0.0000  0.0000]
                     J:   72.4097 along [0.0000  1.0000  0.0000]
                     K:   72.4097 along [0.0000  0.0000  1.0000]

Write to a file? <N>:
```

HINT:

You may want to print the mass properties so you can refer to them later in this exercise. Highlight the text, select Copy from the Edit pull-down menu in the AutoCAD Text Window, and paste it into Notepad for printing.

4 Enter **Yes** in reply to Write to a file? and press **ENTER** or pick the **Save** button to use the default half.mpr file name.

AutoCAD creates an ASCII (text) file, assigning the MPR file extension to it automatically. The MPR file extension stands for "Mass Properties Report." Let's locate and review the file.

5 Close the AutoCAD Text Window if you haven't already, and minimize AutoCAD.

6 Start Notepad.

7 Select **Open...** from the **File** pull-down menu.

8 Locate and open the file named **half.mpr**.

HINT: Change the file name extension from txt to mpr to list only those files that have the MPR extension.

The contents of the file should be identical to the information displayed by the MASSPROP command.

9 If you printed the information displayed by the MASSPROP command, compare it to the information in Notepad.

10 Exit Notepad and maximize AutoCAD.

STLOUT Command

STLOUT creates an STL file from a solid model. STL is the file type required by most rapid prototyping (RP) systems, such as stereolithography and Fused Deposition Modeling (FDM). RP systems enable you to create a physical prototype part from a solid model. The part can serve many purposes, such as a pattern for prototype tooling.

Before creating the STL file, you must orient the part in the position that is best for the RP system. In most cases, the part should be lying flat on the WCS. You can accomplish this in the current drawing by rotating the part around the Y axis.

1 Enter the **ROTATE3D** command, select the solid, and press **ENTER**.

2 Enter the **Yaxis** option because you want to rotate the part around the Y axis.

3 Pick a point anywhere on the left edge of the part (shown in the illustration on page 650) to define the Y axis.

4 Enter **–90** for the rotation angle.

NOTE:

It is important that you rotate the part in the negative direction. A positive rotation positions the part upside-down.

⑤ Enter the **PLAN** command to generate the plan view of the WCS, and **ZOOM All**.

⑥ If necessary, move the part so that it lies entirely within the drawing area, as reflected by the grid.

If the part lies outside the positive XYZ octant, the STLOUT command will not work. The *x, y,* and *z* coordinates of the solid *must* be greater than 0.

⑦ Pick a viewpoint that is slightly above the part.

A view similar to the one in the following illustration appears.

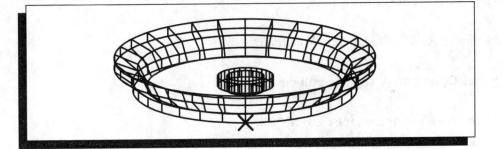

The part should appear to lie on the WCS. Let's check to be sure.

⑧ Using the **ID** command, pick a point at the bottom of the part (indicated by the × in the previous illustration).

HINT:
Use the Nearest object snap mode to snap to a point at the bottom of the part.

The value of the *z* coordinate should be .0000. If it is .0000 or a value less than 0, you must move the solid upward into the positive XYZ octant.

Standard

⑨ Select the **Front** preset view and save your work.

⑩ With **MOVE** command, move the solid upward a short distance, such as **.1**.

11 Enter the **PLAN** command to view the plan view of the WCS and **ZOOM All**.

12 Enter the **STLOUT** command, pick the solid, and press **ENTER**.

AutoCAD wants to know whether you want to create a binary STL file. Normally, you would create a binary file because they are much smaller than ASCII STL files. However, you cannot view the contents of a binary STL file, and we want to view its contents.

13 Enter **No** to create an ASCII STL file.

14 Pick the **Save** button or press **ENTER** to accept half.stl for the name of the file.

Reviewing the STL File

Let's review the contents of the STL file.

1 Minimize AutoCAD and start WordPad.

2 Select **Open...** from the **File** pull-down menu.

3 In the File name box, change *.txt to ***.stl**.

4 Locate the file named **half.stl**, select it, and pick the **Open** button.

The contents of the file appear. The text listing on the next page shows the first part of the file.

```
solid AutoCAD
    facet normal 0.0000000e+000  0.0000000e+000  -1.0000000e+000
        outer loop
            vertex 8.4399540e+000  4.8596573e+000  1.0000000e-001
            vertex 8.5000000e+000  4.2500000e+000  1.0000000e-001
            vertex 5.8750000e+000  4.2500000e+000  1.0000000e-001
        endloop
    endfacet
    facet normal 0.0000000e+000  0.0000000e+000  -1.0000000e+000
        outer loop
            vertex 4.8750000e+000  4.2500000e+000  1.0000000e-001
            vertex 2.2500000e+000  4.2500000e+000  1.0000000e-001
            vertex 2.3100460e+000  4.8596573e+000  1.0000000e-001
        endloop
    endfacet
    facet normal 0.0000000e+000  0.0000000e+000  -1.0000000e+000
        outer loop
            vertex 4.9130602e+000  4.4413417e+000  1.0000000e-001
            vertex 4.8750000e+000  4.2500000e+000  1.0000000e-001
            vertex 2.3100460e+000  4.8596573e+000  1.0000000e-001
        endloop
    endfacet
    facet normal 0.0000000e+000  0.0000000e+000  -1.0000000e+000
```

STL files consist mainly of groups of x, y, and z coordinates. Each group of three defines a triangle.

5 After you've reviewed the file, pick **Exit** from the **File** pull-down menu, and maximize AutoCAD.

6 Close the Solids toolbar, as well as any other toolbars you may have opened.

7 Save your work and exit AutoCAD.

The following view was created using a special utility called stlview.lsp.
This utility reads and displays the contents of an ASCII STL file.

NOTE:

See the optional *Applying AutoCAD Diskette* for a copy of stlview.lsp
and the instructions needed to run it.

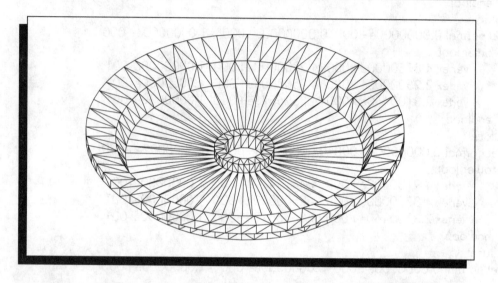

As you can see, an STL file is indeed made up of triangular facets. You can
adjust the size of the triangles using the FACETRES (short for facet
resolution) system variable by changing its value before you create the
STL file. A high value creates a more accurate part with a smoother
surface finish, but the file size and the time required to process the file
increase.

Creating the Prototype Part

More than 200 RP service bureaus around the world can create plastic
prototype parts from STL files. If you were to send the half.stl file to one
of them, they could create a part for you for a fee.

The following are pictures of a completed plastic prototype part. It was
created from the half.stl file using the SLA-250 stereolithography system
from 3D Systems, Inc. (Valencia, CA). The SLA-250 is one of the most
widely used RP systems in the world.

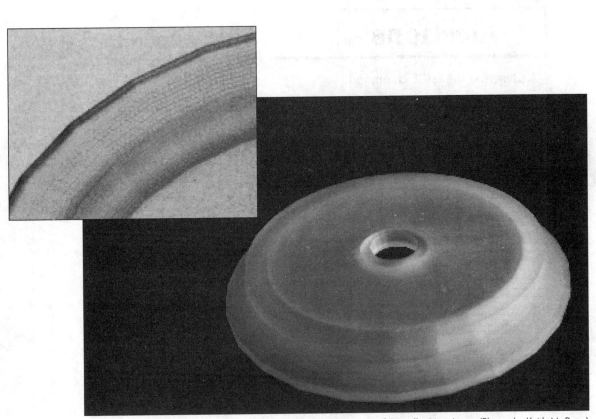

Prototype part courtesy of Laser Prototypes, Inc. of Denville, New Jersey (Photos by Keith M. Berry)

Notice the flat segments that make up the large diameter of the pulley. They are caused by the triangular facets. You could reduce their size by increasing the value of FACETRES before creating the STL file. You could also reduce their visibility by sanding the part, but this could change the accuracy of the part and mold. The part pictured above was built from a version of half.stl with FACETRES set at a relatively low value to show the undesirable effect of large facets.

Questions

1. Describe the SLICE command.

_____

_____

2. When would the SLICE command be useful?

_____

_____

3. AutoCAD can create files with an MPR file extension. How are these files created, and what information do they provide?

_____

_____

4. Name at least three mass properties for which AutoCAD makes information available about a solid model.

_____

_____

5. Explain the purpose of the STLOUT command.

_____

_____

6. What is the purpose of creating an STL file?

_____

_____

7. What AutoCAD system variable affects the size of the triangles in an STL file?

_____

_____

8. What is the primary advantage of creating a binary STL file? An ASCII STL file?

_____

_____

■ *Challenge Your Thinking: Questions for Discussion*

1. Write a paragraph comparing and contrasting the function and purpose of the SECTION and SLICE commands.

2. Describe a situation in which you might need to slice an object using the 3points option.

3. Explore the different rapid prototyping systems that are commercially available. Write an essay that compares each of them.

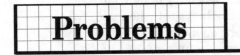

Problems

1. Create ASCII and binary STL files from the solid model stored in compos.dwg. Compare the sizes of the two files and then view the contents of the ASCII STL file.

2. If you know of a rapid prototyping service bureau in your area, request a demonstration of their RP system(s). Explore the possibility of having them create a prototype part for you from an STL file.

3. Open the file named shaft.dwg. Use the SLICE command to split the model down the middle to create a pattern. Use STLOUT to create a binary STL file.

Unit 58 External References

■ OBJECTIVE:

To apply AutoCAD's external reference feature using the XREF and XBIND commands

External references (also called *xrefs*) provide you the option of attaching or overlaying drawings in the current drawing, similar to inserting a drawing as a block. However, xrefs do not become part of the drawing. Instead, the xrefs are loaded automatically each time the drawing file is loaded. In addition to viewing the xrefs, you can make use of the xref objects by, for example, snapping to them.

External references are helpful if you want to view the assembly of individual components as a master drawing. Xrefs are particularly useful when you are working with other AutoCAD users in a network environment.

■ *Applying External References*

Drawing files must be available to apply external references. Therefore, let's create files to use in the following sections.

1 Start AutoCAD and pick **Use a Wizard**, **Quick Setup**, and **OK**.

2 Pick the **Architectural** radio button and the **Next** button.

3 Enter **80′** for the width and **60′** for the length, and pick the **Done** button.

The drawing area is based on a scale of $1/8'' = 1'$ on a standard A-size sheet. The active drawing area on the sheet is $10'' \times 7.5''$.

4 **ZOOM All.**

5 Set the grid at **10′** and snap at **2′**.

6 Save your work in a template file named **xref.dwt**. Enter **Base drawing** for the template description.

7 Using **Save As...**, create a new drawing named **property.dwg**.

8 Create and make current a new layer named **Property**. Assign the color red to it.

Most property lines are not a perfect rectangle. However, to keep things simple, follow the next step.

9 Draw a rectangular property line using the **PLINE** command. Make it almost as large as possible (76′ × 56′), but stay inside the drawing limits.

10 Save your work.

11 Using the **xref.dwt** template file, begin a new drawing.

12 Create and make current a new layer named **Building**. Assign the color magenta to it.

13 Approximate the following building outline using a polyline. Allow space for trees and shrubs around the building, but do not draw any at this time.

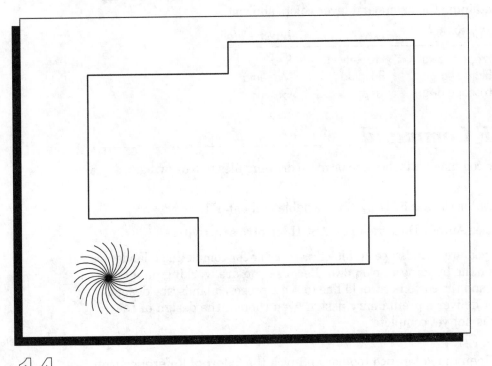

14 Save your work in a file named **bldg.dwg**.

15 Create another drawing using the **xref.dwt** template file.

16 Create and make current a new layer named **Trees**, and assign the color green to it.

17 Create a simple tree symbol, as shown in the previous illustration. Approximate the size and shape of the tree.

HINT:

Use the ARC and ARRAY commands to create the tree.

Draw

18 Create a block of the tree. Name it **Tree**, and use the tree's center as the insertion base point. Do not retain the tree on the screen.

The screen should currently be blank.

19 Save your work in a file named **ldscape.dwg**.

To review our last 19 steps, you should now have the following drawing files, containing the layers and layer colors shown.

File Name	Layers	Layer Color
Property.dwg	Property	Red
Bldg.dwg	Building	Magenta
Ldscape.dwg	Trees	Green

XREF Command

The XREF command attaches one or more drawing files to a drawing.

1 Enter the **XREFCTL** system variable and enter **1**.

This causes AutoCAD to create an ASCII log of xref activity.

Suppose you are a landscape architect responsible for completing a landscape design for the building in bldg.dwg. Because you are working on a tight schedule and the customer would like to see a proposed landscape design, you must deliver a preliminary design even though the design of the building is not yet complete.

2 Open the **Reference** toolbar and pick the **External Reference** icon.

Reference

The following dialog box appears.

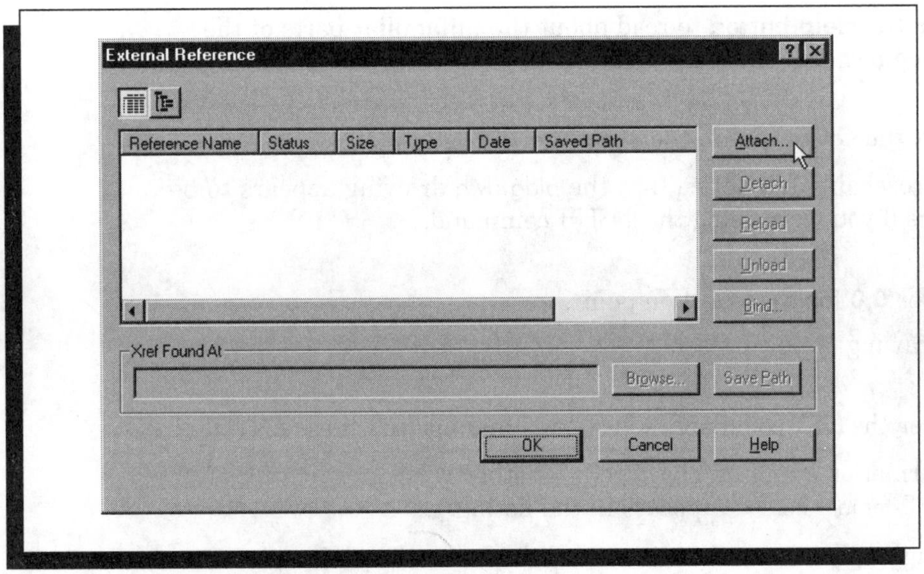

③ Pick the **Attach...** button.

This displays the Select file to attach dialog box.

④ Find and double-click the **bldg.dwg** drawing file.

This displays the Attach Xref dialog box, as shown in the following illustration.

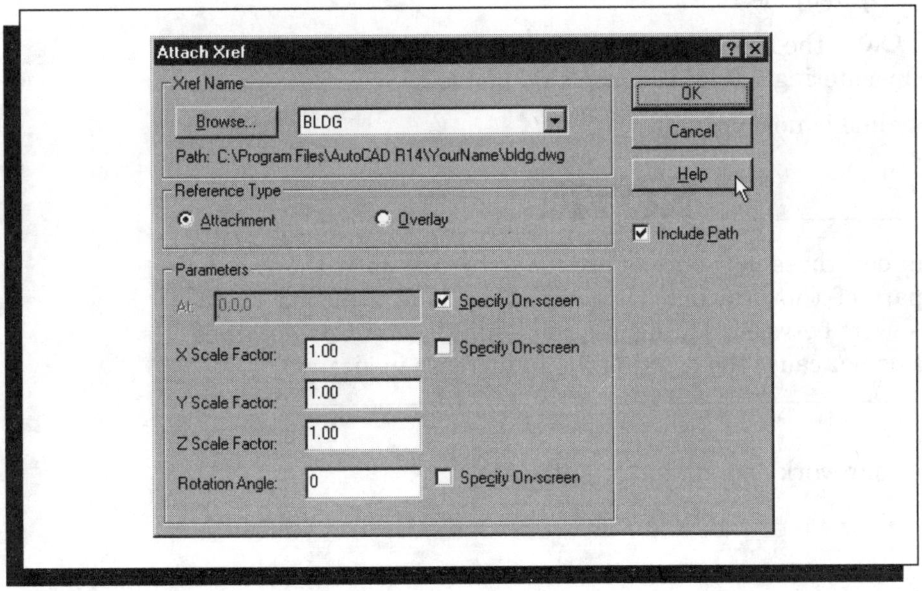

Study the information in the dialog box. Much of it is self-explanatory.

5 Pick the **Help** button to read about the unfamiliar parts of the dialog box, and then exit help.

6 Pick the **OK** button in the Attach Xref dialog box.

Move the crosshairs and notice that the bldg.dwg drawing appears to be inserting, as if you were using the INSERT command.

7 Enter **0,0** for the insertion point.

The bldg drawing is now present.

8 Enter the **LIST** command, select the building, and press **ENTER**.

Listing information about an object tells whether it belongs to an xref. Notice that External reference appears in the listing.

We are going to place several trees around the house. Therefore, it is important to see the property line so that none of the trees extend outside the property line.

9 Close the AutoCAD Text Window.

10 Pick the **External Reference Attach** icon.

The Attach Xref dialog box appears.

Reference

11 Pick the **Browse...** button and find and double-click the **property.dwg** file.

12 Pick **OK** in the Attach Xref dialog box and position the red property line by entering **0,0** for the insertion point.

The property line is now present.

NOTE:

As a reminder, these new objects are for reference only; they do not become part of the drawing. However, they stay attached to the drawing—even between editing sessions—until they are detached. The xrefs do not cause the drawing file to increase in size.

13 Save your work and minimize AutoCAD.

14 Using Notepad, review the contents of the file named **ldscape.xlg**.

This is an ASCII file that AutoCAD creates to maintain a log of xref activity. You may choose to delete it if it consumes disk space that you need to use. Also, you can set the XREFCTL system variable back to 0 to avoid creating logs in the future.

15 Exit Notepad and return to AutoCAD.

16 Insert the block named **Tree** into the current drawing. Place the tree close to the east side of the building, and accept the default values.

17 Place several other trees (of various sizes) around the building.

18 Save your work.

Changing an Xref

Suppose the customer asks the building architect to expand the east side of the building.

1 Open the **bldg.dwg** drawing.

2 Stretch the east part of the building **10′** to the east.

3 Save your work.

The bldg drawing has changed. Must the building architect notify the landscape architect of the change? Suppose the landscape architect loads ldscape to continue working on it.

4 Open **ldscape.dwg**.

The latest change in bldg.dwg is reflected in the ldscape.dwg drawing. As you can see in the following illustration, the trees now interfere with the building. Being linked to related drawings that may change is a beneficial feature of an xref.

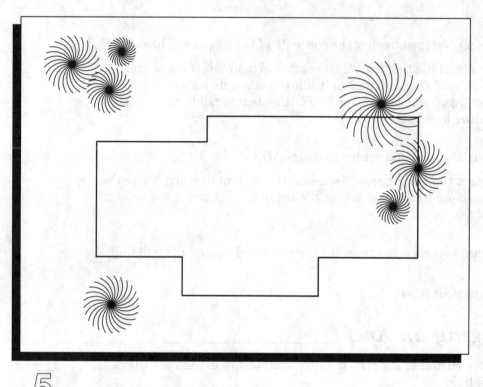

5 Move the tree(s) away from the building.

Xref Layers

Object Properties

1 Display the Layer & Linetype Properties dialog box.

Notice that ldscape.dwg contains two additional layers as a result of attaching the two xrefs. Each of the new layer names is preceded by its parent xref drawing file name and is separated by the | (vertical bar) character. You can control the visibility, color, and linetype of the xref layers.

NOTE:

You may need to expand the Name column in the dialog box to see the full name of each layer.

2 Save your work.

XREF Options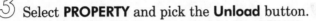

Xref offers several options.

1 Enter **UNDO** and **Mark**.

2 Pick the **External Reference** icon.

Notice the new information in the External Reference dialog box. As you can see, AutoCAD lists the xrefs in the current drawing.

3 Select **PROPERTY** and pick the **Unload** button.

Note that Unload appears under the Status heading. This means that AutoCAD will suppress the display and regeneration of the PROPERTY xref definition. Unloading xrefs that are not currently needed helps current session editing and improves performance. Unloaded xrefs remain attached to the drawing and can be reloaded as necessary.

4 Pick the **OK** button.

The property line disappears.

5 Press the space bar or **ENTER** to redisplay the External Reference dialog box, select **PROPERTY**, and pick the **Reload** button.

The Reload button re-reads and displays the most recently saved version of the drawing.

6 Pick the **OK** button.

The property line reappears.

7 Redisplay the **External Reference** dialog box.

Above the Reload button is the Detach button. Use this button only if you want to detach an xref permanently.

8 Pick the **Help** button and read about the Bind... button.

9 After exiting help, select **PROPERTY** and pick the **Bind...** button.

10 In the Bind Xrefs dialog box, pick the **Bind** radio button, unless it is already picked, and pick **OK**.

PROPERTY disappears from the list of external references in the External Reference dialog box because it is now a permanent part of the drawing.

11 Pick the **OK** button.

NOTE:

In the future, you may not want to bind the entire xref. To bind only a part of an xref, use the XBIND command. See the following section for more information.

12 Enter **UNDO** and **Back**.

Property.dwg is an xref once again.

XBIND Command

The XBIND command permanently binds a subset of an xref's dependent symbols to the current drawing. Dependent symbols are named objects in an xref, such as blocks, layers, and text styles.

1 Pick the **External Reference Bind** icon from the Reference toolbar.

The following Xbind dialog box appears.

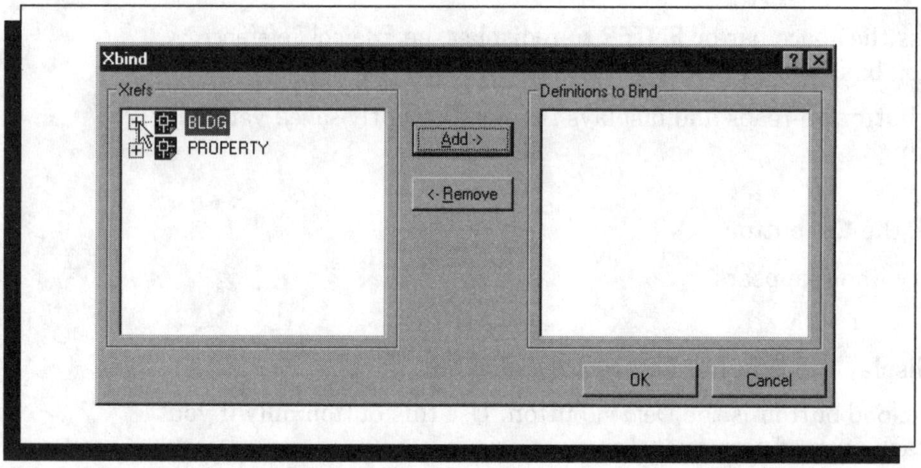

Reference

2 Pick the white box with a plus (+) sign in it located to the left of BLDG.

This displays xref's dependent symbols.

3 Pick the white box with a plus (+) sign in it located to the left of Layer.

This displays the layer BLDG|BUILDING.

4 Pick **BLDG|BUILDING** and pick the **Add** button.

AutoCAD adds BLDG|BUILDING under Definitions to Bind.

5 Pick the **OK** button.

When binding a dependent layer to the current drawing, AutoCAD renames the layer by replacing the | (vertical bar) with $#$.

6 Display the **Layer Control** drop-down box.

In this case, AutoCAD renamed BLDG|BUILDING to Bldg0building.

7 Close the Reference toolbar, as well as any other toolbars you may have opened.

8 Save your work and exit AutoCAD.

Questions

1. What is the primary benefit of using external references (xrefs)?

   _____

   _____

2. What is the function of each of the following buttons in the External
   Reference dialog box?

   Attach _____

   Detach _____

   Reload _____

   Unload _____

   Bind _____

3. Attaching an xref is similar to using what popular AutoCAD
   command?

   _____

4. What command tells whether a given object belongs to an xref?

   _____

5. What information is contained in an AutoCAD XLG file?

   _____

   _____

6. Describe the purpose of the XBIND command.

   _____

   _____

 Challenge Your Thinking: Questions for Discussion

1. Experiment with the Attachment and Overlay options in the Attach Xref
   dialog box. What are the differences between the two? In what ways
   are they similar?

Problem

This problem works best as a small group activity, involving two to four individuals. The group should identify a project. Make it simple; otherwise it may be difficult to organize. The project should involve several components that fit together. Each individual on the project should be responsible for completing one or more different components. The project leader must coordinate the effort and be responsible for completing the final assembly made up of the individual component drawings.

While a local area network would aid greatly in the completion of this project, it can be done using individual AutoCAD stations. However, all individuals working on the project must copy their component drawings to a single location (disk and directory) and make them available to others.

AutoCAD® at Work

CAD Helps Steam Engines Keep Chugging

At one time, all locomotives were powered by steam. Today, however, most locomotives are diesel-electric. What, then, would CAD have to do with old-fashioned steam engines? In Lomita, California, there's a shop that sells kits for miniature working steam trains, and the design drawings for those kits are being produced with AutoCAD.

When Moodie Braun, a former Lieutenant Colonel in the U.S. Air Force Space and Missile System Division, bought the Little Engines shop, he found an archive of about 3,000 drawings. The drawings—for sixteen different models of trains in three scales—dated back to the 1930s. Each kit typically included drawings with details for machining, drilling, filing, soldering, and painting.

As materials ceased to be available or if errors were discovered, the drawings had to be revised. It was a tedious job. Braun and his staff were making erasable photocopies, revising the copies, and having new sepias (brown prints) made.

Recently, the shop began using AutoCAD. Drawing time has been cut in half, and it's much easier to update drawings. Instead of altering photocopies, Braun and his staff simply revise on the computer and plot new drawings. Considering the complexity of some kits (the largest train can carry 75 people), the improvements in the accuracy and quality of the drawings are a benefit to the customers as well as the shop.

Courtesy of Little Engines and Autodesk, Inc.

Unit 59 — An Internal Peek at AutoCAD's Menus

OBJECTIVE:

To examine and understand the contents of AutoCAD's acad.mnu menu file

This unit concentrates on the components that make up the acad.mnu file. It reviews the file in its raw form: the parts you don't see when you're working in AutoCAD. Later, you will learn to modify and create your own menu, one as simple or as sophisticated as you'd like.

The Raw Menu

Let's use Microsoft WordPad to review acad.mnu. First, let's make a backup copy of acad.mnu.

 Produce a copy of acad.mnu. Name the copy **acadback.mnu**.

HINT: This file is located in the \Program Files\AutoCAD R14\Support directory.

 Open WordPad by picking the Windows Start button and selecting Programs, Accessories, and then WordPad.

Pick the **Open** icon in WordPad and find and open **acadback.mnu**.

HINT: In WordPad's Open dialog box, enter *.mnu to see only the MNU files.

The file contains the source code for AutoCAD's menus.

Review the first part of the file.

You should see information similar to that shown on the following pages.

```
//
//   AutoCAD Menu - Release 14.0
//   28 April 1997
//
//   Copyright (C) 1986, 1987, 1988, 1989, 1990, 1991, 1992, 1994, 1996,
//   1997 by Autodesk, Inc.
//
//   Permission to use, copy, modify, and distribute this software
//   for any purpose and without fee is hereby granted, provided that
//   the above copyright notice appears in all copies and that both
//   the copyright notice and the limited warranty and restricted rights
//   notice below appear in all supporting documentation.
//
//   AUTODESK, INC. PROVIDES THIS PROGRAM "AS IS" AND WITH ALL FAULTS.
//   AUTODESK, INC. SPECIFICALLY DISCLAIMS ANY IMPLIED WARRANTY OF
//   MERCHANTABILITY OR FITNESS FOR A PARTICULAR USE.  AUTODESK, INC.
//   DOES NOT WARRANT THAT THE OPERATION OF THE PROGRAM WILL BE
//   UNINTERRUPTED OR ERROR FREE.
//
//   Use, duplication, or disclosure by the U.S. Government is subject to
//   restrictions set forth in FAR 52.227-19 (Commercial Computer
//   Software - Restricted Rights) and DFAR 252.227-7013(c)(1)(ii)
//   (Rights in Technical Data and Computer Software), as applicable.
//
//
//   NOTE:  AutoCAD looks for an ".mnl" (Menu Lisp) file whose name is
//         the same as that of the menu file, and loads it if
//         found.  If you modify this menu and change its name, you
//         should copy acad.mnl to <yourname>.mnl, since the menu
//         relies on AutoLISP routines found there.
//


//
//   Default AutoCAD NAMESPACE declaration:
//
***MENUGROUP=ACAD


//
//   Begin AutoCAD Digitizer Button Menus
//
***BUTTONS1
//   Simple + button
//   if a grip is hot bring up the Grips Cursor Menu (POP 17), else send a carriage
//   return
$M=$(if,$(eq,$(substr,$(getvar,cmdnames),1,5),GRIP_),$P0=ACAD.GRIPS
$P0=*);
$P0=SNAP $p0=*
```

```
^C^C
^B
^O
^G
^D
^E
^T


***BUTTONS2
// Shift + button
$P0=SNAP $p0=*


***BUTTONS3
// Control + button


***BUTTONS4
// Control + shift + button


//
//   Begin System Pointing Device Menus
//
***AUX1
// Simple button
// if a grip is hot bring up the Grips Cursor Menu (POP 17), else send a
carriage return
$M=$(if,$(eq,$(substr,$(getvar,cmdnames),1,5),GRIP_),$P0=ACAD.GRIPS
$P0=*);
$P0=SNAP $p0=*
^C^C
^B
^O
^G
^D
^E
^T


***AUX2
// Shift + button
$P0=SNAP $p0=*
$P0=SNAP $p0=*


***AUX3
// Control + button
$P0=SNAP $p0=*
```

The ASCII file contains as many as 4,566 lines of code, including the blank lines. Much of the code consists of AutoCAD's pull-down menu items. Many examples, such as ***POP2, are presented in this unit.

Individual Menu Elements

AutoCAD uses several menu section labels and special menu characters in acad.mnu. Each of them serves a specific purpose.

1 Attempt to locate each of the following items in the menu file.

HINT: Pick Find... from WordPad's Edit pull-down menu.

Menu Sections

***BUTTONS1	— specifies a buttons menu 1 for the buttons on a mouse or digitizer cursor control
***AUX1	— specifies auxiliary device menu 1
***TABLET1	— specifies tablet menu area 1
***POP2	— pull-down menu sections are defined using ***POP1, ***POP2, etc.
***MENUGROUP=	— specifies a menu file group name
***TOOLBARS	— specifies a toolbar
***HELPSTRINGS	— specifies help comments that appear in place of the status bar when pointing at an icon
***ACCELERATORS	— specifies accelerator keys

Pull-Down Menus

[&Grips...]	— displays Grips... in the menu; the ampersand (&) character denotes the mnemonic key (*i.e.*, underlines the G in the word Grips, indicating that you can enter G at the keyboard to activate this menu item)
->	— as a prefix, indicates that the pull-down or cursor menu has a submenu
[->In&quiry]	— indicates that Inquiry has a submenu; the ampersand (&) character denotes the mnemonic key
<-	— as a prefix, indicates that the pull-down or cursor menu item is the last in the submenu
[<-&Save...]	— indicates that Save... is the last item in a submenu
~	— "grays out" a prompt (displays it in half-tone), disabling the menu item
[--]	— specifies a separator line

Image Tile Menus

***image	— specifies an image tile menu
**image_poly	— specifies an image tile submenu named image poly
$I=	— addresses an image tile menu
[acad(Cone,Cone)]	— addresses the cone slide image contained in the acad slide library for display as an image tile; also displays Cone in the list box

Special Characters

;	— issues ENTER
^M	— issues ENTER
^I	— issues a tab
'	— an apostrophe specifies transparent command entry
\	— the backslash stops the computer, and the computer expects input from the user
(a space)	— an empty space is the same as pressing the space bar
+	— menu item continues on the next line
=*	— displays the current top-level image, pull-down, or cursor menu
$	— special character code that loads a menu section or introduces a conditional DIESEL macro expression ($M=)
^	— this character (called a *caret*) automatically presses the CTRL key
^B	— toggles snap on or off
^C	— issues a cancel
*^C^C	— prefix for a repeating item
^D	— toggles coordinates on or off
^E	— sets the next isometric plane
^G	— toggles grid on or off
^H	— issues a backspace
^O	— issues the CTRL and O keys to toggle the ortho mode; ^O is used in the BUTTONS menu to assign button #6 to this function
^P	— toggles MENUECHO on or off
^Q	— echoes all prompts, status listings, and input to the printer
^T	— toggles tablet on or off
^V	— changes current viewport
^Z	— null character that suppresses the automatic addition of SPACEBAR at the end of a menu item
ID_	— AutoCAD uses this prefix to begin all name tags in acad.mnu
//	— specifies a comment (as opposed to program code)

Other menu elements exist, although the previous list provides enough to get you started with menu development and customization.

Menu Items

You can combine the individual items described in the previous section to create menu items (also referred to as *macros*) that perform specific AutoCAD functions. For example, [Redo]^C^C_redo displays Redo in the menu, issues cancel twice, and enters the REDO command.

1 Study each of the following menu items from the menu file.

***POP5

— specifies the fifth (Format) pull-down menu

ID_ZoomWindo   [&Window]'_zoom_w

— ID_ZoomWindo is the name tag; Windows appears in the menu; the W is underlined; the apostrophe allows for transparent command entry; AutoCAD ignores the underscore and then enters the ZOOM command; then enters W for Window

ID_CircleDia   [Center, &Diameter] ^C^C_circle _d

— ID_CircleDia is the name tag; Center, Diameter appears in the menu; the D is underlined; issues cancel twice; enters the CIRCLE command; presses the space bar, pauses for user input; enters D for Diameter

2 Locate each of the previous menu items in the menu file.

3 Close WordPad (*do not* save any changes), and start AutoCAD. Begin a new drawing from scratch.

4 Locate each of the previous items in AutoCAD's menus and pick each of them.

HINT:

Find the second menu item by picking Zoom from the View pull-down menu. You can find the third menu item by picking Circle from the Draw pull-down menu.

5 Exit AutoCAD.

⑥ Locate the following files in the Support directory.

acad.mns   An ASCII file that looks very similar to acad.mnu; this
           file stores changes to the toolbars or menus made
           from within the AutoCAD graphics screen

acad.mnc   Compiled version of the acad.mns file; AutoCAD
           loads this binary version of the menu file.

acad.mnr   Binary file containing the bitmaps for the icons; this
           file updates as you modify an icon using the icon
           editor in AutoCAD

acad.mnl   AutoCAD menu LISP file, which contains AutoLISP
           expressions that are used by acad.mnu; you will learn
           more about AutoLISP later in the book.

The acad.mnu file is referred to as the *template* menu file. You will
work with the acad menu files in the following units.

Questions

Describe what each of the following menu items will do.

1. [&Utilities] ^C^C_files

_____

_____

2. [Dimstyle]_dimstyle

_____

_____

3. [->&Zoom]

_____

_____

4. **LINE 3

_____

_____

5. [New...]^C^C_new

_____

_____

6. [Window]_w

_____

_____

7. [Text &Style...]'_style

_____

_____

8. [&Start, Center, End]^C^C_arc _c

_____

_____

9. [done];

_____

■ Challenge Your Thinking: Questions for Discussion

Note: To answer all questions and problems that require opening acad.mnu, use the acadback.mnu file you created instead. This will help prevent any unintentional changes to acad.mnu.

1. In AutoCAD, identify a menu item and then execute it. Try to envision what the menu item would look like in the acad.mnu file. What individual menu elements would AutoCAD use?

2. In acad.mnu, locate the menu item that you identified in the previous question. Is it similar to what you had envisioned?

Problems

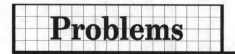

1. Print a portion of the acad.mnu file and then start AutoCAD. Experiment with different command and submenu sequences and attempt to locate the sequences on the printout.

2. Locate items in the printout that you cannot fully visualize. Attempt to find the corresponding items in the menus and execute them.

AUTOCAD® AT WORK

Mining Industry Uses 3D AutoCAD

Modern-day mining operations take advantage of the power of computer technology to define potential deposits of copper, gold, and other minerals. Working with AutoCAD and Autodesk 3D Studio, mining engineers at Noranda, Inc., build 3D computer models of potential ore zones based on

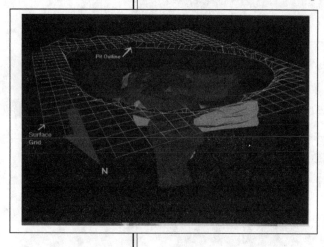

data obtained from test drilling. These models provide accurate representations of ore bodies hundreds of feet below ground and serve as a tool for guiding efficient, productive mining operations. The image shown here is a geological interpretation of one mining site that shows the layers of ore zones hidden underground. When information is needed about a specific section, the mining engineer can quickly generate a 2D slice through the model for use in ore reserve calculations or in the design of underground workings.

Based on a story provided by Autodesk, Inc.

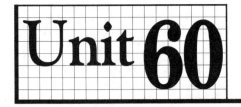

Unit 60 — Creating Custom Menus and Toolbars

■ OBJECTIVE:

To create an AutoCAD pull-down menu and toolbar and to apply the MENU and MENULOAD commands

AutoCAD users can create pull-down menus and toolbars that can include a wide range of AutoCAD commands. Users can develop custom *macros,* which automatically execute any series of inputs. For example, a simple two-item macro can enter ZOOM Previous in one step. Sophisticated macros can activate numerous AutoCAD commands and functions in a single step.

Everything that you can enter at the keyboard can be entered automatically using macros. Thus, you have the flexibility to develop a menu at any level of sophistication.

Developing a Pull-Down Menu

Let's create a simple pull-down menu named pull.mnu.

1 Using Notepad, store the following in a file named **pull.mnu**. Skip the first line so that the line above ***POP1 is blank.

NOTE:

See also the optional *Applying AutoCAD Diskette.* It includes the pull.mnu file, which contains the following code.

```
***POP1
[Construct]
[Line]^C^CLINE
[Circle]^C^CCIRCLE
[Arc]^C^CARC
[--]
[->Display]
 [Pan]'PAN
 [->Zoom]
  [Window]^C^CZOOM W
  [Previous]^C^CZOOM P
  [All]^C^CZOOM A
  [Extents]^C^CZOOM E
```

2 Save the file and exit Notepad.

3 Make a backup copy of the file.

MENU Command

The MENU command permits you to load menus.

1 Start AutoCAD and begin a new drawing from scratch.

2 Enter the **MENU** command.

This displays the Select Menu File dialog box.

3 In Files of type, select **Menu Template (*.mnu)**.

4 Locate and select **pull.mnu**.

AutoCAD displays a warning message indicating that loading a template menu (MNU) file overwrites the menu source (MNS) file if one exists. In this case, pull.mns does not exist.

NOTE:

If the computer you are using is also used by others, check with your supervisor or instructor to make sure you are not overwriting someone else's file.

5 Pick the **Yes** button to continue loading the file.

The new Construct pull-down menu appears, replacing all of the others, including the docked toolbars, as shown on the following page.

6 Try each of the options in the Construct menu.

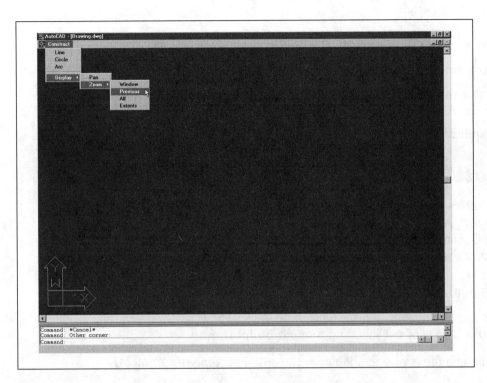

AutoCAD automatically compiles the pull.mnu file into pull.mnc and pull.mnr files. As mentioned in the previous unit, the MNC file is the compiled version, while the MNR file contains the bitmaps, such as the icons, used by the menu. Both are binary files.

When AutoCAD creates the MNC file, it also creates an MNS file. Initially, it is identical to the MNU file, without comments or special formatting. AutoCAD changes it each time you change the contents of the MNU file through the standard AutoCAD interface, such as changing the contents of a toolbar. You will create and edit a toolbar later in this unit. See the on-line *AutoCAD Customization Guide* for more information on the MNU, MNC, MNR, and MNS files.

7 Minimize AutoCAD and review these files.

8 Maximize AutoCAD.

NOTE:

Even though the menus and toolbars are gone, you still have full access to all AutoCAD commands. Just enter them at the keyboard. Notice, however, that the buttons menu is no longer available.

9 Try each of the buttons on the mouse or cursor control.

They do not function the same as before because acad.mnc is not loaded.

10 Enter the **MENU** command and select **acad.mnc**, which is located in the AutoCAD Support directory.

All the standard menus and toolbars should be present once again.

Partial Menus

AutoCAD offers base and partial menus. A base menu is loaded when you first start AutoCAD, or when you use the MENU command. The MENULOAD command loads partial menus. Let's convert pull.mnu to a partial menu that we can use with the standard acad menu.

1 Minimize AutoCAD.

2 Make a copy of pull.mnu and name it **pullpart.mnu**.

The pullpart.mnu file is also available on the optional *Applying AutoCAD Diskette*.

3 Using Notepad, open **pullpart.mnu**.

4 At the top of the file, add *****MENUGROUP=PULLPART**, with a blank line after it.

NOTE:

The menu group name can be different than the file name. We used PULLPART for both to make it easier to remember the names.

5 Select **Save** from the **File** pull-down menu and exit Notepad.

6 Maximize AutoCAD.

7 From the **Tools** pull-down menu, select **Customize Menus...** or enter the **MENULOAD** command.

The Menu Customization dialog box appears. If necessary, pick the Menu Groups tab to display it.

8 Pick the **Browse...** button and find and select the new **pullpart.mnu** file.

⑨ After you've selected this file, pick the **Load** button.

⑩ In reply to the warning message, pick the **Yes** button.

PULLPART appears in the Menu Groups list box.

⑪ Select **PULLPART** in this box.

⑫ Pick the **Menu Bar** tab to change the contents of the dialog box.

⑬ In the Menu Bar list box, select **Tools**.

⑭ Pick the **Insert** button.

This inserts Construct between Format and Tools, as shown in the following illustration.

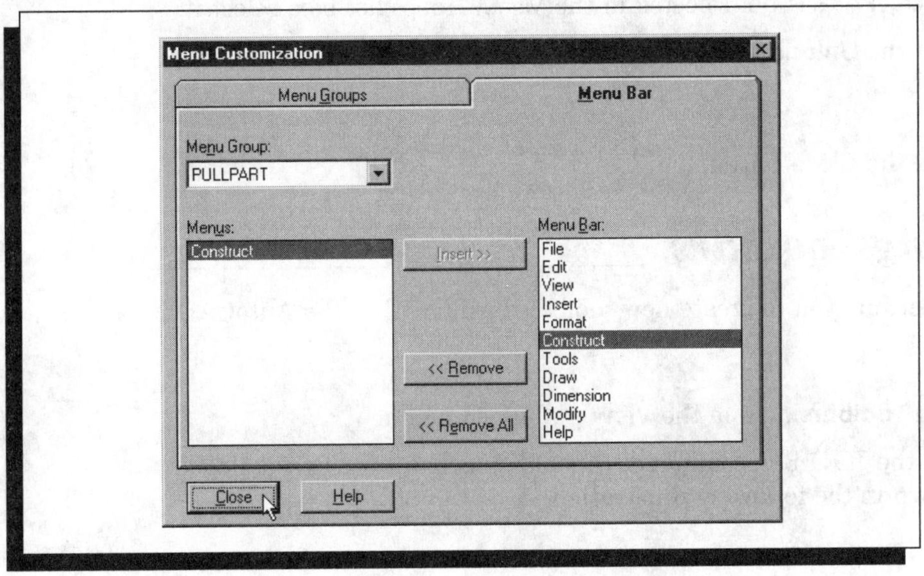

⑮ Pick the **Close** button.

AutoCAD adds the Construct menu to the pull-down menus.

⑯ Try the selections in the Construct menu.

Removing Partial Menus

The Menu Customization dialog box allows you to remove partial menus.

1 Select **Customize Menus...** from the **Tools** pull-down menu.

2 In the Menu Groups list box, pick **PULLPART**.

3 Pick the **Menu Bar** tab to change the contents of the dialog box.

4 In the Menu Bar list box, select **Construct**, and pick the **Remove** button.

This removes the Construct pull-down menu from AutoCAD's menu bar.

5 Pick the **Menu Groups** tab.

6 If **PULLPART** is not selected in the Menu Groups list box, select it.

7 Pick the **Unload** button to remove PULLPART from the Menu Groups list box.

8 Pick the **Close** button.

Creating Toolbars

AutoCAD permits you to create new toolbars easily within the AutoCAD environment.

1 Pick **Toolbars...** from the **View** pull-down menu.

This enters the TOOLBAR command, which displays the Toolbars dialog box, as shown in the following illustration.

2 Pick the **New...** button.

3 Enter **Custom** for the toolbar name, and pick the **OK** button.

This adds Custom to the Toolbars list box. Notice also that the beginning of a toolbar appears near the top of the graphics screen.

4 Pick the **Customize...** button.

5 In the Categories drop-down box, pick **Standard** from the list.

All of the icons from the Standard toolbar and its flyouts appear.

6 Review the icons using the scrollbar.

7 From the dialog box, drag and drop the **New** icon (the first icon in the group) into the new Custom toolbar near the top of the screen.

8 Drag the **Open** icon (second icon) and then the **Save** icon (third) to the Custom toolbar.

As you can see, the toolbar grows as you drag icons to it. You can also copy and drag icons from AutoCAD's toolbars.

9 Press and hold the **CTRL** key. From the Draw toolbar, drag the **Line** icon to the Custom toolbar while holding down the **CTRL** key.

10 Press and hold the **CTRL** key. From the Modify toolbar, drag the **Erase** icon to the Custom toolbar.

11 Press and hold the **CTRL** key. From the Object Properties toolbar, drag the **Layers** icon to the Custom toolbar.

The Custom toolbar should look similar to the following.

NOTE:

Your toolbar may look different than the one shown above, depending on where you drop the icons. For example, you can create the toolbar as two rows of three icons each. After you have finished the toolbar, you can change its shape in the same way you do other toolbars.

12 Delete the **Erase** icon from the Custom toolbar by dragging and dropping it to an open area in the graphics screen.

13 Pick the **Close** button to close the Customize Toolbars dialog box; pick **Close** again to close the Toolbars dialog box.

AutoCAD updates its acad.mns, acad.mnc, and acad.mnr files.

14 Try each of the icons in the new toolbar.

15 Reshape the toolbar and then dock it along the right edge.

Whenever you change the position or visibility of a toolbar, AutoCAD stores the change in the system registry. This is why the screen may look different after someone else uses AutoCAD, even if this person did not create or save any drawing files.

Deleting Toolbars

AutoCAD also makes it easy to delete toolbars. Be careful not to delete toolbars that you should keep.

1. Pick **Toolbars...** from the **View** pull-down menu.

2. In the Toolbars list box, pick **Custom**. Be sure that this is the one you've selected.

3. Pick the **Delete** button. If you're certain that you have selected Custom, pick the **Yes** button.

NOTE:

The Properties... button displays the Toolbar Properties dialog box. It provides information about the toolbar and permits you to change the toolbar name and help text.

4. Pick the **Close** button.

AutoCAD updates its acad.mns, acad.mnc, and acad.mnr files once again.

5. Exit AutoCAD. Do not save.

Questions

1. Briefly define an AutoCAD macro.

_____

_____

2. Why are custom macros useful?

_____

_____

3. Explain the concept of base and partial menus.

_____

_____

4. Which AutoCAD command allows you to load a base menu?

_____

5. AutoCAD compiles MNU files into what two binary file types?

_____

6. What must you do to convert a simple base menu, such as pull.mnu, to a partial menu?

_____

_____

7. Explain how you would add the Polyline icon from the Draw toolbar to a new custom toolbar.

_____

_____

8. Where does AutoCAD store changes to the position and visibility of toolbars?

_____

◼ *Challenge Your Thinking: Questions for Discussion*

1. The INSERT command can be included in a macro like any other command. In conjunction with a drawing file name, how could this be useful?

2. When creating a custom toolbar, is it possible to copy an icon from a flyout to the custom toolbar? Explain.

Problems

1. Create the following pull-down menu. Name it mine.mnu. Position the new partial menu between the Modify and Help pull-down menus. Try each of the menu selections. If available, refer to the optional *Applying AutoCAD Diskette*. It contains the following code.

```
***MENUGROUP=MINE

***POP1
[MINE]
[LINE] ^c^cline
[ERASE W] ^c^cerase w
[Flip Snap] ^b
[--]
[->Display]
  [Pan]'pan
  [->Zoom]
    [ZOOM W] ^c^czoom w
    [ZOOM P] ^c^czoom p
    [<-ZOOM All] ^c^czoom a
  [<-Viewpoint] ^c^cvpoint;;
[--]
[->Text]
  [COMP S] ^c^cstyle comp complex;;;;;;;
  [His Name] ^c^ctext 6,2 .2 0 John Doe;;;Mechanical Engineer;
  [<-My Name] ^c^ctext s comp 6,3 .2 0;
[--]
[Arch Units] ^c^cunits 4;;;;;;
[--]
[->Object Snap]
  [Mid Point]mid
  [Cen Point]cen
  [<-Nearest]near
[--]
[*Cancel*] ^c^c
```

2. Remove and unload the partial menu that you created in Problem 1.

3. Create the following custom toolbar named Frequent. Be sure to press the CTRL key when copying icons from the toolbars.

4. Delete the Frequent toolbar that you created in Problem 3.

 **Creating Tablet Menus**

■ **OBJECTIVE:**

To apply tablet menu development, the TABLET command, and the tablet configuration steps

This unit focuses on creating tablet menus and configuring a tablet menu. You must have a digitizing tablet to complete this unit.

Areas of the digitizing tablet can be designated for *tablet menus,* enabling you to enter a wide variety of AutoCAD commands and functions quickly and conveniently.

Developing a Tablet Menu

The first step in designing a new tablet menu is to ask yourself what you would like to include in the menu. The best way to answer this question is to sketch the tablet menu overlay on paper so that you gain some sense of the placement of each menu component. After the sketch is refined to your liking, you can use it to develop the actual menu file. In this unit, an example tablet menu overlay has been provided. Later, after you've learned the procedures, you'll be able to design your own.

The example on the next page is a relatively simple, but functional, tablet menu overlay. Let's use it as the model for the following steps.

1 Make an enlarged photocopy of the tablet menu overlay so that it comes close to fitting your digitizing tablet. If you do not have access to a copier that has enlargement capability, use the overlay at its existing size.

_____ NOTE: _____

See also the optional *Applying AutoCAD Diskette.*

The overlay must not extend outside the active area on the digitizer. For instance, if the active area is 11″ × 11″, then the overlay must not be larger than 11″ × 11″.

Next, let's create the menu file that holds the menu items. Notice how these menu items correspond to the items on the tablet menu overlay, starting with tablet menu 1.

2 Start AutoCAD and begin a new drawing from scratch.

3 Using Notepad, enter the items on pages 697-698. Name the file **tab.mnu**. Be sure to enter the items exactly as shown, and press **ENTER** after each entry.

_____ NOTE: _____

See also the optional *Applying AutoCAD Diskette.*

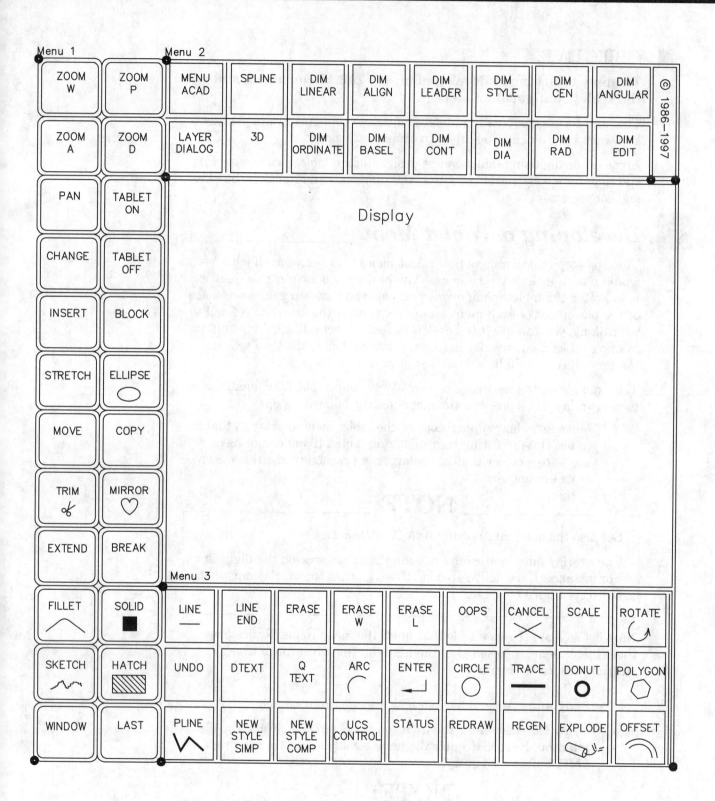

Menu 1

Menu 2

| ZOOM W | ZOOM P | MENU ACAD | SPLINE | DIM LINEAR | DIM ALIGN | DIM LEADER | DIM STYLE | DIM CEN | DIM ANGULAR |
| ZOOM A | ZOOM D | LAYER DIALOG | 3D | DIM ORDINATE | DIM BASEL | DIM CONT | DIM DIA | DIM RAD | DIM EDIT |

© 1986–1997

Display

PAN	TABLET ON
CHANGE	TABLET OFF
INSERT	BLOCK
STRETCH	ELLIPSE
MOVE	COPY
TRIM	MIRROR
EXTEND	BREAK

Menu 3

FILLET	SOLID	LINE	LINE END	ERASE	ERASE W	ERASE L	OOPS	CANCEL	SCALE	ROTATE
SKETCH	HATCH	UNDO	DTEXT	Q TEXT	ARC	ENTER	CIRCLE	TRACE	DONUT	POLYGON
WINDOW	LAST	PLINE	NEW STYLE SIMP	NEW STYLE COMP	UCS CONTROL	STATUS	REDRAW	REGEN	EXPLODE	OFFSET

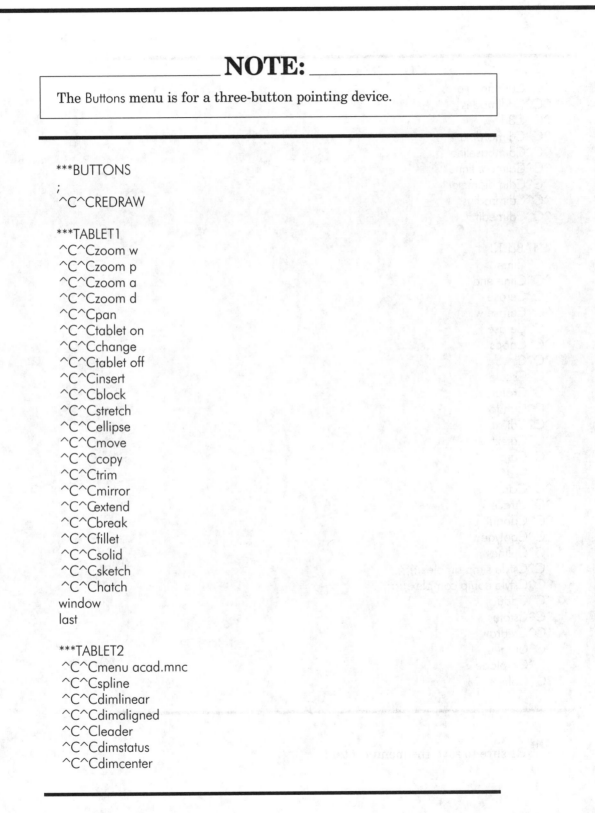

NOTE:

The Buttons menu is for a three-button pointing device.

```
***BUTTONS
;
^C^CREDRAW

***TABLET1
^C^Czoom w
^C^Czoom p
^C^Czoom a
^C^Czoom d
^C^Cpan
^C^Ctablet on
^C^Cchange
^C^Ctablet off
^C^Cinsert
^C^Cblock
^C^Cstretch
^C^Cellipse
^C^Cmove
^C^Ccopy
^C^Ctrim
^C^Cmirror
^C^Cextend
^C^Cbreak
^C^Cfillet
^C^Csolid
^C^Csketch
^C^Chatch
window
last

***TABLET2
^C^Cmenu acad.mnc
^C^Cspline
^C^Cdimlinear
^C^Cdimaligned
^C^Cleader
^C^Cdimstatus
^C^Cdimcenter
```

```
^C^Cdimangular
^C^Cddlmodes
^C^C3d
^C^Cdimordinate
^C^Cdimbaseline
^C^Cdimcontinue
^C^Cdimdiameter
^C^Cdimradius
^C^Cdimedit

***TABLET3
^C^Cline
^C^Cline end
^C^Cerase
^C^Cerase w
^C^Cerase l
^C^Coops
^C^C
^C^Cscale
^C^Crotate
^C^Cundo
^C^Cdtext
^C^Cqtext
^C^Carc
;
^C^Ccircle
^C^Ctrace
^C^Cdonut
^C^Cpolygon
^C^Cpline
^C^Cstyle simp simplex;;;;;;;
^C^Cstyle comp complex;;;;;;;
^C^Cdducs
^C^Cstatus
^C^Credraw
^C^Cregen
^C^Cexplode
^C^Coffset
```

4 Be sure to save the menu contents.

Make a backup copy of the file.

■ *TABLET Command* _____

1 Secure the menu overlay to the digitizer tablet with tape.

NOTE: _____

Be sure that all of the overlay is inside the active pointing area of the digitizing tablet. If it is not, tablet configuration will not work.

2 Enter the **TABLET** command, and then enter the **CFG** option.

NOTE: _____

If you receive the message Your pointing device cannot be used as a tablet, you need to specify a new device in AutoCAD. You can do this by selecting Preferences... from the Tools pull-down menu and picking the Pointer tab.

The following prompt appears on the screen.

_____

Enter number of tablet menus desired (0-4) <0>:

_____

3 Enter **3**.

The following prompt appears on the screen.

_____

Digitize upper left corner of menu area 1:

_____

4 Locate the upper left corner of menu 1, shown in the following illustration, and pick that point. (The point is indicated on the overlay by a small donut.)

NOTE:

In this overlay, menu 1 is comprised of the first two columns.

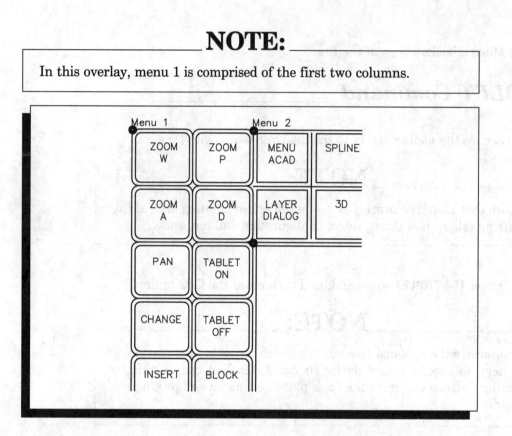

5 Pick the lower left corner of menu 1 (also indicated by a small donut) . . .

6 . . . and the lower right corner of menu 1 (the small donut two cells to the right of the preceding point).

You have just defined the boundaries of menu 1.

7 Enter **2** for the number of columns in menu 1 . . .

8 . . . and **12** for the number of rows in menu 1.

Now AutoCAD knows the exact size and location of all twenty-four cells in tablet menu 1.

AutoCAD now prompts you for the upper left corner of menu 2.

9 Locate menu 2 and its upper left corner, and pick that point.

NOTE:

Menu 2 is comprised of the two upper rows, beginning with the cell called MENU ACAD. Menu 2 consists of eight columns and two rows.

10 Proceed exactly as you did with menu 1 until you are finished with menu 2. Be sure you select the leftmost donut when you pick the lower right corner of menu 2.

11 Proceed with menu 3. It contains nine columns and three rows.

After you are finished, the following prompt appears on the screen.

Do you want to respecify the Fixed Screen Pointing Area? <N>

---

12 Enter **Y** for Yes.

13 Digitize the lower left and the upper right corners of the display pointing area (the square bounded by the three menus). For the upper right corner, be sure to pick the rightmost donut.

14 Enter **No** in reply to the last question.

You are finished with the tablet configuration.

Loading the Menu File

Now let's load the tablet menu called tab.mnu.

1 Enter the **MENU** command and locate and select **tab.mnu** and pick the **Yes** button.

If you correctly completed the above steps, you should now have full access to the new tablet menu.

2 Experiment with the tablet menu by picking each of the cells on the overlay, but do not yet pick the item named MENU ACAD located in the upper left corner.

3 To bring back the standard acad menu, pick the tablet menu item called **MENU ACAD**. If this doesn't work, enter **MENU** and select **acad.mnc** from the Support directory.

You no longer have access to the tablet menu.

4 Exit AutoCAD.

Questions

1. What AutoCAD command and command option are used to configure a digitizing tablet menu?

_____

_____

2. Explain the purpose of tablet configuration.

_____

_____

3. What is the minimum and maximum number of tablet menus that can be included on a digitizing tablet?

_____

4. What command is used to load a tablet menu file?

_____

Challenge Your Thinking: Questions for Discussion

1. Discuss the advantages and disadvantages of using a digitizing tablet instead of or in addition to the pull-down menus and the keyboard.

2. Explore the various options available for digitizing tablets. What should you consider when buying a tablet? Describe the distinguishing features of the various tablets. Tell which features would be most important to you and why.

Problems

1. Develop a new tablet menu and include a symbol library in one section of the menu. Use the previously created library called lib1.dwg or create a new one. The following example should help you get started. Name the menu file tab2.mnu. The tab2.mnu file is available on the optional *Applying AutoCAD Diskette*.

```
***TABLET1
^C^Cinsert LIB ^C
^C^Cinsert TSAW drag \drag \drag
^C^Cinsert DRILLP drag \drag \drag
^C^Cinsert JOINT drag \drag \drag


***TABLET2
^C^Cline
^C^Cerase
^C^Czoom w
```

2. Develop a new tablet menu using the steps outlined in this unit. Make the menu as sophisticated and powerful as possible. Use AutoCAD's macro development capability as much as possible.

Unit 62

Configuring and Customizing AutoCAD's Tablet Menu

■ **OBJECTIVE:**

To configure and customize the standard AutoCAD tablet menu

This unit steps you through the process of configuring AutoCAD's tablet menu. Also, you will discover how to develop tablet menu area 1, located at the top of the tablet menu. Techniques are included for positioning new AutoCAD command macros and symbol libraries in this area of the tablet menu. The AutoCAD tablet menu template (overlay) is shown below.

The top area has been reserved for you to customize. A total of 225 cells are available. This area is referred to as tablet menu 1.

There is more than one way to develop menu area 1. We will use a basic method that allows you to follow each step of the development easily. This method also shows you how the other portions of the tablet menu were designed.

Configuring the Standard Tablet Menu

Be sure the tablet menu overlay is securely fastened to the digitizing tablet.

1 Start AutoCAD and begin a new drawing from scratch.

2 Enter the **TABLET** command and the **CFG** option.

3 Enter **4** for the number of desired tablet menus.

4 As AutoCAD now requests, digitize the upper left corner of menu area 1 as shown in the following illustration.

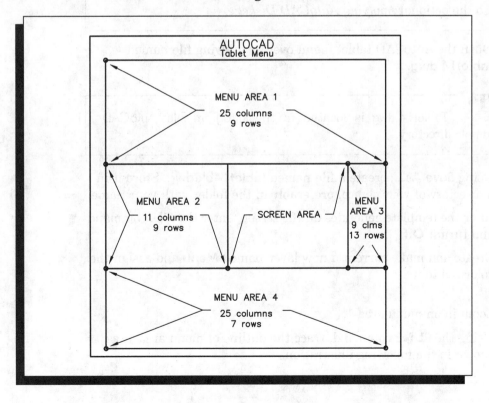

5 Pick each of the remaining corners as instructed by AutoCAD. Be precise. Also enter the correct number of columns and rows as indicated below. Be sure also to specify the fixed screen pointing area. Do not specify a floating screen pointing area.

Menu Area	Columns	Rows
1	25	9
2	11	9
3	9	13
4	25	7

When you are finished, AutoCAD stores this information and the Command prompt becomes available.

6 Pick several items on the tablet menu to make sure that it is configured properly.

Customizing the Tablet Menu

We will begin by creating the rectangular shape of area 1.

NOTE:

See also the optional *Applying AutoCAD Diskette*.

1 Open the AutoCAD tablet menu overlay drawing file named **tablet14.dwg**.

HINT:

Tablet14.dwg is located in the \Program Files\AutoCAD R14\Sample directory.

2 Using **Save As...**, create a file named **tablet14bk.dwg**. Store it in a directory of your choice, preferably in the folder with your name.

3 After the template generates on the screen, enter the **FILL** command and turn it **Off**.

3 Create and make current a new layer named **Menu** and assign the color red to it.

4 Zoom in on menu area 1.

5 Using the **PLINE** command, trace the outline of menu area 1, as shown in the following illustration.

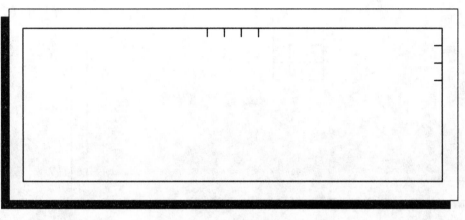

6 Add four short vertical lines beginning at column 12, as shown, and add three short horizontal lines beginning at row A. Zoom closely to these areas prior to drawing the lines.

7 Create a block named **frame** and enter **0,0** for the insertion base point. Select each of the lines and the polyline outline you created.

8 Enter the **WBLOCK** command, enter **frame** for the file name, and enter **frame** for the block name.

9 Save your work.

Designing Area 1

You now have a drawing file, frame.dwg, that matches the size of tablet area 1. Let's open this file and add command macros and symbols. When it is finished, you will be able to plot frame.dwg and position it under the template's transparent area 1.

1 Open the **frame.dwg** file.

The rectangular menu area identical to the one in the preceding illustration appears.

2 Enter **ZOOM Extents**, and then enter **ZOOM** and **.95x**.

3 Use horizontal and vertical lines to create the following nine cells.

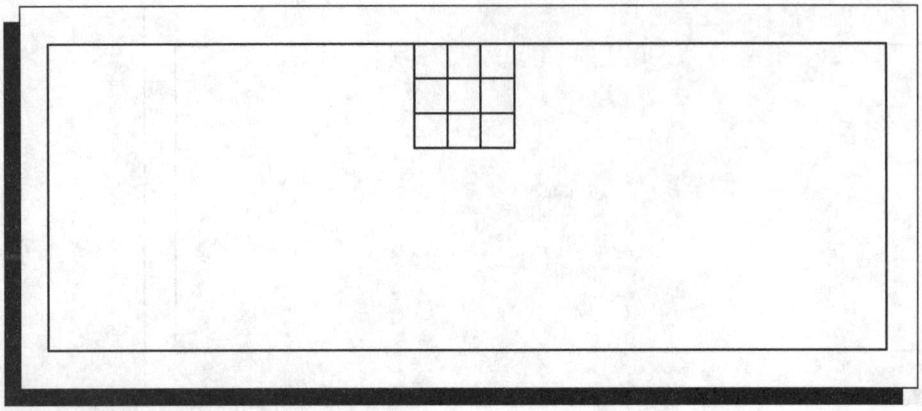

HINT:

Use the RAY, EXTEND and TRIM commands to construct the nine cells as quickly as possible.

4 Save your work.

Now, let's bring in the symbol library named lib1.dwg that you created in Unit 33, and place each of the tools in one of the nine cells.

5 Insert **lib1.dwg**, but be sure to cancel when the Insertion point prompt appears.

Draw

6 Review the block definitions now contained in the drawing.

7 Place each of the tool symbols as shown here. You will need to reduce their size; try **.15** unit.

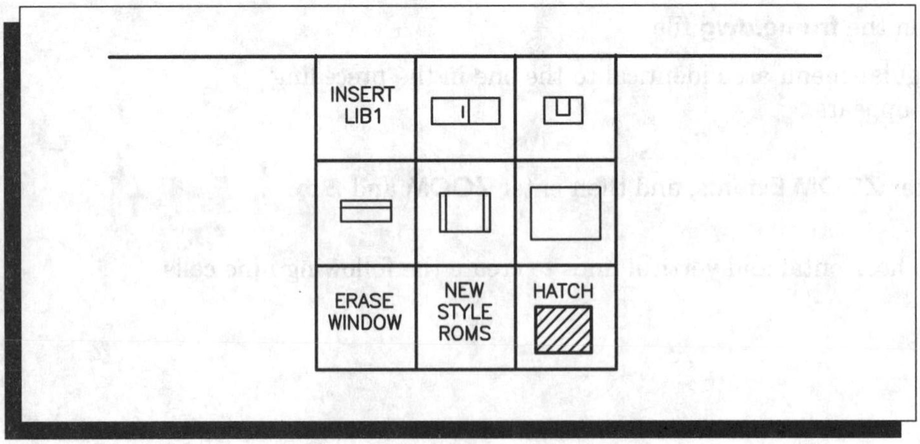

⑧ Place the text in the cells as shown above, and include a small hatch pattern under the word HATCH.

⑨ **ZOOM All** and save your work.

⑩ Plot the drawing extents at a scale of **1=1**.

⑪ After plotting, trim around the menu area with scissors, and secure the menu under the template.

⑫ Exit AutoCAD.

Writing the Code

This part becomes a bit more involved.

① Locate the **acad.mnu** file, which is located in the Support directory. Make a copy of it, and name it **acadnew.mnu**.

② Using WordPad, open **acadnew.mnu**.

③ Using the Find capability, locate the item *****TABLET1** in the file.

After you locate ***TABLET1 in the file, notice the similar items (A-1 through A-25, B-1 through B-25, and so on) that follow it. The first part of this section is shown below.

```
***TABLET1
**TABLET1STD
[A-1]\
[A-2]\
[A-3]\
[A-4]\
[A-5]\
```

This is where you enter the macros that correspond to the items in menu area 1. Likewise, each of these numbers corresponds to cells in menu area 1. To illustrate this, the upper middle portion of area 1 is shown on the following page.

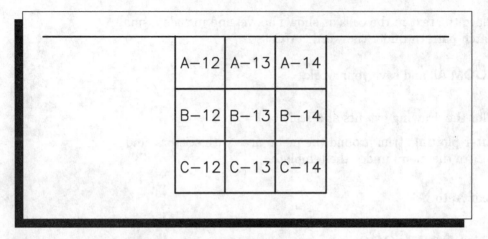

The numbering sequence of the cells begins at the upper left corner of the tablet menu and proceeds to the right. Tablet area 1 contains 225 cells arranged into 25 columns and 9 rows.

4 Type the menu items over the numbers as shown below. Notice that their placement corresponds directly to the cells in the upper middle portion of tablet area 1.

NOTE:

In the following code, text was omitted where you see blank lines. This was done to conserve space on the page.

```
***TABLET1
[A-1]
[A-2]
[A-3]
[A-4]
[A-5]
[A-6]
[A-7]
[A-8]
[A-9]
[A-10]
[A-11]
^C^CINSERT LIB1;^C
^C^CINSERT TSAW DRAG \DRAG \DRAG
^C^CINSERT DRILLP DRAG \DRAG \DRAG
[A-15]
[A-16]
[A-17]
```

```
[A-18]
[A-19]
[A-20]
[A-21]
[A-22]
[A-23]
[A-24]
[A-25]

[B-11]
^C^CINSERT JOINT DRAG \DRAG \DRAG
^C^CINSERT PLANER DRAG \DRAG \DRAG
^C^CINSERT BENCH DRAG \DRAG \DRAG
[B-15]
[B-16]
[B-17]
[B-18]
[B-19]
[B-20]
[B-21]
[B-22]
[B-23]
[B-24]
[B-25]

[C-11]
^C^CERASE W
^C^CSTYLE ROMS ROMANS;;;;;;;
^C^CBHATCH
[C-15]
[C-16]
[C-17]
[C-18]
[C-19]
[C-20]
[C-21]
[C-22]
[C-23]
[C-24]
[C-25]
```

⑤ Save your changes in the **acadnew.mnu** file, being sure to save the file as a text file (*not* as a Word document), and exit WordPad.

6 Copy **lib1.dwg** to the AutoCAD R14 folder.

Using the New Menu

1 Start AutoCAD and begin a new drawing from scratch.

2 Load **acadnew.mnu** using the **MENU** command.

3 Pick the new tablet menu item called **INSERT LIB1**.

This item, as you may have noticed when you typed the macro, inserts the small symbol library, lib1.dwg, into the current drawing.

4 Pick each of the tool symbols and place them one at a time.

5 Pick the remaining three menu items and notice what each of them does.

This gives you a taste of what can be developed with a tablet menu.

6 Load the **acad.mnc** file, and exit AutoCAD without saving.

Questions

1. How do you define the four standard tablet areas when configuring the tablet menu?

_____

_____

2. What is the purpose of the AutoCAD tablet menu area 1?

_____

_____

3. What is the benefit of customizing the tablet menu?

_____

_____

4. Explain the numbering sequence of cells contained in tablet menus.

_____

_____

5. Explain why a text editor such as WordPad is required to modify the file acad.mnu.

_____

_____

 ### *Challenge Your Thinking: Questions for Discussion*

1. Consider how a custom tablet menu could help you use AutoCAD. Discuss your ideas with others. Then design a custom tablet menu to help you do your work more easily.

Problems

1. Experiment with different portions of the AutoCAD tablet menu areas 2, 3, and 4 to discover how these menus were designed.

2. In the acadnew.mnu file, place additional menu items in menu area 1. Categorize the items in area 1 so that related items are grouped together. You may wish to add headings above each group of related items.

3. Create the custom menu you designed in the "Challenge Your Thinking" question and try it. Does it work the way you expected? If not, revise it until it becomes a helpful tool.

Unit 63 Exploring AutoLISP

■ OBJECTIVE:

To explore AutoLISP applications and capabilities by loading and invoking AutoLISP programs

AutoLISP is AutoCAD's variation of the LISP programming language. LISP, short for "LISt Processing," is a powerful programming language sometimes used for artificial intelligence (AI) applications. AutoLISP is embedded into AutoCAD so that you can apply its functions at any time.

AutoLISP, an interpreted language, provides programmers with a high-level language well suited to graphics applications. For instance, you can use AutoLISP to write a program that automates the creation of a staircase in a building. A properly written program prompts you for the distance between the upper and lower floors, asks you for the size and/or number of steps (risers and treads), and draws the detailed staircase for you.

AutoLISP programs, also called *functions* or *routines,* can be stored and used in at least two different formats:

1) as files containing an LSP file extension

2) as pull-down or tablet menu items

Lengthy and sophisticated AutoLISP routines are most often stored as AutoLISP files (with the LSP extension), while shorter and simpler routines are typically stored as menu items.

■ *AutoLISP Example*

Let's take a look at an AutoLISP program.

1 Start Notepad and open the file named **3darray.lsp** located in the \Program Files\AutoCAD R14\Support directory.

This is an AutoLISP file.

2 Review the contents of the file.

The first portion of 3darray.lsp is shown on the following page.

```
; Next available MSG number is    29
; MODULE_ID LSP_3DARRAY_LSP_
;;;
;;;    3darray.lsp
;;;
;;;    Copyright 1987, 1988, 1990, 1992, 1994, 1996 by Autodesk, Inc.
;;;
;;;    Permission to use, copy, modify, and distribute this software
;;;    for any purpose and without fee is hereby granted, provided
;;;    that the above copyright notice appears in all copies and
;;;    that both that copyright notice and the limited warranty and
;;;    restricted rights notice below appear in all supporting
;;;    documentation.
;;;
;;;    AUTODESK PROVIDES THIS PROGRAM "AS IS" AND WITH ALL FAULTS.
;;;    AUTODESK SPECIFICALLY DISCLAIMS ANY IMPLIED WARRANTY OF
;;;    MERCHANTABILITY OR FITNESS FOR A PARTICULAR USE.  AUTODESK, INC.
;;;    DOES NOT WARRANT THAT THE OPERATION OF THE PROGRAM WILL BE
;;;    UNINTERRUPTED OR ERROR FREE.
;;;
;;;    Use, duplication, or disclosure by the U.S. Government is subject to
;;;    restrictions set forth in FAR 52.227-19 (Commercial Computer
;;;    Software - Restricted Rights) and DFAR 252.227-7013(c)(1)(ii)
;;;    (Rights in Technical Data and Computer Software), as applicable.
;;;
;;;==========================================
;;; Functions included:
;;;      1) Rectangular ARRAYS (rows, columns & levels)
;;;      2) Circular ARRAYS around any axis
;;;
;;; All are loaded by: (load "3darray")
;;;
;;; And run by:
;;;      Command: 3darray
;;;          Select objects:
;;;          Rectangular or Polar array (R/P): (select type of array)


;;; =============== load-time error checking ============

(defun ai_abort (app msg)
  (defun *error* (s)
    (if old_error (setq *error* old_error))
    (princ)
  )
```

The beginning of many AutoLISP programs includes copyright information and statements explaining the purpose of the program.

Now that you have seen the contents of 3darray.lsp, let's load it into AutoCAD.

Loading/Invoking AutoLISP Files _____

AutoLISP files, such as 3darray.lsp, can be loaded and invoked by entering the AutoLISP load function at the Command prompt.

1 Exit Notepad. Do not save changes to the file.

2 Start AutoCAD and begin a new drawing from scratch.

3 At the Command prompt, enter the following text exactly as you see it, including the parentheses. (You may use upper- or lowercase letters.) Be sure to press **ENTER**.

(load "3darray")

If the program loads properly, you will receive the message 3DARRAY loaded, and you will have access to a new command called 3DARRAY. If you receive an error message, refer to the section titled "AutoLISP Error Messages" in this unit.

4 Create a 3D cube 1 unit in size, and select a viewpoint that is above, in front, and to the right of the object.

5 Enter **ZOOM** and **1** to reduce the current zoom magnification.

6 Enter **3DARRAY**, select the object and press **ENTER**, and choose to perform a rectangular array.

7 Enter **5** for the number of rows, **4** for the number of columns, and **3** for the number of levels. Specify a distance of **1.4** between rows, columns, and levels, as you would using the ARRAY command.

The 3D array generates on the screen.

8 Enter **ZOOM Extents** and **HIDE**.

The screen should look similar to the following illustration.

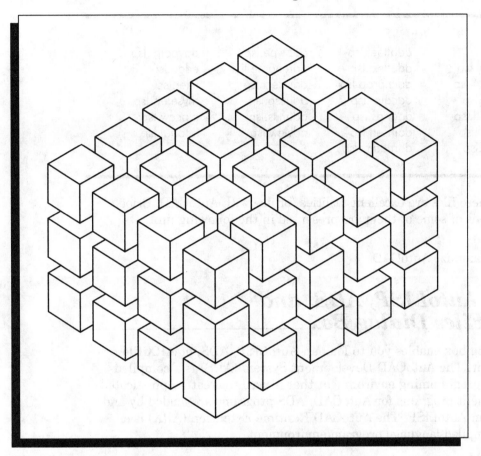

As indicated in the first part of this unit, AutoLISP code can be stored as pull-down or tablet menu items instead of as LSP files. These menu items are stored in a menu file and are invoked by picking the corresponding item from the menu.

Additional AutoLISP Examples

The AutoLISP files that come with AutoCAD are located in the Support folder. They provide examples of good programming practice, and they illustrate the power of AutoLISP.

1 Minimize AutoCAD.

2 Review the AutoLISP files in the Support folder.

The AutoLISP files are shown here.

3d.lsp	ddattdef.lsp	ddptype.lsp	ddvpoint.lsp
3darray.lsp	ddattext.lsp	ddrename.lsp	edge.lsp
acadr14.lsp	ddchprop.lsp	ddselect.lsp	filter.lsp
ai_utils.lsp	ddcolor.lsp	dducsp.lsp	mvsetup.lsp
appload.lsp	ddgrips.lsp	ddunits.lsp	plpccw.lsp
attredef.lsp	ddinsert.lsp	ddview.lsp	tutorial.lsp
bmake.lsp	ddmodify.lsp		

Among these files are excellent utilities for use with AutoCAD. Brief descriptions of some of them are presented in the following unit.

3 Maximize AutoCAD.

Load AutoLISP, ADS, and ARX Files Dialog Box

This dialog box enables you to load an AutoLISP, ADS, or ARX file by selecting it. The AutoCAD Development System (ADS) is a compiled-language programming environment that is used to create sophisticated enhancement programs for AutoCAD. ADS programs are loaded by and called from AutoLISP. The AutoCAD Runtime Extension (ARX) is a newer compiled-language program environment.

1 Select **Load Application...** from the **Tools** pull-down menu.

The Load AutoLISP, ADS, and ARX Files dialog box appears, as shown in the following illustration.

2 Pick the **File...** button.

3 Select **3d.lsp** from the Support folder and pick the **Open** button.

The file appears in the list box.

4 Pick the **Load** button to load the file.

This loads the AutoLISP program as if you had used the AutoLISP load command.

5 Exit AutoCAD. Do not save.

AutoLISP Error Messages

At one time or another, you may receive one of many AutoLISP error messages. Here are three examples.

Insufficient node space

Insufficient string space

LOAD failed.

NOTE:

The on-line *AutoCAD Customization Guide*, Part II, Chapter 16, provides a complete list of AutoLISP error messages.

The error messages insufficient node space and insufficient string space appear when AutoCAD runs out of "heap" space. The heap is an area of memory set aside for storage of all AutoLISP functions and symbols (also called *nodes*). Elaborate AutoLISP programs require greater amounts of heap and stack space.

The stack is also an area of memory set aside by AutoLISP. Stack holds function arguments and partial results; the deeper you "nest" functions, the more stack space is used.

NOTE:

Chapter 15 of the on-line *AutoCAD Customization Guide* provides details on the use of heap and stack and memory management.

The error message LOAD failed appears when the file named in the load function cannot be found, or when the user does not have read access to the file. This message also appears when the file name is misspelled or the incorrect file location (directory) is specified.

Questions

1. What is AutoLISP?

_____

_____

2. Name two applications for AutoLISP.

_____

3. What file extension do AutoLISP programs use?

_____

4. Why is the amount of heap space important when you load AutoLISP programs?

_____

_____

5. What purpose does the 3darray.lsp program serve?

_____

_____

6. How would you load a file named project.lsp?

_____

_____

■ *Challenge Your Thinking: Questions for Discussion*

1. AutoCAD is a very complex CAD program that allows you to do most drafting tasks easily. In addition, you can customize the program by creating your own menu items. With these facts in mind, explain why separate AutoLISP programs may be desirable for some AutoCAD applications.

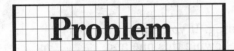

Problem

1. Load and invoke the AutoLISP programs supplied with AutoCAD.

AUTOCAD® AT WORK

STL: The Key to RP for Edmar Engineering

Ernie Marine of Edmar Engineering in Denver has had nothing but good results from his work with STL files and prototype part production. He created five parts with AutoCAD, converted them to STL, and had them built on an SLA-250 machine. The SLA-250 from 3D Systems in Valencia, California, is the most widely installed RP system. (SLA stands for "StereoLithography Apparatus.")

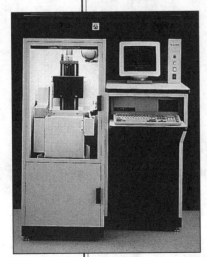

One of the parts was a complex bevel gear that fits in precision dental equipment. Protogenic, a rapid prototyping service bureau in Boulder, Colorado, created the plastic prototype part at 10 times the original size using an SLA. A machinist traced the prototype gear using a pantograph machine and milled an EDM graphite electrode from it at one-tenth the size of the prototype gear. The entire process saved four weeks in prototyping a new line of dental equipment.

The new design, which uses the gear, consists of a disposable prophy cup, which dentists and dental hygienists use to clean patients' teeth. Disposing of the prophy cup rather than cleaning it will avoid the chance of transmitting diseases from one patient to another.

Based on a story by Terry Wohlers in *CADENCE*, November 1992, Copyright © 1992, Miller Freeman, Inc.

Unit 64 — Easing into AutoLISP Programming†

OBJECTIVE:

To practice fundamental AutoLISP programming techniques and to create new AutoLISP files and AutoCAD commands

The significance of AutoLISP embedded into AutoCAD has captured the interest of many. This unit will introduce AutoLISP programming so that you can decide whether or not it's for you. Like many others, you may remain satisfied with applying the power of already-developed AutoLISP programs.

You may find AutoLISP programming very intriguing. If you enjoy programming, you should continue with the unit following this one and then explore and learn other AutoLISP commands and functions on your own.

AutoLISP Arithmetic

As illustrated in the previous unit, you can enter AutoLISP functions, such as load, directly at the Command prompt. Let's enter an AutoLISP arithmetic expression.

1 Start AutoCAD and begin a new drawing from scratch.

2 At the Command prompt, type **(* 5 6)** and press **RETURN**.

The number 30 appears at the bottom of the screen because $5 \times 6 = 30$.

3 Enter **(/ 15 3)**.

The number 5 appears because $15 \div 3 = 5$.

4 Try **(+ 25 4)**.

Did 29 appear?

Setq Function

The setq function is used to assign values to a variable.

1 Type **(setq A 10)** and press **ENTER**. (You can type A in upper- or lowercase letters.)

The value of variable A is now 10.

†Acknowledgment: Some of the basic principles presented here are from the *CADalyst* Journal AutoLISP tutorial series printed in Vol. 3, No. 2 and Vol. 3, No. 3.

2 Enter **(setq B (– 20 4))**.

Notice the parentheses.

The variable B now holds the value of 16.

3 To list the value of A, enter **!A**.

Did 10 appear?

4 Enter **!B**.

Let's try something a bit more complex.

5 Enter **(setq CAT (– B A))**.

Now CAT is equal to the value of B minus the value of A.

Let's try one more.

6 Enter **(setq DOG (* (* 2 CAT)(/ A 5)))**.

The variable DOG is now equal to 24.

NOTE:

The total number of left parentheses must equal the total number of right parentheses. If they are not equal, you will receive a message such as 1>. This means you lack one right parenthesis. Add one by typing another right parenthesis and pressing ENTER.

Storing AutoLISP Routines As Files

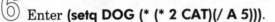

As you may remember, AutoLISP code can be stored in an LSP file and subsequently loaded and invoked at any time. The routines executed in the previous unit are good examples. Let's store the above functions to illustrate this capability.

1 Minimize AutoCAD.

2 Load Notepad and store the following exactly as you see it. Name the file **first.lsp**, save it, exit Notepad, and return to AutoCAD.

```
(defun c:FIRST ()
(setq a 10)
(setq b (– 20 4))
(setq cat (– b a))
(setq dog (* (* 2 cat)(/ a 5)))
)
```

Notice the AutoLISP function defun. The defun function allows you to define a new function or AutoCAD command and to invoke the AutoLISP file. Once loaded, the above program can be invoked by entering FIRST at the Command prompt.

NOTE:

Notice the inclusion of the right parenthesis at the program's last line. This right parenthesis evens the number of left and right parentheses. And, in conjunction with the left parenthesis in front of defun, the right parenthesis encloses the defun function.

3 At the Command prompt, enter **(load "first")** or use the **Load AutoLISP, ADS, and ARX Files** dialog box to load first.lsp.

HINT:

If the first.lsp file is located in a directory other than the default AutoCAD directory, specify the path when entering the load function. For example, enter (load "user/first") where USER is the name of a directory. Notice the slash mark (not a backslash). An alternative to this approach is to open a drawing from the directory that contains first.lsp. AutoCAD will then find the file.

4 Enter **FIRST**, now a new AutoCAD command, to invoke the program.

The number 24 appears.

NOTE:

The preceding code can also be stored without the defun function. Also, setq can be entered just once. It would look like this.

```
(setq
a 10
b (- 20 4)
cat (- b a)
dog (* (* 2 cat)(/ a 5))
)
```

Entering (load "first") would then load and invoke the routine in one step.

Storing AutoLISP Routines As Menu Items

You can store and invoke AutoLISP routines as pull-down menu items.

1 Minimize AutoCAD.

2 Produce a copy of **pullpart.mnu** (from Unit 60). Name it **pullprt2.mnu**.

The pullprt2.mnu file is available on the optional *Applying AutoCAD Diskette.*

3 Using Notepad, open **pullprt2.mnu**.

4 Edit the file so that it matches the following text exactly.

```
***MENUGROUP=PULLPRT2

***POP1
[Construct]
[Line] ^C^CLINE
[Circle] ^C^CCIRCLE
[Arc] ^C^CARC
[--]
[Pick This] ^C^C(setq a 10) (setq b (- 20 4)) (- b a)
[--]
[->Display]
 [Pan]'PAN
 [->Zoom]
  [Window] ^C^CZOOM W
  [Previous] ^C^CZOOM P
  [All] ^C^CZOOM A
  [<-Extents] ^C^CZOOM E
```

5 Save the **pullprt2.mnu** file and exit Notepad.

6 Maximize AutoCAD.

7 Load the partial menu file named **pullprt2.mnu**. Place it between the Insert and Format pull-down menus.

HINT:

Pick Customize Menu... from the Tools pull-down menu. Follow Steps 8 through 15 on pages 673-674.

---

8 After the Construct pull-down menu appears, select **Pick This** from it.

The AutoLISP code should return 6.

9 From the Menu Customization dialog box, unload **PULLPRT2**.

The Construct pull-down menu disappears.

List Function

Setq by itself can assign only one value to a variable. The list function can be used to string together multiple values, such as *x* and *y* coordinates, to form a point. Setq, used with list, can then assign a list of coordinates to a single variable. Let's step through an example.

 Type **(setq fish (list 7 6))** and press **ENTER**.

Did (7 6) appear on the screen? This is now the value of fish.

2 Enter **!fish**.

(7 6) appears again.

3 Enter the **LINE** command and pick a point near the lower left corner of the screen.

4 In reply to To point, enter **!fish**.

A line appears.

5 Press **ENTER**.

6 Enter **LIST** and pick the line. Notice that the coordinates of the second point are 7,6—the value of fish.

7 Close the AutoCAD Text Window.

HINT: This function is especially valuable when you need to reach a point off the screen. For instance, if the upper right corner of the screen is 15,10 and the LINE command has been issued, you can reach point 25,20 (or any point, for that matter) if a variable is assigned to that point. You just enter the variable preceded by the ! as you did in the fish example. Let's try it.

 Enter **(setq trout (list 35 28))**.

2 Enter the **LINE** command and pick a point anywhere on the screen.

3 In reply to To point, enter **!trout**.

Did a line appear? Does it look as though it runs off the screen? It should.

Press **ENTER** and enter **ZOOM Extents**.

You should now see the endpoint 35,28 of the line.

Use **LIST** to verify the coordinates of the point, and then close the AutoCAD Text Window.

Car Function

The car function is used to obtain the first item in a list, such as the *x* coordinate. So, what would be the car of fish? Let's enter it.

Type **(car fish)** and press **ENTER**.

The value 7 appears.

Cadr Function

Cadr is like car, only cadr gives you the *second* item of a list—the *y* coordinate, in this case.

Enter **(cadr fish)**.

The value 6 appears.

So now you see we can obtain either the *x* coordinate or the *y* coordinate from a list.

Combining Several Functions

We can also assign a new variable to a set of coordinates that contains the cadr of fish (the *y* coordinate) and 0 as the *x* coordinate.

First, we must create a new list containing 0 and the cadr of fish. Then we need to use setq to assign the new list to a variable (we'll call it bird).

Enter **(setq bird (list 0 (cadr fish)))**.

AutoCAD displays (0 6) at the prompt line.

Try a similar function using **car** and **0** (for the *y* coordinate). Use **bug** for the variable name.

AutoCAD displays (7 0).

Did you enter (setq bug (list (car fish) 0))? You should have.

728

③ Enter **!bird** and then enter **!bug**.

The values (0 6) and (7 0) should return.

The preceding steps gave you a taste of basic AutoLISP programming. The next unit will pick up from here and will apply most of the preceding AutoLISP programming techniques.

Few AutoCAD users will write sophisticated AutoLISP programs. Typically, drafting, design, and engineering professionals have neither the time nor the interest to learn AutoLISP fully. They are likely to seek ready-made AutoLISP programs, such as those presented in the following section.

AutoLISP Programming Examples

Several AutoLISP programs are supplied with AutoCAD, as explained in the previous unit. They are good examples of AutoLISP programming techniques, so you are encouraged to review them. The following list provides brief descriptions of some of the AutoLISP files distributed with AutoCAD.

These files are among those provided in the Support directory:

3d.lsp	creates various three-dimensional objects, including a pyramid, box, cone, dome/dish, wedge, torus, and 3D mesh
3darray.lsp	creates 3D rectangular arrays by specifying rows, columns, and levels; also creates polar arrays around a specified axis
mvsetup.lsp	sets the drawing units and limits of a new drawing based on the paper size and drawing scale
edge.lsp	lets you interactively change the visibility of the edges of a 3D face
ddunits.lsp	provides a quick and easy interface to the existing UNITS command
ddinsert.lsp	AutoLISP implementation of the INSERT command with a dialog interface.

① Exit AutoCAD. Do not save.

1. What will be returned if you enter (* 4 5) at the Command prompt?

2. Explain the purpose of the setq function.

3. If the value of variable XYZ is 129.5, what will be returned when you enter !XYZ?

4. How do you load and invoke an AutoLISP file named red.lsp that contains (defun c:RED ()?

5. What purpose do the car and cadr functions serve?

6. What is the purpose of the list function?

 Challenge Your Thinking: Questions for Discussion

1. You may have noticed that you can use many of the AutoLISP programs provided with AutoCAD without first loading them. AutoCAD provides a way to load AutoLISP functions automatically. Find out how to do this. Then back up the necessary AutoCAD files and experiment. See if you can get first.lsp, which you created in this unit, to load automatically. Write a short description of the procedure you used.

2. Describe a way to add first.lsp permanently to the menu so that it appears automatically each time you start AutoCAD.

Problems

1. Create an AutoLISP function that assigns a specific coordinate point to variable centerpoint and a specific value to variable radius. Then use these variables to create a circle. Change the values of centerpoint and radius and create another circle. Does the second circle reflect the changes you made?

2. Assign the values indicated to the variables listed below.

Variable	Value
tree	52
flower	174
leaf	89
bud	248

   Perform the following operations in AutoCAD using AutoLISP. Record the keystrokes you entered to perform the operations as well as your answers.

   a. Subtract tree from flower.

   b. Add tree, flower, leaf, and bud.

   c. Subtract bud from the product of flower and tree.

   d. Divide bud by leaf.

3. Using the same variables you assigned in the previous problem, create a screen menu item that sets the variables to the values indicated. Record your keystrokes. Add the screen menu item to pullprt2.mnu and save the file as test.mnu. Load test.mnu and confirm that it works as expected.

Unit 65 — Applying AutoLISP Programming Techniques

OBJECTIVE:

To introduce and combine several AutoLISP functions and apply them to the development of a new program, and to apply parametric programming techniques

This unit continues the lesson begun in the preceding unit. You will apply the techniques you have learned so far, as well as a few new ones.

Applying Your Knowledge

Let's apply several AutoLISP functions to the creation of a border for a drawing. At this point, we'll step through the process. Later, we'll store the function as a routine and add the routine to the screen menu called pullprt2.mnu.

We'll use the setq, list, car, and cadr functions to define a rectangular border for a 36″ × 24″ sheet. In order to place the border 1″ from the outer edge of the sheet, we'll define a drawing area of 34″ × 22″.

1 Start AutoCAD and begin a new drawing from scratch.

Let's begin by assigning a variable to each corner of the border. We'll call the lower left corner variable LL, the lower right variable LR, and so on.

2 Enter **(setq LL (list 0 0))**.

That takes care of the lower left corner.

3 For the upper right corner, enter **(setq UR (list 34 22))**.

Now let's use car and cadr for the remaining two corners.

4 Enter **(setq LR (list (car UR)(cadr LL)))**.

5 Enter **(setq UL (list (car LL)(cadr UR)))**.

Let's try out the new variables.

6 One at a time, enter **!LL**, then **!UR**, then **!LR**, and last **!UL**.

The correct coordinates for each corner appear.

Using the LIMITS and LINE commands, and the above variables, let's establish the new border format.

7 Enter the **LIMITS** command.

8 In reply to Lower left corner, enter **!LL**.

9 In reply to Upper right corner, enter **!UR**.

10 **ZOOM All**.

11 Set the grid to **1** unit.

12 Enter the **LINE** command and enter **!LL** for the first point.

13 For the second point, enter **!LR**; for the third point, **!UR**; for the fourth point, **!UL**; and then close by entering **C**.

14 Enter **ZOOM** and **.9x** so that you can more easily see the border.

This is all very interesting, but it took a lot of steps. Let's combine all the steps into a single routine and store it.

Creating a New Routine

1 Minimize AutoCAD.

2 Using Notepad, create and store the following text in a file named **bord.lsp**.

This file is available on the optional *Applying AutoCAD Diskette*.

```
(defun C:bord ()
(setq LL (list 0 0))
(setq UR (list 34 22))
(setq LR (list (car UR)(cadr LL)))
(setq UL (list (car LL)(cadr UR)))
)
```

3 After saving the file and exiting Notepad, maximize AutoCAD.

4 Begin a new drawing from scratch. Do not save your work in the current drawing.

5 Pick **Load Application...** from the **Tools** pull-down menu.

6 Locate and load the file named **bord.lsp**.

HINT: See pages 705-706 for information on how to use the Load AutoLISP, ADS, and ARX Files dialog box.

7 At the Command prompt, enter **BORD** in upper- or lowercase letters.

This executes the file, although it doesn't do much, at least not yet.

AutoLISP Command Function

The AutoLISP command function executes standard AutoCAD commands from within AutoLISP.

1 Using Notepad, open **bord.lsp**.

2 Using **Save As...**, create a new file named **border.lsp**.

The border.lsp file is available on the optional *Applying AutoCAD Diskette*.

```
(defun C:border ()
    (setq LL (list 0 0))
    (setq UR (list 34 22))
    (setq LR (list (car UR)(cadr LL)))
    (setq UL (list (car LL)(cadr UR)))

    (command "LIMITS" LL UR)
    (command "ZOOM" "A")
    (command "GRID" "1")
    (command "LINE" LL LR UR UL LL "")
    )
```

AutoCAD commands and command options are enclosed by double quotes ("). The two consecutive double quotes ("") are equivalent to pressing the space bar.

3 In AutoCAD, load the **border.lsp** file and enter **BORDER** at the Command prompt.

The border appears.

4 Enter **ZOOM** and **.9x** so that you can more easily see the border.

Before drawing the border, the AutoLISP routine created a new drawing area with the LIMITS command, zoomed all, and set the grid to 1.

Adding Remarks

It's good practice to provide explanatory remarks in AutoLISP files.

1 Using Notepad, store the contents of **border.lsp** in a new file named **34x22.lsp**.

2 Edit the file to include the following changes.

This file is available on the optional *Applying AutoCAD Diskette.*

```
; This routine establishes a drawing area
; for a 34" x 22" format (36" x 24" sheet size)
; and draws a border line.
;

(defun C:34x22 ()
    (setq LL (list 0 0))
    (setq UR (list 34 22))
    (setq LR (list (car UR)(cadr LL)))
    (setq UL (list (car LL)(cadr UR)))

    (command "LIMITS" LL UR)        ; sets drawing limits
    (command "ZOOM" "A")            ; zooms all
    (command "GRID" "1")            ; sets grid
    (command "LINE" LL LR UR UL LL "")
    )
```

A semicolon (;) permits you to include a remark. The explanation can be helpful to you and others that use the program. The indentations make it easier to read.

3 In AutoCAD, begin a new drawing from scratch. Do not save your work in the current drawing.

4 Load the **34x22.lsp** file and enter **34x22** at the Command prompt.

Other AutoLISP Functions

Here is a list of other commonly used AutoLISP functions. Experiment with them and discover ways of including them into 34x22.lsp and other files.

setvar   sets an AutoCAD system variable to a given value and returns that value. The variable name must be enclosed in double quotes.

         Example: (setvar "CHAMFERA" 1.5) sets the first chamfer distance to 1.5

getvar   retrieves the value of an AutoCAD system variable. The variable name must be enclosed in double quotes.

         Example: (getvar "CHAMFERA") returns 1.5, assuming the first chamfer distance specified most recently was 1.5.

getpoint   pauses for user input of a point. You may specify a point by pointing or by typing a coordinate in the current units format.

         Example: (setq xyz (getpoint "Where? "))

getreal   pauses for user input of a real number.

         Example: (setq sf (getreal "Scale factor: "))

getdist   pauses for user input of a distance. You may specify a distance by typing a number in AutoCAD's current units format, or you may enter the distance by pointing to two locations on the screen.

         Example: (setq dist (getdist "How far? "))

getstring   pauses for user input of a string.

         Example: (setq str (getstring "Your name? "))

You are encouraged to explore AutoLISP's potential further. Review the on-line *AutoCAD Customization Guide* to learn more about AutoLISP.

Parametric Programming _____

The AutoCAD INSERT command lets you insert blocks at any height and width. An architectural window symbol, for example, may be inserted into an elevation drawing at .6 on the X axis and .9 on the Y axis. *Parametrics* function similarly, but with far more flexibility and automation.

Unlike blocks, interior and exterior dimensions of parametrics-based objects remain variable. This enables you to specify not only the height and width of an object, but also different sizes of geometry inside the object.

Consider a bicycle design. With parametrics, you can adjust all the elements that make up the frame to user-specified sizes. The wheels and tires can be another size, the sprockets yet another, and so on. This reduces the potentially large number of variations of a design to just one, because each variation shares the same basic geometry. Hence, the use of parametrics reduces the number of files and the amount of disk storage space needed, while increasing flexibility and speed.

The following door and window variations were created using a parametric routine called dwelev.lsp, written by Bruce Chase.

The dwelev.lsp code is printed on the following pages.

```
;Simple parametric DOOR/WINDOW ELEVATION drawing program

; Copywritten by Bruce R Chase, Chase Systems.
; May be copied for non-commercial use.


(setq hpi (* pi 0.5))
(defun d_we1 (p1 x y off / tp)              ;draws the rectang & offsets
  (command "pline" p1 "w" 0.0 0.0
     (setq tp (polar p1 angl x))
     (setq tp (polar tp (+ angl hpi) y))
           (polar tp (- angl pi)  x)
     "cl"
  )
  (setq e (entlast))
  (if off (command "offset" "t"
        (cons (entlast)(list p1))
        (polar (polar p1 angl off)(+ angl hpi) off) ""))
  (setq ee (entlast))
)


(defun d_we2 (spt offbase offside offtop x y numx numy sx sy offin trim /
        p1 p2 p3 p4 xx yy e ee)
  (d_we1 spt x y (if trim (* -1 trim) nil))    ; base d/w w\trim
  (if (and numx numy)(progn                 ; set base of panels
     (setq p1 (polar
              (if offside (polar spt angl offside) spt)
              (+ angl hpi)
              (if offbase offbase 0.0)))
    (d_we1 p1                               ; build panels
       (setq xx (if numy (/ (- x offside offside (* (- numy 1) sy)) numy) x))
       (setq yy (if numy (/ (- y offtop  offbase (* (- numx 1) sx)) numx) y))
       (if offin offin nil)                 ; raised panel or glass trim
    )
    (command "array" e ee "" "R" numx numy  p1 ; array the base panel
         (polar (polar p1 angl (+ sy xx))
                 (+ angl hpi)(+ sx yy)))
))
)




(defun drwdr2 (spt xx x / tp)               ; getdist or default program
  (terpri)(terpri)
  (prompt (strcat xx " <"))
```

```
    (princ (rtos x (getvar "lunits")(getvar "luprec")))
    (if (Null (setq tp (if spt (getdist spt ">: ")(getdist ">: ")))) x tp)
)
(defun d_we3 ()
    (while (null (setq spt (getpoint "\n \nLower left corner of door/window:
"))))
    (setq angl (if (null
        (setq tp (getorient spt "\nBase angle of door/window <0.0>: ")))
        0.0 tp))
)

(defun c:dwelev ( / spt offbase offside offtop        ; gather all the info
                x y numx numy sx sy offin trim tp angl)
    (d_we3)
    (setq x (drwdr2 spt "Width of door/window" 36.0))
    (setq y (drwdr2 spt "Height of door/window" 80.0))
    (setq trim (if (zerop (setq tp (drwdr2 spt "Trim width" 0.0))) nil tp))

    (if (setq numy (getint "\nNumber of panels rows <none>: "))(progn
        (setq numx (if (null (setq numx
            (getint "\nNumber of panel columns <1>: "))) 1 numx))
        (setq offbase (drwdr2 spt "Bottom rail width" 10.0))
        (setq offtop  (drwdr2 spt "Top rail width"    6.0))
        (setq offside (drwdr2 spt "Side rail width"   4.0))
        (if (> numx 1)(setq sx
            (drwdr2 nil "Spacing between panel rows" 1.0)))
        (if (> numy 1)(setq sy
            (drwdr2 nil "Spacing between panel columns"  1.0)))
        (setq offin   (if (zerop
                (setq tp (drwdr2 nil "Offset distance for raised panel"
0)))
                    nil tp))
    ))
    (d_we2 spt offbase offside offtop x y numx numy sx sy offin trim)
(princ)
)
(prompt "\nCommand: DWELEV \n")


(c:dwelev)                ;call up program with this command
                          ;or use within another procedure with actual sizes:

;(progn (d_we3)
; ------ spt offbase offside offtop x   y   numx numy sx  sy  offin trim ------
;   (d_we2 spt 10.0   4.0   6.0 40.0 84.0 2    4 2.0 3.0  1.5 4.0)
;)
```

Let's create and use the dwelev.lsp routine.

1 Using Notepad, accurately enter this AutoLISP routine. Name it **dwelev.lsp**.

HINT: This file is available on the *Applying AutoCAD Diskette,* available from Glencoe/McGraw-Hill.

2 Begin a new drawing (do not save the current drawing), and pick **Use a Wizard**, **Quick Setup**, and **OK**.

3 Prepare an architectural working environment based on a scale of $\frac{1}{4}'' = 1'$ and a C-size (24" × 18") sheet

HINT: Produce a drawing area of 22" × 16" and multiply these numbers by 4 to determine the width and length.

4 **ZOOM All**, set snap at **2'**, and turn off the grid.

5 Load the **dwelev.lsp** file.

Loading the file automatically enters the new DWELEV command created by the program.

NOTE: If the routine does not appear to load, compare your program code to the dwelev.lsp code printed in this unit. They must be identical.

The program is mostly self-explanatory because it employs easy-to-understand prompts.

7 Use the following as a guide as you enter your responses.

Lower left corner of door/window: *(Pick a point at any location.)*
Base angle of door/window <0.0>: *(Press* **ENTER***)*
Width of door/window <3'>: **4'**
Height of door/window <6'8">: *(Press* **ENTER.***)*
Trim width <0">: **1.5"**
Number of panel rows <none>: **4**
Number of panel columns <1>: **2**
Bottom rail width <10">: **8"**
Top rail width <6">: *(Press* **ENTER.***)*
Side rail width <4">: **5"**
Spacing between panel rows <1">: **2"**
Spacing between panel columns <1">: **3"**
Offset distance for raised panel <0">: **1"**

8 Zoom in on the door and examine it.

The door should look identical to the one in the following illustration.

You can also create windows using dwelev.lsp. Just specify 0 in reply to Number of panel rows and Number of panel columns. To create the casement style windows shown earlier, link two or more windows together (using COPY or ARRAY).

9 Create additional doors and windows by entering **DWELEV** at the Command prompt, save your work, and exit AutoCAD.

The dwelev.lsp routine is intentionally basic. This makes it easier for you to understand its operation. If you are an accomplished programmer, you may choose to embellish the routine by including doorknobs, window molding, and other details normally included in door and window symbology.

Questions

1. Explain the purpose of the AutoLISP command function.

_____

_____

2. For what reason are semicolons used in AutoLISP routines?

_____

_____

3. Briefly explain the purpose of the following functions:

getvar _____

_____

getpoint _____

_____

getdist _____

_____

4. Explain the benefits of applying parametric programming techniques.

_____

_____

◼ *Challenge Your Thinking: Questions for Discussion*

1. Explain what makes an application a good choice for parametric programming. Give examples of parametric applications you think could or should be accomplished within AutoCAD using AutoLISP.

2. In this unit, you created a screen menu item that automatically places a border on a 36″ × 24″ sheet. Explain how you could revise this menu item to create a parametric AutoLISP program that allows you to draw a border for any size sheet of paper.

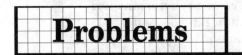
Problems

1. Write the AutoLISP code you would use to create a 4- by 6-unit rectangle on the screen with corners at points pt1, pt2, pt3, and pt4. Enter the code and confirm that it works as expected.

2. On your own, learn other AutoLISP functions, such as getreal and getdist, and develop them into new AutoLISP routines.

3. Using dwelev.lsp and AutoCAD commands, create an architectural elevation drawing similar to the one here. This elevation, minus doors and windows, is available in the elev.dwg file contained on the optional *Applying AutoCAD Diskette.*

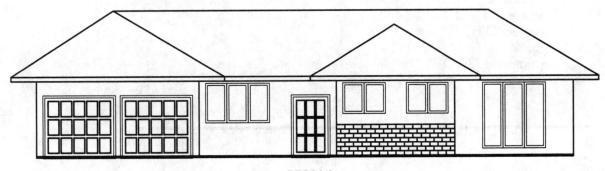

PRB64-3

4. Try to create the parametric program discussed in the second "Challenge Your Thinking" question. Refer to the dwelev.lsp program and the documentation provided with AutoCAD if you need help. Do not be discouraged if your program does not work the first time you try to run it. Most programs, even those written by professionals, have to be "debugged" before they will run properly.

5. The following AutoLISP program, titled v1, opens two viewports with horizontal orientation; makes the upper one the top view and the lower one the front view; and zooms both to 80% of the current view. The second program, v2, resets to a single viewport with an 80% zoom. Write a program named v3 which will create three viewports, with top, front, and isometric orientations. Write a second program, named v4, which creates four viewports with top, front, right-side, and isometric view orientations. Create a three-dimensional object to check your programs. They should generate viewports like the ones shown on the next page. The v1.lsp and v2.lsp files are available on the optional *Applying AutoCAD Diskette*.

```
(setvar "CMDECHO" 0)
(defun c:V1 ()
     (command "VPORTS" "2" "H")
     (setq cvpt (getvar "CVPORT"))
     (IF (= (getvar "CVPORT") 2) (setvar "CVPORT" 3) (setvar "CVPORT" 2))
     (command "VPOINT" "0,-1,0")
     (command "ZOOM" ".8x")
     (setvar "CVPORT" cvpt)
     (command "PLAN" "W")
     (command "ZOOM" ".8x")
     (princ)
)
```

Program v1

```
(setvar "CMDECHO" 0)
(defun c:V2 ()
        (command "VPORTS" "SI")
        (command "ZOOM" "E")
        (command "ZOOM" ".8x")
        (princ)

)
```

Program v2

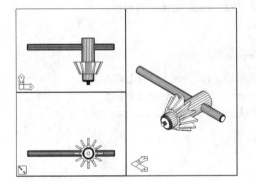

Expected result
after invoking v3

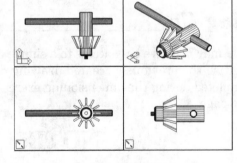

Expected result
after invoking v4

Problem 5 courtesy of Gary J. Hordemann, Gonzaga University

Unit 66 Digitizing Hard-Copy Drawings

■ OBJECTIVE:

To input a hard-copy drawing into AutoCAD using the TABLET command

The intent of this unit is to step through the process of digitizing. Note that you must have a digitizing tablet connected to your CAD system in order to complete this unit.

There will be times, especially in a business environment, when you'll wish your hand-completed drawings were stored in AutoCAD. Suppose your firm has recently implemented CAD. All of your previous drawings were completed by hand, and you need to revise one or more of them. As you know, it's very time-consuming to redraw them by hand.

Fortunately, most CAD systems, including AutoCAD, offer a method of transferring those drawings onto disk. It is not always practical to digitize drawings, but it is often faster than starting the drawings from scratch.

Setting Up

Since you may not have easy access to a simple drawing not yet in AutoCAD, let's digitize the previously created drawing dimen2.dwg. This drawing was completed during the dimensioning exercise (Unit 25) and is shown on the next page.

NOTE:

If you do not have a hard copy of dimen2.dwg, plot the drawing now or make a photocopy of the drawing on the next page.

1 Start AutoCAD and open **dimen2.dwg**.

2 Using **Save As...**, create a new drawing file named **digit.dwg**.

3 Erase the entire drawing so that the screen is blank.

NOTE:

Be sure the snap resolution is set at 6″ and is turned on. Also be sure to display the entire drawing area by entering ZOOM All.

4 Make **Objects** the current layer, if it is not already.

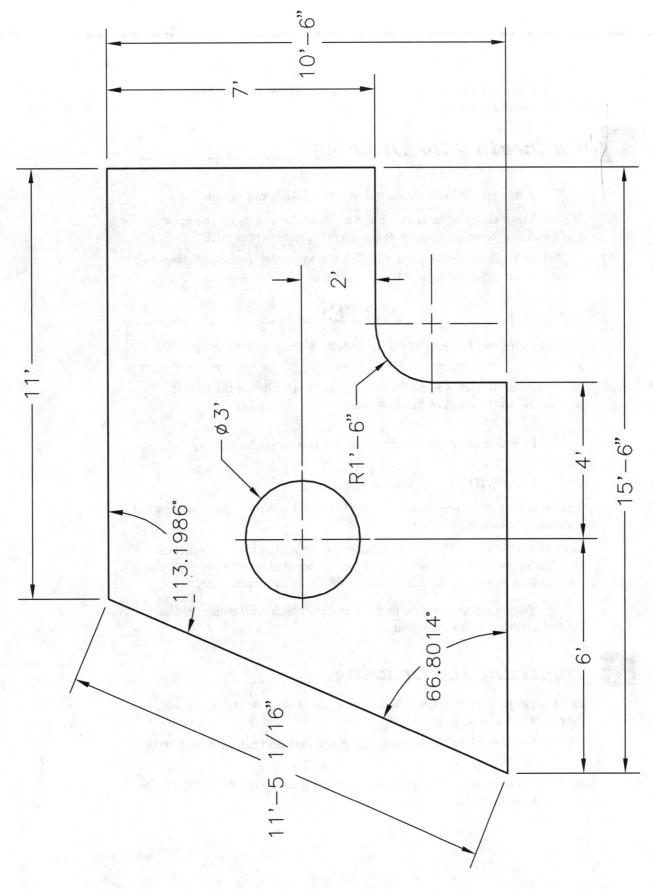

5 Fasten the hard copy of dimen2.dwg onto the center of the digitizing tablet.

Calibrating the Drawing _____

1 Enter the **TABLET** command and the **CALibrate** option.

We need to identify at least two known (absolute) points on the drawing. Let's call the lower left corner of the object absolute point 6',6'.

2 In reply to Digitize point #1, pick the lower left corner of the object (be precise) and enter the coordinates **6',6'**.

NOTE: _____

Crosshairs will not appear on the screen when you pick the point.

Now we need to pick a second known point. Let's choose the corner located 10' to the right of the first point.

3 Pick this point and enter **16',6'** for the coordinates.

4 Press **ENTER** to end the calibration.

You have just calibrated the drawing. AutoCAD can now size the drawing according to this calibration.

Note that the word TABLET now appears in the status bar. This means that tablet mode is on. Tablet mode can be toggled on and off with the F4 function key or by double-clicking TABLET in the status bar.

5 Toggle tablet mode on and off and notice the difference in the position of the crosshairs.

Digitizing the Drawing _____

Let's begin to trace (digitize) the drawing. Let's start at the lower left corner of the object, at point 6',6'.

1 Enter the **LINE** command, and make certain that snap and ortho are on.

2 Turn on the tablet mode and digitize (pick) point **6',6'**. (Read the following hint.)

Whenever you digitize points from the drawing, tablet mode must be turned *on*. Whenever you select icons or pull-down menus, tablet mode must be turned *off*.

HINT: If you encounter difficulty in reaching absolute point 6',6' or any other portion of the screen with the pointing device, it may be that the fixed screen pointing area is too small. If this is the case, enter the TABLET command and the CFG option, specify 0 tablet menus, and enlarge the screen pointing area.

3 With tablet mode on, digitize the next corner point, working counterclockwise.

This completes the first line segment.

4 When digitizing the third point, ignore the fillet and pick the approximate location of the corner. With snap on, your selection of the corner will be accurate. You will insert the fillet later.

5 Continue around the object until you close the polygon, and save your work.

6 Using the **FILLET** command, specify a fillet radius of **1'6"** and place the fillet at the proper location. Remember to turn tablet mode off if you want to use the icons or pull-down menus.

7 Using the **CIRCLE** command, place the hole. (See the following hint.)

HINT: When picking the hole's center and diameter, be sure that tablet mode is on and digitize the center and radius from the hard-copy drawing.

8 Make **DIM** the current layer.

9 Using AutoCAD's dimensioning commands, fully dimension the object with tablet mode off.

You're finished.

10 Save your work and exit AutoCAD.

Questions

1. What command and command option are used to calibrate a drawing to be digitized?

_____

2. Why is the calibration process necessary?

_____

3. Briefly explain the process of calibrating a drawing to be digitized.

_____

_____

_____

_____

4. Why is the snap resolution important when digitizing?

_____

_____

5. Explain when tablet mode should be turned on and when it should be turned off.

_____

_____

 ### *Challenge Your Thinking: Questions for Discussion*

1. If you have a mouse, explain whether you can digitize a drawing using only a conventional mouse.

2. Digitizing can make the task of entering drawings into AutoCAD a much simpler process. However, the digitizing process is not perfect. Discuss any problems you encountered as you digitized dimen2.dwg in this unit. What other problems do you think might occur? List the problems and describe ways to minimize each.

Problems

1. Obtain two hand-completed drawings and digitize each. Save the drawing files as prb65-1A.dwg and prb65-1B.dwg.

NOTE:

It is possible to digitize drawings that are larger than the digitizing tablet. When you complete one part of the drawing, move a new part of the drawing into the active area on the tablet and recalibrate.

2. The following is a drawing of the Great Lakes, showing the net flow for each lake. Reproduce this figure by digitizing it. Use the SKETCH command with SKPOLY set to 1. Be sure to close the polylines in order to make hatching easy. The hatch pattern is FLEX.

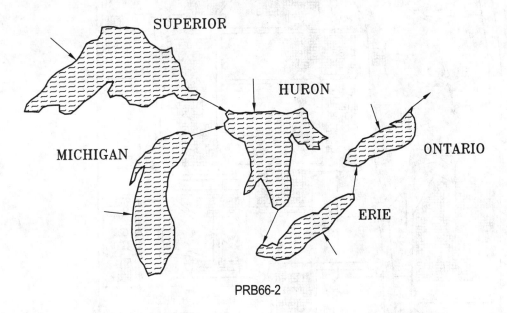

PRB66-2

3. Use an enlarged copy of the map on the next page to practice digitizing. Use the PLINE command to digitize the streets. Use the LINE command to digitize the heavy dashed lines, which represent buried conduit. Use the SKETCH command to digitize the lake and the river. Determine the compass bearing of one of the streets by changing to surveyor's units and digitizing the north arrow. Use the DONUT command to digitize the small dots (street lights).

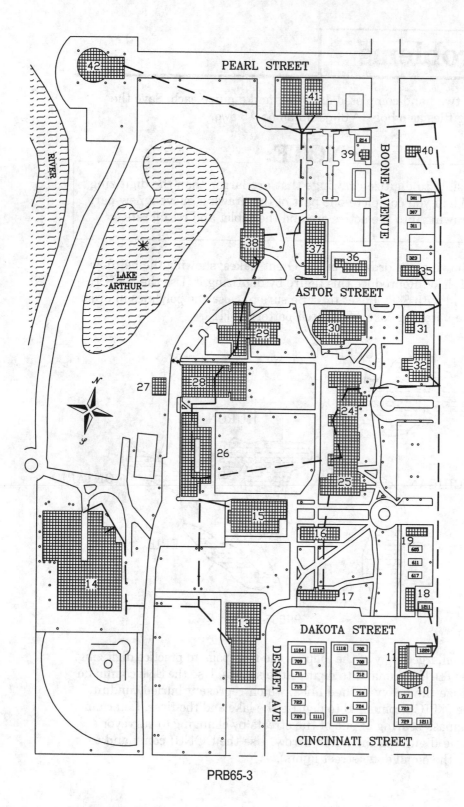

PEARL STREET

RIVER

LAKE ARTHUR

BOONE AVENUE

ASTOR STREET

N S

DAKOTA STREET

DESMET AVE

CINCINNATI STREET

PRB65-3

Problems 2 and 3 courtesy of Gary J. Hordemann, Gonzaga University

752

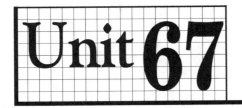

Unit 67 — Importing and Exporting Files

■ OBJECTIVE:

**To practice importing and exporting DXF, PostScript (EPS),
3D Studio (3DS), and DWF files**

Standard file formats are important for moving graphics from one system
to another. This unit steps through the translation process using several
industry standard formats.

AutoCAD's drawing interchange format (DXF) is used frequently to
exchange drawings among users of different CAD systems. IGES and SAT
are also popular file formats for design data exchange. PostScript, with its
EPS file format, is a popular language used in desktop publishing,
illustration, and presentation programs. 3D Studio, with its 3DS file
format, is a popular presentation and animation product. AutoCAD's
Drawing Web Format (DWF) allows you to publish drawings on the World
Wide Web.

■ *Exporting DXF Files*

The DXF file format is a de facto standard for translating files from one
CAD system, such as AutoCAD, to another, such as CADKEY. Autodesk,
Inc., makers of AutoCAD, created the DXF format, also referred to as a
drawing interchange file format. You can translate DXF files to other DXF-
compatible CAD systems or to programs for specialized applications. For
example, certain manufacturing software uses DXF files to generate tool
path code for computer numerical control (CNC) mills and lathes.

You can easily generate a DXF file from an existing AutoCAD drawing.

1 Start AutoCAD and open **dimen.dwg**.

2 Select **Export...** from the **File** pull-down menu.

This displays the Export Data dialog box.

3 Display the **Save as type** drop-down box and review the list of
options.

Note that AutoCAD offers three variations of DXF.

4 Select **AutoCAD R14 DXF (*.dxf)**.

5 Pick the **Options...** button.

This displays the Export Options dialog box, as shown in the following illustration.

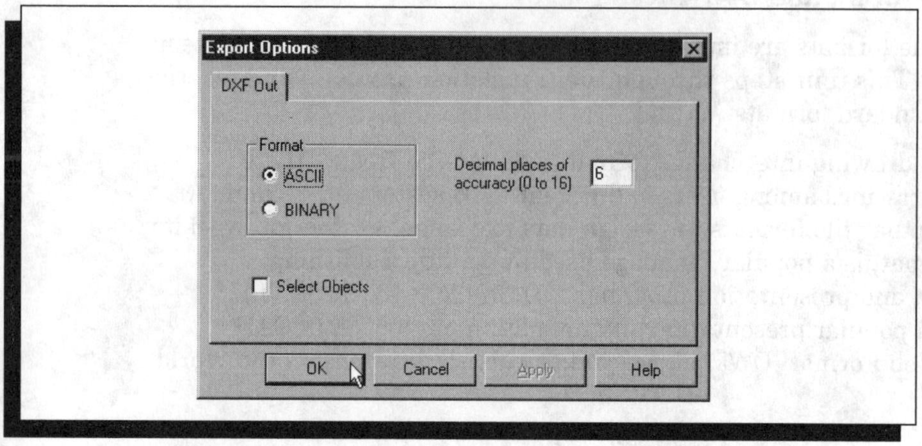

AutoCAD permits you to produce either an ASCII or binary file. The Select Objects check box lets you select one or more objects to translate instead of the entire drawing. AutoCAD also gives you the option of entering decimal places of accuracy. A high number produces a more accurate file, although the file size increases.

6 Pick the **OK** button to accept the default settings.

7 In the File name box, accept the suggested **dimen.dxf** name and pick the **Save** button.

The translation can take a while for complex drawings. When the Command prompt returns, the translation is complete.

8 Minimize AutoCAD.

9 Find the file named **dimen.dxf**.

How does the file size of the DXF file compare to its DWG counterpart?

10 Using WordPad, review the contents of the ASCII DXF file.

The beginning of the DXF text file should look similar to the following.

```
   0
SECTION
   2
HEADER
   9
$ACADVER
   1
AC1009
   9
$INSBASE
  10
0.0
  20
0.0
  30
0.0
   9
$EXTMIN
  10
```

11 Exit WordPad and maximize AutoCAD.

Binary DXF Files

Binary DXF files contain all of the information in an ASCII DXF file, but they are much more compact. Their file size is approximately 25 percent smaller, and they can be written and read by AutoCAD much faster.

1 Select **Export...** from the **File** pull-down menu.

2 Select **AutoCAD R14 DXF (*.dxf)** from the Save as type drop-down box.

3 Pick the **Options...** button, pick the **BINARY** radio button, and pick **OK**.

4 In the File name box, change the name to **dim.dxf** and pick the **Save** button.

5 Compare the size of the ASCII DXF file with that of the binary DXF file.

The binary DXF file should be at least 25 percent smaller. Also, binary DXF files preserve all of the floating-point accuracy in the drawing database. ASCII DXF files, instead, increase in size as you increase the decimal places of accuracy.

6 Return to AutoCAD.

Importing DXF Files

Both ASCII and binary DXF files can be imported into AutoCAD.

1 Begin a new drawing from scratch.

2 Pick the **Open** icon from the docked Standard toolbar.

3 From the Files of type drop-down box, select **DXF (*.dxf)**.

4 Select a DXF file currently on disk. If one is not available, select one of the two DXF files you just created and pick the **Open** button.

NOTE:

The optional *Applying AutoCAD Diskette* includes example DXF files.

The DXF file generates on the screen. At this point, you could save the drawing as a drawing (DWG) file, but don't.

IGES

IGES stands for "Initial Graphics Exchange Specification." IGES is an industry standard approved by the American National Standards Institute (ANSI) for interchange of graphic files between small- and large-scale CAD systems.

Translating files from one CAD system to another using IGES is useful. However, each CAD system is unique. Consequently, certain characteristics, such as layers, blocks, linetypes, colors, text, and dimensions, are potential problem areas as a result of the translation.

For example, some CAD systems use numbers for layer names and do not accept names such as Object or Dimension. If an AutoCAD drawing file is translated to a system using layer numbers, all of the AutoCAD layer names are changed to numbers. These types of problems are also present when you translate files using the DXF file format.

Translator software for importing and exporting IGES is available as an option from Autodesk, Inc.

ACIS

AutoCAD permits you to export objects representing NURBS surfaces, regions, and solids to an ACIS file in ASCII (SAT) format. ACIS is the 3D geometry engine that AutoCAD uses to create solid models. You can import SAT files into other ACIS-based products. Importing and exporting SAT files is covered in Unit 53.

PostScript Files

AutoCAD allows you to output a PostScript (EPS) file.

1. Open the drawing named **lamp.dwg** and fill the screen with the lamp.

2. Select **Export...** from the **File** pull-down menu.

3. Display the **Save as type** drop-down box and select **Encapsulated PS (*.eps)**.

4. Pick the **Options...** button.

5. Pick the **Help** button and read about the options available in the dialog box.

6. After exiting help, pick the **OK** button to accept all of the default settings.

7. Pick the **Save** button to accept the lamp.eps name.

AutoCAD creates a file named lamp.eps.

8. Check to see if the file exists.

9 Begin a new drawing from scratch. Do not save any changes.

10 Select **Encapsulated PostScript...** from the **Insert** pull-down menu.

11 Unless you have another EPS file, select **lamp.eps** and pick **Open**.

12 Press **ENTER** to accept the 0,0,0 insertion point.

13 In reply to Scale factor, drag the crosshairs to the right until the box fills most of the screen and pick a point.

The PostScript image appears.

14 View it from different angles in space.

Notice that it is now a flat image instead of a 3D model. That's because PostScript does not recognize 3D objects.

AutoCAD offers the PSDRAG command to control the appearance of a PostScript image as you are dragging it into position. The PSQUALITY system variable is available if you want to adjust the rendering quality of PostScript images. The PSFILL command allows you to fill 3D polyline outlines using a PostScript fill pattern.

3D Studio Files

AutoCAD creates a 3D Studio (3DS) file from selected objects.

1 Open the **shaft.dwg** drawing file of the solid model. Do not save the current drawing.

2 Select **Export...** from the **File** pull-down menu.

3 Display the **Save as type** drop-down box and select **3D Studio (*.3ds)**.

4 Pick the **Save** button to accept the shaft.3ds name.

5 Select the shaft model and press **ENTER**.

AutoCAD displays the 3D Studio File Export Options dialog box, as shown in the following illustration.

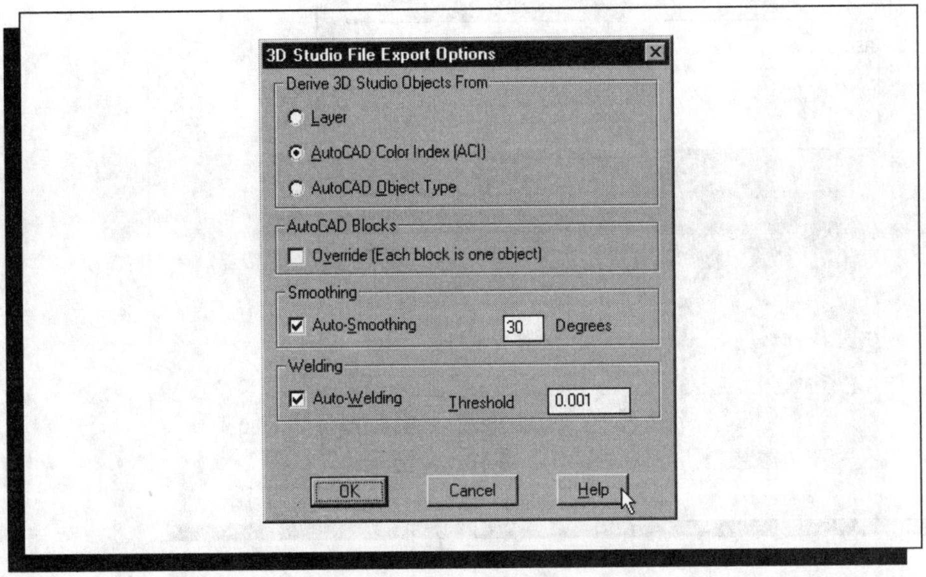

6 Review the options and then pick the **Help** button to read about them.

7 Pick **OK** to accept the default settings.

AutoCAD creates a file named shaft.3ds. You can read this file into the 3D Studio software available from Autodesk's Kinetix multimedia division.

1 Begin a new drawing from scratch. Do not save any changes to shaft.dwg.

2 Select **3D Studio...** from the **Insert** pull-down menu.

3 Unless you have another 3D Studio file, select **shaft.3ds**, and pick **Open**.

NOTE:

The optional *Applying AutoCAD Diskette* includes example 3DS files.

This displays the 3D Studio File Import Options dialog box, as shown in the following illustration.

4 Review the options in the dialog box and pick the **Help** button to read about them.

5 After exiting help, pick the **Add All** button to select all available objects, and pick the **OK** button.

AutoCAD imports the 3D Studio file. At this point, you could save this as a DWG file, but don't.

6 View the file from any angle in space and shade the drawing.

DWF Files

You can export a drawing to AutoCAD's Drawing Web Format (DWF). A DWF file is a highly compressed 2D vector file that you can use to publish AutoCAD drawings on the World Wide Web. Using a Web browser, such as Netscape Navigator or Microsoft Internet Explorer, and the WHIP! plug-in provided by Autodesk, you can view DWF files directly or while they are embedded in a Web (HTML) page. The WHIP! plug-in provides dynamic pan and zoom capabilities.

1 Open the drawing file named **pulley2.dwg**. Do not save the current drawing.

2 Switch to model space.

HINT:

Double-click PAPER on the status bar.

3 Pick **Export...** from the **File** pull-down menu.

4 From the Save as type drop-down box, select **Drawing Web Format (*.dwf)**.

5 Pick the **Options** button and pick the **Help** button to read about the options in this dialog box.

6 After exiting help, pick the **OK** button to accept the default settings in the DWF Export Options dialog box.

7 Pick the **Save** button to accept the pulley2.dwf name.

AutoCAD creates a file named pulley2.dwf.

The icon named Launch Browser, located in the docked Standard toolbar, enters the BROWSER command. Ths launches Netscape Navigator or Microsoft Internet Explorer.

The AutoCAD Internet Utilities let you open, insert, and save drawings from anywhere on the Web. You can also embed Universal Resource Locations (URLs) into AutoCAD drawings to provide hyperlinks to other Web pages.

8 Exit AutoCAD. Do not save your changes.

Questions

1. What is a DXF file, and what is its purpose?

   _____

   _____

2. List below the 12 file formats that AutoCAD can export.

   _____

   _____

3. Explain the advantages of using binary DXF files instead of ASCII DXF files.

   _____

   _____

   _____

4. EPS stands for what?

   _____

   _____

5. Can you store 3D data in a 3DS file?

   _____

   _____

6. Briefly describe the purpose of AutoCAD's Drawing Web Format (DWF).

   _____

   _____

7. What is IGES, and what is the purpose of an IGES file?

   _____

   _____

   _____

8. What are the potential problem areas associated with translating DXF and IGES files from one CAD system to another?

   _____

   _____

   _____

■ Challenge Your Thinking: Questions for Discussion

1. Experiment further with importing and exporting DXF files. What happens if you try to import a DXF file into a drawing that already has objects in it?

2. Experiment further with importing and exporting 3DS files. Try each of the options under Save to Layers in the 3D Studio File Import Options dialog box. What, if any, information is lost when you translate an AutoCAD file to a 3DS format and then translate it back to an AutoCAD (DWG) format?

3. Discuss applications for AutoCAD's Drawing Web Format (DWF). In a company with operations inside and outside the United States, how might they take advantage of DWF?

Problems

1. With AutoCAD drawings, create ASCII DXF files using the capabilities provided by AutoCAD. After the translations are complete, review the contents of the DXF files.

2. If you have access to another CAD system, load the DXF files from Problem 1 into that system. (Note: The CAD system must be able to import a DXF file. If you are successful, review characteristics of the drawings, such as layers, colors, linetypes, blocks, text, and dimensions. Note the differences between the CAD systems.

3. Create or obtain DXF files from another CAD system and import them into AutoCAD. Note the differences between AutoCAD and the systems used to create the files you imported into AutoCAD. If available, refer to the optional *Applying AutoCAD Diskette* for example files.

4. Create EPS and 3DS files from any of the drawing files you have created in the past.

5. Obtain or create EPS and 3DS files and import them into AutoCAD. If available, refer to the optional *Applying AutoCAD Diskette* for example files.

6. Download the WHIP! plug-in from Autodesk's Web site. The Universal Resource Locator (URL) is http://www.autodesk.com.

Unit 68 Raster Image Files

◼ OBJECTIVE:

To import, display, and edit raster images

AutoCAD allows you to insert a raster file (also referred to as a bitmapped image), as an 8-bit grayscale, 8-bit color, or 24-bit color object. You can combine them in a variety of formats with AutoCAD drawings. Also, you can display more than one image in any viewport, and you are not limited to any number or size of images in a single file. Raster files reside in AutoCAD as unique object types.

◼ *Attaching an Image File*

1 Start AutoCAD and begin a new drawing from scratch.

2 Display the **Reference** toolbar.

3 Pick the **Image Attach** icon from the Reference toolbar.

Reference

This displays the Attach Image File dialog box.

4 Display the **Files of type** drop-down box and review the list of raster file types supported by AutoCAD.

5 Select the **All image files** default.

6 Find and open AutoCAD's **Sample** folder.

As you can see, AutoCAD's Sample folder contains six raster image files in JPG and TIF formats.

7 Pick the **Details** button located in the upper area of the screen to display a detailed listing of the files.

Notice the size of each file.

8 Single-click the first file in the list.

A preview of the raster image appears in the dialog box.

HINT:

If a preview does not appear, pick the Show preview button below the Preview box. When the preview appears, notice the Hide Preview below it.

9 Preview each of the remaining raster files.

10 Select the file named **downtown.jpg** and pick the **Open** button.

The Attach Image dialog box appears, as shown in the following illustration.

Pick the **Details** button to display more information.

11 Pick the **Details** button to display more information.

12 Review the options in the dialog box and then pick the **Help** button to read about them.

13 After exiting help, pick the **OK** button to accept the default settings.

Notice that a small box is attached to the crosshairs.

14 In reply to Insertion point, pick a point in the lower left area of the screen.

15 In reply to Scale factor, drag the box up and to the right, filling about half of the screen, and pick a point.

As you can see, the raster image is an aerial photograph of a city's downtown area.

Mixing Raster and Vector Objects

With AutoCAD, you can mix raster and vector objects. First, let's resize the raster image.

1 Using AutoCAD's realtime zoom and pan capabilities, review the detail in the image.

Fill the screen with the lower right quarter of the image.

Create a new layer named **City**, assign the color yellow to it, and make it the current layer.

Using the **PLINE** command, trace the edges of a couple of city blocks.

As you can see, you can add vector objects into a file that contains raster objects.

Freeze layer 0.

AutoCAD displays only the vector information. Using this method of mixing raster and vector objects, you can digitize photographs and other raster images.

Thaw layer **0** and make it the current layer. Freeze layer City.

Save your work in a file named **raster.dwg**.

Editing the Image

You can copy, move, rotate, resize, and clip imported raster images.

ZOOM All and pick the border of the image.

AutoCAD hatches the image to show that you have selected it.

Pick one of the four grips.

Drag to increase the size of the image, and press **ESC** twice.

Reference

Pick the **Image Clip** icon from the Reference toolbar.

Select the outline of the image.

In reply to ON/OFF/Delete/<New Boundary>, press **ENTER** to accept the default.

Form a rectangle of any size by picking two points inside the image.

AutoCAD crops the image.

Reference

8 Pick the **Image Frame** icon from the Reference toolbar.

9 Enter **Off**.

This removes the border line.

10 Make the border reappear.

11 Increase the size of the image using its grips, and cancel twice.

Adjusting the Image Quality

You can also adjust the image color, contrast, brightness, and transparency of raster images.

1 Using the steps outlined earlier, import into the current drawing the file named **r300-20.jpg** located in the Sample folder. Use all of the default settings and position the image at any size and location. It's okay if it covers part of the downtown.jpg image.

HINT:

When the Attach Image dialog box appears, pick the Browse... button.

Reference

2 After the r300-20.jpg image appears, pick the **Image Quality** button from the Reference toolbar.

AutoCAD presents High/Draft options on the Command line. These quality settings do not affect the actual image; they only affect display performance. High-quality images take longer to display.

3 Press **ESC** to cancel.

Reference

4 Pick the **Image Transparency** icon from the Reference toolbar.

5 Pick one of the two images and press **ENTER**.

AutoCAD presents ON/OFF on the Command line. Several raster file formats allow images with transparent pixels. If you turn on image transparency, AutoCAD recognizes the transparent pixels so that graphics on the screen show through those pixels.

6 Enter **On**.

Reference

7 Pick the **Image Adjust** icon from the Reference toolbar and pick the first (black and white) image.

This displays the Image Adjust dialog box, as shown in the following illustration.

8 Adjust the brightness, contrast, and fade controls and pick the **OK** button.

Reference

9 Pick the **Image Adjust** icon again (or press the space bar or **ENTER**) and pick the second color image.

10 Adjust the brightness, contrast, and fade controls and pick the **OK** button.

Image Dialog Box

The Image dialog box allows you to unload, reload, attach, and detach raster image files.

Reference

1 Pick the **Image** icon from the Reference toolbar.

This displays the Image dialog box, as shown in the following illustration.

Image name	Status	Size	Type	Date	Saved path
DOWNTOWN	Loaded	847kB	JFIF	05/06/97 ...	C:\Program Files

Attach...
Detach
Reload
Unload
Details...

Image Found At

Browse...   Save Path

OK   Cancel   Help

Notice that several buttons are not available.

2 Click **DOWNTOWN**.

These buttons become available.

3 Pick the **Unload** button.

Unload appears under the Status heading. Unload removes the image data
from working memory without erasing the image object from the drawing.
When the image is no longer needed for editing, Unload is recommended
because it increases performance by reducing the memory requirement
for AutoCAD without detaching the raster file.

4 Pick the **OK** button.

The outline of the image remains.

Reference

5 Redisplay the image dialog box.

6 Click **DOWNTOWN**, pick the **Reload** command, and pick **OK**.

This reloads the image.

Reference

7 Redisplay the **Image** dialog box and click **DOWNTOWN**.

Picking the Attach button is equivalent to picking the Image Attach icon from the Reference toolbar. The Detach button removes the selected image definitions from the drawing database and erases all the associated image objects from the drawing and display.

⑧ Pick the **Details...** button.

This displays details on the image, as shown in the following illustration.

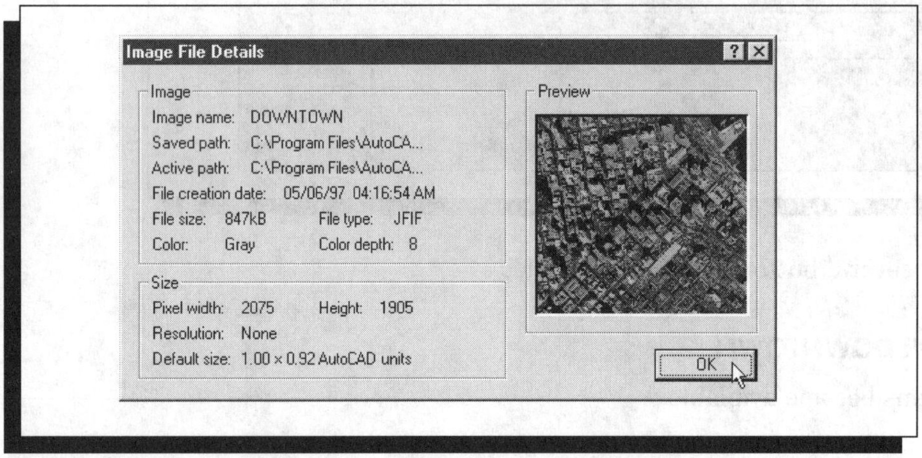

⑨ Pick the **OK** button.

⑩ In the upper left area of the **Image** dialog box, pick the **Tree view** button.

This illustrates the images that are attached to the current drawing. You can also press the F4 function key to display this information.

⑪ Pick the **OK** button.

⑫ Minimize AutoCAD and review the size of the **raster.dwg** file.

The file is very small (probably around 30Kb) compared to the 848Kb for downtown.jpg and 1,013Kb for r300-20.jpg. Raster.dwg remains small because AutoCAD attaches (externally references) the raster files instead of importing their contents.

⑬ Maximize AutoCAD and save your work.

Exporting Raster Image Files _____

AutoCAD can export raster files in Microsoft Windows (BMP), TrueVision (TGA), Targa, PCX, and Tagged image file format (TIF).

1 Open the file named **hatch.dwg**.

2 Select **Export...** from the **File** pull-down menu.

3 From the **Save as type** drop-down box, select **Bitmap (*.bmp)**.

4 Pick the **Save** button to accept the hatch.bmp name.

5 Select the entire drawing and press **ENTER**.

AutoCAD creates a raster image file named hatch.bmp.

6 Close the Reference toolbar, do not save your work, and exit AutoCAD.

Questions

1. List six raster image file formats that AutoCAD can import.

   _____

   _____

2. How do you resize an attached raster image in AutoCAD?

   _____

   _____

3. What is the purpose of the Unload and Reload buttons in the Image dialog box?

   _____

   _____

4. When would you choose to use the Detach button in the Image dialog box?

   _____

   _____

5. Why do AutoCAD drawing files with "imported" raster files remain small?

   _____

   _____

6. What raster image file types can AutoCAD export?

   _____

   _____

■ *Challenge Your Thinking: Questions for Discussion*

1. Discuss potential applications of mixing raster and vector objects. When might this be useful to a company or government agency?

2. An Image Quality icon is available on the Reference toolbar. When might you need to use it? How does it affect the quality of a printed image?

Problems

1. Begin a new drawing from scratch. Attach the hatch.bmp file you created in this unit to the new drawing file. Insert the image to fill most of the screen. Save your work in a file named prb68-1.dwg.

2. Begin a new drawing from scratch. Attach the file named jblake.jpg located in AutoCAD Sample folder. Insert the image to fill most of the screen. Increase the brightness of the image. Crop the image to remove most of the water, sky, and trees. On a new layer named Roof, trace part of the cabin's roof. Zoom and pan as necessary, and save your work in a file named prb68-2.dwg.

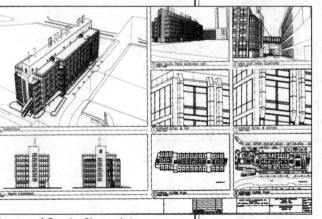

Courtesy of Goody, Clancy, & Associates, Architects, Inc.

AUTOCAD® AT WORK

MIT Biology Building

The new headquarters for the Massachusetts Institute of Technology biology department contains 250,000 sq. ft. of research laboratories and associated teaching and administrative facilities. In developing the design for this dramatic structure, the architectural firm, Goody, Clancy, & Associates, chose to take full advantage of the 3D capabilities of AutoCAD—working in a 3D format throughout the architectural process, rather than using traditional 2D design. As a result, the designers and consulting engineers were able to gain a reliable, practical understanding of every part of the building as it was developed. They were able to make design changes quickly and construct a variety of views of not only the broad aesthetic effects but also the relationships between building materials. The result is an exceptional new laboratory design and mechanical system. MIT uses the architects' AutoCAD files for facility management, so the drawings continue to have a useful life although construction has been completed.

Based on a story provided by Autodesk, Inc.

■ **OBJECTIVE:**

To exercise AutoCAD's Object Linking and Embedding (OLE) capabilities

As you know, you can copy an AutoCAD drawing and paste it into another Windows program. Using OLE, you can maintain a link between AutoCAD and the other program, if the program supports OLE. When you edit the AutoCAD drawing, the change occurs automatically in the second program. This can save time and ensure accuracy if you want to maintain the same version of a drawing in another document.

You can also link information from another program to AutoCAD. For example, you can paste a table from Microsoft Word® or a spreadsheet from Microsoft Excel™ into AutoCAD. When you change the table in Word, or the spreadsheet in Excel, AutoCAD updates automatically to reflect the change.

Linking AutoCAD to Another Document

When you link an AutoCAD view to a document or file in another program, such as Microsoft WordPad, the AutoCAD drawing becomes the server document. The document in WordPad becomes the client document.

1 Start AutoCAD and open the drawing named **region.dwg**.

2 Using **Save As...**, create a new file named **ole.dwg**.

3 Pick **Copy Link** from the **Edit** pull-down menu.

This enters the COPYLINK command.

4 Exit AutoCAD and start Microsoft WordPad.

HINT:

From the Windows Start menu, pick Programs, Accessories, and WordPad.

5 In WordPad, select **Object...** from the **Insert** pull-down menu.

The Insert Object dialog box appears, as shown in the following illustration.

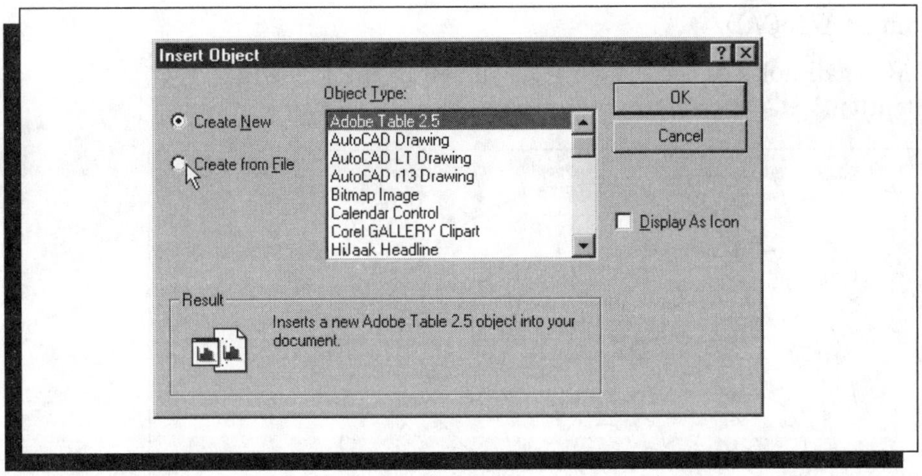

6 Pick the **Create from File** radio button and check the **Link** check box.

7 Pick the **Browse...** button and find and double-click the **ole.dwg** file.

8 Pick the **OK** button in the Insert Object dialog box.

This opens the file in both AutoCAD and WordPad.

9 Minimize AutoCAD.

10 In WordPad, pick **Save** or **Save As...** from the **File** pull-down menu and enter **link.doc** for the file name. Be sure to select the folder with your name.

11 Minimize WordPad.

Editing the Server Document _____

Let's change the drawing in AutoCAD—the server document.

1 Maximize AutoCAD.

2 Add 10 small holes to the drawing, as shown in the following illustration.

3 Save your work and minimize AutoCAD.

4 Maximize WordPad.

The change is reflected automatically in the WordPad file.

5 Pick **Links...** from the **Edit** pull-down menu.

The Links dialog box appears, as shown in the following illustration.

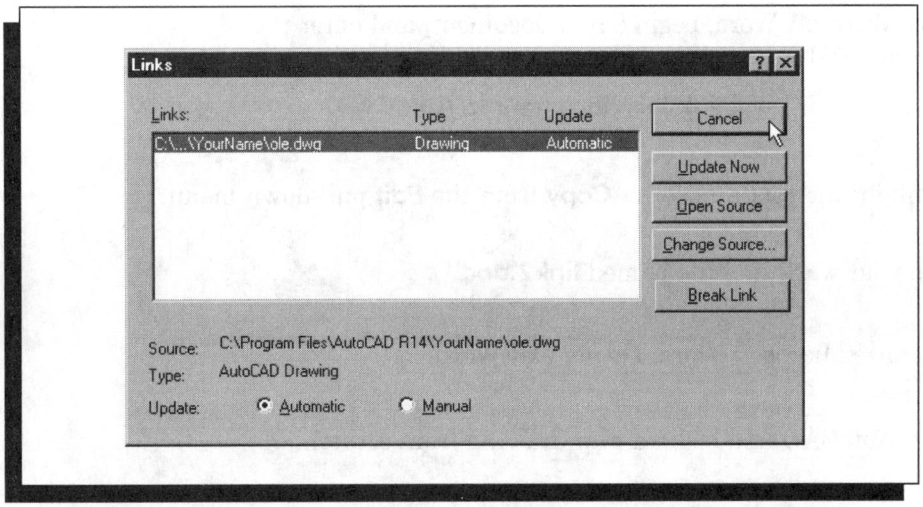

The buttons in this dialog box permit you to update a link, open or change a source, and break a link. As you can see at the bottom, Update is set to Automatic, which means that the link updates automatically, as it did. Also, the link is listed as Automatic in the Links list box. If Update were set to Manual, you would need to pick the Update Now button to update the link.

6 Pick the **Cancel** button.

7 Save your work and exit WordPad.

8 Maximize and exit AutoCAD.

Linking a Document to AutoCAD (Optional)

The beginning of this unit mentioned that you can link a document to AutoCAD. If you have a product such as Microsoft Word or Excel, try it. The steps are similar to the previous ones, except Word or Excel becomes the server and AutoCAD becomes the client.

_____ NOTE: _____

The following steps assume that you are using Microsoft Word. If you are using a different OLE-capable application, use these steps as a guide.

1 Start Microsoft Word, begin a new document, and enter the following text:

I will link this document to AutoCAD.

2 Highlight the text and select **Copy** from the **Edit** pull-down menu.

3 Save your work in a file named **link2.doc**.

4 Minimize Microsoft Word. *Do not* exit Word.

5 Start AutoCAD and begin a new drawing from scratch.

6 Select **Paste Special...** from AutoCAD's **Edit** pull-down menu.

The Paste Special dialog box appears. Notice that Source is the Word document you just copied to the Clipboard.

7 Pick the **Paste Link** radio button, and pick **OK**.

The contents of the Word document appear in the AutoCAD drawing area, bounded by a box.

8 Click and drag the box to reposition it in the drawing.

9 Save your work in a file named **link.dwg**.

10 Minimize AutoCAD and maximize Microsoft Word.

11 Edit the sentence to read:

This document is linked to AutoCAD.

12 Save your changes and minimize Microsoft Word.

13 Maximize AutoCAD.

Notice that the changes you made in Microsoft Word have updated automatically in AutoCAD. This is because both the client and the server are currently open.

14 Save the change and exit AutoCAD.

15 Maximize Word and exit it also.

Questions

1. What is the overall purpose of Object Linking and Embedding (OLE)?

   _____

   _____

2. When you link an AutoCAD drawing to a WordPad document, is AutoCAD the server or the client? Why?

   _____

   _____

3. If WordPad is a client, describe what you would do in WordPad to update, change, or cancel the link between the server and client.

   _____

   _____

4. Suppose you want to link a spreadsheet from Microsoft Excel to AutoCAD. Is the spreadsheet document the server or the client?

   _____

Challenge Your Thinking: Questions for Discussion

1. The procedures described in this unit focus on linking an AutoCAD drawing to files from other applications and vice versa. Find out how linking is different from embedding. Write a short report explaining the difference and describing at least one application in which each would be useful.

Problems

1. Link information from a program such as Microsoft Word or Excel to AutoCAD. Change the server document and then update the client document.

2. Use Microsoft WordPad to create a document named shop.doc. In the document, write two or three paragraphs describing the best way to set up a workshop. Link workshop.dwg (from Unit 33) to the shop.doc file to illustrate the paragraphs you have written.

3. Make changes to workshop.dwg in AutoCAD and then update the link you created in Problem 2.

4. In AutoCAD, save workshop.dwg with a new file name of traffic.dwg. Add arrows showing the projected traffic patterns around the tools in the workshop. Save the changes to traffic.dwg. In the shop.doc file (created in Problem 2), add a paragraph or two discussing the effect of foot traffic through the workshop on worker efficiency. Link traffic.dwg to the shop.doc file to illustrate the point. View and update all the links to the shop.doc file.

Unit 70 — Lights, Camera, . . .

■ **OBJECTIVE:**

To develop a slide show as a script file, to create a slide library, and to develop an image tile menu

AutoCAD permits you to develop a slide show by making slides and including them in a script file. Though it may sound complicated, it is really very simple.

The following is an example script file. It's nothing more than an ASCII text file with an SCR file extension. You can execute a script when you start AutoCAD or from within AutoCAD using the SCRIPT command.

```
UNITS 4 4 1 0 0 N
LIMITS 0,0 15',10'
ZOOM A
GRID 1'
SNAP ON
```

This script file is available on the optional *Applying AutoCAD Diskette.*

With earlier versions of AutoCAD, people used script files such as the one above to store drawing parameters and settings to expedite the setup process. The use of drawing template files has largely replaced this practice.

Script files can be used for other purposes, too, such as showing a continuous sequence of drawings in a sort of electronic flipchart. AutoCAD calls this a slide show.

This unit also focuses on image tile menus. You can use small pictures, called *image tiles,* to create menus and symbol libraries. Image tiles make it easier for you to use the blocks you've created. Instead of having to remember block names or referring to hard copy, you can access image tile menus (that represent your blocks) on the screen. This enables you to review and choose the blocks easily, and it eliminates the need to enter the INSERT or DDINSERT command and the block name repeatedly.

NOTE:

This unit requires knowledge of AutoCAD's acad.mnu menu file, which was presented in Units 59-62.

MSLIDE Command

The first step in creating a slide show is to create slides. You can create slides from existing drawings using the MSLIDE command. Let's create slides from a couple of drawings.

1. Begin a new drawing from scratch, and name it **show.dwg**. Save it in the folder with your name.

2. Pick the **Insert Block** icon from the docked Draw toolbar, and insert the drawing named **lamp.dwg**. Use **0,0,0** for the insertion point, and accept the remaining default settings. (You may need to zoom in on the drawing.)

Draw

3. View the drawing from any point in space and enter **HIDE**.

4. Enter the **MSLIDE** command, enter **lamp.sld** for the file name, and pick the **Save** button.

AutoCAD creates a slide of the lamp.

NOTE:

The new slide file cannot be edited. The drawing files used to create the slides remain untouched and can be edited.

5. Erase all objects on the screen so that it is blank.

HINT:

Use the All object selection option.

6. Insert **compos.dwg**.

7. Zoom and pan so that it fits on the screen, and enter **HIDE**.

8. With the **MSLIDE** command, create another slide. Name it **compos.sld**.

9. Erase the compos.dwg model so that the screen is blank.

VSLIDE Command

Let's apply the VSLIDE command to look at the first slide we created.

1. Enter **VSLIDE**, select **lamp.sld**, and pick **Open**.

The lamp.sld slide appears on the screen.

2 View the **compos.sld** slide.

3 To restore the original screen, enter **R** (for REDRAW).

4 If you'd like to make additional slides from other drawings, create them now.

Creating a Script File _____

Now let's create a script file (slide show). It's going to be a short one!

1 Minimize AutoCAD and start Notepad.

2 Enter the following using upper- or lowercase letters. Be sure to press **ENTER** after typing redraw.

vslide lamp
delay 1000
vslide compos
delay 1500
redraw

_____ NOTE: _____

If you've created additional slides, you can include them also.

3 Name the file **show.scr**, and be sure to store it in the directory that contains the slides you created.

4 Exit Notepad and maximize AutoCAD.

You have just created a simple slide show stored as a script file. The DELAY command tells AutoCAD to hold the slide on the screen for X number of milliseconds. Although computer clocks run at different speeds, 1000 milliseconds is approximately a one-second delay.

Showtime

Now let's try the slide show.

1. Select **Run Script...** from the **Tools** pull-down menu to enter the SCRIPT command, select **show.scr**, and pick **Open**.

AutoCAD displays each slide on the screen.

2. To repeat the slide show, enter the **RSCRIPT** command.

You can include this command at the end of a script file to repeat the slide show automatically.

NOTE:

The backspace key will interrupt a running script. This allows you to issue other AutoCAD commands. If you wish to return to the script, enter the RESUME command.

3. Create new slides from other drawings and include them in the show.scr file. Include the **RSCRIPT** command at the end of the script.

4. Run the revised slide show.

Creating a Slide Library

The slide library facility lets you store slide files in a single file, similar to filling a carousel tray of 35-mm slides. Once the slide library is complete, you can use it in conjunction with an image tile menu. The individual slide files need not be present.

First, we must identify (or create) several blocks to include in the slide library. Let's use the small symbol library of blocks from Unit 33. The blocks are printed here for your reference.

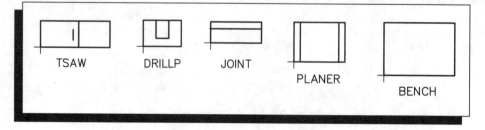

These five blocks will later become image tiles.

1. Open the file named **lib1.dwg** that you created in Unit 33. (It contains the five blocks shown on the previous page.) Do not save your changes.

2. Using **Save As...**, create a new file named **equip.dwg** and store it in the folder with your name.

NOTE:

You will be creating several new files in this unit. Store all of them in the folder with your name. If you do not, the image tile menu may not work.

3. If the attribute display is on, turn it off using the **ATTDISP** command.

4. Using the **MSLIDE** command, produce a slide of each block. (First, read the following hint.) Use the block name for the slide name, and store the slides in the folder with your name.

HINT:

Using the ZOOM command, make each block as large as possible on the display before creating the slide. Keep them simple. Complex image tiles take longer to display on the screen, and their complexity may confuse more than communicate.

5. After you are confident that you have five slides that correspond to the five blocks, minimize AutoCAD.

6. Locate the AutoCAD file named **slidelib.exe** (contained in AutoCAD's Support folder) and copy it to the folder with your name and the slide files.

7. From the Windows **Start** menu, pick **Programs** and open a DOS window.

8. At the DOS prompt, change to the directory with your name. An example is shown below.

cd \Program Files\AutoCAD R14\Jody

NOTE:

If you receive the message Too many parameters - Files\AutoCAD, enter the following text instead, exactly as you see it here:

cd \Progra~1\AutoCA~1\Jody

(Substitute Jody with the name of your folder.)

⑨ At the DOS prompt, enter **slidelib tools** to begin a slide library file named tools.slb.

This starts the slidelib.exe program and displays a copyright statement on the screen.

⑩ Enter the following text exactly as you see it here.

tsaw
drillp
joint
planer
bench (press **ENTER** twice)

Press the **F6** function key and **ENTER**.

You should now have a slide library file named tools.slb in the folder with your name.

⑪ Close the DOS window.

Creating the Image Tile Menu

First let's look at an already-developed image tile menu in AutoCAD.

① Maximize AutoCAD.

② Select **Tiled Viewports** and **Layout...** from the **View** pull-down menu.

The Tiled Viewports Layout dialog box is an example of an image tile menu.

Without the use of image tiles, the selection of a tiled viewport layout would not be so easy.

③ Pick **Cancel**.

④ Minimize AutoCAD.

Creation of an image tile menu involves the following steps.

① Copy **acad.mnu** from the AutoCAD Support folder to the folder with your name.

② Rename the copy to **acadrev.mnu**.

③ Using WordPad, open **acadrev.mnu**.

④ Using WordPad's Find feature, find *****image**.

⑤ Below ***image, as shown in the following, insert the menu items that are in boldface print.

These menu items are stored in a file available on the optional *Applying AutoCAD Diskette.*

```
//        Begin AutoCAD Image Menus
//

***image
**tools
[Select Tool]
[tools(tsaw,Table Saw)]^cinsert tsaw \\\\
[tools(drillp,Drill Press)]^cinsert drillp \\\\
[tools(joint,Jointer)]^cinsert joint \\\\
[tools(planer,Surface Planer)]^cinsert planer \\\\
[tools(bench,Work Bench)]^cinsert bench \\\\


**image_3DObjects
[3D Objects]
```

Each of the five consistently structured menu items will: (1) display the slide image as an image tile on the screen, (2) display the symbol name in the list box, (3) allow you to choose it, and (4) insert the block you chose.

If you want to learn more about the specific components within these menu items, review Unit 59.

⑥ Save your work, but do not exit WordPad.

NOTE:

When saving the file in WordPad, be sure to save it as a text (TXT) document. If you save it as a Word (DOC) document, the procedure will not work.

Making the Image Tile Menu Accessible

It is possible to access an image tile menu using the screen, pull-down, or tablet menu. In any case, you must add a new menu item in acadrev.mnu.

1. Using WordPad's Find feature, find the ***POP7** menu item.

2. Above ***POP7, as shown in the following, insert the menu item that is in boldface print.

ID_Preferenc [&Preferences...] ^C^C_preferences
[Tools...]^c^cinsert equip ^c$i=tools $i=*

***POP7
**DRAW

This menu item will display Tools... at the bottom of the Tools pull-down menu. Also, it will insert the five block definitions contained in equip.dwg. The graphics in equip.dwg will not insert because a cancel is automatically issued at the Insertion point step of the INSERT command. The last part of the macro addresses the image tile menu and displays it on the screen.

3. Save your work and exit WordPad.

4. Make sure that acadrev.mnu and equip.dwg are located in the folder with your name.

Using the Image Tile Menu

1. Maximize AutoCAD and begin a new drawing from scratch. Do not save any changes to equip.dwg.

2. Save the drawing file as **job.dwg** in the folder with your name.

3. Enter the **MENU** command, select **acadrev.mnu** in the folder with your name, and pick **Open**.

HINT:

In the File of type drop-down box, select Menu Template (*.mnu) so that you can see the MNU files in the folder.

788

4️⃣ When the warning appears, pick the **Yes** button, indicating that you want to proceed.

AutoCAD compiles the MNU file and creates MNC, MNS, and MNR files.

5️⃣ Display the **Tools** pull-down menu.

You should see the new Tools... item at the bottom.

5️⃣ Pick the **Tools...** item.

The new image tile menu appears, as shown in the following illustration.

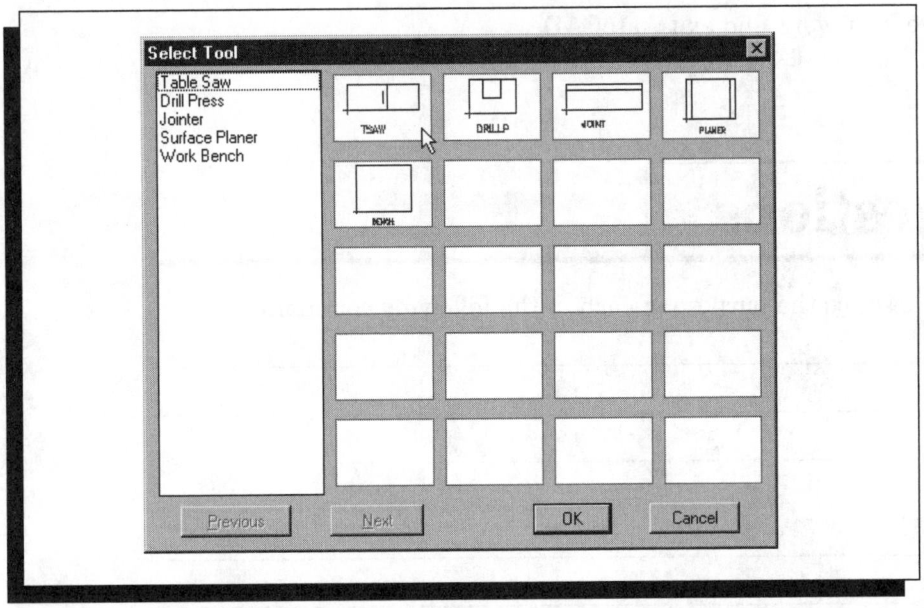

If the image tile menu does not appear, or if one or more of the image tiles is missing, a typing error may exist in the information you inserted into acadrev.mnu. Review the file and make the necessary corrections.

7️⃣ If the image tile menu appears as shown, select one of the tools and place it on the screen.

HINT: Double-click the image tile or single-click it and pick the **OK** button.

8️⃣ Select and place the remaining tools.

NOTE:

An image tile menu can be of any length, although AutoCAD can display only 20 image tiles at one time. If you include more than 20 image tile slides in the image tile menu, AutoCAD makes the Next and Previous buttons available automatically.

⑨ Using the **MENU** command, select the **acad.mnc** file from the Support folder and pick **Open**.

This reloads the standard AutoCAD menu.

⑩ Save your work and exit AutoCAD.

Questions

1. Briefly describe the purpose of each of the following commands.

   MSLIDE _____

   VSLIDE _____

   SCRIPT _____

   RSCRIPT _____

   DELAY _____

   RESUME _____

2. Describe the purpose of an AutoCAD script file.

   _____

   _____

3. What is the file extension of a script file?

   _____

4. What does the number following the DELAY command indicate?

   _____

   _____

5. Why are image tile menus beneficial?

   _____

   _____

6. Name at least one rule you should consider before creating slides for use in an image tile menu.

   _____

   _____

7. From what type of files does the slide library utility create a slide library file?

   _____

8. In what AutoCAD file should you enter the image tile menu information necessary for displaying the image tile menu on the screen?

   _____

9. What does the ***image menu item specify?

   _____

10. Explain the following:

    [Tools...] ^c^cinsert equip ^c$i=tools $i=*

    _____

    _____

11. Suppose you want to include 50 selections in an image tile menu. How do you go about it when AutoCAD can only display 20 selections?

    _____

    _____

 Challenge Your Thinking: Questions for Discussion

1. Is it more practical to store a drawing setup in a script file or in a drawing template file? Explain why.

2. Brainstorm possible uses for an AutoCAD slide show. Be creative. Then choose one possibility and develop a plan for the slide show.

3. Design an image tile menu from which you can choose a text style. Describe the steps you would take.

Problems

1. Create a dozen or so slides of previously created drawings. Include them in a slide show stored as a script file. Run the show.

2. Develop a script file that includes several AutoCAD commands. Make it elaborate. When you're finished, print the file so that you can work out the bugs as you run it.

3. Create the slide show for which you developed a plan in the second "Challenge Your Thinking" question.

4. Create the image tile menu you designed in the "Challenge Your Thinking" question. First, group several blocks and create a slide of each. Next, create a slide library using the slide files. Then use the slide library to create an image tile menu. Use the steps outlined in this unit to complete this problem.

5. List other applications for which you think image tile menus would be useful. Then form a group in which each member is responsible for developing one image tile menu. When the image tile menus have been developed, each member of the group should make his or her image tile menu available to the rest of the group or class. In this way, it is possible to build a library of image tile menus that everyone can use.

AUTOCAD® AT WORK

An Atlas of Applications:
The AutoCAD Resource Guide

From Imagine Nursery™ to Cemetery Master®, the list of applications software in the *AutoCAD Resource Guide* tells a story of diversity—of the myriad ways people in different walks of life are adapting AutoCAD to their workplace. Cemetery Master Information System is a planning tool for the mortuary business. Nursery, which has nothing to do with babies, offers a graphics library of 58 trees catalogued by height, time of year, and leaf density. Both are specialty programs within the broader field of facilities management. Another specialty within the facilities management field is DentaCAD, a program that helps in optimizing the layout of a dentist's office. When you think about it, dentists have unusual requirements for walls, plumbing, and power, and some decidedly strange furniture.

Architecture, hydrologic analysis, mapping, modeling . . . 37 applications categories in the *AutoCAD Resource Guide* testify to the range of specialties within which AutoCAD users perform. Published four times a year by Autodesk, Inc., the guide itemizes more than 1,600 software and hardware products, many of which can only be categorized as "Other."

Anthropologists tell us that a stick becomes a tool when an intelligent being conceives of the stick as a means to achieve some particular purpose, such as digging in soil, extending one's reach into a tree, or bonking a rival on the head. Thus does AutoCAD become a tool, giving rise to applications

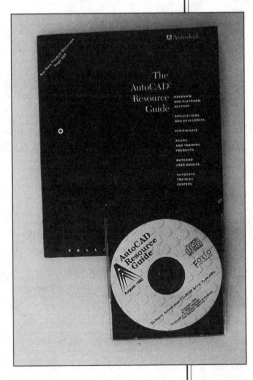

such as the DataSketch Law Enforcement Symbols Libraries, which include "highly detailed predrawn symbols for sketching crime scenes and traffic accidents." To help prevent accidents, look to the Traffic Sign Graphic Computer System which, as its name implies, is for use by government agencies in the manufacture of traffic signs. Lumberjack is a design system for logging roads and other access roads that don't need a lot of traffic signs.

Some of the intelligent beings using AutoCAD want everybody to have access to AutoCAD, and have produced programs such as MAX, which supports Korean display and rendering, and TransCYRILLIC® for Windows™ and AutoCAD, which provides fonts for Russian and 37 other Cyrillic-alphabet languages. Arabesque offers 13 Arabic and 30 Islamic patterns, while Shacham provides support for Hebrew. Pinnacle Chinese Handwriting and Rubon Chinese System bring a big part of Asia into the world of AutoCAD.

Among the whimsical software names that are bound to turn up in an applications list as long as AutoCAD's are CadZooks, which provides 3D detailing of precast concrete structures, ACD Bath POD Max Pack, showcasing 225 bathroom designs with a selection of 50 tubs, 50 showers, and 100 whirlpools, and, finally, Easydie—no relation to Cemetery Master—which simplifies die design.

793

Optional Problems

Introduction

The following problems provide additional practice with AutoCAD. These problems will help you expand your knowledge and ability, and they will offer you new and challenging experiences.

The problems encompass a variety of disciplines and have been sequenced from simple to advanced. Your instructor may ask you to complete the problems in a different order.

The key to successful completion is to *plan before beginning*. Review previously learned commands and techniques and ask yourself how you can apply them to the problem. For example, when laying out rectangular objects, plan to use grid, snap, and ortho. When drawing lines of specific lengths and angles, consider using the relative and polar methods of specifying endpoints. Plan how to use COPY, MIRROR, and ARRAY to simplify and speed your work.

As you discover new and easier methods of creating drawings, apply these methods to solving the problems. Since there is usually more than one way to complete a drawing, experiment with alternative methods. Discuss these alternatives with other users and create strategies for efficient completion of the problems.

Remember, there is no substitute for practice. The expertise you gain is proportionate to the time you spend on the system. Set aside blocks of time to work with AutoCAD, think through your approach, and enjoy this fascinating technology.

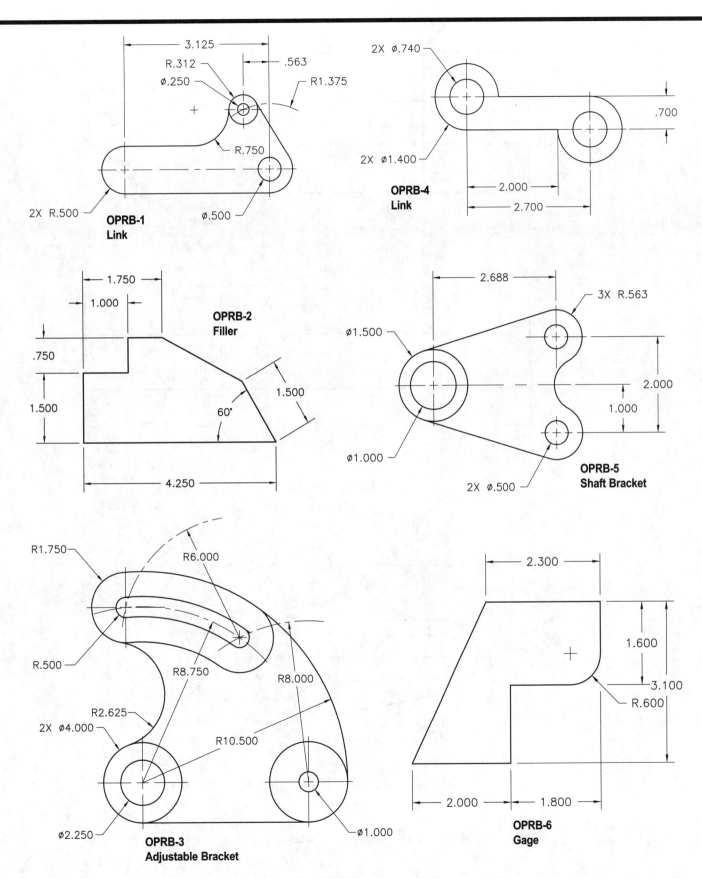

OPRB-1
Link

OPRB-2
Filler

OPRB-3
Adjustable Bracket

OPRB-4
Link

OPRB-5
Shaft Bracket

OPRB-6
Gage

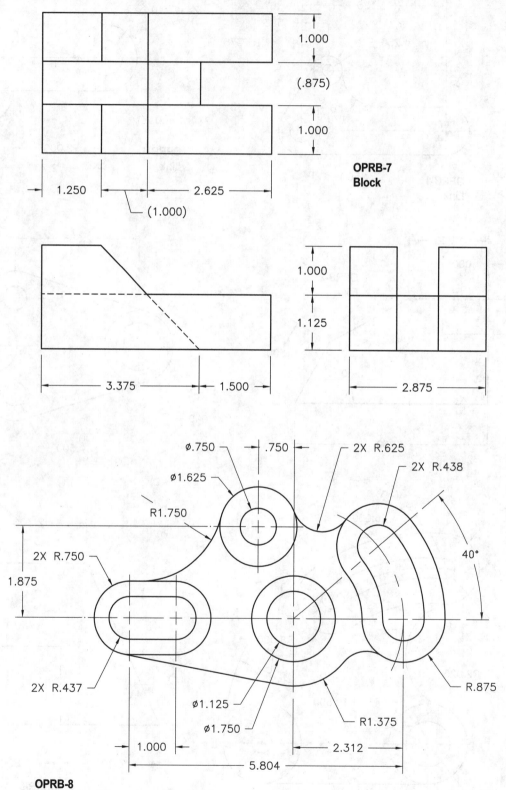

OPRB-7
Block

ø.750 .750 2X R.625
ø1.625 2X R.438
R1.750
2X R.750
1.875
40°
2X R.437
2X R.437
ø1.125
ø1.750
R1.375
R.875
1.000
2.312
5.804

OPRB-8
Idler Plate (Courtesy of Steve Huycke, Lake Michigan College)

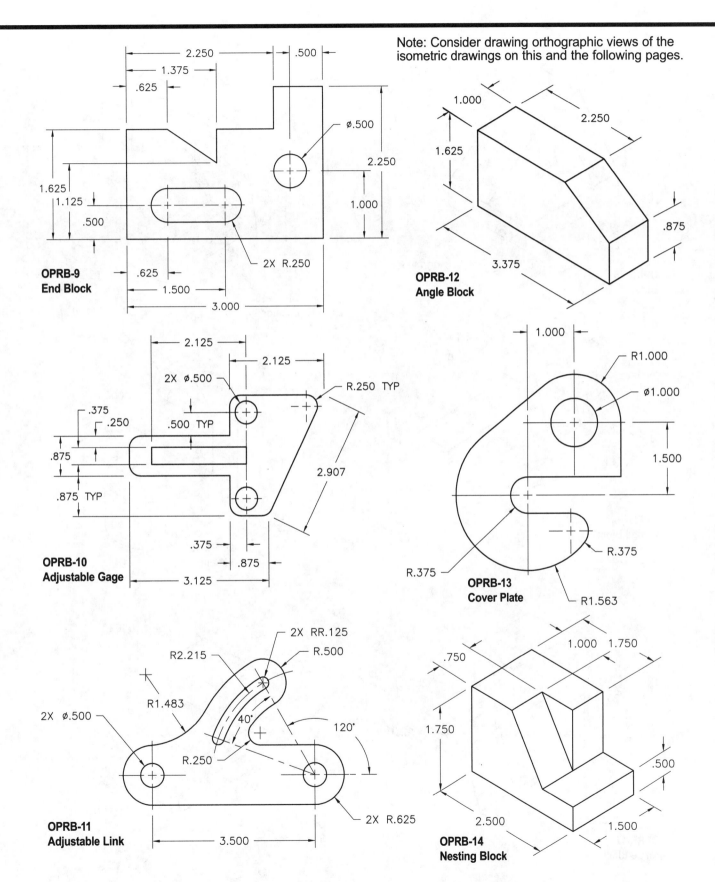

Note: Consider drawing orthographic views of the isometric drawings on this and the following pages.

OPRB-9
End Block

2.250
1.375
.625
Ø.500
2.250
1.000
1.625
1.125
.500
2X R.250
.625
1.500
3.000
.500

OPRB-12
Angle Block

1.000
2.250
1.625
.875
3.375

OPRB-10
Adjustable Gage

2.125
2.125
2X Ø.500
R.250 TYP
.375
.250
.500 TYP
.875
2.907
.875 TYP
.375
.875
3.125

OPRB-13
Cover Plate

1.000
R1.000
Ø1.000
1.500
R.375
R.375
R1.563

OPRB-11
Adjustable Link

2X RR.125
R2.215
R.500
R1.483
2X Ø.500
40°
120°
R.250
2X R.625
3.500

OPRB-14
Nesting Block

1.000 1.750
.750
1.750
.500
2.500
1.500

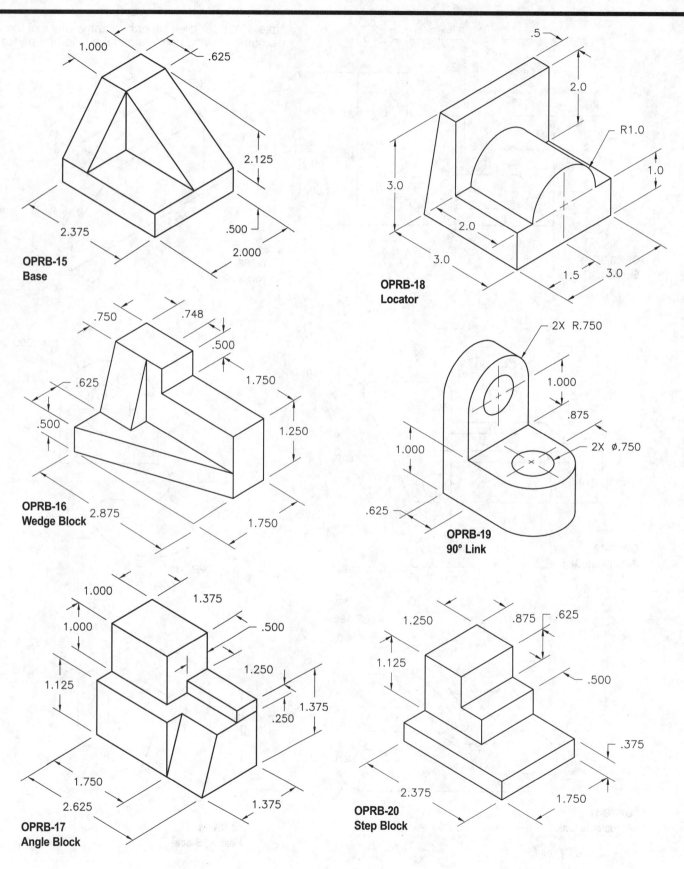

OPRB-15
Base

OPRB-16
Wedge Block

OPRB-17
Angle Block

OPRB-18
Locator

OPRB-19
90° Link

OPRB-20
Step Block

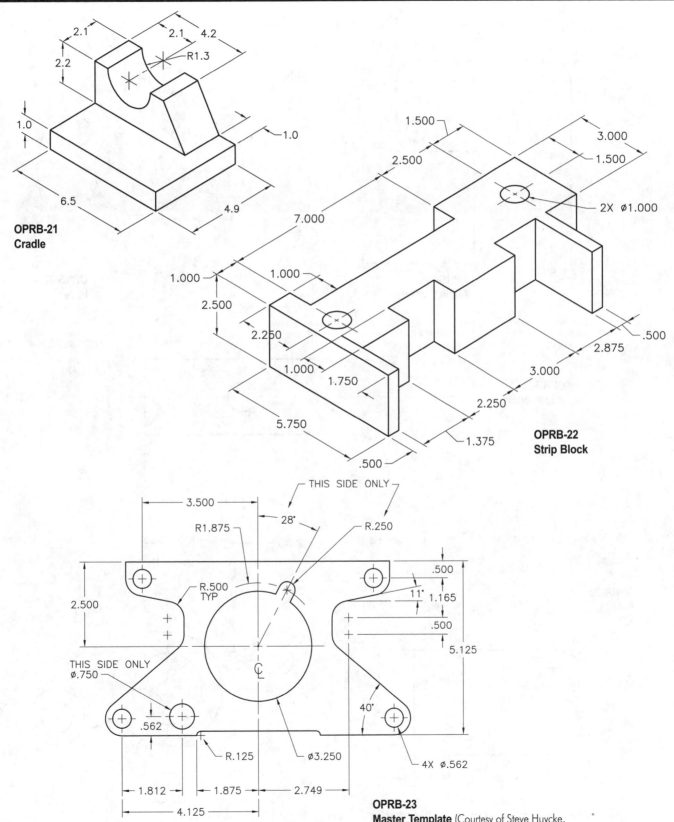

OPRB-21
Cradle

2.1   2.1   4.2
2.2
R1.3
1.0
1.0
6.5
4.9

OPRB-22
Strip Block

1.500
3.000
2.500
1.500
7.000
2X ⌀1.000
1.000   1.000
2.500
.500
2.250
2.875
1.000
3.000
1.750
2.250
5.750
.500   1.375

OPRB-23
Master Template (Courtesy of Steve Huycke,
Lake Michigan College)

THIS SIDE ONLY
3.500
28°
R.250
R1.875
.500
R.500
TYP
11°   1.165
2.500
.500
5.125
THIS SIDE ONLY
⌀.750
40°
.562
4X ⌀.562
R.125   ⌀3.250
1.812   1.875   2.749
4.125

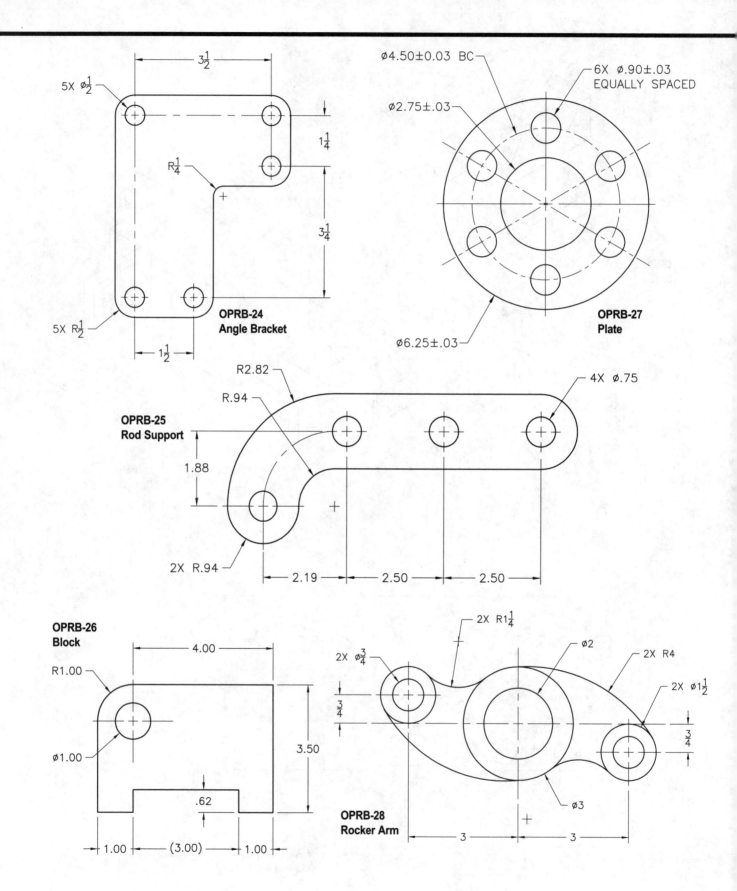

OPRB-24
Angle Bracket

OPRB-27
Plate

OPRB-25
Rod Support

OPRB-26
Block

OPRB-28
Rocker Arm

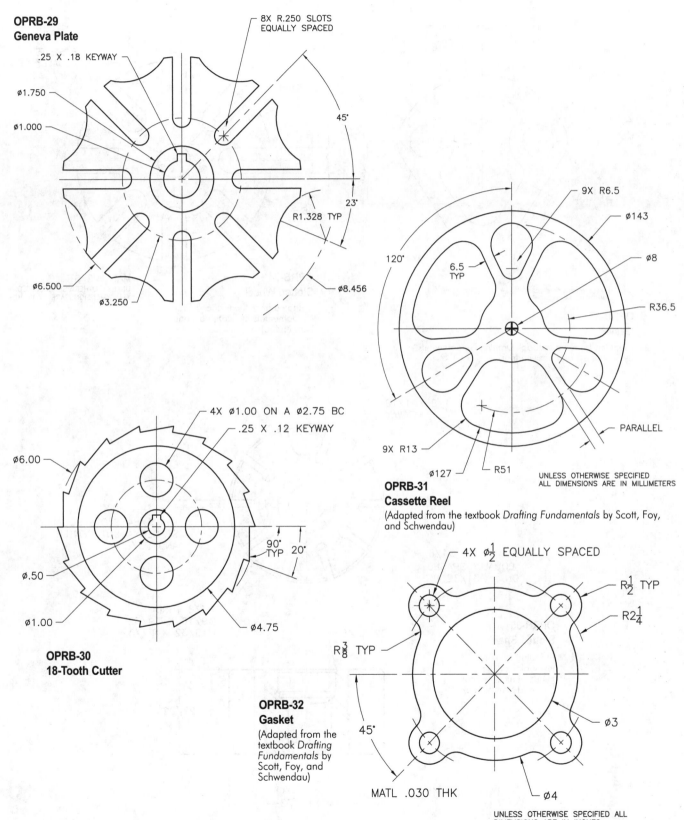

OPRB-29
Geneva Plate

.25 X .18 KEYWAY

ø1.750

ø1.000

8X R.250 SLOTS
EQUALLY SPACED

45°

23°

R1.328 TYP

ø8.456

ø6.500

ø3.250

9X R6.5

ø143

6.5
TYP

ø8

120°

R36.5

9X R13

ø127

R51

PARALLEL

UNLESS OTHERWISE SPECIFIED
ALL DIMENSIONS ARE IN MILLIMETERS

OPRB-31
Cassette Reel
(Adapted from the textbook *Drafting Fundamentals* by Scott, Foy, and Schwendau)

4X ø1.00 ON A ø2.75 BC

.25 X .12 KEYWAY

ø6.00

90°
TYP 20°

ø.50

ø1.00

ø4.75

OPRB-30
18-Tooth Cutter

4X ø$\frac{1}{2}$ EQUALLY SPACED

R$\frac{1}{2}$ TYP

R2$\frac{1}{4}$

R$\frac{3}{8}$ TYP

ø3

45°

ø4

MATL .030 THK

OPRB-32
Gasket
(Adapted from the textbook *Drafting Fundamentals* by Scott, Foy, and Schwendau)

UNLESS OTHERWISE SPECIFIED ALL
DIMENSIONS ARE IN INCHES

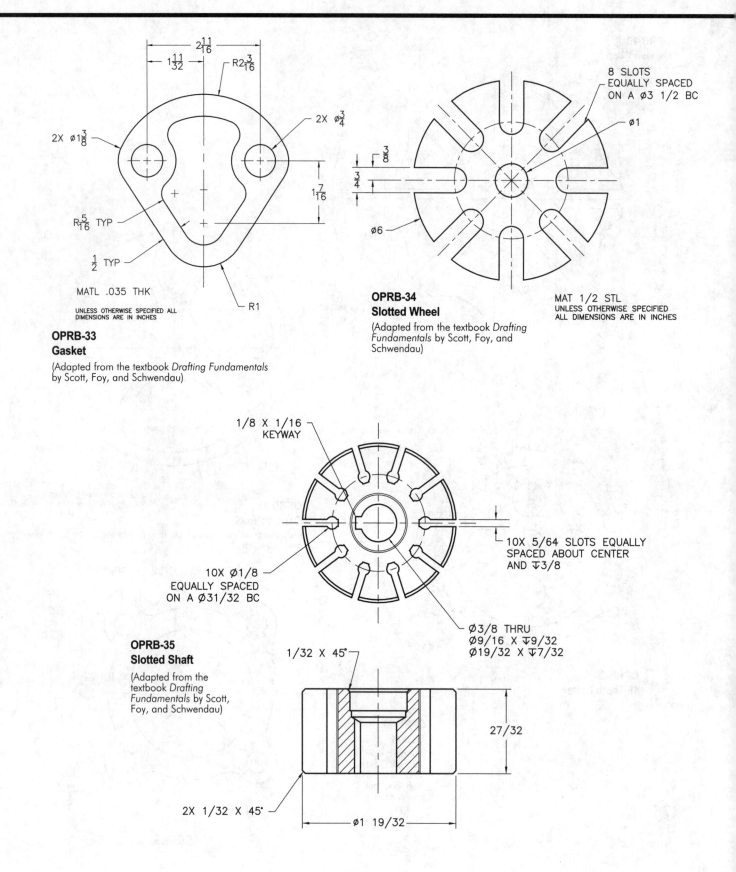

OPRB-33
Gasket

(Adapted from the textbook *Drafting Fundamentals* by Scott, Foy, and Schwendau)

MATL .035 THK

UNLESS OTHERWISE SPECIFIED ALL
DIMENSIONS ARE IN INCHES

2X ⌀1 3/8

R 5/16 TYP

1/2 TYP

2 11/16

1 11/32

R2 3/16

2X ⌀ 3/4

1 7/16

R1

OPRB-34
Slotted Wheel

(Adapted from the textbook *Drafting Fundamentals* by Scott, Foy, and Schwendau)

8 SLOTS
EQUALLY SPACED
ON A ⌀3 1/2 BC

⌀1

⌀6

3/4

3/32

MAT 1/2 STL
UNLESS OTHERWISE SPECIFIED
ALL DIMENSIONS ARE IN INCHES

OPRB-35
Slotted Shaft

(Adapted from the textbook *Drafting Fundamentals* by Scott, Foy, and Schwendau)

1/8 X 1/16
KEYWAY

10X ⌀1/8
EQUALLY SPACED
ON A ⌀31/32 BC

10X 5/64 SLOTS EQUALLY
SPACED ABOUT CENTER
AND ⌵3/8

⌀3/8 THRU
⌀9/16 X ⌵9/32
⌀19/32 X ⌵7/32

1/32 X 45°

2X 1/32 X 45°

27/32

⌀1 19/32

802

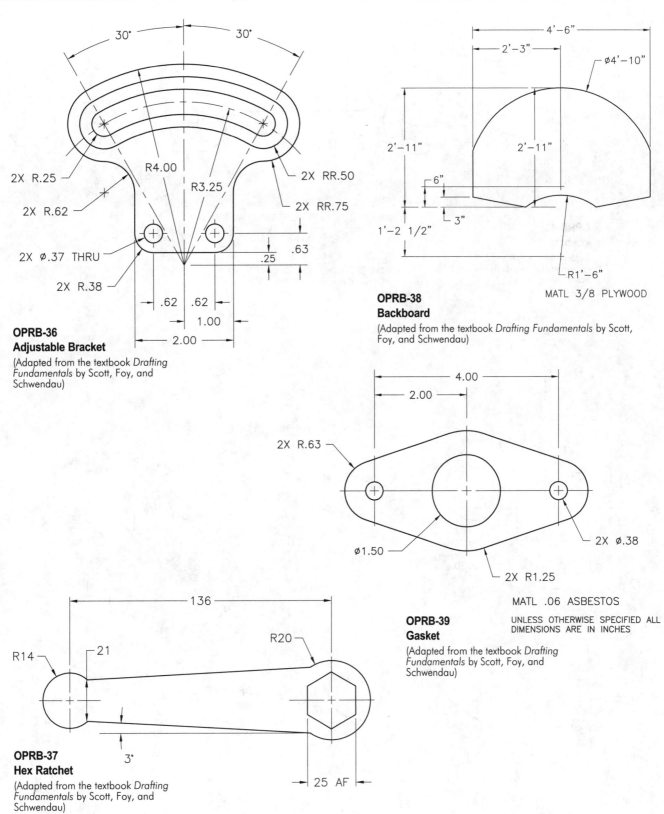

OPRB-36
Adjustable Bracket

(Adapted from the textbook *Drafting Fundamentals* by Scott, Foy, and Schwendau)

OPRB-38
Backboard

(Adapted from the textbook *Drafting Fundamentals* by Scott, Foy, and Schwendau)

MATL 3/8 PLYWOOD

OPRB-39
Gasket

(Adapted from the textbook *Drafting Fundamentals* by Scott, Foy, and Schwendau)

MATL .06 ASBESTOS

UNLESS OTHERWISE SPECIFIED ALL DIMENSIONS ARE IN INCHES

OPRB-37
Hex Ratchet

(Adapted from the textbook *Drafting Fundamentals* by Scott, Foy, and Schwendau)

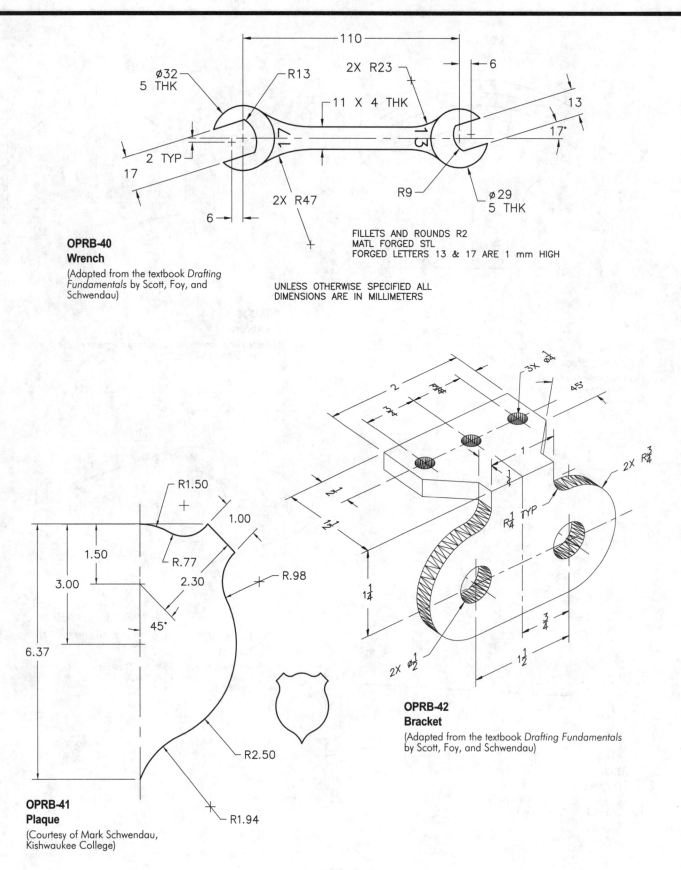

110

Ø32
5 THK

R13

2X R23

6

11 X 4 THK

13

17°

17

2 TYP

17

13

6

2X R47

R9

Ø29
5 THK

FILLETS AND ROUNDS R2
MATL FORGED STL
FORGED LETTERS 13 & 17 ARE 1 mm HIGH

UNLESS OTHERWISE SPECIFIED ALL
DIMENSIONS ARE IN MILLIMETERS

OPRB-40
Wrench

(Adapted from the textbook *Drafting Fundamentals* by Scott, Foy, and Schwendau)

R1.50

1.00

1.50

R.77

3.00

2.30

R.98

45°

6.37

R2.50

R1.94

OPRB-41
Plaque

(Courtesy of Mark Schwendau,
Kishwaukee College)

2

3X Ø¼

45°

3/4

1

2X R¾

1¼

R¼ TYP

1¼

1¼

3/4

2X Ø½

1½

OPRB-42
Bracket

(Adapted from the textbook *Drafting Fundamentals*
by Scott, Foy, and Schwendau)

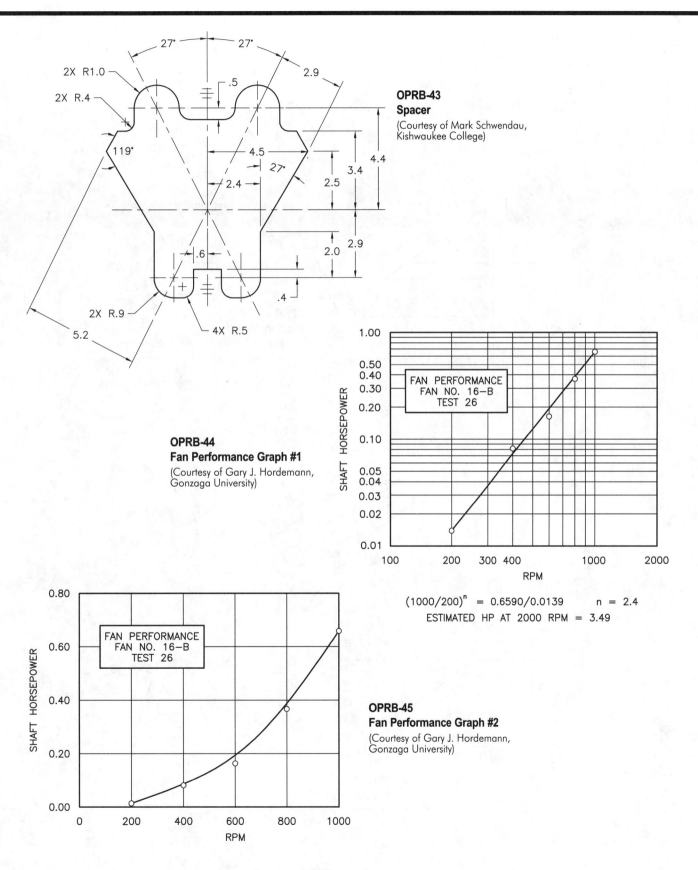

OPRB-43
Spacer
(Courtesy of Mark Schwendau,
Kishwaukee College)

OPRB-44
Fan Performance Graph #1
(Courtesy of Gary J. Hordemann,
Gonzaga University)

FAN PERFORMANCE
FAN NO. 16-B
TEST 26

$(1000/200)^n = 0.6590/0.0139 \qquad n = 2.4$
ESTIMATED HP AT 2000 RPM = 3.49

FAN PERFORMANCE
FAN NO. 16-B
TEST 26

OPRB-45
Fan Performance Graph #2
(Courtesy of Gary J. Hordemann,
Gonzaga University)

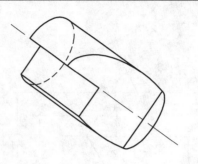

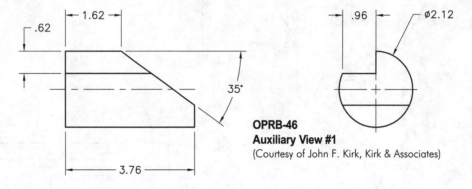

OPRB-46
Auxiliary View #1
(Courtesy of John F. Kirk, Kirk & Associates)

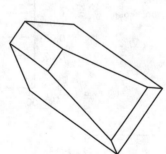

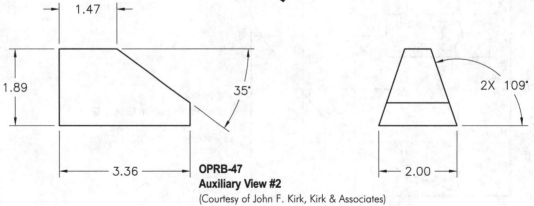

OPRB-47
Auxiliary View #2
(Courtesy of John F. Kirk, Kirk & Associates)

OPRB-48
Casting
(Courtesy of Gary J. Hordemann,
Gonzaga University)

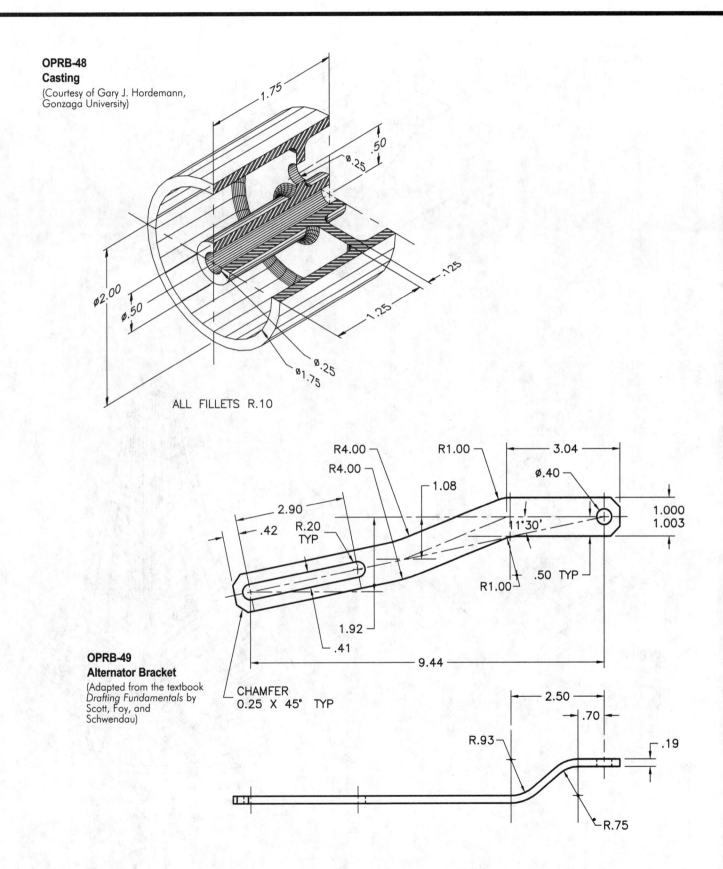

1.75

.50

ø.25

.125

1.25

ø2.00

ø.50

ø.25
ø1.75

ALL FILLETS R.10

R4.00

R4.00

R1.00

3.04

ø.40

1.08

2.90

1.000
1.003

.42

R.20
TYP

11°30'

.50 TYP

R1.00

1.92

.41

9.44

OPRB-49
Alternator Bracket
(Adapted from the textbook
Drafting Fundamentals by
Scott, Foy, and
Schwendau)

CHAMFER
0.25 X 45° TYP

2.50

.70

.19

R.93

R.75

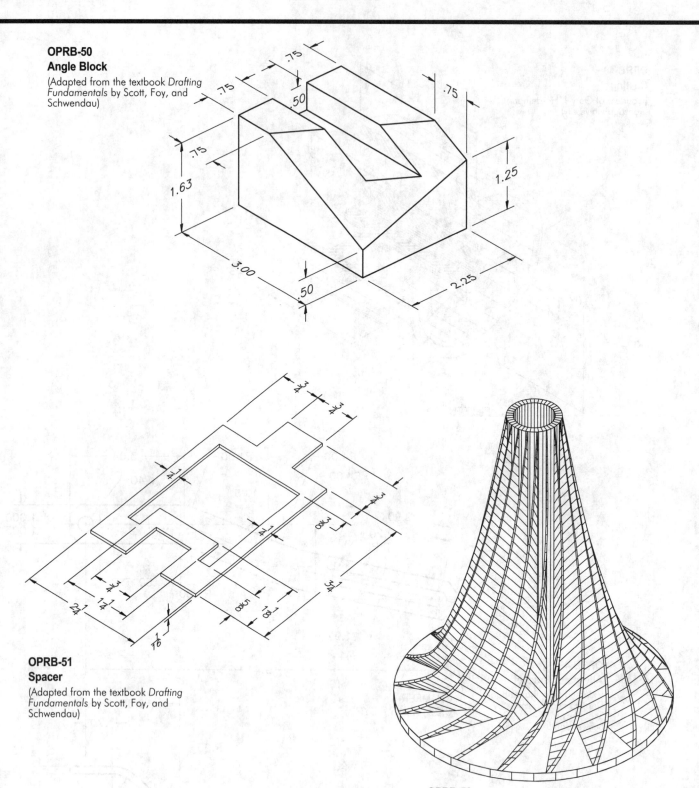

OPRB-50
Angle Block

(Adapted from the textbook *Drafting Fundamentals* by Scott, Foy, and Schwendau)

.75
.75
.50
.75
.75
.75
1.63
1.25
3.00
.50
2.25

OPRB-51
Spacer

(Adapted from the textbook *Drafting Fundamentals* by Scott, Foy, and Schwendau)

OPRB-52
Impeller

(Courtesy of Gary J. Hordemann, Gonzaga University)

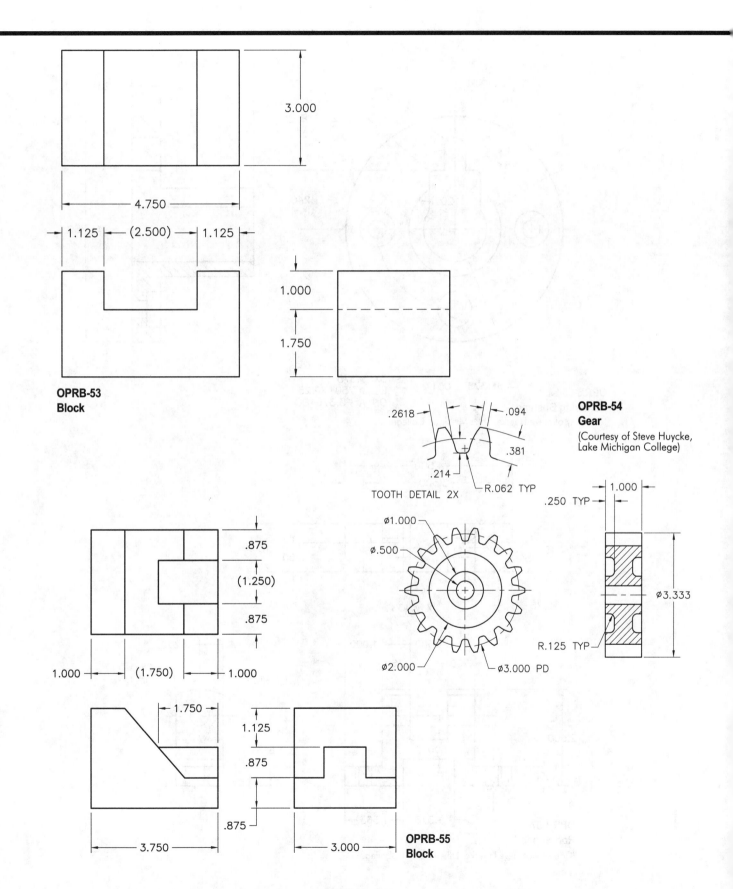

OPRB-53
Block

TOOTH DETAIL 2X

.2618   .094
.381
.214
R.062 TYP

OPRB-54
Gear
(Courtesy of Steve Huycke,
Lake Michigan College)

Ø1.000
Ø.500
Ø2.000
Ø3.000 PD

.250 TYP
1.000
Ø3.333
R.125 TYP

OPRB-55
Block

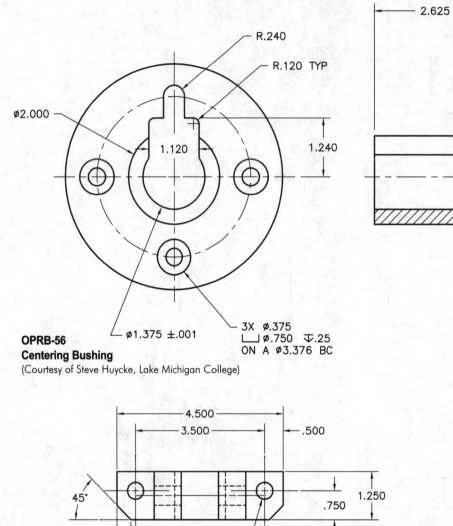

R.240

R.120 TYP

Ø2.000

1.120

1.240

Ø1.375 ±.001

3X Ø.375
⌴ Ø.750 ⌵.25
ON A Ø3.376 BC

OPRB-56
Centering Bushing
(Courtesy of Steve Huycke, Lake Michigan College)

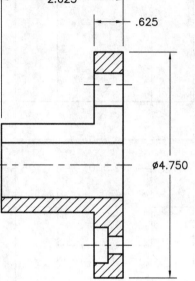

2.625

.625

Ø4.750

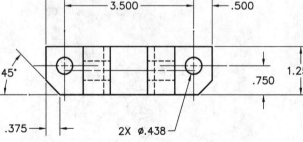

4.500

3.500

.500

45°

1.250

.750

.375

2X Ø.438

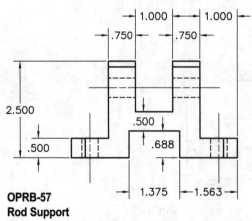

1.000

1.000

.750

.750

2.500

.500

.500

.688

1.375

1.563

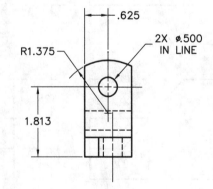

.625

2X Ø.500
IN LINE

R1.375

1.813

OPRB-57
Rod Support
(Courtesy of Steve Huycke, Lake Michigan College)

OPRB-58
45° Elbow
(Courtesy of Steve Huycke,
Lake Michigan College)

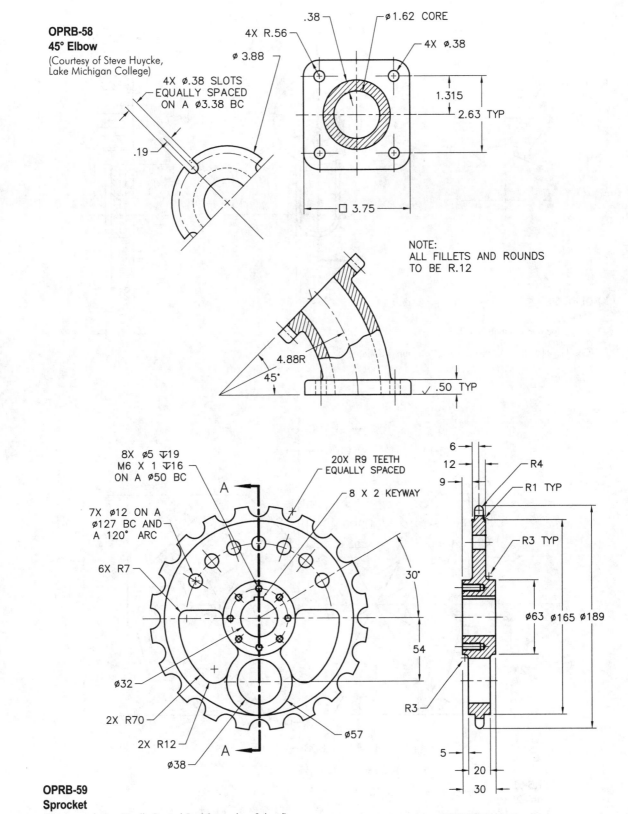

NOTE:
ALL FILLETS AND ROUNDS
TO BE R.12

OPRB-59
Sprocket
(Courtesy of Alan Fitzell, Central Peel Secondary School)

SECTION A—A

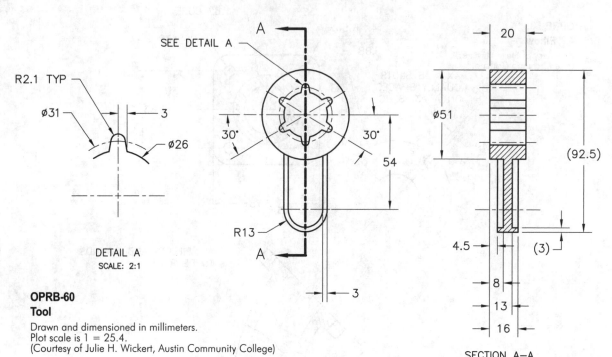

R2.1 TYP

Ø31

3

Ø26

DETAIL A
SCALE: 2:1

SEE DETAIL A

A

30° 30°

54

A

R13

3

20

Ø51

(92.5)

4.5

(3)

8

13

16

SECTION A–A

OPRB-60
Tool

Drawn and dimensioned in millimeters.
Plot scale is 1 = 25.4.
(Courtesy of Julie H. Wickert, Austin Community College)

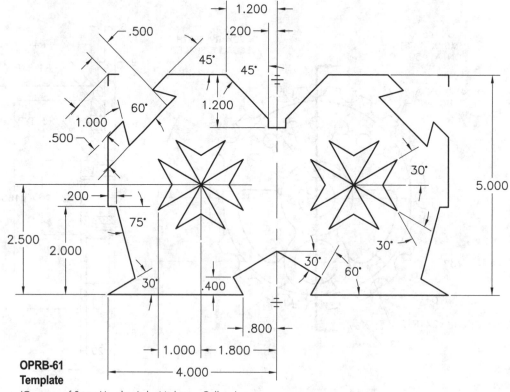

.500

1.200

.200

45°

45°

1.200

60°

1.000

.500

30°

30°

.200

75°

5.000

2.500

2.000

30°

30°

60°

.400

30°

.800

1.000 1.800

4.000

OPRB-61
Template

(Courtesy of Steve Huycke, Lake Michigan College)

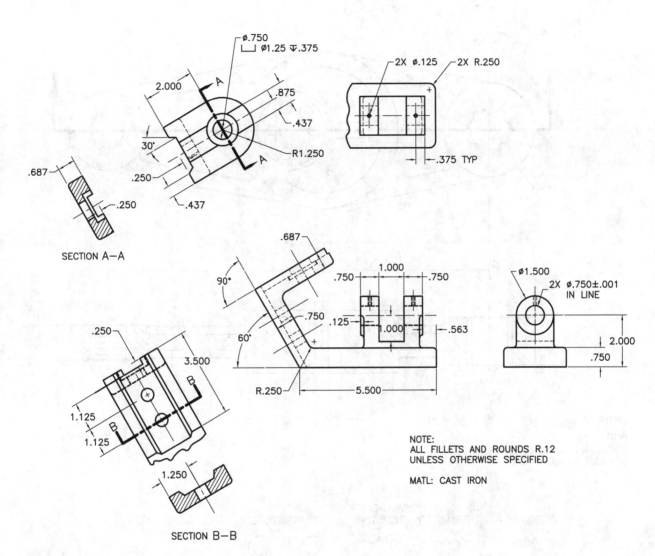

SECTION A—A

SECTION B—B

NOTE:
ALL FILLETS AND ROUNDS R.12
UNLESS OTHERWISE SPECIFIED

MATL: CAST IRON

OPRB-62
Control Bracket
(Courtesy of Steve Huycke, Lake Michigan College)

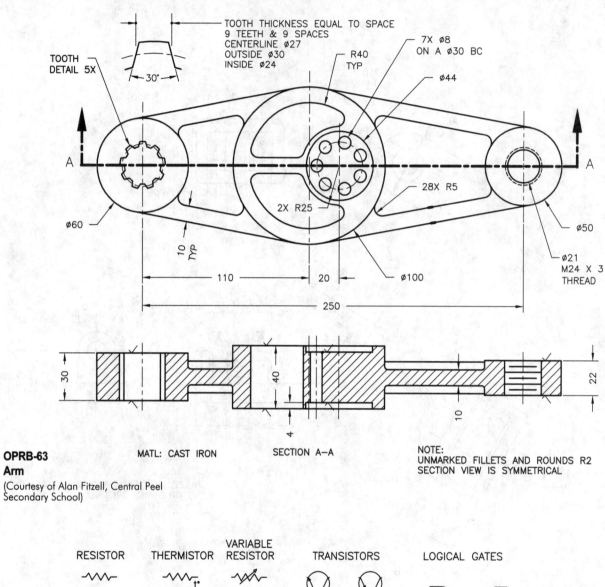

TOOTH THICKNESS EQUAL TO SPACE
9 TEETH & 9 SPACES
CENTERLINE ⌀27
OUTSIDE ⌀30
INSIDE ⌀24

TOOTH
DETAIL 5X

30°

A — A

7X ⌀8
ON A ⌀30 BC

R40
TYP

⌀44

⌀60

2X R25

28X R5

10
TYP

⌀50

⌀21
M24 X 3
THREAD

110   20   ⌀100

250

30   40   22

4   10

OPRB-63
Arm

(Courtesy of Alan Fitzell, Central Peel
Secondary School)

MATL: CAST IRON   SECTION A–A

NOTE:
UNMARKED FILLETS AND ROUNDS R2
SECTION VIEW IS SYMMETRICAL

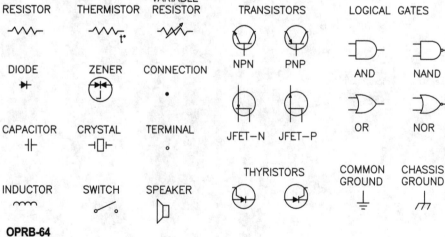

RESISTOR   THERMISTOR   VARIABLE
RESISTOR   TRANSISTORS   LOGICAL GATES

NPN   PNP   AND   NAND

DIODE   ZENER   CONNECTION

CAPACITOR   CRYSTAL   TERMINAL   JFET-N   JFET-P   OR   NOR

THYRISTORS   COMMON   CHASSIS
GROUND   GROUND

INDUCTOR   SWITCH   SPEAKER

OPRB-64
Electronic Symbols

(Courtesy of Robert Pruse, Fort Wayne Community Schools)

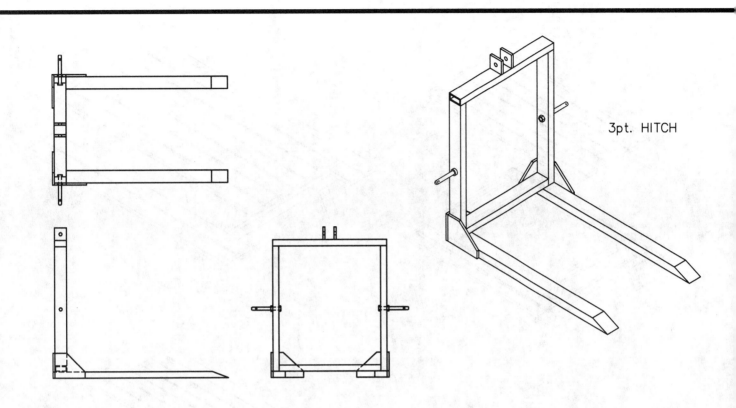

3pt. HITCH

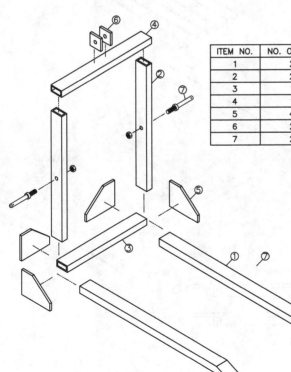

ITEM NO.	NO. OF PCS	SIZE	DESCRIPTION
1	2	42.75	1/8-1.5-3 TUBING
2	2	30	1/8-1.5-3 TUBING
3	1	22.5	1/8-1.5-3 TUBING
4	1	25.5	1/8-1.5-3 TUBING
5	4	3/8-6-8	ANGLE BRACES
6	2	1/2-3-3	UPPER ARM SUPPORTS
7	2	7.25 - 3/4 HEX. NUT	PIN HITCH-STANDARD PART

OPRB-65
Fork Lift
(Courtesy of Craig Pelate and Ron Weseloh,
Red Bud High School)

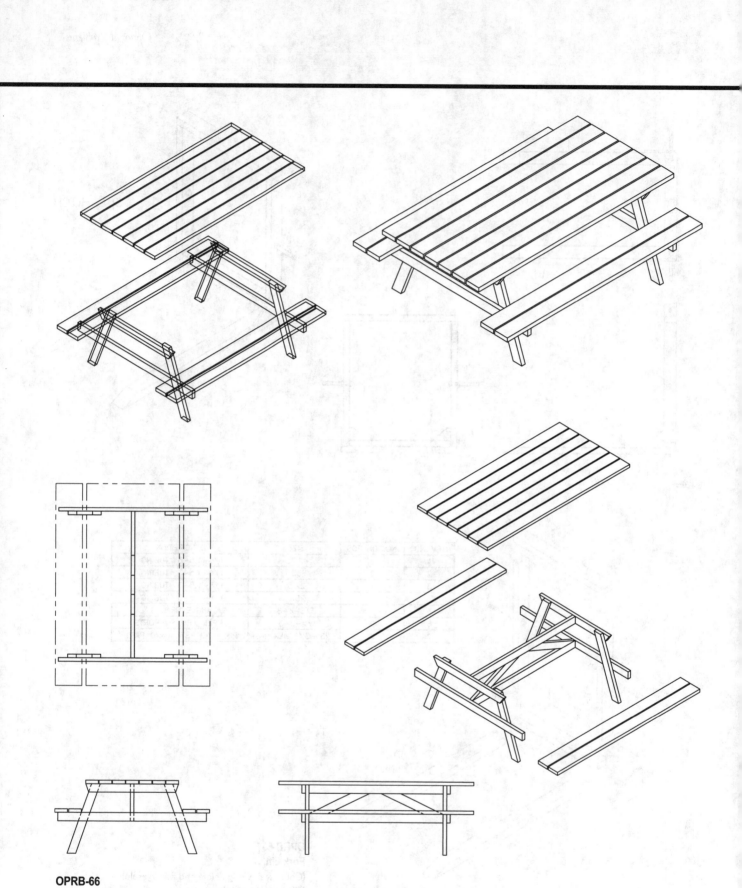

OPRB-66
Picnic Table
(Courtesy of Dan Cowell and Ron Weseloh, Red Bud High School)

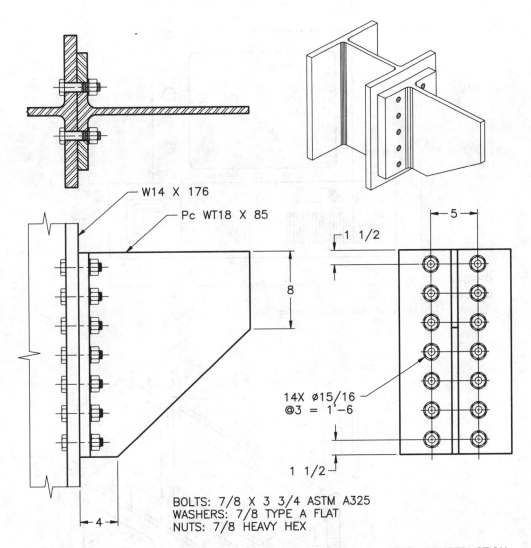

W14 X 176

Pc WT18 X 85

8

1 1/2

5

14X ⌀15/16
@3 = 1'-6

1 1/2

4

BOLTS: 7/8 X 3 3/4 ASTM A325
WASHERS: 7/8 TYPE A FLAT
NUTS: 7/8 HEAVY HEX

FOR DIMENSIONS, SEE: MANUAL OF STEEL CONSTRUCTION,
AMERICAN INSTITUTE OF STEEL CONSTRUCTION

OPRB-67
Structural Bracket
(Courtesy of Gary J. Hordemann, Gonzaga University)

Approximate the missing dimensions.

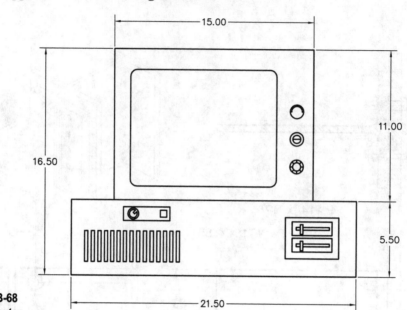

OPRB-68
Computer
(Courtesy of Dan Myers, Informance Computer Services)

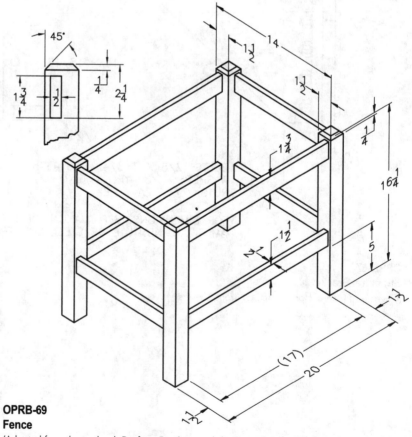

OPRB-69
Fence
(Adapted from the textbook *Drafting Fundamentals* by Scott, Foy, and Schwendau)

Approximate the missing dimensions.

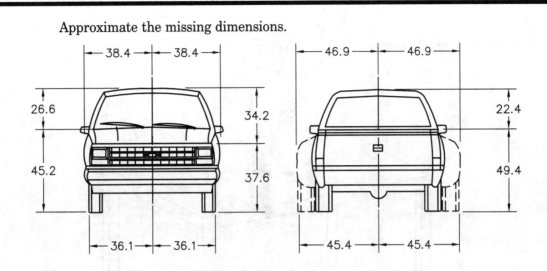

Scale: 1" = 3'

OPRB-70
Pickup (End Views)

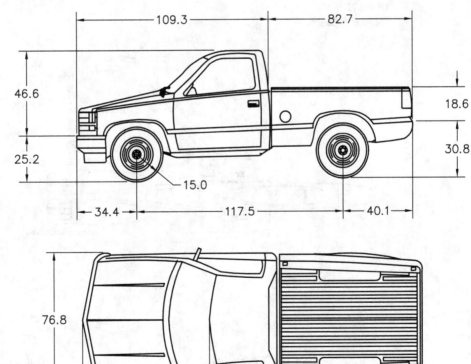

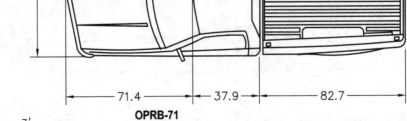

Scale: 1" = 3'

OPRB-71
Pickup (Side and Top Views)

Approximate all dimensions.

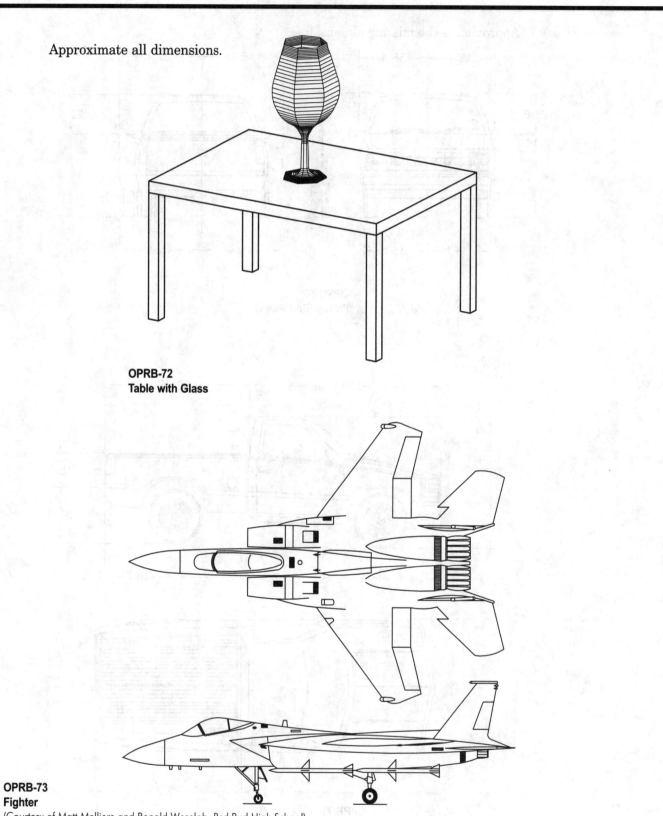

OPRB-72
Table with Glass

OPRB-73
Fighter
(Courtesy of Matt Melliere and Ronald Weseloh, Red Bud High School)

Approximate all dimensions.

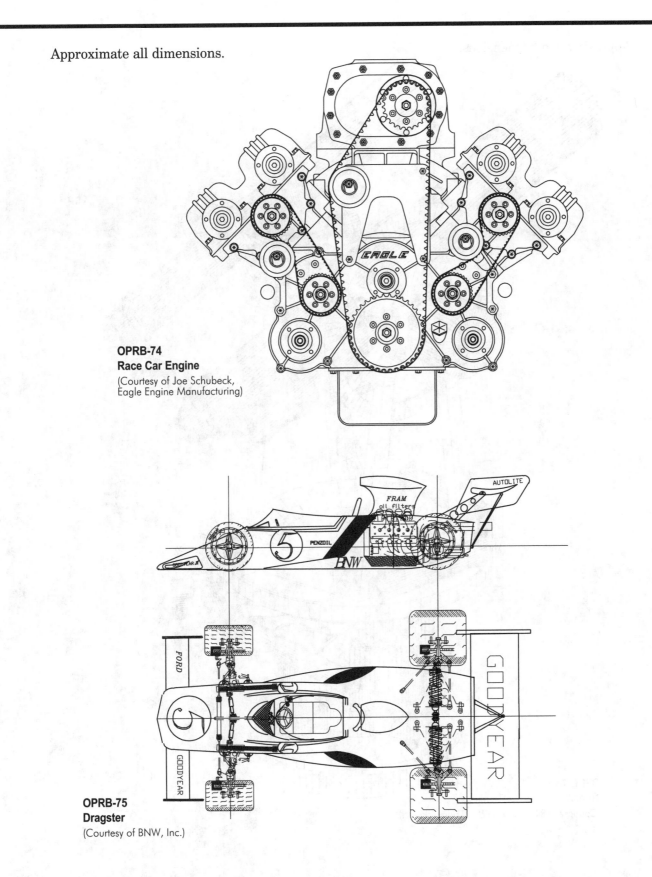

OPRB-74
Race Car Engine
(Courtesy of Joe Schubeck,
Eagle Engine Manufacturing)

OPRB-75
Dragster
(Courtesy of BNW, Inc.)

Approximate all dimensions.

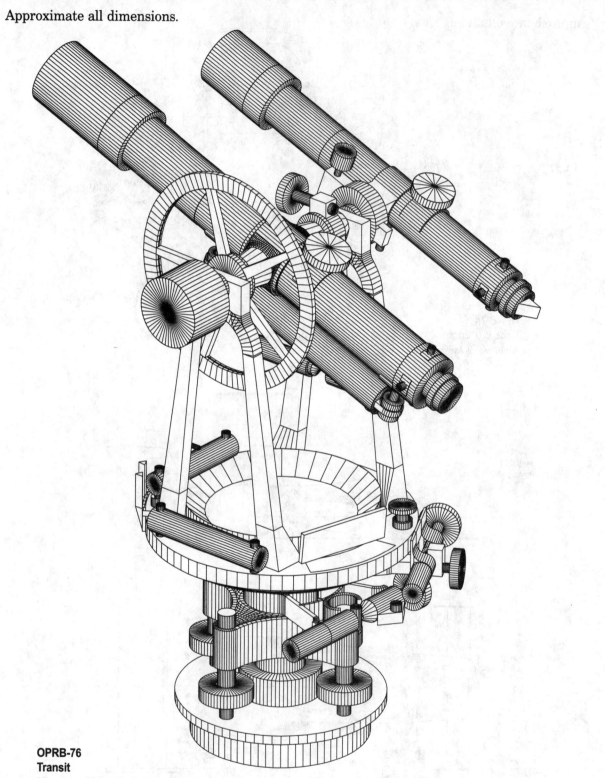

OPRB-76
Transit
(Courtesy of Riley Clark, Hicks & Hartwick, Inc.)

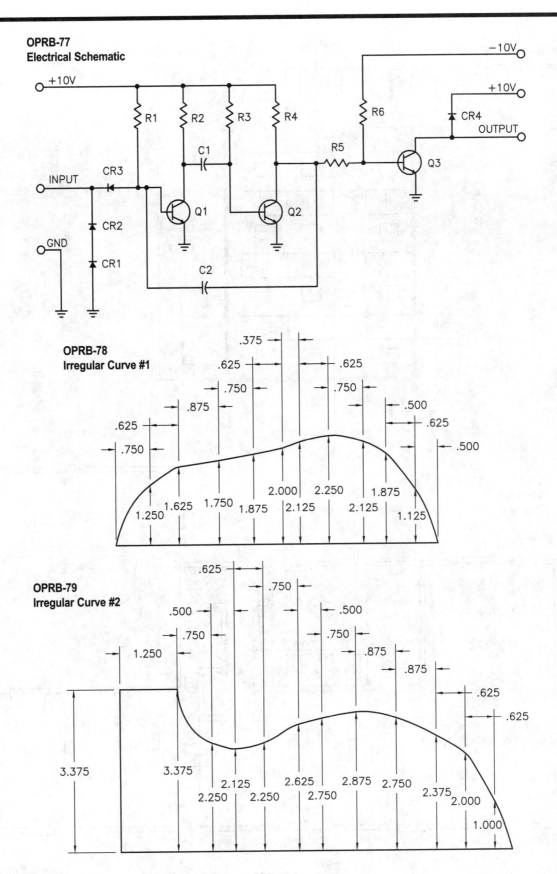

OPRB-77
Electrical Schematic

−10V

+10V

+10V
OUTPUT

R1   R2   R3   R4   R6

CR4

INPUT   CR3   C1   R5   Q3

CR2   Q1   Q2

GND   CR1   C2

OPRB-78
Irregular Curve #1

.375

.625   .625

.750   .750

.875   .500

.625   .625

.750   .500

2.000   2.250   1.875

1.250   1.625   1.750   1.875   2.125   2.125   1.125

OPRB-79
Irregular Curve #2

.625

.750

.500   .500

.750   .750

1.250   .875

.875

.625

.625

3.375   3.375

2.125   2.625   2.875   2.750

2.250   2.250   2.750   2.375   2.000

1.000

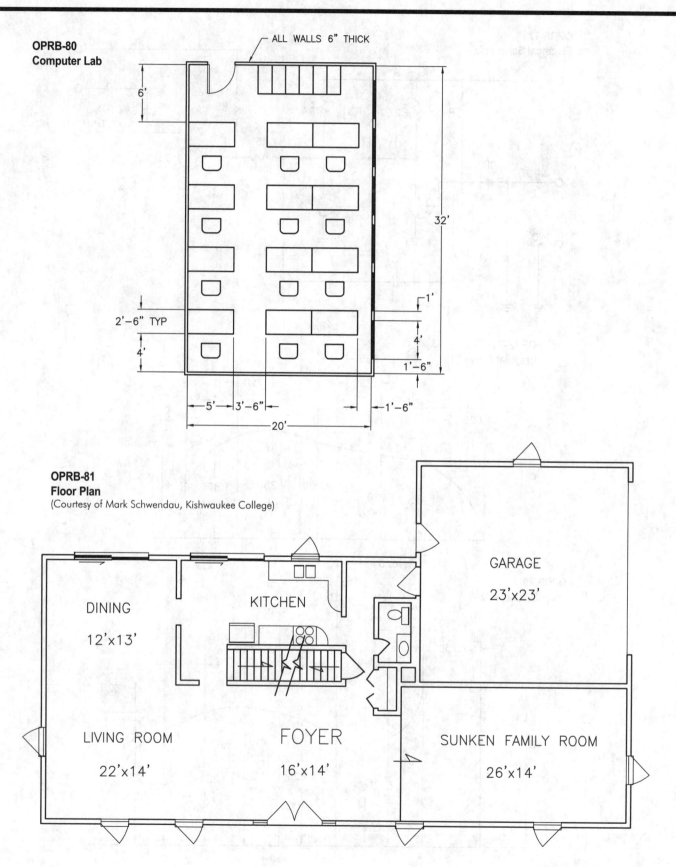

OPRB-80
Computer Lab

ALL WALLS 6" THICK

6'

32'

2'-6" TYP

4'

1'

4'

1'-6"

5' 3'-6" 1'-6"

20'

OPRB-81
Floor Plan
(Courtesy of Mark Schwendau, Kishwaukee College)

GARAGE
23'x23'

DINING
12'x13'

KITCHEN

LIVING ROOM
22'x14'

FOYER
16'x14'

SUNKEN FAMILY ROOM
26'x14'

OPRB-82
Site Plan
(Courtesy of Mill Brothers Landscape
and Nursery, Inc.)

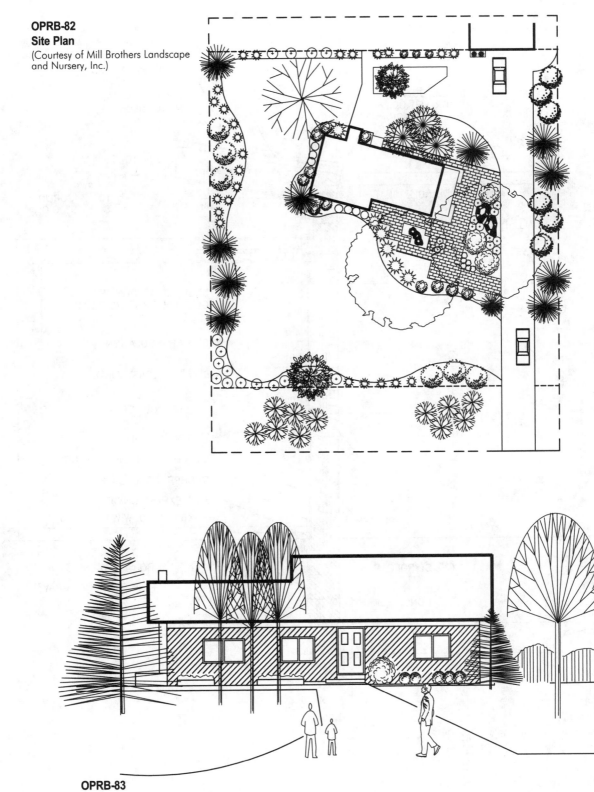

OPRB-83
Elevation
(Courtesy of Mill Brothers Landscape and Nursery, Inc.)

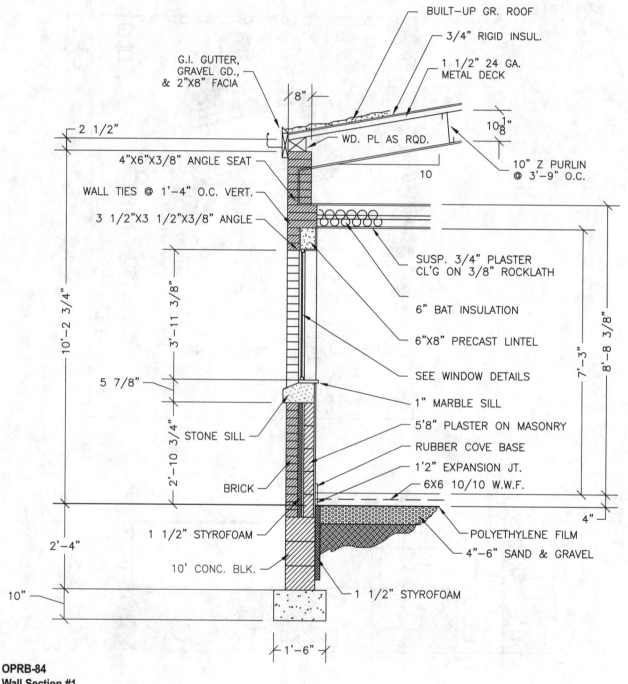

BUILT-UP GR. ROOF

3/4" RIGID INSUL.

1 1/2" 24 GA. METAL DECK

G.I. GUTTER, GRAVEL GD., & 2"X8" FACIA

8"

2 1/2"

WD. PL AS RQD.

10 1/8"

10

10" Z PURLIN @ 3'-9" O.C.

4"X6"X3/8" ANGLE SEAT

WALL TIES @ 1'-4" O.C. VERT.

3 1/2"X3 1/2"X3/8" ANGLE

SUSP. 3/4" PLASTER CL'G ON 3/8" ROCKLATH

6" BAT INSULATION

6"X8" PRECAST LINTEL

SEE WINDOW DETAILS

10'-2 3/4"

3'-11 3/8"

5 7/8"

STONE SILL

7'-3"

8'-8 3/8"

1" MARBLE SILL

5'8" PLASTER ON MASONRY

RUBBER COVE BASE

1'2" EXPANSION JT.

6X6 10/10 W.W.F.

2'-10 3/4"

BRICK

4"

2'-4"

1 1/2" STYROFOAM

POLYETHYLENE FILM

4"-6" SAND & GRAVEL

10"

10' CONC. BLK.

1 1/2" STYROFOAM

1'-6"

OPRB-84
Wall Section #1
(Adapted from a drawing by Paul Driscoll)

826

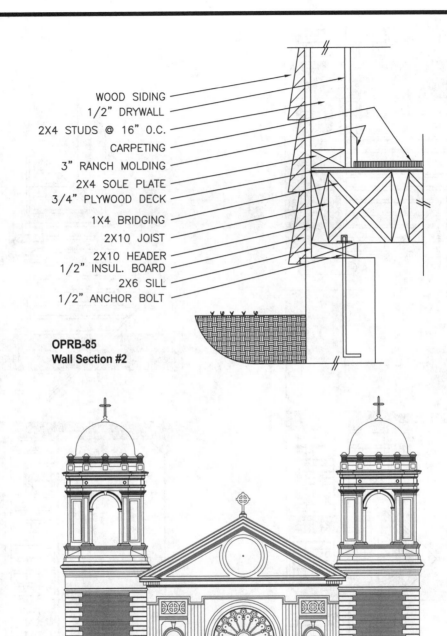

WOOD SIDING
1/2" DRYWALL
2X4 STUDS @ 16" O.C.
CARPETING
3" RANCH MOLDING
2X4 SOLE PLATE
3/4" PLYWOOD DECK
1X4 BRIDGING
2X10 JOIST
2X10 HEADER
1/2" INSUL. BOARD
2X6 SILL
1/2" ANCHOR BOLT

OPRB-85
Wall Section #2

OPRB-86
Musteadt Cathedral

(Courtesy of David Sala, Forsgren
Associates, from *World Atlas of Architecture*
(G.K. Hall and Company))

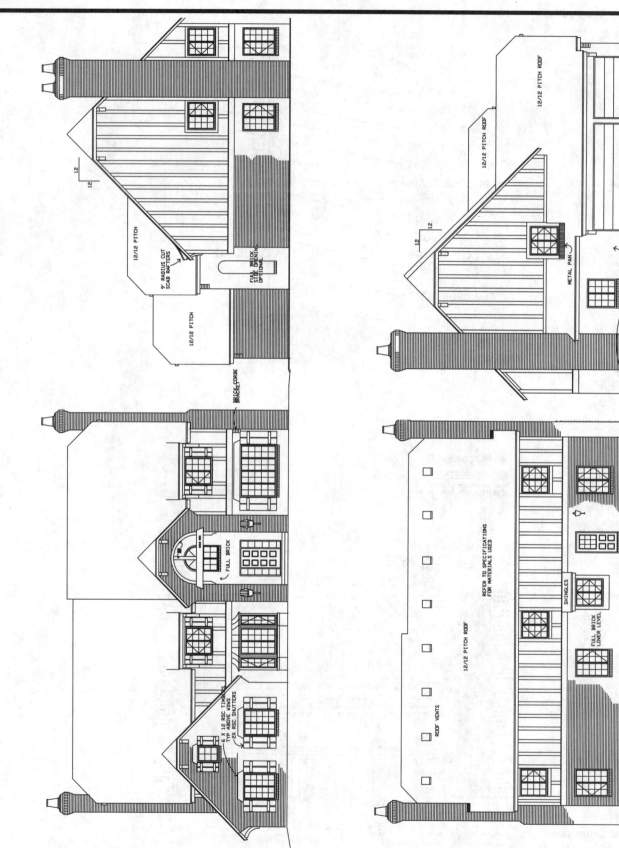

12/12 PITCH

9' RADIUS CUT
SCAB RAFTERS

FULL BRICK
SIDE OPENING
OPTIONAL

12/12 PITCH

12/12 PITCH ROOF

12/12 PITCH ROOF

METAL PAN

FULL BRICK

BRICK CORBE

FULL BRICK

6 X 12 RSC TIMBERS
TYP ABOVE WDWS
2X RSC SHUTTERS

ROOF VENTS

12/12 PITCH ROOF

REFER TO SPECIFICATIONS
FOR MATERIALS USED

SHINGLES

FULL BRICK
LOWER LEVEL

OPRB-87

Elevations (Courtesy of Rodger A. Brooks, Architect)

OPRB-88
Elevation

(Courtesy of Gary J. Hordemann,
Gonzaga University)

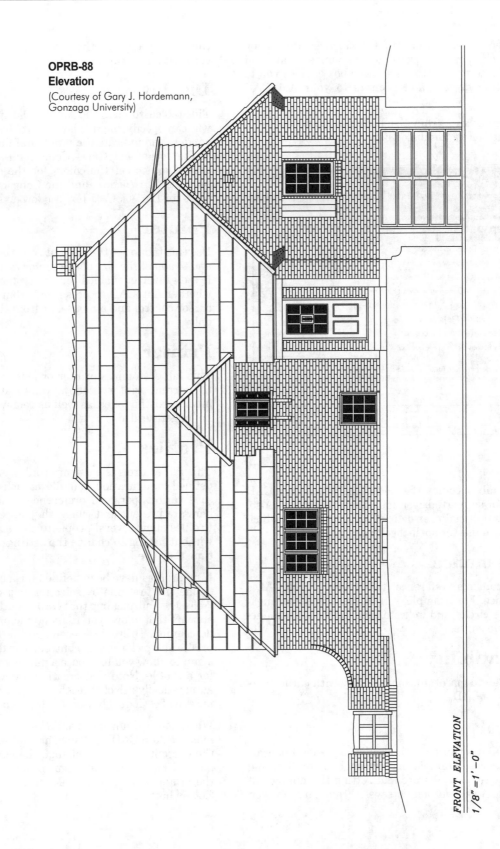

FRONT ELEVATION
1/8" = 1' -0"

829

Appendix A: Preferences Dialog Box

This appendix focuses on AutoCAD's Preferences dialog box, which permits you to customize many of AutoCAD's settings. Also, the dialog box enables you to add pointing devices and printers to your AutoCAD system.

Select Preferences... from the Tools pull-down menu to display the Preferences dialog box.

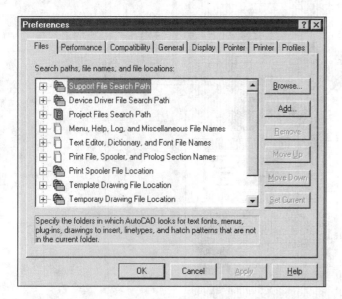

Files

The Files tab specifies the folders in which AutoCAD searches for support, driver, menu, and other files. Also, it specifies optional user-defined settings such as which dictionary to use for spell-checking.

Performance

This tab controls preferences that relate to AutoCAD's performance. For example, you can set the smoothness of arcs and circles and indicate the number of segments per polyline curve.

Compatibility

Use this tab to maintain compatibility with earlier versions of AutoCAD.

General

This tab allows you to set a number of general operating preferences. Examples include enabling or disabling the automatic save feature and setting the number of minutes between automatic saves. From this tab, you can also specify whether you want to create a backup copy with each save.

Display

This tab customizes the AutoCAD display. For instance, you can specify the display of scrollbars in the drawing window and indicate the number of lines of text to show in the docked AutoCAD - Command Line window. It also allows you to set the colors for the drawing area, the AutoCAD Text Window, and the Command line, as well as fonts for the AutoCAD Text Window and Command line.

Pointer

Use this tab to set the current AutoCAD pointing device and to adjust the size of the cursor relative to the screen. If a digitizer is enabled for the system, you can specify whether AutoCAD accepts input from both the digitizer cursor control and a mouse, or from the digitizer cursor only.

Printer

This tab customizes printer or plotter settings and sets the current printer. It allows you to add new devices and modify existing ones, as well as remove devices that are no longer needed.

Profiles

This tab controls the use of profiles. A profile is a user-defined configuration that allows each user to set up and save various personal preferences, such as which tool-bars are present and where they appear on the screen. Profiles allow several people to share a single computer without having to change the settings individually each time they use it.

If no profiles have been defined for a computer, you can define the first one by selecting (clicking) <<Unnamed Profile>> and picking the Copy... button. A dialog box appears that allows you to name the profile and give it a description. To create a second profile, highlight the first profile and pick the Copy button. The dialog box appears again to allow you to choose a new name and description for a profile. Note that creating a new profile does not automatically select it as the current profile. You must pick the Set Current button to set the active profile.

When more than one profile exists, any changes you make to AutoCAD's settings are automatically saved to the current profile. To change the current profile, display the Preferences dialog box, pick the Profiles tab, select the profile you want to make current, and pick the Set Current button.

Appendix B: AutoCAD Management Tips

Organizing an AutoCAD installation requires several considerations. For example, you should store files in specific directories (also called *folders*). Name the files and directories using standard naming conventions. If you don't, you may not be able to find the files the next time you need to edit or plot them. It is also very important that you produce backup files regularly. If you overlook these and other system management tasks, AutoCAD may not meet your expectations. You may lose days of work, causing frustration that you could otherwise avoid.

AutoCAD users and managers alike will find the following information helpful. If you manage an AutoCAD system with care, you should never have to create the same drawing twice. This comes only with careful planning and cooperation among those who will be using the system.

Bear in mind that the process of implementing, managing, and expanding an AutoCAD system evolves over time. As you and others become more familiar with AutoCAD and the importance of managing files and directories, refer back to this appendix.

AutoCAD System Manager

One person (or possibly two, but no more) within the organization should have the responsibility for managing the system and overseeing its use. This person should be the resident CAD authority and should answer questions and provide directions to other users of the system. The manager should oversee the components of the system, including software, documentation, and hardware. The manager should work with the AutoCAD users to establish procedural standards for use with the system.

Key Management Considerations

The system manager should consider questions such as those listed below when installing and organizing AutoCAD. The questions are intended to help guide your thinking, from a management perspective, as you become familiar with the various components of AutoCAD.

- How can I best categorize the files so that each directory does not grow to more than 75 files total?

- If I plan to install three or more AutoCAD stations, should I centralize the storage of user-created files and plotting by using a network and file server?

After AutoCAD is in place and you are familiar with the system, you will create many new files. The following questions address the efficiency with which you create and store these files.

- Are there template files (or existing drawings) on file that may serve as a starting point for new drawings?

- Where should new drawing files be stored, what should they be named, and how can users easily locate them?

- Are there predefined libraries of symbols and details that I can use while I develop a drawing?

- Is a custom pull-down, tablet, or image tile menu available that lends itself to my drawing application?

- Are AutoLISP routines available that would help me perform certain drafting operations more easily?

- As I create the drawings, am I using time-saving techniques, such as freezing layers and using the QTEXT command?

If you feel uncomfortable about your answers to these questions, there is probably room for improvement. The following discussion is provided to help you organize and manage your CAD system more effectively.

NOTE:

Generally, the following applies to all AutoCAD users and files. However, there are inevitable differences among users (backgrounds/interests), drawing applications, and the specific hardware and software which make up the system. Take these differences into consideration.

Software/Documentation

The AutoCAD system manager spends considerable time organizing and documenting files, establishing rules and guidelines, and tracking new software and hardware developments.

File Management—Know where files are located and the purpose of each. Understand which ones are AutoCAD system files and which are not. Create a system for making backup files, and back up regularly. Emphasize this to all users. Delete "junk" files.

Template Files—Create a simple system for the development, storage, and retrieval of AutoCAD template files. Allow for ongoing correction and development of each template. Store the template files in a directory dedicated to templates so they are accessible by other users.

Document the contents of each template file by printing the drawing status information, layers, text styles, linetype scale, status of the dimensioning variables, and other relevant information. On the first page of this information, write the name of the template file, its location, sheet size, and plot scale. Keep this information in a three-ring binder for future reference to other users.

User Drawing Files—Store these in separate directories. Place drawing components on the proper layers. Assign standard colors, linetypes, and line thicknesses to the standard layer names. Make a backup copy of each drawing and store it on a separate disk or tape backup system. Plot the drawings most likely to be used by others and store them in a three-ring binder or similar holder for future reference.

Symbol Libraries—Develop a system for ongoing library development. (See Unit 33 for details on creating symbol libraries.) Plot each symbol library drawing file, and place the library drawings on the wall near the system(s) or in a binder. Encourage users to contribute to the libraries.

Menu Files—Develop, set up, and make available custom pull-down, tablet, and image tile menu files and tablet overlays. (See Units 59-62 and 70 for details on creating pull-down, tablet, and image tile menus.) Store the menus in a directory dedicated to menus so they are accessible to others.

AutoCAD Upgrades—Handle the acquisition and installation of AutoCAD software upgrades. Inform users of the new features and changes contained in the new software. Coordinate upgrade training.

AutoCAD Third-Party Software—Handle the acquisition and installation of third-party software developed for specific applications and utility purposes. Inform users of its availability and use.

Hardware

Oversee the use and maintenance of the hardware components that make up the system. Consider hardware upgrades as user and software requirements change.

Procedural Standards

Develop clear, practical standards in the organization to minimize inconsistency and confusion. Each template file should have a standard set of drawing layers, with a specific color and linetype dedicated to each layer. For example, you may reserve a layer called Dimension, with color 2 (yellow), and a continuous linetype, for all dimensions. Then, when plotting, users assign color 2 to pen 2, which could be a .3-mm black pen. That way, whenever dimensions are on a drawing, they'll be yellow on the screen and will be plotted with a .3-mm black pen. Also, assign a specific pen to each stall on your pen plotter (if you are using one) and make this information available to others. This will avoid confusion and improve consistency within your organization. Develop similar standards for other AutoCAD-related practices.

In summary, take seriously the management of your AutoCAD system. Set up subsystems so that users can contribute to the system's ongoing development. Encourage users to experiment and to be creative by making software and hardware available to them. Make a team effort out of learning, developing, and managing the AutoCAD system so that everyone can learn and benefit from its tremendous power and capability. and utility purposes. Inform users of its availability and use.

Appendix C: Hard Disk Organization

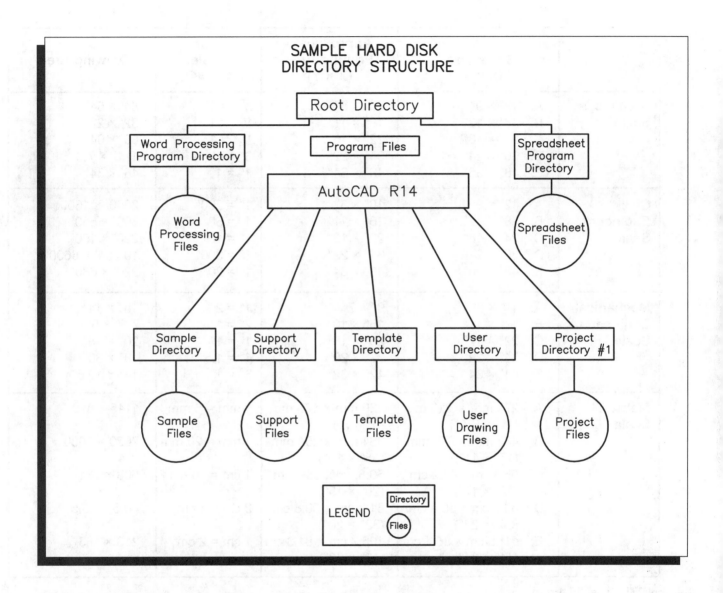

SAMPLE HARD DISK DIRECTORY STRUCTURE

- Root Directory
  - Word Processing Program Directory
    - Word Processing Files
  - Program Files
    - AutoCAD R14
      - Sample Directory
        - Sample Files
      - Support Directory
        - Support Files
      - Template Directory
        - Template Files
      - User Directory
        - User Drawing Files
      - Project Directory #1
        - Project Files
  - Spreadsheet Program Directory
    - Spreadsheet Files

LEGEND
- Directory
- Files

Hard Disk Organization

A directory, also called a folder, is a collection of related files. The diagram above shows a sample directory structure for the hard disk. Notice that all the directories (sometimes called *subdirectories*) grow from the root directory, like branches on a tree. That's why it is called a tree-structured (or *hierarchical*) directory system. Note the AutoCAD directory and the files and directories contained within it.

It is important to have a tree structure similar to the one here. The benefits of such a structure include proper categorization of files, faster retrieval of files, and better overall organization of the system.

Keep the AutoCAD directory clean of user drawing files. Use the AutoCAD directory mainly for the files illustrated above; otherwise it will grow too large and cumbersome to use effectively. Store drawing files in a subdirectory within the AutoCAD directory, as shown above. You may even want to devote subdirectories to AutoLISP files, menu files, symbol libraries, and prototype drawings. In any case, attempt to keep directories small (*i.e.,* fewer that 75 files).

Store all backup files on a separate disk or tape backup system. That way, if files are lost or damaged or if the hard disk crashes, you will have a copy of the files.

Appendix D: Drawing Area Guidelines

	Sheet Size (X × Y)	Approximate Printing Area (X × Y)	Scale	Drawing Area
Architect's Scale	A: 12″ × 9″ B: 18″ × 12″ C: 24″ × 18″ D: 36″ × 24″ E: 48″ × 36″	10″ × 8″ 16″ × 11″ 22″ × 16″ 34″ × 22″ 46″ × 34″	$1/_8″$ = 1′ $1/_2″$ = 1′ $1/_4″$ = 1′ 3″ = 1′ 1″ = 1′	80′ × 64′ 32′ × 22′ 88′ × 64′ 11.3′ × 7.3′ 46′ × 34′
Civil Engineer's Scale	A: 12″ × 9″ B: 18″ × 12″ C: 24″ × 18″ D: 36″ × 24″ E: 48″ × 36″	10″ × 8″ 16″ × 11″ 22″ × 16″ 34″ × 22″ 46″ × 34″	1″ = 200′ 1″ = 50′ 1″ = 10′ 1″ = 300′ 1″ = 20′	2000′ × 1600′ 800′ × 550′ 220′ × 160′ 10,200′ × 6600′ 920′ × 680′
Mechanical Engineer's Scale	A: 11″ × 8$1/_2$″ B: 17″ × 11″ C: 22″ × 17″ D: 34″ × 22″ E: 44″ × 34″	9″ × 7″ 15″ × 10″ 20″ × 15″ 32″ × 20″ 42″ × 32″	1″ = 2″ 2″ = 1″ 1″ = 1″ 1″ = 1.5″ 3″ = 1″	18″ × 14″ 7.5″ × 5″ 20″ × 15″ 48″ × 30″ 14″ × 10.6″
Metric Scale	A: 279 mm × 216 mm (11″ × 8$1/_2$″) B: 432 mm × 279 mm (17″ × 11″) C: 55.9 cm × 43.2 cm (22″ × 17″) D: 86.4 cm × 55.9 cm (34″ × 22″) E: 111.8 cm × 86.4 cm (44″ × 34″)	229 mm × 178 mm (9″ × 7″) 381 mm x 254 mm (15″ × 10″) 50.8 cm x 38.1 cm (20″ × 15″) 81.3 cm × 50.8 cm (32″ × 20″) 106.7 cm × 81.3 cm (42″ × 32″)	1 mm = 5 mm 1 mm = 20 mm 1 cm = 10 cm 2 cm = 1 cm 1 cm = 2 cm	1145 × 890 7620 × 5080 508 × 381 40.5 × 25.5 213 × 163

NOTE: 1″ = 25.4 mm

Appendix E: Dimensioning Symbols

Geometric Characteristic Symbols		
Type of Tolerance	**Symbol**	**Name**
Location	◎	Position
	⌖	Concentricity/coaxiality
	≡	Symmetry
Orientation	∠	Parallelism
	//	Perpendicularity
	⊥	Angularity
Form	⌭	Cylindricity
	▱	Flatness
	○	Circularity (roundness)
	—	Straightness
Profile	⌒	Profile of surface
	◠	Profile of line
Runout	↗	Circular runout
	↗↗	Total runout
Supplementary	Ⓜ	Maximum material condition (MMC)
	Ⓛ	Least material condition (LMC)
	Ⓟ	Projected tolerance zone
	⌾	All around

Dimensioning Symbols			
Symbol	**Type of Dimension**	**Symbol**	**Type of Dimension**
⌀	Diameter	∨	Countersink
R	Radius	⊔	Counterbore/Spotface
SR	Spherical radius (ISO name)	↧	Deep
S⌀	Spherical diameter (ISO name)	X	Places, times, or by
( )	Reference		

Appendix F: AutoCAD Fonts

AutoCAD provides several standard fonts, which have file extensions of SHX. You can use the STYLE command to apply expansion, compression, or obliquing to any of these fonts, thereby tailoring the characters to your needs. (See Unit 20 for details on the STYLE command.) You can draw characters of any desired height using any of the fonts.

The standard fonts supplied with AutoCAD are listed below, along with samples of their appearance. With the exception of monotxt.shx (not included in the examples below), each font's characters are proportionately spaced. Hence, the space needed for the letter "i," for example, is narrower than that needed for the letter "m."

Each font resides in a separate file with a name such as txt.shx. This is the "compiled" form of the font, for direct use by AutoCAD. Examples of standard and TrueType fonts are shown in this appendix.

Standard Fonts

AutoCAD's standard SHX fonts include both text and symbol files that have been created as AutoCAD shapes. You can change the appearance of these fonts by expanding, compressing, or slanting the characters.

Name	Description	Appearance
txt.shx	Basic AutoCAD font; this font is very simple and generates quickly on the screen	ABCDEFGHIJKLMNOPQRSTUVWXYZ abcdefghijklmnopqrstuvwxyz
romans.shx	A "simplex" roman font drawn by means of many short line segments; produces smoother characters than txt.shx, but takes longer to generate on the screen	ABCDEFGHIJKLMNOPQRSTUVWXYZ abcdefghijklmnopqrstuvwxyz
romand.shx	Similar to romans.shx, but instead of a single stroke, it uses a double stroke technique to produce darker, thicker lines	ABCDEFGHIJKLMNOPQRSTUVWXYZ abcdefghijklmnopqrstuvwxyz0123456789
itallicc.shx	Complex italic font using double stroke and serifs	*ABCDEFGHIJKLMNOPQRSTUVWXYZ abcdefghijklmnopqrstuvwxyz0123456789*
scripts.shx	A single-stroke (simplex) script font	*ABCDEFGHIJKLMNOPQRSTUVWXYZ abcdefghijklmnopqrstuvwxyz0123456789*
gothice.shx	Gothic English font	𝕬𝖁𝕮𝕯𝕰𝕱𝕲𝕳𝕴𝕵𝕶𝕷𝕸𝕹𝕺𝕻𝕼𝕽𝕾𝕿𝖀𝖁𝖂𝖃𝖄𝖅 abcdefghijklmnopqrstuvwxyz0123456789
syastro.shx	A symbol font that includes common astronomical symbols	☉♀♀⊕♂♃♄♅Ψⅇℂ☿✳︎♈♉♊♋♌♍♎♏♐♑♒♓≈ ✳︎"∪∪∩∈→↑←↓∂∇∧^´`✗§†‡ℲℲℒ®©
symusic.shx	A symbol font that includes common music symbols	(common music symbols)

836

TrueType Fonts

AutoCAD uses several TrueType fonts and font "families" (groups of related fonts). In the TrueType fonts, each font in a family has its own font file. The examples below are shown in outline form, as they appear by default in AutoCAD. To display solid characters, set the TEXTFILL system variable to 1. You may also change the print quality by changing the value of the TEXTQLTY system variable.

Name	Appearance
Arial Narrow	ABCDEFGHIJKLMNOPQRSTUVWXYZ abcdefghijklmnopqrstuvwxyz0123456789
Dutch801 RmBT	*ABCDEFGHIJKLMNOPQRSTUVWXYZ abcdefghijklmnopqrstuvwxyz0123456789*
Lucida Console	ABCDEFGHIJKLMNOPQRSTUVWXYZ abcdefghijklmnopqrstuvwxyz0123456789
Swis721 BdOul BT	ABCDEFGHIJKLMNOPQRSTUVWXYZ abcdefghijklmnopqrstuvwxyz0123456789
UniversalMath1 BT	ΑΒΨΔΕΦΓΗΙΞΚΛΜΝΟΠΘΡΣΤΘΩϭΧΥΖ αβψδεφγηιξκλμνοπϑρστθωφχυζ″ + − × ÷ = ± ∓°′
Wingdings	(symbol characters)

Appendix G: Standard Hatch Patterns

Shown here are the standard hatch patterns supplied in the file acad.pat.

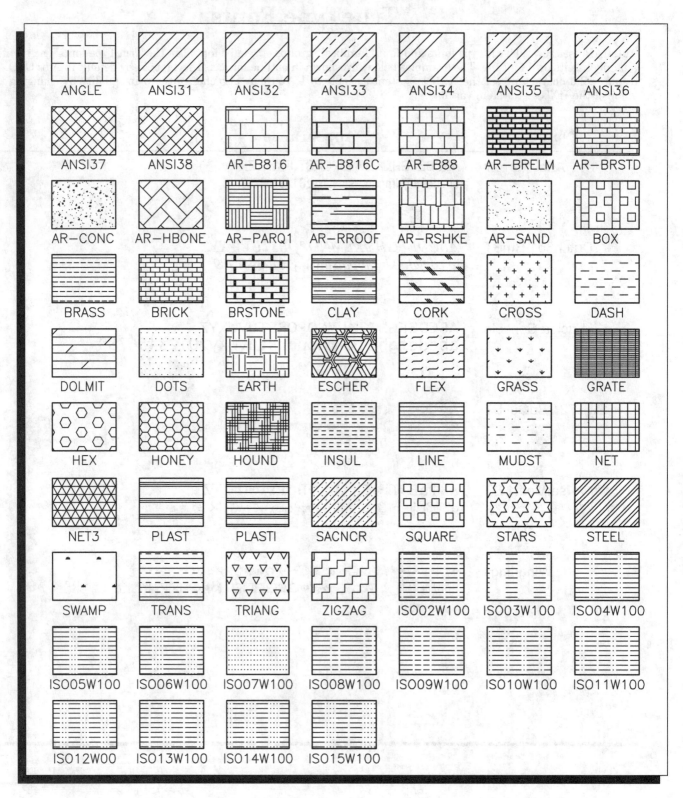

Appendix H: Toolbars

AutoCAD comes with a collection of 16 toolbars, each consisting of several icons that enter commands and options. Some of the toolbars also contain flyouts—sets of additional icons that you can access by holding down the pick button on the pointing device over certain icons. Icons that display flyouts contain a small black triangle in the lower right corner.

This appendix lists, in alphabetical order, the AutoCAD toolbars and the icons they contain. The Insert flyout is shown to the right of the Draw toolbar, from which it originates. The other flyouts, all of which originate from the Standard toolbar, also occur as standalone toolbars. For these flyouts, a note has been placed at the bottom of the toolbar indicating which icon in the Standard toolbar presents the flyout.

For more information about each icon, refer to the index.

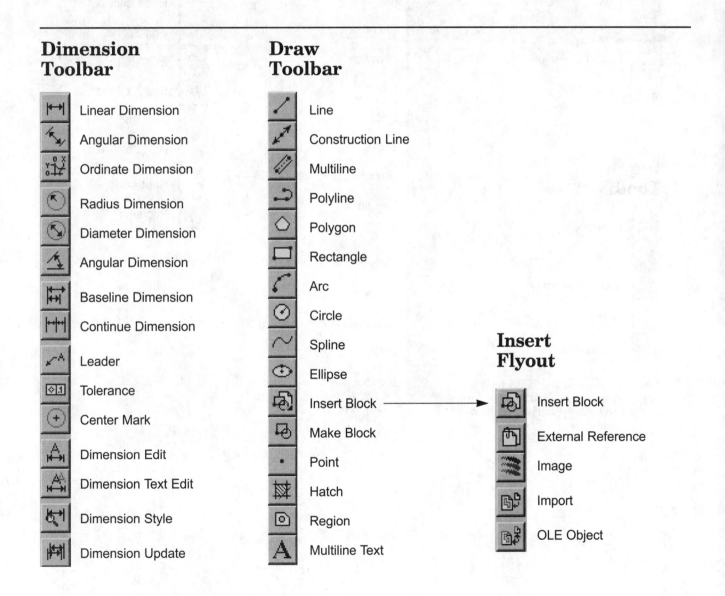

Dimension Toolbar

- Linear Dimension
- Angular Dimension
- Ordinate Dimension
- Radius Dimension
- Diameter Dimension
- Angular Dimension
- Baseline Dimension
- Continue Dimension
- Leader
- Tolerance
- Center Mark
- Dimension Edit
- Dimension Text Edit
- Dimension Style
- Dimension Update

Draw Toolbar

- Line
- Construction Line
- Multiline
- Polyline
- Polygon
- Rectangle
- Arc
- Circle
- Spline
- Ellipse
- Insert Block
- Make Block
- Point
- Hatch
- Region
- Multiline Text

Insert Flyout

- Insert Block
- External Reference
- Image
- Import
- OLE Object

External Database Toolbar

	Administration
	Rows
	Links
	Select Objects
	Export Links
	SQL Editor

Inquiry Toolbar

	Distance
	Area
	Mass Properties
	List
	Locate Point

Note: The icons on the Inquiry toolbar also appear as a flyout from the Distance icon on the Standard toolbar.

Modify Toolbar

	Erase
	Copy Object
	Mirror
	Offset
	Array
	Move
	Rotate
	Scale
	Stretch
	Lengthen
	Trim
	Extend
	Break
	Chamfer
	Fillet
	Explode

Modify II Toolbar

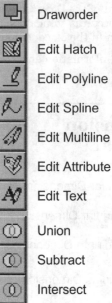

	Draworder
	Edit Hatch
	Edit Polyline
	Edit Spline
	Edit Multiline
	Edit Attribute
	Edit Text
	Union
	Subtract
	Intersect

Object Properties Toolbar

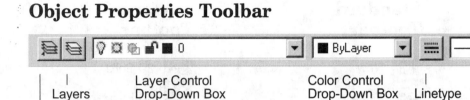

	Layer Control	Color Control		Linetype Control	
Layers	Drop-Down Box	Drop-Down Box	Linetype	Drop-Down Box	Properties

Make Object's Layer Current

Object Snap Toolbar

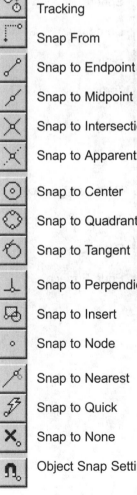

Tracking

Snap From

Snap to Endpoint

Snap to Midpoint

Snap to Intersection

Snap to Apparent Intersect

Snap to Center

Snap to Quadrant

Snap to Tangent

Snap to Perpendicular

Snap to Insert

Snap to Node

Snap to Nearest

Snap to Quick

Snap to None

Object Snap Settings

Note: The icons on the Object Snap toolbar also appear as a flyout from the Tracking icon on the Standard toolbar.

Reference Toolbar

External Reference

External Reference Attach

External Reference Clip

External Reference Bind

External Reference Clip Frame

Image

Image Attach

Image Clip

Image Adjust

Image Quality

Image Transparency

Image Frame

Render Toolbar

Hide

Shade

Render

Scenes

Lights

Materials

Materials Library

Mapping

Background

Fog

Landscape New

Landscape Edit

Landscape Library

Render Preferences

Statistics

Solids
Toolbar

Box
Sphere
Cylinder
Cone
Wedge
Torus
Extrude
Revolve
Slice
Section
Interfere
Setup Drawing
Setup View
Setup Profile

Standard
Toolbar

New
Open
Save
Print
Print Preview
Spelling
Cut
Copy
Paste
Match Properties
Undo
Redo
Launch Browser
Tracking
UCS
Distance
Redraw All
Aerial View
Named Views
Pan Realtime
Zoom Realtime
Zoom Window
Zoom Previous
Help

Surfaces
Toolbar

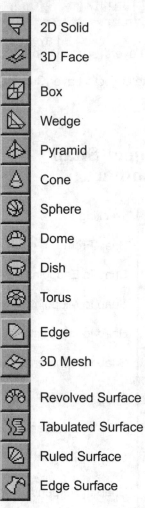

2D Solid
3D Face
Box
Wedge
Pyramid
Cone
Sphere
Dome
Dish
Torus
Edge
3D Mesh
Revolved Surface
Tabulated Surface
Ruled Surface
Edge Surface

UCS
Toolbar

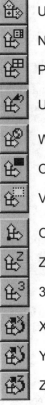

- UCS
- Named UCS
- Preset UCS
- UCS Previous
- World UCS
- Object UCS
- View UCS
- Origin UCS
- Z-Axis Vector UCS
- 3-Point UCS
- X Axis Rotate UCS
- Y Axis Rotate UCS
- Z Axis Rotate UCS

Note: The icons on the UCS toolbar also appear as a flyout from the UCS icon on the Standard toolbar.

Viewpoint
Toolbar

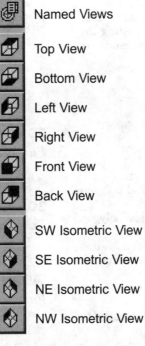

- Named Views
- Top View
- Bottom View
- Left View
- Right View
- Front View
- Back View
- SW Isometric View
- SE Isometric View
- NE Isometric View
- NW Isometric View

Note: The icons on the Viewpoint toolbar also appear as a flyout from the Named Views icon on the Standard toolbar.

Zoom
Toolbar

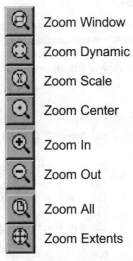

- Zoom Window
- Zoom Dynamic
- Zoom Scale
- Zoom Center
- Zoom In
- Zoom Out
- Zoom All
- Zoom Extents

Note: The icons on the Zoom toolbar also appear as a flyout from the Zoom Window icon on the Standard toolbar.

Appendix I: AutoCAD Command Glossary

This appendix lists brief descriptions of the AutoCAD commands. For the locations of more detailed descriptions and instructions on how to apply them, refer to the index.

Some commands can be used transparently (that is, used while another command is in progress) by preceding the command name with an apostrophe. Such commands are listed here with an apostrophe.

3D

Purpose

Creates a three-dimensional polygon mesh object

Options

B	Creates a 3D box polygon mesh
C	Creates a cone-shaped polygon mesh
D	Creates the lower half of a spherical polygon mesh
D	Creates the upper half of a spherical polygon mesh
M	Creates a polygon mesh whose *M* and *N* sizes determine the number of lines drawn in each direction along the mesh
P	Creates a pyramid or a tetrahedron
S	Creates a spherical polygon mesh
T	Creates a toroidal polygon mesh parallel to the XY plane of the current UCS
W	Creates a right-angle wedge-shaped polygon mesh with the sloped face tapering along the X axis.

3DARRAY

Purpose

Creates a three-dimensional array

Options

| R | Creates a rectangular array |
| P | Creates a polar array |

3DFACE

Purpose

Creates a three-dimensional face

3DMESH

Purpose

Creates a free-form polygon mesh

3DPOLY

Purpose

Creates a polyline of straight line segments in 3D space

Options

E	Draws a straight line from the previous point to the specified new point
C	Draws a closing line back to the first point and ends the command
U	Deletes the last line and allows you to continue drawing from the previous point

3DSIN

Purpose

Presents a dialog box to control the import of a 3D Studio (3DS) file

3DSOUT

Purpose

Presents a dialog box to control the export of a 3D Studio (3DS) file

'ABOUT

Purpose

Displays the AutoCAD version and serial number, license information, and the contents of the acad.msg file

ACISIN

Purpose

Presents a dialog box from which you can select an ACIS file to import

ACISOUT

Purpose

Presents a dialog box from which you can export an AutoCAD solid, body, or region to an ACIS file

ALIGN

Purpose

Moves and rotates objects in two or three dimensions to align with other objects using one, two, or three sets of points

AMECONVERT

Purpose

Converts AME solid models to AutoCAD solid objects

'APERTURE

Purpose

Controls the size of the object snap target box

'APPLOAD

Purpose

Presents a dialog box to load and unload AutoLISP, ADS, and ARX applications

ARC

Purpose

Creates an arc

Options

C	Center point
ENTER	(as reply to Start point) sets start point and direction tangent to last line or arc

AREA

Purpose

Calculates the area and perimeter of objects or defined areas

Options

A	Sets Add mode
S	Sets Subtract mode
O	Computes area of a selected object

ARRAY

Purpose

Makes multiple copies of selected objects in a specified pattern

Options

P	Polar (circular) array
R	Rectangular array

ARX

Purpose

Loads, unloads, and provides information about ARX applications

Options

L	Loads an ARX application
U	Unloads an ARX application
O	Options (for developers of ARX applications)

ASEADMIN

Purpose

Displays a dialog to perform administrative functions for external database commands

ASEEXPORT

Purpose

Presents a dialog box to export link information for selected objects to external database files

ASELINKS

Purpose

Presents a dialog box to manipulate links between objects and an external database

ASEROWS

Purpose

Presents a dialog box that displays and edits table data and creates links and selection sets

ASESELECT

Purpose

Presents a dialog box to create a selection set from rows linked to textual and graphic selection sets

ASESQLED

Purpose

Presents a dialog box from which you can execute Structured Query Language (SQL) statements

ATTDEF

Purpose

Creates an attribute definition entity

Options

I	Controls attribute visibility
C	Controls constant/variable mode
V	Controls verify mode
P	Controls preset mode

'ATTDISP

Purpose

Controls the visibility of attribute entities on a global basis

Options

ON	Makes all attributes visible
OFF	Makes all attributes invisible
N	(Normal) Keeps the current visibility of each attribute (visible attributes are displayed, but invisible attributes are not)

ATTEDIT

Purpose

Permits editing of attributes

ATTEXT

Purpose

Extracts attribute data from a drawing

Options

C	CDF (comma-delimited) format extract
D	DXF format extract
S	SDF (space-delimited) format extract
O	Extracts attributes from selected objects

ATTREDEF

Purpose

Redefines a block and updates associated attributes

AUDIT

Purpose

Evaluates the integrity of a drawing

Options

Y	Fixes errors encountered
N	Reports, but does not fix, errors encountered

BACKGROUND

Purpose

Sets up the background for a scene

'BASE

Purpose

Sets the insertion base point for the current drawing

BHATCH

Purpose

Presents a dialog box from which you can create an associative hatch pattern within an automatically defined boundary; allows you to preview the hatch and make repeated adjustments without starting over each time

'BLIPMODE

Purpose

Controls the display of marker blips

Options

ON	Enables temporary marker blips
OFF	Disables temporary marker blips

BLOCK

Purpose

Creates a block definition from a group of selected objects

Option

?	Lists specified names of defined blocks

BMAKE

Purpose

Defines a block using a dialog box

BMPOUT

Purpose

Saves selected objects to a file in device-independent bitmap format

BOUNDARY

Purpose

Presents a dialog box from which you can create a region or polyline of a boundary enclosed by overlapping objects

BOX

Purpose

Creates a three-dimensional solid box

Options

C	Creates a box using a specified center point
ENTER	Defines the first corner of the box

BREAK

Purpose

Erases part of an object or splits an object in two

Option

F	Respecifies first point

BROWSER

Purpose

Launches the default Web browser defined in the system registry

'CAL

Purpose

Evaluates mathematical and geometric expressions

CHAMFER

Purpose

Creates a chamfer at the intersection of two lines

Options

P	Chamfers an entire 2D polyline
D	Sets the chamfer distances from the selected edge
A	Sets the chamfer distances using a specified distance and an angle
T	Controls whether AutoCAD trims the selected edges to the chamfer line endpoints
M	Controls whether AutoCAD uses two distances or a distance and an angle to create the chamfer

CHANGE

Purpose

Changes the properties of existing objects

Options

P	Changes common properties of objects
C	Color
E	Elevation
LA	Layer
LT	Linetype
S	Linetype scale
T	Thickness

CHPROP

Purpose

Changes the color, layer, linetype, linetype scale factor, and thickness of an object

Options

C	Color
LA	Layer
LT	Linetype
S	Linetype scale
T	Thickness

CIRCLE

Purpose

Creates a circle

Options

C	Draws a circle based on a center point and diameter or radius
3P	Draws a circle based on three points on the circumference
2P	Draws a circle based on two endpoints of the diameter
TTR	Draws a circle tangent to two objects with a specified radius

'COLOR

Purpose

Sets the color for subsequently drawn objects

Options

(number)	Sets entity color number
(name)	Sets entity color to standard color name
BYBLOCK	Sets floating entity color
BYLAYER	Uses layer's color for entities

COMPILE

Purpose

Presents a dialog box from which you can compile shape files and PostScript font files

CONE

Purpose

Creates a three-dimensional solid cone

Option

E	Creates a cone with an elliptical base

CONVERT

Purpose

Converts 2D polylines and associative hatches to the optimized R14 format

COPY

Purpose

Draws a copy of selected objects

Option

M	Makes multiple copies of the selected objects

COPYCLIP

Purpose

Copies objects to the Windows Clipboard

COPYHIST

Purpose

Copies text from the command line history window to the Windows Clipboard

COPYLINK

Purpose

Copies the current view to the Windows Clipboard for linking the AutoCAD view to another OLE-capable application

CUTCLIP

Purpose

Copies objects to the Windows Clipboard and erases them from the drawing

CYLINDER

Purpose

Creates a three-dimensional solid cylinder

Option

E	Creates a cylinder with an elliptical base

DBLIST

Purpose

Lists database information for every object in the drawing

DDATTDEF

Purpose

Presents a dialog box from which you can create an attribute definition

DDATTE

Purpose

Allows attribute editing via a dialog box

DDATTEXT

Purpose

Displays a dialog box from which you can extract data from a drawing; available formats are DXF, CDF, SDF, or selected objects

DDCHPROP

Purpose

Presents a dialog box from which you can change the color, layer, linetype, linetype scale, and thickness of an object

DDCOLOR

Purpose

Displays a dialog box from which you can select a new color for subsequently drawn objects

DDEDIT

Purpose

Presents a dialog box from which you can edit text and attribute definitions

'DDGRIPS

Purpose

Displays a dialog box from which you can enable grips and set their color and size

DDIM

Purpose

Presents a dialog box and series of subdialog boxes from which you can create and modify dimension styles

DDINSERT

Purpose

Presents a dialog box from which you can insert a block or another drawing and set an insertion point and scale; also allows you to rotate or explode the part

DDMODIFY

Purpose

Presents a dialog box from which you can control object properties; the dialog box presented is specific to the type of object you select to modify

'DDPTYPE

Purpose

Presents a dialog box from which you can specify the display mode and size of point objects

DDRENAME

Purpose

Presents a dialog box from which you can change the names of named objects such as blocks and layers

'DDRMODES

Purpose

Presents a dialog box from which you can set drawing aids such as ortho, grid, and snap

'DDSELECT

Purpose

Displays a dialog box from which you can set object selection modes, size of the pick box, and the object sort method

DDUCS

Purpose

Presents a dialog box from which you can manage defined user coordinate systems in the current space

DDUCSP

Purpose

Presents a dialog box from which you can select a preset user coordinate system

'DDUNITS

Purpose

Displays a dialog box that allows you to set coordinate and angle display formats and precision

DDVIEW

Purpose

Displays a dialog box from which you can create and restore views

'DDVPOINT

Purpose

Presents a dialog box that allows you to set the three-dimensional viewing direction

DELAY

Purpose

Provides a timed pause within a script

DIM

Purpose

Provides compatibility with previous releases of AutoCAD; enters the Dimensioning mode, in which dimensioning subcommands can be used to dimension objects.

DIMALIGNED

Purpose

Creates an aligned linear dimension

DIMANGULAR

Purpose

Creates an angular dimension

DIMBASELINE

Purpose

Continues a linear, angular, or ordinate dimension from the baseline of the previous or selected dimension

DIMCENTER

Purpose

Creates the center mark or the center lines of circles and arcs

DIMCONTINUE

Purpose

Continues a linear, angular, or ordinate dimension from the second extension line of the previous or a selected dimension

DIMDIAMETER

Purpose

Creates diameter dimensions for circles and arcs

DIMEDIT

Purpose

Edits dimensions

Options

H	Moves dimension text that has been moved back to its default position
N	Changes dimension text
R	Rotates dimension text
O	Adjusts the obliquing angle of the extension lines for linear dimensions

DIMLINEAR

Purpose

Creates linear (vertical and horizontal) dimensions

DIMORDINATE

Purpose

Creates ordinate point dimensions

Options

X	Measures the x coordinate and determines orientation of the leader line and dimension text
Y	Measures the y coordinate and determines orientation of the leader line and dimension text
M	Allows you to customize mtext objects
T	Allows you to customize the text

DIMOVERRIDE

Purpose

Overrides dimension system variables

Option

C	Clears previous overrides

DIMRADIUS

Purpose

Creates radial dimensions for circles and arcs

DIMSTYLE

Purpose

Creates and modifies dimension styles at the command line

Options

R	Changes the dimensioning system variable settings by reading new settings from an existing dimension style
S	Saves the current settings of dimensioning system variables to a dimension style
ST	Displays the current values of all dimensioning system variables
V	Lists the dimensioning system variable settings of a dimension style without modifying the current settings
A	Updates the dimension objects you select so that they use the current settings of the dimensioning system variables

DIMTEDIT

Purpose

Moves and rotates dimension text

Options

L	Left justifies the dimension text along the dimension line for linear, radial, and diameter dimensions

R	Right justifies the dimension text along the dimension line for linear, radial, and diameter dimensions
H	Moves dimension text that has been moved back to its default position
A	Changes the angle of the dimension text

'DIST

Purpose

Measures the distance and angle between two points

DIVIDE

Purpose

Places evenly spaced point objects or blocks along the length or perimeter of an object

Option

| B | Places blocks at a specified interval along the selected object |

DONUT

Purpose

Draws filled circles and rings

'DRAGMODE

Purpose

Controls the way dragged objects are displayed

Options

ON	Permits dragging, but you must enter DRAG where appropriate to initiate dragging
OFF	Ignores all dragging requests, including those embedded in menu items
Auto	Turns on dragging for every command that supports it and performs drags automatically so that you do not have to enter DRAG each time

DRAWORDER

Purpose

Changes the display order of objects and images

Options

A	Moves an object above a specified reference object
U	Moves an object below a specified reference object
F	Object moves to the top of the drawing order
B	Object moves to the bottom of the drawing order

DSVIEWER

Purpose

Opens the Aerial View

DTEXT

Purpose

Draws text items dynamically (displays the text on screen as it is entered)

Options

| J | Presents justification options to control text alignment |
| S | Sets the text style |

DVIEW

Purpose

Defines parallel projection or perspective views

Options

X	Ends the DVIEW command
CA	Specifies a new camera position by rotating the camera about the target point
TA	Specifies a new position for the target by rotating it around the camera
D	Moves the camera in or out along the line of sight relative to the target; turns on perspective viewing
PO	Locates the camera and target points using x,y,z coordinates
PA	Shifts the image without changing the level of magnification
Z	If perspective viewing is off, performs the equivalent of a ZOOM Center; if perspective viewing is on, adjusts the camera lens length, which changes the field of view and causes more or less of the drawing to be visible at a given camera and target distance
TW	Twists or tilts the view around the line of sight
CL	Clips the view, obscuring portions of the drawing that are behind or in front of the front clipping plane
H	Performs hidden line suppression on selected objects
O	Turns off perspective viewing
U	Reverses the effects of the last DVIEW operation

DWFOUT

Purpose

Exports a drawing in web format file in user-specified precision; allows use of file compression at user's discretion

DXBIN

Purpose

Displays a dialog box from which you can import specially coded binary files

DXFIN

Purpose

Displays a dialog box from which you can import DXF (drawing interchange) files

DXFOUT

Purpose

Displays a dialog box from which you can export DXF (drawing interchange) files

EDGE

Purpose

Changes the visibility of three-dimensional face edges

Options

D	Highlights invisible edges of 3D faces so that you can edit them
S	Hides selected edges

EDGESURF

Purpose

Creates a 3D polygon mesh approximating a Coons surface patch (a bicubic surface interpolated between four adjoining edges)

'ELEV

Purpose

Sets elevation and extrusion thickness of new objects

ELLIPSE

Purpose

Creates an ellipse or an elliptical arc

Options

A	Creates an elliptical arc
I	Creates an isometric circle in the current isometric drawing plane
C	Creates the ellipse using a specified center point

ERASE

Purpose

Removes objects from a drawing

EXPLODE

Purpose

Breaks a block, polyline, or other compound object into its component parts

EXPORT

Purpose

Saves objects to other file formats

EXTEND

Purpose

Extends a line, arc, elliptical arc, open 2D and 3D polyline, or ray to meet another object

Options

P	Specifies the projection mode AutoCAD uses when extending objects (none, UCS, or view)
E	Determines whether the object is extended to another object's implied edge or only to an object that actually intersects it in 3D space
U	Reverses the most recent change made by EXTEND

EXTRUDE

Purpose

Creates unique solid primitives by extruding existing two-dimensional objects

Options

H	Specifies height of extrusion
P	Selects the extrusion path based on a specified curved object

'FILL

Purpose

Controls the filling of multilines, traces, solids, and wide polylines

Options

ON	Enables fill mode
OFF	Disables fill mode

FILLET

Purpose

Rounds and fillets the edges of objects

Options

P	Inserts fillet arcs at each vertex of a 2D polyline
R	Defines the radius of the fillet arc
T	Controls whether AutoCAD trims the selected edges to the fillet arc endpoints

'FILTER

Purpose

Presents a dialog box from which you can create lists to select objects based on properties

FOG

Purpose

Provides visual cues for the apparent distance of objects; allows use of fog and depth cueing, in which white is "fog" and traditional depth cuing is black. With the FOG command, the user can specify any color in between these two extremes to provide visual depth cues to a drawing

'GRAPHSCR

Purpose

Switches from the text screen to the graphics screen; used in command scripts and menus

'GRID

Purpose

Displays a grid of dots at specified spacing on the screen

Options

ON	Turns on the grid at the current spacing
OFF	Turns off the grid
S	Sets the grid spacing to the current snap interval (as set by the SNAP command)
A	Sets the grid to a different spacing on the X and Y axes

GROUP

Purpose

Presents a dialog box from which you can create a named selection set of objects

HATCH

Purpose

Fills an area with a hatch pattern

Options

?	Lists pattern names with brief descriptions
S	Specifies a solid fill
U	Allows you to specify a user-defined pattern name

HATCHEDIT

Purpose

Presents a dialog box from which you can modify an existing associative hatch block

'HELP

Purpose

Displays on-line help. (The F1 function key also performs this function.)

HIDE

Purpose

Regenerates a three-dimensional model with hidden lines suppressed

'ID

Purpose

Displays the coordinates of a location

IMAGE

Purpose

Inserts images in various formats into an AutoCAD drawing file

Options

These options can be entered by picking buttons in the dialog box or by entering them at the keyboard

?	Lists image defined in the drawing database
A	Displays the Attach Image dialog box, from which various formats can be attached to the drawing
D	Detaches an image from an AutoCAD drawing file
U	Unloads an image from working memory, but image remains in the drawing database
T	Shows a preview of the image (if one is available) and lists details such as paths, creation date, file size, file type, color, and so on.
B	Allows the user to browse other directories to find the specific image to load
P	(Path) Allows the user to update the path name associated with an image

IMAGEADJUST

Purpose

Opens a dialog box from which the user can control brightness, contrast, and fade values of an image in the drawing database

Options

C	(Contrast) Controls contrast in the image on a scale of 0 through 100, with 100 being the brightest
F	(Fade) Controls how an image fades into the background on a scale of 0 through 100; the higher the value, the more each pixel is forced to its primary or secondary color
B	(Brightness) Controls brightness of an image on a scale of 0 through 100, with 100 being the brightest
Note:	From the dialog box, you can also preview the image and reset the above options to their default settings.

IMAGEATTACH

Purpose

Opens the Attach Image dialog box directly from the keyboard; allows the user to specify the image name, as well as parameters such as scale factor, rotation angle, and the option to specify parameters on-screen

IMAGECLIP

Purpose

Creates new clipping boundaries for single-image objects

Options

OFF	Turns off image clipping and shows the entire drawing

| ON | Turns on image clipping and clips to a previously defined boundary |
| Delete | Deletes a clipping boundary and displays the entire image |

IMAGEFRAME

Purpose

Controls whether the image frame is displayed on the screen or hidden from view

Options

| ON | Turns on the frame so it can be seen on the screen |
| OFF | Removes the frame from the screen |

IMAGEQUALITY

Purpose

Controls display quality of images

Options

| High | Produces high-quality images, but they are displayed and plotted more slowly than draft-quality images |
| Draft | Produces lower-quality images, but is faster than using the High option |

IMPORT

Purpose

Displays a dialog box that allows you to import various file formats into AutoCAD

INSERT

Purpose

Inserts a named block or drawing into the current drawing

Options

| B | Allows you to enter the name of a block or drawing |
| ? | Lists named blocks in the current drawing |

INSERTOBJ

Purpose

Inserts a linked or embedded object into AutoCAD

INTERFERE

Purpose

Finds the interferences of two or more solids and creates a composite solid from their common volume

INTERSECT

Purpose

Creates composite solids or regions from the intersection of two or more solids or regions

'ISOPLANE

Purpose

Specifies the current isometric plane

Options

RETURN	Toggles to the next plane in a clockwise fashion from left to top to right
L	Selects the left isometric plane
T	Selects the top isometric plane
R	Selects the right isometric plane

'LAYER

Purpose

Displays a dialog box that permits you to create named drawing layers and assigns color and linetype properties to those layers

LEADER

Purpose

Creates a line that connects annotation to a feature

Options

F	Controls the way the leader is drawn and whether it has an arrowhead
A	Inserts annotation at the end of the leader line; the annotation can be text, a feature control frame, a block or an mtext object
U	Undoes the last vertex point on the leader line

LENGTHEN

Purpose

Changes the length of objects and the included angle of arcs

Options

DE	Changes the length of an object or the included angle of an arc by a specified incremental length; a positive value extends the entity, and a negative value trims the entity
P	Sets the length of an object by a specified percentage of its total length; sets the included angle of an arc by a specified percentage of the total angle of the selected arc
T	Sets the length of a selected object by specifying the total absolute length; sets the total angle of a selected arc by a specified total included angle
DY	Enters dynamic dragging mode; changes the length of a selected object based on where its endpoint is dragged

LIGHT

Purpose

Presents a dialog box from which you can manage lights and lighting effects

'LIMITS

Purpose

Sets and controls the drawing area

Options

ON   Turns on limits checking; causes AutoCAD to reject attempts to enter points outside the drawing limits

OFF   Turns off limits checking but maintains the current values for the next time limits checking is set to ON

LINE

Purpose

Creates straight line segments

Options

RETURN   Begins the current line at the last endpoint of the most recently drawn line

U   Undoes the most recent line segment

C   Draws a line segment from the most recent endpoint of a line to the first point, creating a closed polygon

'LINETYPE

Purpose

Displays a dialog box that enables you to create, load, and set linetypes

LIST

Purpose

Displays database information for selected objects

LOAD

Purpose

Presents a dialog box from which you can make shapes available for use by the SHAPE command

LOGFILEOFF

Purpose

Closes the log file opened by LOGFILEON

LOGFILEON

Purpose

Writes the text window contents to a file; the default log file name is acad.log

LSEDIT

Purpose

Allows you to edit a landscape object

LSLIB

Purpose

Allows you to maintain libraries of landscapes by modifying or deleting landscape options

LSNEW

Purpose

Adds realistic landscape items such as trees to your drawings

'LTSCALE

Purpose

Sets the linetype scale factor

MASSPROP

Purpose

Calculates and displays the mass properties of regions and solids

MATCHPROP

Purpose

Copies the properties from one object to one or more objects

MATLIB

Purpose

Presents a dialog box from which you can import and export materials to and from a library of materials

MEASURE

Purpose

Places point objects or blocks at measured intervals on an object

Option

B   Places blocks at a specified interval along the selected object

MENU

Purpose

Presents a dialog box from which you can load a menu file

MENULOAD

Purpose

Loads partial menu files

MENUUNLOAD

Purpose

Unloads partial menu files

MINSERT

Purpose

Inserts multiple instances of a block in a rectangular array

Options

?   Lists currently defined block definitions in the drawing

~   Displays the Select Drawing File dialog box

MIRROR

Purpose

Creates a mirror-image copy of objects

MIRROR3D

Purpose

Creates a mirror-image copy of objects about a plane

Options

3	Defines the mirroring plane by three points
O	Uses the plane of a selected planar object as the mirroring plane
L	Mirrors the selected objects about the last defined mirroring plane
Z	Defines the mirroring plane by a point on the plane and a point normal to the plane (on the Z axis)
V	Aligns the mirroring plane to the viewing plane of the current viewport through a point
XY	Aligns the mirroring plane with the standard XY plane through a specified point
YZ	Aligns the mirroring plane with the standard YZ plane through a specified point
ZX	Aligns the mirroring plane with the standard ZX plane through a specified point

MLEDIT

Purpose

Presents a dialog box from which you can edit multiple parallel lines

MLINE

Purpose

Creates multiple parallel lines

Options

J	Determines how the multiline is drawn between the points you specify; available choices are Top (draws the multiline below the cursor so that the top line aligns with the specified points), Zero (draws the multiline with its origin entered at the cursor so that the multiline is centered around the specified points), and Bottom (draws the multiline above the cursor so that the bottom line aligns with the specified points)
S	Controls the overall width of the multiline; does not affect linetype scale
ST	Specifies a style to use for the multiline

MLSTYLE

Purpose

Presents a dialog box from which you can create, load, or rename multiline styles and control the properties of each style

MOVE

Purpose

Displaces objects a specified distance in a specified direction

MSLIDE

Purpose

Presents a dialog box from which you can create a slide file of the current viewport

MSPACE

Purpose

Switches from paper space to a model space viewport

MTEXT

Purpose

Creates paragraph text

Options

H	Specifies the height of uppercase text
J	Determines both justification and text flow, for new or selected text, in relation to the text boundary
R	Specifies the rotation angle of the text boundary
S	Specifies the text style to use for paragraph text
W	Specifies the width of the multiline text object

MULTIPLE

Purpose

Repeats the next command until canceled

MVIEW

Purpose

Creates floating viewports and turns on existing floating viewports

Options

ON	Turns on a viewport
OFF	Turns off a viewport
H	Removes hidden lines from a viewport during plotting from paper space
F	Creates one viewport that fills the available display area
2	Divides the specified area horizontally or vertically into two viewports
3	Divides the specified area into three viewports
4	Divides the specified area horizontally and vertically into four viewports of equal size
R	Translates viewport configurations saved with the VPORTS command into individual viewports in paper space

MVSETUP

Purpose

Sets up the specifications of a drawing; available only in paper space (when the TILEMODE system variable is set to 0)

Options

A   Pans the view in a viewport so that it aligns with a basepoint in another viewport

C   Creates viewports

S   Adjusts the scale factor (the ratio between the scale of the border in paper space and the scale of the drawing objects displayed in the viewports) of the objects displayed in the viewports

O   Sets MVSETUP preferences before you change the drawing; you can set the layer on which to insert the title block, specify whether to reset the limits to the drawing extents after a title block has been inserted, specify paper space units, and specify whether the title block is to be inserted or externally referenced

T   Prepares paper space, orients the drawing by setting the origin, and creates a drawing border and title block

U   Reverses operations performed in the current MVSETUP session

NEW

Purpose

Presents a dialog box from which you can create a new drawing file

OFFSET

Purpose

Creates concentric circles, parallel lines, and parallel curves

Option

T   Creates an object passing through a specified point

OLELINKS

Purpose

Updates, changes, and cancels OLE links

OOPS

Purpose

Restores erased objects

OPEN

Purpose

Presents a dialog box from which you can specify a drawing file to open

'ORTHO

Purpose

Constrains cursor movement

Options

ON   Turns on the ortho mode

OFF   Turns off the ortho mode

'OSNAP

Purpose

Displays a dialog box that allows you to set running object snap modes

Options

END   Closest endpoint of arc, elliptical arc, ray, mline, or line, or to the closest corner of a trace, solid, or 3D face

MID   Midpoint of arc, elliptical arc, spline, ellipse, ray, solid, xline, mline, or line

CEN   Center of arc, circle, elliptical arc, ellipse, or solid

NOD   Snaps to a point object

QUA   Snaps to quadrant point of an arc, elliptical arc, ellipse, solid, or circle

INT   Intersection of line, arc, circle, spline, elliptical arc, ellipse, ray, xline, or mline

INS   Insertion point of text, block, attribute, or shape

PER   Perpendicular to arc, elliptical arc, ellipse, spline, ray, xline, mline, line, solid, or circle

TAN   Tangent to an arc, circle, ellipse, or elliptical arc

NEA   Nearest point of arc, elliptical arc, ellipse, spline, ray, xline, mline, circle, line, or point

APPINT   Apparent intersection of two objects which may or may not actually intersect in 3D space

QUI   Snaps to the first snap point found

'PAN

Purpose

Moves the drawing display in the current viewport

PASTECLIP

Purpose

Inserts data from the Windows Clipboard

PASTESPEC

Purpose

Inserts data from the Windows Clipboard and controls the format of the data; used with OLE

PEDIT

Purpose

Edits polylines and three-dimensional polygon meshes

Options for 2D Polylines

C	Closes an open polyline
O	Opens a closed polyline
J	Joins to polyline
W	Specifies a new uniform width for the entire polyline
E	Edits vertices
F	Fits curve to polyline
S	Uses the polyline vertices as the frame for a spline curve (type set by SPLINETYPE)
D	Removes extra vertices inserted by a fit or spline curve and straightens all segments of the polyline
L	Toggles linetype generation to be either a continuous pattern at vertices, or with dashes generated at the start and end of vertices
U	Reverses operations one at a time as far back as the beginning of the PEDIT session
X	Exits the PEDIT command

Options for 3D Polylines

C	Closes an open polyline
O	Opens a closed polyline
E	Edits vertices
S	Uses the polyline vertices as the frame for a spline curve (type set by SPLINETYPE)
D	Decurves, or returns a spline curve to its control frame
U	Reverses operations one at a time as far back as the beginning of the PEDIT session
X	Exits the PEDIT command

Options for Polygon Meshes

E	Edits mesh vertices
S	Fits a smooth surface (type set by SPLINETYPE)
D	Restores the original control-point polygon mesh
M	Opens or closes *M*-direction polylines
N	Opens or closes *N*-direction polylines
U	Reverses operations one at a time as far back as the beginning of the PEDIT session
X	Exits the PEDIT command

PFACE

Purpose

Creates a three-dimensional polyface mesh vertex by vertex

PLAN

Purpose

Displays the plan view of a user coordinate system

Options

C	Establishes a plan view of the current UCS
U	Establishes a plan view of the specified UCS
W	Establishes a plan view of the WCS

PLINE

Purpose

Creates two-dimensional polylines

Options

A	Changes PLINE to arc mode so that you can add arc segments to the polyline
C	Closes the polyline by creating a segment from the last endpoint to the first
H	Specifies the width from the center of a wide polyline line segment to one of its edges
L	Draws a line segment of a specified length at the same angle as the previous segment (or tangent to the arc if the preceding segment was an arc)
U	Removes the most recent line segment added to the polyline
W	Specifies the width of the next line segment

PLOT

Purpose

Presents a dialog box from which you can plot a drawing to a plotter, printer, or file

POINT

Purpose

Creates a point object

POLYGON

Purpose

Creates an equilateral closed polyline

Options

C	Defines the center of the polygon
E	Defines a polygon by specifying the endpoints of the first edge

PREFERENCES

Purpose

Presents a dialog box from which you can customize AutoCAD settings

857

PREVIEW

Purpose

Shows how a drawing will look when it is printed or plotted

PSDRAG

Purpose

Controls the appearance of a PostScript image as it is dragged into position with PSIN

Options

0	Displays only the image's bounding box as you drag it into place
1	Displays the rendered PostScript image as you drag it into place

PSFILL

Purpose

Fills a 2D polyline outline with a PostScript pattern

PSIN

Purpose

Imports a PostScript file

PSOUT

Purpose

Displays a dialog box that allows you to create an Encapsulated PostScript (EPS) file

PSPACE

Purpose

Switches from a model space viewport to paper space

PURGE

Purpose

Removes unused named references, such as unused blocks or layers, from the database

QSAVE

Purpose

Saves the current drawing

QTEXT

Purpose

Controls the display and plotting of text and attribute objects

Options

ON	Displays all existing text and attribute objects as bounding boxes only
OFF	Displays the actual text and attribute objects

QUIT

Purpose

Exits AutoCAD without saving the current drawing

RAY

Purpose

Creates a line that extends infinitely in one direction only

RECOVER

Purpose

Repairs a damaged drawing

RECTANG

Purpose

Draws a rectangular polyline

REDEFINE

Purpose

Restores AutoCAD internal commands overridden by UNDEFINE

REDO

Purpose

Reverses the effects of the previous UNDO or U command

'REDRAW

Purpose

Refreshes the display of the current viewport

'REDRAWALL

Purpose

Refreshes the display of all viewports

REGEN

Purpose

Regenerates the drawing and refreshes the current viewport

REGENALL

Purpose

Regenerates the drawing and refreshes all viewports

'REGENAUTO

Purpose

Controls automatic regeneration of a drawing

Options

ON	Drawing regenerates automatically as needed
OFF	When drawing needs to be regenerated, AutoCAD prompts the user instead of regenerating automatically

REGION

Purpose

Creates a region object from a selection set of existing objects

REINIT

Purpose

Displays a dialog box from which you can reinitialize the input/output ports, digitizer, display, and program parameters file

RENAME

Purpose

Changes the names of objects including blocks, dimension styles, layers, linetypes, styles, UCSs, views, and viewports

RENDER

Purpose

Displays a dialog box from which you can create a realistically shaded image of a 3D wireframe or solid model

REPLAY

Purpose

Presents a dialog box from which you can specify images to be displayed in the GIF, TGA, or TIFF image formats

'RESUME

Purpose

Continues an interrupted script

REVOLVE

Purpose

Creates a solid by revolving a two-dimensional object about an axis

Options

O	Selects an existing line or single-segment polyline that defines the axis about which to revolve the object
X	Uses the positive X axis of the current UCS as the positive axis direction
Y	Uses the positive Y axis of the current UCS as the positive axis direction

REVSURF

Purpose

Creates a rotated surface about a selected axis

RMAT

Purpose

Presents a dialog box from which you can preview, select, modify, duplicate, create, attach, or detach rendering materials

ROTATE

Purpose

Moves objects about a base point

Option

R	Specifies the absolute current rotation angle and desired new rotation angle

ROTATE3D

Purpose

Moves objects about a three-dimensional axis

Options

2	Uses two points to define the axis of rotation
L	Uses the last axis of rotation
V	Aligns the axis of rotation with the viewing direction of the current viewport that passes through the selected point
X	Aligns the axis of rotation with the X axis that passes through the selected point
Y	Aligns the axis of rotation with the Y axis that passes through the selected point
Z	Aligns the axis of rotation with the Z axis that passes through the selected point

RPREF

Purpose

Displays a dialog box that allows you to set rendering preferences

RSCRIPT

Purpose

Creates a script that repeats continuously

RULESURF

Purpose

Creates a ruled surface between two curves

SAVE

Purpose

Saves the drawing with the current file name or a specified name

SAVEAS

Purpose

Saves an unnamed drawing with a file name or renames the current drawing

SAVEIMG

Purpose

Presents a dialog box from which you can save a rendered image to a BMP, TIFF, or TGA file

SCALE

Purpose

Enlarges or reduces objects equally in the X, Y, and Z directions

Options

S	Sets the scale factor by which the selected objects are multiplied
R	Scales the selected objects based on a reference length and a specified new length

SCENE

Purpose

Presents a dialog box from which you can manage scenes in model space

'SCRIPT

Purpose

Executes a sequence of commands from a script

SECTION

Purpose

Uses the intersection of a plane and solids to create a region

Options

3	Defines three points on the sectioning plane
O	Aligns the sectioning plane with a circle, ellipse, circular or elliptical arc, 2D spline, or 2D polyline segment
Z	Defines the sectioning plane by a specified origin point on the Z axis of the plane
V	Aligns the sectioning plane with the current viewport's viewing plane
XY	Aligns the sectioning plane with the XY plane of the current UCS
YZ	Aligns the sectioning plane with the YZ plane of the current UCS
ZX	Aligns the sectioning plane with the ZX plane of the current UCS

SELECT

Purpose

Places selected objects in a selection set for use with subsequent commands

Options

AU	Chooses automatic selection
A	Switches to Add mode
ALL	Selects all objects on thawed layers
BOX	Selects all objects inside or crossing a rectangle specified by two points
C	Selects objects within and crossing an area defined by two points, specified from right to left
CP	Selects objects within and crossing a polygon defined by specifying points around the objects to be selected
F	Selects all objects crossing a selection fence
G	Selects all objects within a specified group
L	Selects the most recently created visible object
M	Allows specification of multiple points without highlighting the objects
P	Selects the most recent selection set
R	Switches to the Remove mode
SI	Places object selection in Single mode and selects the first object or set of objects designated, then ends the selection process
U	Cancels the selection of the object most recently added to the selection set
W	Selects all objects completely inside a rectangle defined by two points, specified from left to right
WP	Selects objects within a polygon defined by points indicated around the objects to be selected

SETUV

Purpose

Lets you map materials onto geometry

'SETVAR

Purpose

Lists or changes values of system variables

Option

?	Lists current system variable settings

SHADE

Purpose

Displays a flat-shaded image of the drawing in the current viewport

SHAPE

Purpose

Inserts a shape

Option

?	Lists shape names

SHELL

Purpose

Accesses operating system commands

Option

ENTER	Shells out to DOS until you enter EXIT to return to AutoCAD

SHOWMAT

Purpose

Lists material type and attachment method for a selected object

SKETCH

Purpose

Creates a series of freehand line segments

Options

P	Raises and lowers the sketching pen
X	Records and reports the number of temporary lines sketched and exits Sketch mode
Q	Discards all temporary lines sketched since the start of SKETCH or the last use of the Record option, and exits Sketch mode
R	Records temporary lines as permanent and does not change the pen's position
E	Erases any portion of a temporary line and raises the pen if it is down
C	Lowers the pen to continue a sketch sequence from the endpoint of the last sketched line
.	(period) Lowers the pen, draws a straight line from the endpoint of the last sketched line to the pen's current location, and returns the pen to the "up" position

SLICE

Purpose

Slices a set of solids with a plane

Options

3	Defines three points on the cutting plane
O	Aligns the cutting plane with a circle, ellipse, circular or elliptical arc, 2D spline, or 2D polyline segment
Z	Defines the cutting plane by a specified origin point on the Z axis of the XY plane
V	Aligns the cutting plane with the current viewport's viewing plane
XY	Aligns the cutting plane with the XY plane of the current UCS
YZ	Aligns the cutting plane with the YZ plane of the current UCS
ZX	Aligns the cutting plane with the ZX plane of the current UCS

'SNAP

Purpose

Restricts cursor movement to specified intervals

Options

ON	Activates snap
OFF	Deactivates snap
A	Specifies differing X and Y spacings for the snap grid
R	Sets the rotation of the snap grid with respect to the drawing and the display screen
S	Selects the format of the snap grid (standard or isometric)

SOLDRAW

Purpose

Works with SOLVIEW to create profiles and sections based on 3D solid models

SOLID

Purpose

Creates solid-filled polygons

SOLPROF

Purpose

Creates profile images of solid models

SOLVIEW

Purpose

Creates floating viewports using orthographic projection to lay out multi- and sectional-view drawings of 3D solids

Options

U	Creates a profile view relative to a specified UCS
O	Creates a folded orthographic view from an existing view
A	Creates an auxiliary view from an existing view (see Glossary)
S	Creates a sectional view of solids, complete with crosshatching as appropriate

'SPELL

Purpose

Displays a dialog box from which you can spell-check text created with TEXT, DTEXT, and MTEXT

SPHERE

Purpose

Creates a three-dimensional solid sphere

Options

D	Defines the diameter of the sphere
R	Defines the radius of the sphere

SPLINE

Purpose

Creates a quadratic or cubic spline (NURBS) curve

Option

O	Converts 2D and 3D spline-fit polygons to equivalent spline entities

SPLINEDIT

Purpose

Edits a spline object

Options

F	Edits fit data
C	Closes an open spline
O	Opens a closed spline
M	Relocates a spline's control vertices
R	Fine-tunes a spline definition
E	Reverses the spline's direction
U	Cancels the last editing operation
X	Ends SPLINEDIT

STATS

Purpose

Presents a dialog box that displays rendering statistics

'STATUS

Purpose

Displays drawing statistics, modes, and extents

STLOUT

Purpose

Stores a solid in an ASCII or binary file

STRETCH

Purpose

Moves or stretches objects

STYLE

Purpose

Displays a dialog box that permits you to create named text styles

SUBTRACT

Purpose

Creates a composite region or solid by subtracting the area of one set of regions from another or subtracting the volume of one set of solids from another

SYSWINDOWS

Purpose

Arranges windows

Options

C	Overlaps windows with visible title bars
H	Arranges windows in horizontal, nonoverlapping tiles
V	Arranges windows in vertical, nonoverlapping tiles
A	Arranges the window icons

TABLET

Purpose

Calibrates the tablet with the coordinate system of a paper drawing

Options

ON	Turns on tablet mode
OFF	Turns off tablet mode
CAL	Calibrates tablet for use in the current space
CFG	Configures tablet menus and screen pointing area

TABSURF

Purpose

Creates a tabulated surface from a path curve and direction vector

TEXT

Purpose

Creates a single line of text

Options

J	Allows you to justify the text
S	Sets the text style for the text

'TEXTSCR

Purpose

Switches from the graphics screen to the text screen

'TIME

Purpose

Displays the date and time statistics of a drawing

Options

D	Repeats the display with updated times
ON	Starts the user elapsed timer
OFF	Stops the user elapsed timer
R	Resets the user elapsed timer

TOLERANCE

Purpose

Creates geometric tolerances

TOOLBAR

Purpose

Displays, hides, and positions toolbars

TORUS

Purpose

Creates a donut-shaped solid

Options

D	Defines the radius of the torus; allows you to specify the radius or diameter of the tube
R	Defines the diameter of the torus; allows you to specify the radius or diameter of the tube

TRACE

Purpose

Creates solid lines of a specified width

TRANSPARENCY

Purpose

Controls whether background pixels in an image are transparent or opaque

Options

ON	Turns transparency on so that objects underneath are visible
OFF	Turns transparency off so that objects underneath the image are hidden from view

'TREESTAT

Purpose

Displays information on the drawing's current spatial index

TRIM

Purpose

Trims objects at a cutting edge defined by other objects

Options

P	Specifies the projection mode for trimming objects (none, UCS, or view)
E	Determines whether an object is trimmed at another object's implied edge, or only to an object that intersects it in 3D space
U	Reverses the most recent trim operation

U

Purpose

Reverses the most recent operation

UCS

Purpose

Manages user coordinate systems

Options

W	Sets the current UCS to the WCS
O	Defines a new UCS by shifting the origin of the current UCS, leaving the direction of the X, Y, and Z axes unchanged
ZA	Defines a UCS with a specified positive Z axis
3	Uses three points to specify a new UCS origin and the direction of the positive X and Y axes
OB	Defines a new UCS based on a selected object
V	Establishes a new UCS with the XY plane perpendicular to the viewing direction (or parallel to the screen)
X	Rotates the current UCS about the X axis
Y	Rotates the current UCS about the Y axis
Z	Rotates the current UCS about the Z axis
P	Restores the previous UCS
R	Restores a saved UCS to become the current UCS
S	Saves the current UCS to a specified name
D	Removes the specified UCS from the list of saved coordinate systems
?	Lists currently defined UCSs

UCSICON

Purpose

Controls the visibility and placement of the UCS icon

Options

ON	Enables the coordinate system icon
OFF	Disables the coordinate system icon
All	Applies changes to the icon in all active viewports
N	Displays the icon at the lower left corner of the viewport regardless of the location of the UCS origin
OR	Forces the icon to appear at the origin of the current coordinate system

UNDEFINE

Purpose

Allows an application-defined command to override an internal AutoCAD command

UNDO

Purpose

Reverses the effect of commands

Options

(number)	Undoes the specified number of preceding operations
A	Undoes a menu selection as a single command, reversible by a single U command
C	Limits or turns off the UNDO command
BE	Begins an UNDO group definition from a sequence of operations
E	Ends the UNDO group definition begun by BE
M	Places a mark in the undo information
B	Undoes all operations in memory until it reaches a mark set by the M option

UNION

Purpose

Creates a composite region or solid

'UNITS

Purpose

Selects coordinate and angle display formats and precision

'VIEW

Purpose

Saves and restores named views

Options

?	Lists named views
D	Deletes one or more named views
R	Restores the view you specify to the current viewport
S	Saves the display in the current viewport using the name you supply
W	Saves a portion of the current display as a view

VIEWRES

Purpose

Sets the resolution for object generation in the current viewport

VPLAYER

Purpose

Sets layer visibility within viewports

Options

?	Lists frozen layers
F	Freezes specified layers in selected viewports
T	Thaws specified layers in selected viewports
R	Sets the visibility of layers in specified viewports to their current default setting
N	Creates new layers that are frozen in all viewports
V	Determines whether the specified layers are thawed or frozen in subsequently created viewports

VPOINT

Purpose

Sets the viewing direction for a three-dimensional visualization of the drawing

Options

RETURN	Presents an axis tripod from which you can select the new viewing direction
V	Allows you to enter *x,y,z* coordinates to create a vector that defines a direction from which the drawing can be viewed
R	Specifies a new direction using two angles

VPORTS

Purpose

Divides the graphics area into multiple tiled viewports

Options

S	Saves the current viewport configuration
R	Restores a previously saved viewport configuration
D	Deletes a named viewport configuration

J	Combines two adjacent viewports into one larger viewport
SI	Returns the drawing to a single viewport view, using the view from the active viewport
?	Lists viewport configurations
2	Divides the current viewport in half
3	Divides current viewport into three viewports
4	Divides current viewport into four viewports of equal size

VSLIDE

Purpose

Displays a raster image slide file in the current viewport

WBLOCK

Purpose

Writes objects to a new drawing file

WEDGE

Purpose

Creates a three-dimensional solid with a tapered, sloping face

Option

C	Allows you to specify the center point of the wedge

WMFIN

Purpose

Imports a Windows metafile (WMF format)

WMFOPTS

Purpose

Presents a dialog box that allows you to set options for WMFIN

WMFOUT

Purpose

Saves objects in a Windows Metafile (WMF format)

XATTACH

Purpose

Presents a dialog box from which you can attach an external reference to the current drawing

XBIND

Purpose

Binds dependent symbols of an xref to a drawing; dependent symbols may include blocks, dimension styles, layers, linetypes, and text styles

XCLIP

Purpose

Defines an xref clipping boundary and sets the front or back clipping planes

Options

ON	Displays the clipped portion of the xref or block only
OFF	Displays all of the geometry of the xref or block, ignoring clipping boundaries
C	(Clipdepth) Sets front and back clipping planes on an external reference or block
D	(Delete) Removes a clipping boundary for the selected xref or block. (This is a permanent deletion; to remove the clipping boundary temporarily from the screen, use OFF instead.)
P	(Generate Polyline) Automatically draws a polyline (in the current layer) that coincides with the clipping boundary; this allows the clipping boundary to be edited with the PEDIT command
N	(New boundary) Creates a new clipping boundary

XLINE

Purpose

Creates a line that extends infinitely in both directions

Options

H	Creates a horizontal xline
V	Creates a vertical xline
A	Creates an xline at a specified angle
B	Creates an xline that passes through the selected angle's vertex and then bisects the angle
O	Creates an xline parallel to another object at a specified distance

XREF

Purpose

Displays a dialog box that alows you to control external references to drawing files

'ZOOM

Purpose

Increases or decreases the apparent size of objects in the current viewport

Options

A	Displays the entire drawing in the current viewport; In a plan view, AutoCAD zooms to the drawing limits (area) or current extents, whichever is greater
C	Displays a window defined by a center point and a magnification value or height
D	Displays the generated portion of the drawing with a view box. The view box represents the current viewport, which you can shrink or enlarge and move around the drawing
E	Displays the drawing as large as possible on the screen
P	Displays the previous view
S	Displays at a specified scale factor. The value you enter is relative to the limits (area) of the drawing; for example, entering 3 thriples the apparent display size of objects
W	Displays an area specified by two opposite corners of a rectangular window
R	The cursor changes to a magnifying glass with plus (+) and minus (–) signs, allowing you to zoom interactively to a logical extent

Appendix J: AutoCAD System Variables

This appendix lists the AutoCAD system variables and their meanings. You can change all but read-only system variables at the Command prompt using either the SETVAR command or AutoLISP's getvar and setvar functions. In addition, you can change the values of many of the system variables by simply entering the name of the variable at the Command prompt.

ACADPREFIX

Type: String
Saved in: (Not saved)
Description: (Read-only) Stores the directory path specified by the ACAD environment variable, with path separators appended if necessary

ACADVER

Type: String
Saved in: (Not saved)
Description: (Read-only) Stores the AutoCAD version number; note that this variable differs from the DXF file $ACADVER header variable, which contains the drawing database level number

ACISOUTVER

Type: Integer
Saved in: Drawing
Description: Controls the version of ACIS used to create SAT files using the ACISOUT command; note that Release 14 supports only one version of ACIS, but future releases of AutoCAD will support additional ACIS versions

AFLAGS

Type: Integer
Saved in: (Not saved)
Description: Sets the attribute flags for the ATTDEF command (sum of the following):

0	No attribute mode selected
1	Invisible
2	Constant
4	Verify
8	Preset

ANGBASE

Type: Real
Saved in: Drawing
Description: Sets the base angle 0 with respect to the current UCS

ANGDIR

Type: Integer
Saved in: Drawing
Description: Sets the angle from angle 0 with respect to the current UCS

0	Counterclockwise
1	Clockwise

APBOX

Type: Integer
Saved in: Registry
Description: Turns the AutoSnap aperture box on or off

0	Aperture box is not displayed
1	Aperture box is displayed

APERTURE

Type: Integer
Saved in: Registry
Description: Sets object snap target height, in pixels

AREA

Type: Real
Saved in: (Not saved)
Description: (Read-only) Stores the last area computed by AREA, LIST, or DBLIST

ATTDIA

Type: Integer
Saved in: Drawing
Description: Controls whether INSERT uses a dialog box for attribute value entry

0	Issues prompts on the command line
1	Uses a dialog box

ATTMODE

Type: Integer
Saved in: Drawing
Description: Controls attribute display mode

0	Off
1	Normal
2	On

ATTREQ

Type: Integer
Saved in: Drawing
Description: Determines whether INSERT uses default attribute settings during insertion of blocks

0	Assumes the defaults for the values of all attributes
1	Enables prompts or dialog box for attribute values, as selected by ATTDIA

AUDITCTL
Type: Integer
Saved in: Registry
Description: Controls whether AutoCAD creates an ADT file (audit report)
0   Disables or prevents writing of ADT files
1   Enables the writing of ADT files

AUNITS
Type: Integer
Saved in: Drawing
Description: Sets angular units mode
0   Decimal degrees
1   Degrees/minutes/seconds
2   Gradians
3   Radians
4   Surveyor's units

AUPREC
Type: Integer
Saved in: Drawing
Description: Sets angular units decimal places

AUTOSNAP
Type: Integer
Saved in: Registry
Description: Controls display of the AutoSnap marker and SnapTips; also turns the AutoSnap magnet on and off
0   Turns off the marker, SnapTip, and magnet
1   Turns on the marker
2   Turns on the SnapTip
4   Turns on the magnet

BACKZ
Type: Real
Saved in: Drawing
Description: Stores the back clipping plane offset from the target plane for the current viewport, in drawing units

BLIPMODE
Type: Integer
Saved in: Drawing
Description: Controls whether marker blips are visible
0   Turns off marker blips
1   Turns on marker blips

CDATE
Type: Real
Saved in: (Not saved)
Description: Sets calendar date and time

CECOLOR
Type: String
Saved in: Drawing
Description: Sets the color of new objects

CELTSCALE
Type: Real
Saved in: Drawing
Description: Sets the current global linetype scale for objects

CELTYPE
Type: String
Saved in: Drawing
Description: Sets the linetype of new objects

CHAMFERA
Type: Real
Saved in: Drawing
Description: Sets the first chamfer distance

CHAMFERB
Type: Real
Saved in: Drawing
Description: Sets the second chamfer distance

CHAMFERC
Type: Real
Saved in: Drawing
Description: Sets the chamfer length

CHAMFERD
Type: Real
Saved in: Drawing
Description: Sets the chamfer angle

CHAMMODE
Type: Integer
Saved in: (Not saved)
Description: Sets the input method by which AutoCAD creates chamfers
0   Requires two chamfer distances
1   Requires one chamfer length and an angle

CIRCLERAD
Type: Real
Saved in: (Not saved)
Description: Sets the default circle radius; a zero sets no default

CLAYER
Type: String
Saved in: Drawing
Description: Sets the current layer

CMDACTIVE

Type: Integer
Saved in: (Not saved)
Description: (Read-only) Stores bit-code that indicates whether an ordinary command, transparent command, script, or dialog box is active; the sum of the following:

1   Ordinary command
2   Ordinary command and transparent command
4   Script
8   Dialog box

CMDDIA

Type: Integer
Saved in: Registry
Description: Controls whether dialog boxes are enabled for more than just PLOT and external database commands

0   Disables dialog boxes
1   Enables dialog boxes

CMDECHO

Type: Integer
Saved in: (Not saved)
Description: Controls whether AutoCAD echoes prompts and input during the AutoLISP command function

0   Disables echoing
1   Enables echoing

CMDNAMES

Type: String
Saved in: (Not saved)
Description: (Read-only) Displays the name of the currently active command and transparent command

CMLJUST

Type: Integer
Saved in: Drawing
Description: Specifies multiline justification

0   Top
1   Middle
2   Bottom

CMLSCALE

Type: Real
Saved in: Drawing
Description: Controls the overall width of a multiline as a function of the style definition

CMLSTYLE

Type: String
Saved in: Drawing
Description: Sets the name of the multiline style that AutoCAD uses to draw the multiline

COORDS

Type: Integer
Saved in: Drawing

Description: Controls when coordinates are updated

0   Coordinate display is updated on pick points only
1   Display of absolute coordinates is continuously updated
2   Distance and angle from the last point are displayed when distance or angle is requested

CURSORSIZE

Type: Integer
Saved in: Registry
Description: Determines the size of the crosshairs as a percentage of the screen size

CVPORT

Type: Integer
Saved in: Drawing
Description: Sets the identification number of the current viewport

DATE

Type: Real
Saved in: (Not saved)
Description: (Read-only) Stores the current date and time represented as a Julian data and fraction in a real number: <Julian date>.<Fraction>

DBMOD

Type: Integer
Saved in: (Not saved)
Description: Indicates the drawing modification status using bit-code; the sum of the following:

0   Object database modified
1   Symbol table modified
2   Database variable modified
8   Window modified
16 View modified

DCTCUST

Type: String
Saved in: Registry
Description: Displays the current custom spelling dictionary path and file name

DCTMAIN

Type: String
Saved in: Registry
Description: Displays current main spelling dictionary file name

DELOBJ

Type: Integer
Saved in: Drawing
Description: Controls whether objects used to create other objects are retained or deleted from the drawing database

0   Objects are retained
1   Objects are deleted

DIASTAT

Type: Integer
Saved in: (Not saved)
Description: Stores the exit method of the most recently used dialog box
0	Cancel
1	OK

DIMALT

Type: Switch
Saved in: Drawing
Description: When turned on, enables alternate units dimensioning

DIMADEC

Type: Integer
Saved in: Drawing
Description: Controls the number of places of precision displayed for angular dimension text

DIMALTD

Type: Integer
Saved in: Drawing
Description: Controls number of decimal places used in alternate measurement

DIMALTF

Type: Real
Saved in: Drawing
Description: Controls scale factor of alternate units

DIMALTTD

Type: Integer
Saved in: Drawing
Description: Sets the number of decimal places for the tolerance values of an alternate units dimension

DIMALTTZ

Type: Integer
Saved in: Drawing
Description: Toggles suppression of zeros for tolerance values in alternate units

DIMALTU

Type: Integer
Saved in: Drawing
Description: Sets the units format for alternate units of all dimension style family members except angular
1	Scientific
2	Decimal
3	Engineering
4	Architectural
5	Fractional

DIMALTZ

Type: Integer
Saved in: Drawing
Description: Toggles suppression of zeros for alternate unit dimension values
0	Turns off suppression of zeros
1	Turns on suppression of zeros

DIMAPOST

Type: String
Saved in: Drawing
Description: Specifies a text prefix or suffix (or both) to the alternate dimension measurement for all types of dimensions except angular

DIMASO

Type: Switch
Saved in: Drawing
Description: Controls the creation of associative dimension objects
OFF	Dimensions are not associative
ON	Dimensions are associative

DIMASZ

Type: Real
Saved in: Drawing
Description: Controls size of dimension line and leader line arrowheads and the size of hook lines

DIMAUNIT

Type: Integer
Saved in: Drawing
Description: Sets angle format for angular dimensions
0	Decimal degrees
1	Degrees/minutes/seconds
2	Gradians
3	Radians
4	Surveyor's units

DIMBLK

Type: String
Saved in: Drawing
Description: Sets the name of a block to be drawn instead of the normal arrowhead at the ends of the dimension line or leader line

DIMBLK1

Type: String
Saved in: Drawing
Description: If DIMSAH is on, DIMBLK1 specifies user-defined arrowhead blocks for the first end of the dimension line

DIMBLK2

Type: String
Saved in: Drawing
Description: If DIMSAH is on, DIMBLK2 specifies user-defined arrowhead blocks for the second end of the dimension line

DIMCEN

Type: Real
Saved in: Drawing
Description: Controls drawing of center marks and centerlines by the DIMCENTER, DIMDIAMETER, and DIMRADIUS dimensioning commands

DIMCLRD

Type: Integer
Saved in: Drawing
Description: Assigns colors to leader lines, arrowheads, and dimension lines

DIMCLRE

Type: Integer
Saved in: Drawing
Description: Assigns colors to dimension extension lines

DIMCLRT

Type: Integer
Saved in: Drawing
Description: Assigns colors to dimension text

DIMDEC

Type: Integer
Saved in: Drawing
Description: Sets the number of decimal places for the tolerance values of a primary units dimension

DIMDLE

Type: Real
Saved in: Drawing
Description: Extends the dimension line beyond the extension line when oblique strokes are drawn instead of arrowheads

DIMDLI

Type: Real
Saved in: Drawing
Description: Controls the dimension line spacing for baseline dimensions; each baseline dimension is offset by this amount, if necessary, to avoid drawing over the previous dimension

DIMEXE

Type: Real
Saved in: Drawing
Description: Determines how far to extend the extension line beyond the dimension line

DIMEXO

Type: Real
Saved in: Drawing
Description: Determines how far extension lines are offset from origin points

DIMFIT

Type: Integer
Saved in: Drawing
Description: Controls the placement of text and arrowheads inside or outside extension lines based on available space between extension lines
(DIMFIT options, continued)

0   Places text and arrowheads between extension lines if space is available; otherwise places both text and arrowheads outside extension lines

1   Places text and arrowheads between extension lines when space is available; if not, places text between extension lines and arrowheads outside; if not enough space for text inside extension lines, places text and arrowheads outside extension lines

2   Places text and arrowheads between extension lines when space is available; otherwise places text inside if it fits, or places arrowheads inside if they fit; if neither fits inside extension lines, both are placed outside

3   Places whatever best fits between extension lines

4   Creates leader lines when there is not enough space for text between extension lines

DIMGAP

Type: Real
Saved in: Drawing
Description: Sets the distance around the dimension text when you break the dimension line to accommodate dimension text; sets the gap between annotation and hook line in leaders

DIMJUST

Type: Integer
Saved in: Drawing
Description: Controls horizontal dimension text position
0   Center-justifies text between extension lines
1   Positions text next to first extension line
2   Positions text next to second extension line
3   Positions text above and aligned with first extension line
4   Positions text above and aligned with second extension line

DIMLFAC

Type: Real
Saved in: Drawing
Description: Sets global scale factor for linear dimensioning measurements

DIMLIM

Type: Switch
Saved in: Drawing
Description: When turned on, generates dimension limits as the default text

DIMPOST

Type: String
Saved in: Drawing
Description: Specifies a text prefix or suffix (or both) to the dimension measurement

DIMRND

Type: Real
Saved in: Drawing
Description: Rounds all dimensioning distances to the specified value

DIMSAH

Type: Switch
Saved in: Drawing
Description: Controls the use of user-defined arrowhead blocks at the ends of the dimension line

ON   Normal arrowheads or user-defined blocks set by DIMBLK are used

OFF   User-defined arrowhead blocks are used

DIMSCALE

Type: Real
Saved in: Drawing
Description: Sets the overall scale factor applied to dimensioning variables that specify sizes, distances, or offsets

0.0   AutoCAD computes a reasonable default value based on scaling between current model space viewport and paper space

\>0   AutoCAD computes a scale factor that leads text sizes, arrowhead sizes, and other scaled distances to plot at their face values

DIMSD1

Type: Switch
Saved in: Drawing
Description: When turned on, suppresses drawing of the first dimension line

DIMSD2

Type: Switch
Saved in: Drawing
Description: When turned on, suppresses drawing of the second dimension line

DIMSE1

Type: Switch
Saved in: Drawing
Description: When turned on, suppresses drawing of the first extension line

DIMSE2

Type: Switch
Saved in: Drawing
Description: When turned on, suppresses drawing of the second extension line

DIMSHO

Type: Switch
Saved in: Drawing
Description: When turned on, controls redefinition of dimension objects while dragging

DIMSOXD

Type: Switch
Saved in: Drawing
Description: When turned on, suppresses drawing of dimension lines outside the extension lines

DIMSTYLE

Type: String
Saved in: Drawing
Description: (Read-only) Sets the current dimension style by name

DIMTAD

Type: Integer
Saved in: Drawing
Description: Controls vertical position of text in relation to the dimension line

0   Centers text between extension lines

1   Places text above dimension line except when dimension line is not horizontal and text inside extension lines is forced horizontal

2   Places text on the side of the dimension line farthest away from the defining points

3   Places text to conform to a JIS representation

DIMTDEC

Type: Integer
Saved in: Drawing
Description: Sets the number of decimal places for the tolerance values for a primary units dimension

DIMTFAC

Type: Real
Saved in: Drawing
Description: Specifies a scale factor for text height of tolerance values relative to the dimension text height as set by DIMTXT:

$$DIMTFAC = \frac{\text{Tolerance Height}}{\text{Text Height}}$$

DIMTIH

Type: Switch
Saved in: Drawing
Description: Controls the position of dimension text inside the extension lines for all dimension types except ordinate dimensions

OFF   Aligns text with dimension line

ON   Draws text horizontally

DIMTIX

Type: Switch
Saved in: Drawing
Description: Draws text between extension lines

OFF For linear and angular dimensions, places text inside extension lines if there is sufficient room; for radius and diameter dimensions, forces text outside the circle or arc

ON Draws dimension text between the extension lines even if AutoCAD would ordinarily place it outside those lines

DIMTM

Type: Real
Saved in: Drawing
Description: When DIMTOL or DIMLIM is on, sets the minimum (lower) tolerance limit for dimension text

DIMTOFL

Type: Switch
Saved in: Drawing
Description: When turned on, draws a dimension line between the extension lines even when the text is placed outside the extension lines; for radius and diameter dimensions (while DIMTIX is off), draws a dimension line and arrowheads inside the circle or arc and places the text and leader outside

DIMTOH

Type: Switch
Saved in: Drawing
Description: When turned on, controls the position of dimension text outside the extension lines

0 Aligns text with the dimension line
1 Draws text horizontally

DIMTOL

Type: Switch
Saved in: Drawing
Description: When turned on, appends dimension tolerances to dimension text

DIMTOLJ

Type: Integer
Saved in: Drawing
Description: Sets vertical justification for tolerance values relative to the nominal dimension text

0 Bottom
1 Middle
2 Top

DIMTP

Type: Real
Saved in: Drawing
Description: When DIMTOL or DIMLIM is on, sets the maximum (upper) tolerance limit for dimension text

DIMTSZ

Type: Real
Saved in: Drawing
Description: Specifies size of oblique strokes drawn instead of arrowheads for linear, radius, and diameter dimensioning

0 Draws arrows
>0 Draws oblique strokes; size of strokes is determined by this value multiplied by the DIMSCALE value

DIMTVP

Type: Real
Saved in: Drawing
Description: Adjusts vertical position of dimension text above or below the dimension line when DIMTAD is off

DIMTXSTY

Type: String
Saved in: Drawing
Description: Specifies the text style of the dimension

DIMTXT

Type: Real
Saved in: Drawing
Description: Specifies the height of dimension text, unless the current text style has a fixed height

DIMTZIN

Type: Integer
Saved in: Drawing
Description: Toggles suppression of zeros for tolerance values

DIMUNIT

Type: Integer
Saved in: Drawing
Description: Sets the units format for all dimension style family members except angular

1 Scientific
2 Decimal
3 Engineering
4 Architectural
5 Fractional

DIMUPT

Type: Switch
Saved in: Drawing
Description: Controls cursor functionality for user-positioned text

0 Cursor controls only dimension line location
1 Cursor controls text position as well as dimension line location

DIMZIN

Type: Integer
Saved in: Drawing
Description: Controls suppression of the inches portion of a feet-and-inches dimension when the distance is an integral number of feet, or the feet portion when the distance is less than one foot

0	Suppresses zero feet and precisely zero inches
1	Includes zero feet and precisely zero inches
2	Includes zero feet and suppresses zero inches
3	Includes zero inches and suppresses zero feet

DISPSILH

Type: Integer
Saved in: Drawing
Description: Controls the display of silhouette curves of body objects in wireframe mode

0	Off
1	On

DISTANCE

Type: Real
Saved in: Drawing
Description: (Read-only) Stores the distance computed by the DIST command

DONUTID

Type: Real
Saved in: (Not saved)
Description: Sets the default for the inside diameter of a donut

DONUTOD

Type: Real
Saved in: (Not saved)
Description: Sets the default for the outside diameter of a donut (must be nonzero)

DRAGMODE

Type: Integer
Saved in: Drawing
Description: Sets object drag mode

0	No dragging
1	On (if requested)
2	Auto

DRAGP1

Type: Integer
Saved in: Registry
Description: Sets regen-drag input sampling rate

DRAGP2

Type: Integer
Saved in: Registry
Description: Sets fast-drag input sampling rate

DWGCODEPAGE

Type: String
Saved in: Drawing
Description: (Read-only) Stores the drawing code page

DWGNAME

Type: String
Saved in: (Not saved)
Description: (Read-only) Stores the drawing name as entered by the user

DWGPREFIX

Type: String
Saved in: (Not saved)
Description: Stores the drive/directory prefix for the drawing

DWGTITLED

Type: Integer
Saved in: (Not saved)
Description: Indicates whether the current drawing has been named

0	Not named
1	Named

EDGEMODE

Type: Integer
Saved in: (Not saved)
Description: Controls determination of cutting edges for TRIM and EXTEND commands

ELEVATION

Type: Real
Saved in: Drawing
Description: Stores the current 3D elevation relative to the current UCS for the current space

EXPERT

Type: Integer
Saved in: (Not saved)
Description: Controls issuance of certain prompts; when prompts are suppressed, the operation is performed as though you had entered **y** at the prompt

0	Issues all prompts normally
1	Suppresses About to regen, proceed? and Really want to turn the current layer off?

(ELEVATION options, continued)

2	Suppresses preceding prompts, Block already defined. Redefine it?, and A drawing with this name already exists. Overwrite it?
3	Suppresses preceding prompts and those issued by LINETYPE if you try to load a linetype that's already loaded or create a new linetype in a file that already defines it
4	Suppresses preceding prompts and those issued by UCS Save and VPORTS Save if the name you supply already exists

5   Suppresses preceding prompts and those issued by DIMSTYLE Save and DIMOVERRIDE if the dimension style name you supply already exists (the entries are redefined)

EXPLMODE
Type: Integer
Saved in: Drawing
Description: Controls whether the EXPLODE command supports non-uniformly scaled (NUS) blocks
0   Does not explode NUS blocks
1   Explodes NUS blocks

EXTMAX
Type: 3D Point
Saved in: Drawing
Description: Stores the upper right point of drawing extents

EXTMIN
Type: 3D Point
Saved in: Drawing
Description: Stores the lower left point of drawing extents

FACETRES
Type: Real
Saved in: Drawing
Description: Further adjusts the smoothness of shaded and hidden line-removed objects; valid values are from .01 to 10.0

FILEDIA
Type: Integer
Saved in: Registry
Description: Suppresses display of file dialog boxes
0   Disables file dialog boxes
1   Enables file dialog boxes

FILLETRAD
Type: Real
Saved in: Drawing
Description: Stores the current fillet radius

FILLMODE
Type: Integer
Saved in: Drawing
Description: Specifies whether objects created with SOLID are filled in
0   Not filled
1   Filled

FONTALT
Type: String
Saved in: Registry
Description: Specifies the alternate font to be used when the specified font file cannot be located

FONTMAP
Type: String
Saved in: Registry
Description: Specifies the font mapping file to be used when the specified font cannot be located

FRONTZ
Type: Real
Saved in: Drawing
Description: Stores the front clipping plane offset from the target plane for the current viewport, in drawing units

GRIDMODE
Type: Integer
Saved in: Drawing
Description: Specifies whether the grid is turned on
0   Turns the grid off
1   Turns the grid on

GRIDUNIT
Type: Real
Saved in: Drawing
Description: Specifies the grid spacing (X and Y) for the current viewport

GRIPBLOCK
Type: Integer
Saved in: Registry
Description: Controls assignment of grips in blocks
0   Assigns grip only to insertion point of block
1   Assigns grips to objects within the block

GRIPCOLOR
Type: Integer
Saved in: Registry
Description: Controls the color of nonselected grips

GRIPHOT
Type: Integer
Saved in: Registry
Description: Controls the color of selected grips

GRIPS
Type: Integer
Saved in: Registry
Description: Allows the use of selection set grips for the Stretch, Move, Rotate, Scale, and Mirror grip modes
0   Disables grips
1   Enables grips

GRIPSIZE
Type: Integer
Saved in: Registry
Description: Sets the size of the box drawn to display the grip in pixels

HANDLES

Type: Integer
Saved in: Drawing
Description: (Read-only) Reports that object handles are enabled and can be accessed by applications

HIGHLIGHT

Type: Integer
Saved in: (Not saved)
Description: Controls object highlighting; does not affect objects selected with grips

0	Disables object selection highlighting
1	Enables object selection highlighting

HPANG

Type: Real
Saved in: (Not saved)
Description: Specifies the hatch pattern angle

HPBOUND

Type: Real
Saved in: (Not saved)
Description: Controls the object type created by the BHATCH and BOUNDARY commands

0	Creates a polyline
1	Creates a region

HPDOUBLE

Type: Integer
Saved in: (Not saved)
Description: Specifies hatch pattern doubling for "U" user-defined patterns

0	Disables hatch pattern doubling
1	Enables hatch pattern doubling

HPNAME

Type: String
Saved in: (Not saved)
Description: Sets default hatch pattern name of up to 34 characters, no spaces allowed

HPSCALE

Type: Real
Saved in: (Not saved)
Description: Specifies the hatch pattern scale factor; must be nonzero

HPSPACE

Type: Real
Saved in: (Not saved)
Description: Specifies the hatch pattern line spacing for "U" user-defined hatch patterns; must be nonzero

INDEXCTL

Type: Integer
Saved in: Drawing
Description: Controls whether layer and spatial indexes are created and saved in drawing files

0	No indexes are created
1	Creates layer index
2	Creates spatial index
3	Creates both layer and spatial indexes

INETLOCATION

Type: Real
Saved in: Registry
Description: Stores the Internet location used by BROWSER

INSBASE

Type: 3D point
Saved in: Drawing
Description: Stores insertion base point set by BASE command, expressed in UCS coordinates for the current space

INSNAME

Type: String
Saved in: Not saved
Description: Sets default block name for DDINSERT or INSERT

ISAVEBAK

Type: Integer
Saved in: Registry
Description: Improves the speed of incremental saves, especially for large drawings in Windows

0	No BAK file is created (even for a full save)
1	A BAK file is created

ISAVEPERCENT

Type: Integer
Saved in: Registry
Description: Determines the amount of wasted space tolerated in a drawing file; values are integers between 0 and 100; default value is 50. Value is the percent of total file size wasted in a file. Wasted space is eliminated with each full save, so when AutoCAD's estimate reaches the value set in ISAVEPERCENT, it automatically performs a full save.

ISOLINES

Type: Integer
Saved in: Drawing
Description: Specifies the number of isolines per surface on objects; valid integer values are from 0 to 2047

LASTANGLE

Type: Real
Saved in: (Not saved)
Description: (Read-only) Stores the end angle of the last arc entered, relative to the XY plane of the current UCS for the current space

LASTPOINT

Type: 3D point
Saved in: Drawing
Description: Stores the last point entered, expressed in UCS coordinates for the current space; referenced by @ during keyboard entry

LASTPROMPT

Type: String
Saved in: (Not Saved)
Description: Stores the last string echoed to the Command line; read-only

LENSLENGTH

Type: Real
Saved in: Drawing
Description: Stores lens length (in millimeters) used in perspective viewing for the current viewport

LIMCHECK

Type: Integer
Saved in: Drawing
Description: Controls object creation outside drawing limits
0   Enables object creation
1   Disables object creation

LIMMAX

Type: 2D point
Saved in: Drawing
Description: Stores upper right drawing limits for the current space expressed in world coordinates

LIMMIN

Type: 2D point
Saved in: Drawing
Description: Stores lower left drawing limits for the current space expressed in world coordinates

LISPINIT

Type: Integer
Saved in: Registry
Description: Specifies whether AutoLISP-defined functions and variables are preserved when you open a new drawing
0   AutoLISP functions and variables are preserved from drawing to drawing
1   AutoLISP functions and variables are valid in current drawing only (Release 13 behavior)

LOCALE

Type: String
Saved in: (Not saved)
Description: (Read-only) Displays the ISO language code of the current AutoCAD version

LOGFILEMODE

Type: Integer
Saved in: Registry
Description: Specifies whether the contents of the text window are written to a log file
0   Log file is not maintained
1   Log file is maintained

LOGFILENAME

Type: String
Saved in: Registry
Description: Specifies path for the log file

LOGINNAME

Type: String
Saved in: (Not saved)
Description: Displays the user's name as configured or input when AutoCAD is loaded

LTSCALE

Type: Real
Saved in: Drawing
Description: Sets global linetype scale factor

LUNITS

Type: Integer
Saved in: Drawing
Description: Sets mode for linear units
1   Scientific
2   Decimal
3   Engineering
4   Architectural
5   Fractional

LUPREC

Type: Integer
Saved in: Drawing
Description: Sets linear units decimal places or denominator

MAXACTVP

Type: Integer
Saved in: (Not saved)
Description: Sets the maximum number of viewports to regenerate at one time

MAXOBJMEM

Type: Integer
Saved in: (Not saved)
Description: Controls the object pager (specifies how much virtual memory AutoCAD allows the drawing to use before it starts paging the drawing out to disk into the object pager's swap files)

MAXSORT
Type: Integer
Saved in: Registry
Description: Sets maximum number of symbol names or file names to be sorted by listing commands

MEASUREMENT
Type: Integer
Saved in: Drawing
Description: Sets drawing units as English or metric
0	English; AutoCAD uses the hatch pattern file and linetype file designated by ANSIHatch and ANSILinetype registry settings
1	Metric; AutoCAD uses the hatch pattern file and linetype file designated by the ISOHatch and ISOLinetype registry settings

MENUCTL
Type: Integer
Saved in: Registry
Description: Controls the page switching of the screen menu
0	Screen menu does not switch pages in response to keyboard command entry
1	Screen menu switches pages in response to keyboard command entry

MENUECHO
Type: Integer
Saved in: (Not saved)
Description: Sets menu echo and prompt control bits; the sum of the following:
1	Suppresses echo of menu items (^P in a menu item toggles echoing)
2	Suppresses display of system prompts during menu
4	Disables ^P toggle of menu echoing
8	Displays input/output strings; debugging aid for DIESEL macros

MENUNAME
Type: String
Saved in: (Not saved)
Description: Stores the name and path of the currently loaded base menu file

MIRRTEXT
Type: Integer
Saved in: Drawing
Description: Controls how MIRROR reflects text
0	Retains text direction
1	Mirrors the text

MODEMACRO
Type: String
Saved in: (Not saved)
Description: Displays a text string on the status line, such as the name of the current drawing, time/date stamp, or special modes

MTEXTED
Type: String
Saved in: Registry
Description: Sets the name of the program to use for editing mtext objects

OFFSETDIST
Type: Real
Saved in: (Not saved)
Description: Sets the default offset distance
<0	Changes to Through mode
>0	Sets the default offset distance

OLEHIDE
Type: Integer
Saved in: Registry
Description: Controls the display of OLE objects in AutoCAD both on the screen and the plotted image
0	All OLE objects are visible
1	OLE objects are visible in paper space only
2	OLE objects are visible in model space only
3	No OLE objects are visible

ORTHOMODE
Type: Integer
Saved in: Drawing
Description: Controls orthogonal display of lines or polylines
0	Off
1	On

OSMODE
Type: Integer
Saved in: Drawing
Description: Sets running object snap modes using the following bit-codes; to enter more than one object snap, enter the sum of their values
0	NONe
1	ENDpoint
2	MIDpoint
4	CENter
8	NODe
16	QUAdrant
32	INTersection
64	INSertion
128	PERpendicular
256	TANgent
512	NEArest
1024	QUIck
2048	APPint

OSNAPCOORD

Type: Integer
Saved in: Registry
Description: Controls whether coordinates entered on the command line override running object snaps

0	Running object snap settings override keyboard coordinate entry
1	Keyboard entry overrides object snap settings
2	Keyboard entry overrides object snap settings except in scripts

PDMODE

Type: Integer
Saved in: Drawing
Description: Sets point object display mode

PDSIZE

Type: Real
Saved in: Drawing
Description: Sets point object display size

0	Creates a point at 5% of graphics area height
>0	Specifies an absolute size
<0	Specifies a percentage of the viewport size

PELLIPSE

Type: Integer
Saved in: Drawing
Description: Controls the ellipse type created with ELLIPSE

0	Creates a true ellipse object
1	Creates a polyline representation of an ellipse

PERIMETER

Type: Real
Saved in: (Not saved)
Description: (Read-only) Stores the last perimeter value computed by AREA, LIST, or DBLIST

PCFACEVMAX

Type: Integer
Saved in: (Not saved)
Description: (Read-only) Sets the maximum number of vertices per face

PICKADD

Type: Integer
Saved in: Registry
Description: Controls additive selection of objects

0	Disables PICKADD
1	Enables PICKADD

PICKAUTO

Type: Integer
Saved in: Registry
Description: Controls automatic windowing when the Select objects prompt appears

0	Disables PICKAUTO
1	Draws a selection window (both window and crossing window) automatically at the Select objects prompt

PICKBOX

Type: Integer
Saved in: Registry
Description: Sets object selection target height, in pixels

PICKDRAG

Type: Integer
Saved in: Registry
Description: Controls the method of drawing a selection window

0	Draws window by clicking mouse or digitizer at opposite corners
1	Draws window by clicking at one corner, holding down mouse or digitizer button, dragging, and releasing button at other corner

PICKFIRST

Type: Integer
Saved in: Registry
Description: Controls method of object selection so that you select objects first and then use an edit or inquiry command

0	Disables PICKFIRST
1	Enables PICKFIRST

PICKSTYLE

Type: Integer
Saved in: Drawing
Description: Controls group selection and associative hatch selection

0	No group selection or associative hatch selection
1	Group selection
2	Associative hatch selection
3	Group selection and associative hatch selection

PLATFORM

Type: String
Saved in: (Not saved)
Description: Indicates which platform of AutoCAD is in use

PLINEGEN

Type: Integer
Saved in: Drawing

Description: Sets the linetype pattern generation around the vertices of a 2D polyline; does not apply to polylines with tapered segments

0   Polylines are generated to start and end with a dash at each vertex

1   Generates the linetype in a continuous pattern around the vertices of the polyline

PLINETYPE

Type: Integer
Saved in: Registry
Description: Specifies whether AutoCAD uses optimized 2D polylines; also affects the polyline type used with BOUNDARY, DONUT, ELLIPSE (when value is set to 1), PEDIT, POLYGON, and SKETCH (when SKPOLY is set to 1)

0   Polylines in older drawings are not converted on open; PLINE creates old-format polylines

1   Polylines in older drawings are not converted on open; PLINE creates optimized polylines

2   Polylines in older drawings are converted on open; PLINE creates optimized polylines

PLINEWID

Type: Real
Saved in: Drawing
Description: Stores the default polyline width

PLOTID

Type: String
Saved in: Registry
Description: Changes the default plotter, based on its assigned description, and retains the text string of the current plotter description

PLOTROTMODE

Type: Integer
Saved in: Drawing
Description: Controls the orientation of plots

0   Rotates effective plotting area so that corner with rotation icon aligns with paper at lower left (0), top left (90), top right (180) or lower right (270) corner

1   Aligns lower left corner of effective plotting area with lower left corner of the paper

PLOTTER

Type: Integer
Saved in: Config
Description: Changes the default plotter, based on its assigned integer, and retains an integer number that AutoCAD assigns for each plotter

POLYSIDES

Type: Integer
Saved in: (Not saved)
Description: Sets the default number of sides for POLYGON; the range is 3 to 1024

POPUPS

Type: Integer
Saved in: (Not saved)
Description: (Read-only) Displays the status of the currently configured display driver

0   Does not support dialog boxes, the menu bar, pull-down menus, and image tile menus

1   Supports the above features

PROJECTNAME

Type: String
Saved in: Drawing
Description: Stores the current project name; each project name can contain one or more search paths. Used when an xref or image is not found in the original search path.

PROJMODE

Type: Integer
Saved in: Registry
Description: Sets the current projection mode for TRIM and EXTEND operations

0   True 3D mode (no projection)

1   Project to the XY plane of the current UCS

2   Project to the current view plane

PROXYGRAPHICS

Type: Integer
Saved in: Drawing
Description: Specifies whether images of proxy objects are saved in the drawing

0   Image is not saved with the drawing; a bounding box is displayed instead

1   Image is saved with the drawing

PROXYNOTICE

Type: Integer
Saved in: Registry
Desription: Displays a notice when you open a drawing containing custom objects created by an application that is not present

0   No proxy warning is displayed

1   Proxy warning is displayed

PROXYSHOW

Type: Integer
Saved in: Registry
Description: Controls the display of proxy objects in a drawing

0	Proxy objects are not displayed
1	Graphic images are displayed for al proxy objects
2	Only the bounding box is displayed for all proxy objects

PLTSCALE

Type: Integer
Saved in: Drawing
Description: Controls paper space linetype scaling

0	No special linetype scaling
1	Viewport scaling governs linetype scaling

PSPROLOG

Type: String
Saved in: Registry
Description: Assigns a name for a prologue section to be read from the acad.psf file when using PSOUT

PSQUALITY

Type: Integer
Saved in: Drawing
Description: Controls the rendering quality of PostScript images and whether they are drawn as filled objects or as outlines

0	Disables PostScript image generation
<0	Sets number of pixels per AutoCAD drawing unit for the PostScript resolution
>0	Sets number of pixels per drawing unit, but uses the absolute value; causes AutoCAD to show the PostScript paths as outlines and does not fill them

QTEXTMODE

Type: Integer
Saved in: Drawing
Description: Controls quick text mode

0	Off
1	On

RASTERPREVIEW

Type: Integer
Saved in: Drawing
Description: Controls whether drawing preview images are saved with the drawing and sets the format type

0	BMP only
1	BMP and WMF
2	WMF only
3	No preview image created

REGENMODE

Type: Integer
Saved in: Drawing
Description: Controls automatic regeneration of the drawing

0	Turns REGENAUTO off
1	Turns REGENAUTO on

RE-INIT

Type: Integer
Saved in: (Not saved)
Description: Reinitializes the I/O ports, digitizer display, plotter, and acad.pgp file using the following bit-codes; to specify more than one, enter the sum of their values

0	No initialization
1	Digitizer port reinitialization
2	Plotter port reinitialization
4	Digitizer reinitialization
8	Display reinitialization
16	PGP file reinitialization

RTDDISPLAY

Type: Integer
Saved in: Registry
Description: Controls the display of raster images during realtime zoom or pan

0	Displays raster image content
1	Displays raster image outline only

SAVEFILE

Type: String
Saved in: Registry
Description: (Read-only) Stores current auto-save file name

SAVENAME

Type: String
Saved in: Drawing
Description: (Read-only) Stores the file name you assign to a drawing

SAVETIME

Type: Integer
Saved in: Registry
Description: Sets automatic save interval, in minutes (or 0 to disable automatic saves)

SCREENBOXES

Type: Integer
Saved in: Registry
Description: (Read-only) Stores the number of boxes in the screen menu area of the graphics area

SCREENMODE

Type: Integer
Saved in: Registry
Description: (Read-only) Stores a bit-code indicating the graphics/text state of the AutoCAD display; the sum of the following values

0	Text screen is displayed
1	Graphics mode is displayed
2	Dual-screen display is configured

SCREENSIZE

Type: 2D point
Saved in: (Not saved)
Description: Stores current viewport size in pixels

SHADEDGE

Type: Integer
Saved in: Drawing
Description: Controls shading of edges in rendering

0	Faces shaded, edges not highlighted
1	Faces shaded, edges drawn in background color
2	Faces not filled, edges in object color
3	Faces in object color, edges in background color

SHADEDIF

Type: Integer
Saved in: Drawing
Description: Sets ratio of diffuse reflective light to ambient light (in percent of diffuse reflective light)

SHPNAME

Type: String
Saved in: (Not saved)
Description: Sets default shape name

SKETCHINC

Type: Real
Saved in: Drawing
Description: Sets SKETCH record increments

SKPOLY

Type: Integer
Saved in: Drawing
Description: Determines whether SKETCH generates lines or polylines

0	Lines
1	Polylines

SNAPANG

Type: Real
Saved in: Drawing
Description: Sets (UCS-relative) snap/grid rotation angle for the current viewport

SNAPBASE

Type: 2D point
Saved in: Drawing
Description: Sets snap/grid origin point for the current viewport

SNAPISOPAIR

Type: Integer
Saved in: Drawing
Description: Controls current isometric plane for the current viewport

0	Left
1	Top
2	Right

SNAPMODE

Type: Integer
Saved in: Drawing
Description: Controls the snap mode

0	Off
1	On (for current viewport)

SNAPSTYL

Type: Integer
Saved in: Drawing
Description: Sets snap style for the current viewport

0	Standard
1	Isometric

SNAPUNIT

Type: 2D point
Saved in: Drawing
Description: Sets snap spacing for current viewport

SORTENTS

Type: Integer
Saved in: Config
Description: Controls the display of object sort order operations using the following codes; to select more than one, enter the sum of their codes

0	Disables SORTENTS
1	Sorts for object selection
2	Sorts for object snap
4	Sorts for redraws
8	Sorts for MSLIDE slide creation
16	Sorts for regenerations
32	Sorts for plotting
64	Sorts for PostScript output

SPLFRAME

Type: Integer
Saved in: Drawing
Description: Controls display of spline-fit polylines

0	Does not display control polygon; displays fit surface of polygon mesh; does not display invisible edges of 3D faces or polyface meshes

1   Displays control polygon; displays defining mesh of surface-fit polygon mesh; displays invisible edges of 3D faces or polyface meshes

SPLINESEGS
Type: Integer
Saved in: Drawing
Description: Sets the number of line segments to be generated for each spline

SPLINETYPE
Type: Integer
Saved in: Drawing
Description: Sets the type of spline curve to be generated by PEDIT Spline
5   Quadratic B-spline
6   Cubic B-spline

SURFTAB1
Type: Integer
Saved in: Drawing
Description: Sets the number of tabulations to be generated for RULESURF and TABSURF; sets the mesh density in the *M* direction for REVSURF and EDGESURF

SURFTAB2
Type: Integer
Saved in: Drawing
Description: Sets the mesh density in the *N* direction for REVSURF and EDGESURF

SURFTYPE
Type: Integer
Saved in: Drawing
Description: Controls the type of surface fitting to be performed by PEDIT Smooth
5   Quadratic B-spline surface
6   Cubic B-spline surface
8   Bezier surface

SURFU
Type: Integer
Saved in: Drawing
Description: Sets the surface density in the *M* direction

SURFV
Type: Integer
Saved in: Drawing
Description: Sets the surface density in the N direction

SYSCODEPAGE
Type: String
Saved in: Drawing
Description: Indicates the system code pages specified in acad.xmf

TABMODE
Type: Integer
Saved in: (Not saved)
Description: Controls use of tablet mode
0   Disables tablet mode
1   Enables tablet mode

TARGET
Type: 3D point
Saved in: Drawing
Description: (Read-only) Stores location of the target point for current viewport (in UCS coordinates)

TDCREATE
Type: Real
Saved in: Drawing
Description: (Read-only) Stores time and date of drawing creation

TDINDWG
Type: Real
Saved in: Drawing
Description: (Read-only) Stores total editing time

TDUPDATE
Type: Real
Saved in: Drawing
Description: (Read-only) Stores time and date of last update/save

TDUSRTIMER
Type: Real
Saved in: Drawing
Description: (Read-only) Stores user-elapsed timer

TEMPPREFIX
Type: String
Saved in: (Not saved)
Description: Contains the directory name configured for placement of temporary files

TEXTEVAL
Type: Integer
Saved in: (Not saved)
Description: Controls method of evaluation of text strings
0   All responses to prompts for text strings and attribute values are taken literally
1   Text starting with "(" or "!" is evaluated as an AutoLISP expression, as for nontextual input

TEXTFILL
Type: Integer
Saved in: Drawing
Description: Controls the filling of Bitstream, TrueType, and Adobe Type 1 fonts
0   Outlines
1   Filled images

TEXTQLTY

Type: Real
Saved in: Drawing
Description: Sets resolution of Bitstream, TrueType, and Adobe Type 1 fonts

TEXTSIZE

Type: Real
Saved in: Drawing
Description: Sets the default height for new text objects drawn with the current text style

TEXTSTYLE

Type: String
Saved in: Drawing
Description: Contains the name of the current text style

THICKNESS

Type: Real
Saved in: Drawing
Description: Sets the current 3D thickness

TILEMODE

Type: Integer
Saved in: Drawing
Description: Controls access to paper space, as well as the behavior of AutoCAD viewports
0	Enables paper space and viewport objects
1	Enables Release 10 compatibility mode (uses VPORTS)

TOOLTIPS

Type: Integer
Saved in: Registry
Description: Controls the display of tooltips

TRACEWID

Type: Real
Saved in: Drawing
Description: Sets default trace width

TREEDEPTH

Type: Integer
Saved in: Drawing
Description: Specifies maximum depth, that is, the number of times the tree-structured spatial index may divide into branches

TREEMAX

Type: Integer
Saved in: Registry
Description: Limits memory consumption during drawing regeneration by limiting the maximum number of nodes in the spatial index (oct-tree)

TRIMMODE

Type: Integer
Saved in: (Not saved)
Description: Controls whether AutoCAD trims selected edges for chamfers and fillets
0	Leaves selected edges intact
1	Trims selected edges to the endpoints of the chamfer lines and fillet arcs

UCSFOLLOW

Type: Integer
Saved in: Drawing
Description: Generates a plan view when you change from one UCS to another; can be set separately for each viewport
0	UCS does not affect the view
1	Any UCS change causes a change to plan view of the new UCS in the current viewport

UCSICON

Type: Integer
Saved in: Drawing
Description: Displays the coordinate system icon using bit-code for the current viewport; the sum of the following
0	Off (disabled)
1	On; icon display is enabled
2	Origin; if icon display is enabled, the icon floats to the UCS origin if possible

UCSNAME

Type: String
Saved in: Drawing
Description: (Read-only) Stores the name of the current coordinate system for the current space

UCSORG

Type: 3D point
Saved in: Drawing
Description: (Read-only) Stores the origin point of the current coordinate system for the current space (in world coordinates)

UCSXDIR

Type: 3D point
Saved in: Drawing
Description: (Read-only) Stores the X direction of the current UCS for the current space

UCSYDIR

Type: 3D point
Saved in: Drawing
Description: (Read-only) Stores the Y direction of the current UCS for the current space

UNDOCTL

Type: Integer
Saved in: (Not saved)
Description: Stores a bit-code indicating the state of the UNDO feature; the sum of the following values

0	UNDO is disabled
1	UNDO is enabled
2	Only one command can be undone
4	Auto-group mode is enabled
8	A group is currently active

UNDOMARKS

Type: Integer
Saved in: (Not saved)
Description: (Read-only) Stores the number of marks placed in the UNDO control stream by the Mark option

UNITMODE

Type: Integer
Saved in: Drawing
Description: Controls units display format

0	Displays as previously set
1	Displays in input format

USERI1-5

Type: Integer
Saved in: Drawing
Description: USERI1, USERI2, USERI3, USERI4, and USERI5 are used to store and retrieve integer values

USERR1-5

Type: Real
Saved in: Drawing
Description: USERR1, USERR2, USERR3, USERR4, and USERR5 are used to store and retrieve real numbers

USERS1-5

Type: String
Saved in: (Not saved)
Description: USERS1, USERS2, USERS3, USERS4, and USERS5 are used to store and retrieve text string data

VIEWCTR

Type: 3D point
Saved in: Drawing
Description: (Read-only) Stores the center of view in the current viewport, expressed in UCS coordinates

VIEWDIR

Type: 3D vector
Saved in: Drawing
Description: (Read-only) Stores viewing direction in the current viewport, expressed in UCS coordinates

VIEWMODE

Type: Integer
Saved in: Drawing
Description: Controls viewing mode for the current viewport using bit-code; sum of the following values

0	Disabled
1	Perspective view active
2	Front clipping on
4	Back clipping on
8	UCS follow mode on
16	Front clip not at eye (if on, FRONTZ determines front clipping plane; if off, front clipping plane is set to pass through camera point)

VIEWSIZE

Type: Real
Saved in: Drawing
Description: Stores height of view in current viewport, expressed in drawing units

VIEWTWIST

Type: Real
Saved in: Drawing
Description: Stores view twist angle for the current viewport

VISRETAIN

Type: Integer
Saved in: Drawing
Description: Controls visibility of layers in xref files

0	Xref layer definition in the current drawing takes precedence over these settings: On/Off, Freeze/Thaw, color, and linetype for xref-dependent layers
1	The above settings for xref-dependent layers take precedence over the xref layer definition in the current drawing

VSMAX

Type: 3D point
Saved in: Drawing
Description: (Read-only) Stores the upper right corner of the current viewport's virtual screen, expressed in UCS coordinates

VSMIN

Type: 3D point
Saved in: Drawing
Description: (Read-only) Stores the lower left corner of the current viewport virtual screen, expressed in UCS coordinates

WORLDUCS

Type: Integer
Saved in: (Not saved)
Description: (Read-only) Indicates whether the UCS is the same as the world coordinate system
0   Current UCS is different from WCS
1   Current UCS is the same as WCS

WORLDVIEW

Type: Integer
Saved in: Drawing
Description: Controls whether UCS changes to WCS during DVIEW or VPOINT
0   Current UCS remains unchanged
1   Current UCS is changed to the WCS for the duration of the DVIEW or VPOINT command; DVIEW and VPOINT command input is relative to the current UCS

XCLIPFRAME

Type: Integer
Saved in: Drawing
Description: Controls visibility of xref clipping boundaries
0   Clipping boundary is not visible
1   Clipping boundary is visible

XLOADCTL

Type: Integer
Saved in: Registry
Description: Turns demand loading on and off and controls whether it loads the original drawing or a copy
0   Turns off demand loading; entire drawing is loaded
1   Turns on demand loading, reference file is kept open
2   Turns on demand loading; a copy of the reference file is loaded

XLOADPATH

Type: String
Saved in: Registry
Description: Creates a path for storing temporary copies of demand-loaded xref files

XREFCTL

Type: Integer
Saved in: Registry
Description: Controls whether AutoCAD writes XLG files (external reference log files)
0   Does not write XLG files
1   Writes XLG files

Appendix K: Glossary of Terms

2D   Two-dimensional.

3D   Three-dimensional.

acad.dwt   AutoCAD's default drawing template file. Unless you specify a unique template file, AutoCAD uses all of the settings and values stored in acad.dwt when you begin a new drawing. Any drawing file can be saved as a template file.

ACIS   3D geometry engine that AutoCAD uses to create solid models. CAD software products that use ACIS can output ASCII files known as SAT files, which are readable directly by other ACIS-based products. ACIS was developed and made commercially available by Spatial Technology, Inc. (Boulder, CO).

ANSI dimensioning   Standardized format for dimensioning and tolerancing of engineering drawings established by the American National Standards Institute (ANSI).

ASCII   American Standard Code for Information Interchange (ASCII) consists of standard text, numbers, and special characters produced by a computer keyboard.

Bezier curve   Smooth curve used for 3D surface models of free-form shapes. A Bezier curve is made up of four control points that influence the shape of the curve. The curve does not necessarily pass through the control points.

binary   Standard two-digit numerical system in which 0 and 1 are the only digits; forms the basis for all arithmetic calculations in computers.

bitmap image   Digital representation of an image in which bits are referenced to pixels. In color graphics, a different value represents each red, green, and blue component of a pixel.

BMP file   Microsoft Windows bitmapped image file format.

Boolean operation   In CAD, combining two solid objects, subtracting one from another, or determining intersecting areas between overlapping solid objects; based on Boolean algebra, a mathematical system designed to analyze symbolic logic.

CAD   Computer-Aided Design or Drafting. Programs used to create and document designs such as manufactured products and commercial buildings. AutoCAD is an example of a CAD program.

CAE   Computer-Aided Engineering. CAE programs offer capabilities for engineering design and analysis such as determining a design's structural integrity and its capacity to transfer heat.

CAM   Computer-Aided Manufacturing. Typically refers to systems that use CAD surface data to drive computer numerical control (CNC) machines such as mills, lathes, and flame cutters to fabricate parts, molds, and dies.

clipping plane   Plane that slices through one or more 3D objects, causing the part of the object on one side of the plane to be omitted.

compiled language program environment   Type of computer programming language that requires you to compile the code before users can execute it. Compiling means to covert the programming code into machine language. AutoCAD's ARX (AutoCAD Runtime Extension) is a compiled language program environment, whereas AutoLISP is an interpreted language, which does not require compilation. Each line of code is interpreted directly, one line at a time.

Coons patch   Technique for creating complex surfaces that interpolate boundary curves derived from Coons mathematics. AutoCAD's EDGESURF command creates a Coons surface patch.

drawing template file   A drawing used as a template for creating new drawings. Template files typically contain frequently used settings and values, such as linetypes and layers, that are inserted into new drawings automatically when the new drawing is created. AutoCAD's default template file is acad.dwt.

facet   Three- or four-sided polygon element that represents a piece of a 3D surface. Polygonal mesh models are sometimes referred to as *faceted models*.

facet shading   See *flat shading*.

fillet   Rounded corner.

flat shading   Method of rendering a polygonal model that fills each polygon with a single shade of color to give the model a faceted look. Also called *facet shading*.

font   Distinctive set of characters consisting of letters and numbers of specific size and design.

Geometric Dimensioning & Tolerancing (GD&T)   Standardized format for dimensioning and showing variations (called *tolerances*) in the manufacturing process. Tolerances specify the largest variation allowable for a given dimension. GD&T uses geometric characteristic symbols and feature control frames that follow industry standard practices. Used by many government contractors, GD&T standards are defined in the American National Standards Institute (ANSI) Y14.5M Dimensioning/Tolerancing handbook.

Gouraud shading   Method of smooth rendering of polygonal models by interpolating (averaging and blending) adjacent color intensities. The Gouraud method applies light to each vertex of a polygon face and interpolates the results to produce a realistic-looking model.

GUI Graphical User Interface. Microsoft Windows, OSF Motif and the Macintosh environment, as well as programs specifically designed for them, are GUI examples.

hidden line removal The deletion of tessellation lines that would not be visible if you were viewing the 3D model as a real object.

icon 1) Graphic symbol typically used in GUI software. When selected, an icon issues a command or performs a specific operation. 2) A small, nonprinting symbol on the AutoCAD graphics screen that provides information about the current state of the drawing. Examples include the UCS icon and the paper space icon.

IGES Initial Graphics Exchange Specification. IGES is an industry standard format for exchanging CAD data between systems.

interpolation Method of averaging and blending.

ISO International Standards Organization.

machine-independent software Software that is not dependent on a specific brand or class of machine.

model Representation of a proposed or real object. A soft model is a computer mock-up of a design, while a hard model is a physical object that was fabricated from wood, clay, plastic, metal or some other material.

NURBS Non-uniform rational B-spline. Smooth curve or surface defined by a series of weighted control points. AutoCAD's SPLINE command creates a NURBS curve.

operating system (OS) Computer software environment normally provided by the computer manufacturer. Microsoft Windows, IBM OS/2, Macintosh System 7, and Silicon Graphics IRIX (a Unix variation), are examples of operating systems.

Phong shading Method of smooth rendering of polygonal models; calculates the light at several points across the model's surface. Phong shading produces more accurate specular highlights than Gouraud shading.

photorealistic rendering High-quality rendering that resembles a photograph.

pixel Picture element. Smallest addressable dot, or element, on a computer screen or a raster device such as an ink jet or laser printer.

prototype tooling Molds used to produce models and protoype parts; sometimes referred to as *soft tooling*.

raster Technique of defining a line or curve using a series of dots or pixels. Computer screens and most printers (ink jet, laser and electrostatic) use raster techniques.

raster printer Printer that defines lines, curves, and solid objects using small dots. See *raster*.

rapid prototyping (RP) Refers to a class of machines used for producing physical models and prototype parts from 3D computer model data such as CAD. Unlike CNC machines, which subtract material, RP processes join together liquid, powder, and sheet materials to form parts. Layer by layer, these machines fabricate plastic, wood, ceramic, and metal objects from thin horizontal cross-sections taken from the 3D computer model. Stereolithography and Fused Deposition Modeling are examples of RP processes.

ruled surface Surface created by linear interpolation between two curves. AutoCAD's RULESURF command creates a ruled surface.

SAT files Machine-independent geometry files that contain topological and surface information about models created using the ACIS engine. AutoCAD can import and export SAT files using the ACISIN and ACISOUT commands, respectively.

solid model 3D model defined using solid modeling techniques, which is somewhat similar to using physical materials such as wood or clay to produce shapes. CAD solid modeling programs that take advantage of Constructive Solid Geometry (CSG) techniques use primitives, such as cylinders, cones, and spheres, and Boolean operations to construct shapes. CAD programs that use the Boundary Representation (B-Rep) solid modeling technique store geometry directly using a mathematical representation of each surface boundary. Some CAD programs use both CSG and B-Rep modeling techniques.

solid primitive Basic solid shape, such as a cylinder or sphere, usually used to produce more complex shapes.

spline curve Smooth curved line. NURBS is one example of a spline curve.

stereolithography Rapid prototyping (RP) process involving the solidification of ultraviolet light-sensitive liquid resin using a laser to harden and produce a plastic part of a 3D computer model.

STL File format for converting CAD models to physical parts using rapid prototyping (RP) such as stereolithography. 3D Systems, Inc. (Valencia, CA) developed and published the STL format, which is available in binary and ASCII forms.

surface model Three-dimensional (3D) model defined by surfaces. The surface consists of a mesh of polygons, such as triangles, or a mathematical description such as a Bezier B-spline surface or non-uniform rational B-spline (NURBS) surface.

surface of revolution Surface created by rotating a line or curve around an axis. AutoCAD's REVSURF command creates a surface of revolution.

system registry A folder in which a computer stores important information about the operating system, hardware and software installations, user preferences, application settings, and other types of information that make a computer run smoothly.

tabulated surface Type of ruled surface defined by a path curve and direction vector. AutoCAD's TABSURF command creates a tabulated surface.

tessellation lines Lines that describe a curved surface or 3D model.

TGA file TrueVision Targa file format

TIF file Tagged image file format (often called *TIFF*).

vector Line segment defined by its endpoints or by a starting point and direction in 3D space.

wireframe model 3D model whose shape is defined by a series of lines and curves.

888

Index

configuring plotter devices, 433-435

construction lines, 94-96

Continue Dimension icon, 281

Control option, of UNDO command, 104-105

controlling views, with View Control dialog box, 186

converting AME models to R14 solids, 606

Coons surface patch, 523, 526

coordinate display, 4, 77
    controlling with CTRL D, 78
    controlling with F6, 77

coordinate entry
    absolute, 55
    direct distance method, 57
    polar, 55
    relative, 55

coordinate pair, 54

coordinate system icon, 4

coordinates, Cartesian, 54

COPY command, 130

Copy History option in AutoCAD Text Window, 82

Copy Object icon, 130

COPYCLIP command, 374

copying and pasting among drawing files, 374

copying files, 204
    in Select File dialog box, 204

copying, with grips, 163

Create new Drawing dialog box, 10

creating a new group, 359

creating a new multiline style, 115-118

creating a script file, 783

creating a solid pulley, 627

creating an image tile menu, 786-790

creating new folders, 205

creating new lights for rendering, 557

cross hatching, patterns for, 838

cross section of solid model, 628

crosshairs
    current position of, 77
    effect of rotating snap on, 92

crossing object selection, 41, 149
    with STRETCH command, 149
    with EXTEND and TRIM commands, 154

CTRL B, to toggle snap, 91

CTRL E, to toggle isometric planes, 458

CTRL G, to toggle grid feature, 90

cubic B-splines, 333

curve fitting, in polyline, 329

curves, controlling resolution of, 177

custom menus
    partial, 686-688
    unloading, 688

custom toolbars, 689-690
    deleting, 691

customizing AutoCAD's tablet menu, 706-712

CYLINDER command, 593
    Elliptical option, 594

Cylinder icon, 593

D

DBLIST command, 347

DCL files, 406

DDATTE command, 399

DDATTEXT command, 402-404

DDCOLOR command, 249

DDEDIT command, 217, 283
    to edit mtext, 226-227
    to edit standard text, 225, 226

DDIM command, 278, 279, 297

DDINSERT command, 371, 402

DDPTYPE command, 350

DDRMODES command, 93

DDSELECT command, 166

DDUCSP command, 499

DDUNITS command, 238

DDVIEW command, 187

DDVPOINT command, 471-473

Default System Printer, used to print drawings, 425

default values, display in AutoCAD, 20

Defined Blocks dialog box, 402

defining boundaries, 576

defining curves for ruled surfaces, 518

defun function, in AutoLISP, 724

Delete key, to erase selected objects, 165

deleting files and folders, using right-click method, 205

deleting views, 186

Delta option, of LENGTHEN command, 155

Detach option, of XREF command, 669

Details about files, in Select File dialog box, 204

Device and Default Selection dialog box, 438-439

Diameter Dimension icon, 267

digitizing hard-copy drawings, 746-749
    calibrating the drawing, 748

DIMALIGNED command, 266

DIMANGULAR command, 267

DIMASO dimensioning system variable, 282

DIMASZ dimensioning system variable, 276

DIMBASELINE command, 268

DIMCEN dimensioning system variable, 277, 303

DIMDIAMETER command

DIMEDIT command, 295-296
    New option, 295
    Oblique option, 295
    Rotate option, 295

dimension lines, suppressing, 301

Dimension Style icon, 278

Dimension Styles dialog box, 278

Dimension Styles icon, 297

dimension styles, 277-279
    and families, 278

Dimension Text Edit icon, 294

dimension text, 279

dimensioning, 264-318
    circles and arcs, 267
    horizontal lines, 265
    inclined lines, 266
    symbols, 835
    system variables, 276-279
    vertical lines, 266
    baseline, 268
    using Continue option, 281
    choosing arrowhead style, 303
    controlling center marks/lines, 303
    controlling extension line characteristics, 302
    editing, 294-303
    limits method, 311
    moving, 271-272
    rotating, 295
    suppressing zeros in, 298
    symmetrical, 309
    using alternate units, 298

DIMLIN, 281

DIMLINEAR command, 265, 281

DIMORDINATE command, 270